INTERSCIENCE MONOGRAPHS AND TEXTS IN PHYSICS AND ASTRONOMY

Edited by **R. E. MARSHAK**

Volume I: E. R. Cohen, K. M. Crowe, and J. W. M. DuMond, **The Fundamental Constants of Physics**

Volume II: G. J. Dienes and G. H. Vineyard, **Radiation Effects in Solids**

Volume III: N. N. Bogalubov and D. V. Shirkov, **Introduction to the Theory of Quantized Fields**

Volume IV: J. B. Marion and J. L. Fowler, *Editors*, **Fast Neutron Physics** *In two parts*
Part I: Techniques **Part II: Experiments and Theory**

Volume V: D. M. Ritson, *Editor*, **Techniques of High Energy Physics**

Volume VI: R. N. Thomas and R. G. Athay, **Physics of the Solar Chromosphere**

Volume VII: Lawrence H. Aller, **The Abundance of the Elements**

Volume VIII: E. N. Parker, **Interplanetary Dynamical Processes**

Volume IX: Conrad L. Longmire, **Elementary Plasma Physics**

Volume X: R. Brout and P. Carruthers, **Lectures on the Many-Electron Problem**

Volume XI: A. I. Akhiezer and V. B. Berestetskii, **Quantum Electrodynamics**

Volume XII: John L. Lumley and Hans A. Panofsky, **The Structure of Atmospheric Turbulence**

Volume XIII: Robert D. Heidenreich, **Fundamentals of Transmission Electron Microscopy**

Volume XIV: G. Rickayzen, **Theory of Superconductivity**

Volume XV: Raymond J. Seeger and G. Temple, *Editors*, **Research Frontiers in Fluid Dynamics**

Volume XVI: C. S. Wu and S. A. Moszkowski, **Beta Decay**

Volume XVII: Klaus G. Steffen, **High Energy Beam Optics**

Volume XVIII: V. M. Agranovich and V. L. Ginzburg, **Spatial Dispersion in Crystal Optics and the Theory of Excitons**

Volume XIX: A. B. Migdal, **Theory of Finite Fermi Systems and Applications to Atomic Nuclei**

Volume XX: B. L. Moiseiwitsch, **Variational Principles**

Volume XXI: L. S. Shklovsky, **Supernovae**

Volume XXII: S. Hayakawa, **Cosmic Ray Physics**

Volume XXIII: J. H. Piddington, **Cosmic Electrodynamics**

Volume XXIV: R. E. Marshak, Riazuddin, and C. P. Ryan, **Theory of Weak Interactions in Particle Physics**

Volume XXV: Norman Austern, **Direct Nuclear Reaction Theories**

INTERSCIENCE MONOGRAPHS AND TEXTS IN PHYSICS AND ASTRONOMY

Edited by R. E. MARSHAK
University of Rochester, Rochester, New York

VOLUME XXV

DIRECT NUCLEAR REACTION THEORIES

NORMAN AUSTERN

Department of Physics
University of Pittsburgh

WILEY-INTERSCIENCE a division of
John Wiley & Sons New York · London · Sydney · Toronto

10 9 8 7 6 5 4 3 2 1

Library of Congress Catalogue Card Number: 78-100327

ISBN 0–471–03770–2

Printed in the United States of America

Preface

My interest in the subject of this book began in 1953, when I collaborated with S. T. Butler and H. McManus in a study of (p, n) reactions localized in the nuclear surface. Naturally, when we later found we had written an early article on "direct reaction" theory, I decided to learn more about that subject. This decision motivated a review of direct reactions, prepared during a visit to Australia in 1957–1958, and published in *Fast Neutron Physics*. In 1962 a series of lectures at the Low Tatra Summer School (see *Selected Topics in Nuclear Theory*) was then intended as a sketch for the present full-length book. I am grateful that Professor R. E. Marshak and John Wiley and Sons have waited for the book so patiently. It was completed in 1969, during a second visit to Australia.

Direct reaction theory has become one of the most popular subjects in nuclear physics, as measured by the numbers of articles published. The present book is intended as a guide and introduction to this vast literature. The presentation centers around the amazingly effective distorted-waves theory, the simplest direct reaction theory that makes any sense. Recondite theories that do not lead to accurate practical calculations are avoided. The discussion of very complicated, specialized reactions has also been avoided. Within these limitations I hope there have not been too many other omissions caused purely by carelessness.

While I hope this book will be of interest to those doing active research in nuclear reactions, it is intended primarily for use by advanced graduate students. I have attempted to keep my experimentalist friends in mind, and to comply with Evans Hayward's request that theoretical books should have "words between the equations."

To keep the book of finite length, severe limitations of subject matter are required. Hence this is in no sense a general text on reaction theory or even on nuclear reactions. On the other hand, enough background material has been inserted to allow the book to be read without previous study of reaction theory. Nonrelativistic Schrödinger quantum mechanics is used throughout. Reactions are not treated as an adjunct to the study of bound states nor, as an end in themselves, but rather as "nuclear structure in the continuum." Relations between reaction theory and structure theory are emphasized at all stages, and explicit dynamical models are used freely.

(Present knowledge about nuclei is too extensive to justify much emphasis on model-independent theories.)

At this stage I would like to express thanks to all those friends who have helped me learn about nuclear reactions, and especially to J. S. Blair, S. T. Butler, R. M. Drisko, and G. R. Satchler. Various sections of the book have been read in manuscript by E. U. Baranger, R. H. Bassel, J. S. Blair, B. L. Cohen, W. Dachnick, R. M. Drisko, H. Feshbach, R. C. Johnson, R. Lipperheide, I. E. McCarthy, M. Macfarlane, V. A. Madsen, K. C. Richards, K. F. Ratcliff, D. Robson, F. Tabakin, T. Tamura, N. S. Wall, H. A. Weidenmüller, and S. Yoshida. The completed manuscript was read by J. B. French and G. R. Satchler. I am grateful for the criticisms provided by all these people. I am grateful to my wife, Wilma, for her patience. Finally, I wish to dedicate the book to my teachers, and particularly to J. Hlavaty, J. E. Mack, R. G. Sachs, E. P. Wigner, and H. A. Bethe.

Norman Austern

Pittsburgh, Pennsylvania
October, 1969

Contents

CHAPTER 1

Wave Packets

The quantum theory of scattering is the theory of the interaction of wave packets; therefore it is appropriate that a book about aspects of nuclear scattering theory should begin with a discussion of this subject. This discussion is of particular benefit to the present book because many qualitative statements about nuclear reaction mechanisms are based on wave-packet ideas.

Wave-packet states are used in scattering analyses because their mathematical properties correspond closely to the physical properties of experimental setups. Energy eigenfunctions, solutions of the time-independent Schrödinger equation, give no equivalent close correspondence with experiment. On the other hand, computation with energy eigenfunctions is more straightforward than with wave packets; therefore these eigenfunctions are used for most practical scattering theories. Expressions for cross sections can be obtained from the stationary-state eigenfunctions by the application of certain simple rules, provided these eigenfunctions are chosen to satisfy *scattering boundary conditions*. The following discussion develops the relations between wave packets and the stationary-state eigenfunctions and makes evident why the stationary-state theory yields results that are sufficiently accurate for nearly all applications. Further discussions of wave-packet ideas may be found in the treatise by Goldberger and Watson [1].

The discussion is limited to the treatment of nonrelativistic motion, as is customary in nuclear physics, and antisymmetrization is not introduced here. It is conveniently included at a later stage by making appropriate linear combinations of the results obtained in the absence of antisymmetrization. Also, we consider only those reactions that are initiated by the collision of two particles and that result in two-particle breakup. However, the requirement of two-body breakup is in no way essential and is introduced only for notational convenience. Some cases of three-body breakup are treated in Sections 1.8 and 2.7 later in this book. All wavefunctions used below are written in the center-of-mass coordinate system.

1.1 The Stationary-state Wavefunction

We begin by writing down the stationary-state wavefunction in the asymptotic region of configuration space. This part of the stationary-state wavefunction is a linear combination of different modes of breakup of the system into two separated nuclei. A given mode of two-particle breakup is specified by a product

$$\psi_\alpha = \psi_a \psi_A \tag{1.1}$$

of the bound internal wavefunctions of the two nuclei. Here A and a denote the nuclei into which the system breaks up and also their states of excitation, their angular momenta, and the projections of their angular momenta. (More detailed notation for these properties of ψ_a and ψ_A are used later on in the book.) A complete specification of the type of two-particle breakup and of the internal states of the two particles, as just described, is called a *channel.* This same word channel is also understood sometimes to include all the properties already mentioned, together with a definite value for the orbital angular momentum of the relative motion of the centers of mass of the two separating nuclei. Both meanings of the word channel are used from time to time in this book. The channel functions ψ_α are considered to be normalized to unity with respect to integrations over the internal variables of A and a.

A given system of nucleons may separate into two nuclei in more than one way; for example, three neutrons and three protons may separate into He^4 plus a deuteron, into H^3 plus He^3, and so forth. Accordingly, the variable $\mathbf{r}_\alpha$, which is the displacement between the centers of mass of the two separating nuclei, must be defined differently for each different mode of breakup. The relative kinetic energy of the separating nuclei E_α must also depend on the mode of breakup, as does the relative momentum $\mathbf{k}_\alpha$, where $E_\alpha = \hbar^2 k_\alpha{}^2/2\mu_\alpha$ and where

$$\mu_\alpha = \frac{M_a M_A}{M_a + M_A} \tag{1.2}$$

is the reduced mass for the indicated mode of breakup.

We consider a set of stationary-state eigenfunctions that obey

$$H\Psi^{(+)}(\mathbf{k}_\alpha) = (E_\alpha + \varepsilon_\alpha)\Psi^{(+)}(\mathbf{k}_\alpha), \tag{1.3}$$

where ε_α is the sum of the internal energies of the particles in channel α. Because the $\Psi^{(+)}(\mathbf{k}_\alpha)$ are energy eigenstates the channel energies all obey $E_\alpha + \varepsilon_\alpha = E_\beta + \varepsilon_\beta$. The functions $\Psi^{(+)}(\mathbf{k}_\alpha)$ are chosen to obey standard asymptotic boundary conditions, such that channel α contains an incident plane wave with relative momentum $\mathbf{k}_\alpha$ and all other channels contain only outgoing scattered waves. Then, in terms of the notation given above, the complete asymptotic form of the wavefunction is

$$\Psi^{(+)}(\mathbf{k}_\alpha) = \psi_\alpha e^{i(\mathbf{k}_\alpha \cdot \mathbf{r}_\alpha)} + \sum_\beta \psi_\beta f(\mathbf{k}_\alpha, \hat{r}_\beta) r_\beta^{-1} e^{ik_\beta r_\beta}. \tag{1.4}$$

The superscript $(+)$ indicates that the wavefunction has been chosen with all the scattered waves radially outgoing as shown. The coefficient $f(\mathbf{k}_\alpha, \hat{r}_\beta)$ is the *scattered amplitude* for channel β. It is a function of the scattering angle, as indicated by the unit vector $\hat{r}_\beta = \mathbf{r}_\beta / r_\beta$. Finally, it may be remarked that (1.4) makes no use of Coulomb wavefunctions. This is no real limitation of the analysis because (1.4) will always be applied at distances r_α, r_β that are so large that the Coulomb fields are fully screened. However, Coulomb effects do enter into the determination of the scattered amplitudes.

The evaluation of the wavefunction $\Psi^{(+)}(\mathbf{k}_\alpha)$ is the principal subject of applied scattering theory. The cross section for a reaction $\alpha \rightarrow \beta$ is easily found from this wavefunction by calculating the radially outgoing flux in channel β and dividing it by the incoming flux per unit area in channel α. Nevertheless, although nearly all the useful results of scattering theory can be extracted from the stationary-state wavefunction and expressed in terms of the amplitudes $f(\mathbf{k}_\alpha, \hat{r}_\beta)$, it is clear that this wavefunction is not a precise description of the physics of a scattering experiment. The wavefunction $\Psi^{(+)}(\mathbf{k}_\alpha)$ pervades all space and is not normalizable. It does not represent the incident beam as a collimated stream of individual particles, which are emitted from a source at definite times, scattered at definite later times, and received by detectors at still later times. It also does not represent the initial localization of the target nucleus. All of these shortcomings are corrected in wave-packet treatments of scattering. Fortunately, the asymptotic $\Psi^{(+)}(\mathbf{k}_\alpha)$ is decisive for building the required packets, and in tracing out this connection we are able to see under what conditions the stationary-state wavefunctions yield adequate physical results.

1.2 Introduction of Packet States

For the closest correspondence with experiment the target nucleus and the incident projectile should be localized separately by the use of two independent wave packets. However, this procedure is very complicated [1]. It leads to a wave-packet treatment of the relative motion of the two collision partners and the motion of their center of mass. It is simpler to use the somewhat artificial procedure of immediately going over to the center-of-mass coordinate system and using a wave packet only to localize the incident channel-displacement variable $\mathbf{r}_\alpha$. A wave-packet function of $\mathbf{r}_\alpha$ expresses simultaneously the initial localization and separation of the two collision partners that initiate the reaction in channel α.

The initial packet function of $\mathbf{r}_\alpha$ is first encountered at a time t_0 centered about a point $\mathbf{z}_0$ on the negative z_α-axis and with an average momentum $\mathbf{k}_0$ directed toward the origin along the z_α-axis. A suitable expression for the packet wavefunction at time t_0 is

$$\Phi(t_0) = G(\mathbf{r}_\alpha - \mathbf{z}_0)e^{i(\mathbf{k}_0\cdot\mathbf{r}_\alpha)}\psi_\alpha. \tag{1.5}$$

The function $G(\mathbf{r}_\alpha - \mathbf{z}_0)$ is the envelope function of the wave packet and describes its shape and its localization about $\mathbf{z}_0$. The properties of this function are determined by the accelerator that produces the packet and by the subsequent collimating system. We have already defined $\mathbf{z}_0$ to be at the center of the wave packet and have therefore tacitly made the reasonable assumption that G is even with respect to reflections in this center. To define t_0 it is necessary to choose an origin for the measurement of time. We choose this origin so that $t = 0$ is the time at which the centers of mass of the two colliding nuclei would coincide if the initial packet were to move undisturbed. By extrapolation back to $\mathbf{z}_0$ the result

$$t_0 = -|\mathbf{z}_0|\left(\frac{\hbar k_0}{\mu_\alpha}\right)^{-1} \tag{1.6}$$

is found.

If we know the wavefunction at time t_0, we can generate the wavefunction at later times from (1.5) by application of the time-development operator of quantum mechanics:

$$\Phi(t) = e^{-(iH/\hbar)(t-t_0)}\Phi(t_0). \tag{1.7}$$

Equation 1.7 is a complete description of the scattering process, from the initial preparation of the packet through all of its subsequent development. The stationary-state eigenfunctions are now used for the further analysis of (1.7).

Much of the subsequent theory hinges on an understanding of the properties of $G(\mathbf{r}_\alpha)$. This envelope function is formed by macroscopic devices: accelerators and collimating systems. Therefore, when examined over distances of the order of the reduced wavelength k_0^{-1}, this function must vary slowly. In particular,

$$k_0^{-1}\,|\nabla G| \ll G. \tag{1.8}$$

Because of (1.8) the significant Fourier components of $\Phi(t_0)$ are grouped very closely about the value $\mathbf{k}_0$. If the coefficients that arise in the Fourier expansion of G are

$$a(\mathbf{p}) = (2\pi)^{-3}\int e^{-i(\mathbf{p}\cdot\mathbf{r}_\alpha)}G(\mathbf{r}_\alpha)\,d^3r_\alpha, \tag{1.9}$$

then those that arise in the expansion of $\Phi(t_0)$ are

$$(2\pi)^{-3}\int e^{-i(\mathbf{p}\cdot\mathbf{r}_\alpha)}\Phi(t_0)\,d^3r_\alpha = a(\mathbf{p}-\mathbf{k}_0)\psi_\alpha e^{i(\mathbf{z}\cdot\mathbf{k}_0)-i(\mathbf{z}_0\cdot\mathbf{p})}. \tag{1.10}$$

In terms of these coefficients

$$\Phi(t_0) = \psi_\alpha\int a(\mathbf{p}-\mathbf{k}_0)e^{i(\mathbf{z}_0\cdot\mathbf{k}_0)-i(\mathbf{z}_0\cdot\mathbf{p})+i(\mathbf{p}\cdot\mathbf{r}_\alpha)}\,d^3p. \tag{1.11}$$

The coefficients $a(\mathbf{p}-\mathbf{k}_0)$ in (1.11) are large only if $\mathbf{p}-\mathbf{k}_0 \approx 0$. In fact, the important values of momenta lie in a range

$$|\mathbf{p}-\mathbf{k}_0| \lesssim w^{-1},$$

where w is some typical *width* over which significant changes of the function G do occur. Equation 1.8 implies $k_0^{-1} \ll w$.

We now replace the Fourier expansion of $\Phi(t_0)$ with an expansion in terms of the stationary eigenstates $\Psi^{(+)}(\mathbf{k}_\alpha)$. By this means the subsequent application of the development operator becomes a simple matter. It is important that the eigenfunction expansion can use the same coefficients that appear in the Fourier expansion, provided the wave packet is initially so far from the scattering center that $|\mathbf{z}_0| \gg w$. Under this condition the initial wave packet overlaps with $\Psi^{(+)}(\mathbf{k}_\alpha)$ in a region in which $\Psi^{(+)}(\mathbf{k}_\alpha)$ is asymptotic and (1.4) may be used. Then in the localized region of overlap with $\Phi(t)$ the scattered waves in (1.4) have well-defined momenta whose directions are opposite that of the incident momentum $\mathbf{k}_0$. The latter property was inserted in $\Psi^{(+)}(\mathbf{k}_\alpha)$ *by our choice of boundary conditions*. We then see that because the important Fourier components of $\Phi(t_0)$ are localized in an interval of radius w^{-1} around $\mathbf{k}_0$ only the incident plane wave part of $\Psi^{(+)}(\mathbf{k}_\alpha)$ can have significant overlap with $\Phi(t_0)$. As a result the overlap gives

$$(2\pi)^{-3}(\Psi^{(+)}(\mathbf{k}_\alpha), \Phi(t_0)) = a(\mathbf{k}_\alpha - \mathbf{k}_0)e^{i(\mathbf{z}_0\cdot\mathbf{k}_0)-i(\mathbf{z}_0\cdot\mathbf{k}_\alpha)}, \tag{1.12}$$

the analog of (1.10), and the initial packet has the expansion

$$\Phi(t_0) = \int a(\mathbf{k}_\alpha - \mathbf{k}_0)e^{i(\mathbf{z}_0\cdot\mathbf{k}_0)-i(\mathbf{z}_0\cdot\mathbf{k}_\alpha)}\Psi^{(+)}(\mathbf{k}_\alpha)\,d^3k_\alpha. \tag{1.13}$$

Equation 1.13 is the analog of (1.11) and is valid provided the initial wave packet fulfills the two conditions that have been discussed:

$$k_0^{-1} \ll w \ll |\mathbf{z}_0|. \tag{1.14}$$

Scattering experiments ordinarily fulfill these conditions very well.

1.3 Properties of the Outgoing Packets

The time development of the incident wave packet is obtained by inserting (1.13) into (1.7). Because (1.13) expresses $\Phi(t_0)$ in terms of eigenstates of H, it follows that

$$\Phi(t) = \int a(\mathbf{k}_\alpha - \mathbf{k}_0) e^{i(\mathbf{z}_0\cdot\mathbf{k}_0) - i(\mathbf{z}_0\cdot\mathbf{k}_\alpha)} e^{-(i/\hbar)(E_\alpha + \varepsilon_\alpha)(t - t_0)} \Psi^{(+)}(\mathbf{k}_\alpha)\, d^3k_\alpha. \quad (1.15)$$

To make this expression for $\Phi(t)$ more explicit and to make evident its wave-packet properties, it is only necessary to perform the integration over $\mathbf{k}_\alpha$ to sufficient accuracy. This integration is performed very easily. The factor $a(\mathbf{k}_\alpha - \mathbf{k}_0)$ constrains $\mathbf{k}_\alpha$ to lie near $\mathbf{k}_0$, within an interval of order w^{-1}. Therefore the variable

$$\mathbf{q} \equiv \mathbf{k}_\alpha - \mathbf{k}_0 \quad (1.16)$$

is small, and the remainder of the integrand may be expanded in terms of this variable. This step is especially easy in the asymptotic region, which is of experimental concern, because there (1.4) may be used for $\Psi^{(+)}(\mathbf{k}_\alpha)$. Of course, in the asymptotic region the individual phase factors in the integrand of (1.15) vary rapidly with respect to $\mathbf{q}$. The variable $\mathbf{q}$ appears in the exponents of these phase factors, and under asymptotic conditions it is multiplied by very large coefficients. Nevertheless, if the exponents are expanded in powers of $\mathbf{q}$, the expansions converge rapidly and only zero-order and linear terms need to be carried. Terms of quadratic or higher order contribute mainly to the *spreading* of the wave packet. Under the circumstances of a collision experiment such spreading has no experimental consequences [1]; therefore the terms of higher order in $\mathbf{q}$ are not required.

Two functions in phase factors in the integrand of (1.15) must be linearized with respect to $\mathbf{q}$. These are the energy $E_\alpha + \varepsilon_\alpha$ that governs the rate at which the phase changes with respect to time and the momentum k_β that governs the rate at which the phase of $\Psi^{(+)}$ changes with respect to displacement from the scattering center. The required expansions are simplified algebraically if the notation

$$\mathbf{v}_\alpha = \frac{\hbar \mathbf{k}_0}{\mu_\alpha}, \qquad v_\beta = \frac{\hbar \bar{k}_\beta}{\mu_\beta}, \quad (1.17)$$

is introduced for the average incoming and outgoing velocities. In (1.17) we understand that $\bar{k}_\beta$ corresponds, by energy conservation, to $k_\alpha = k_0$. Then the expansion of $E_\alpha + \varepsilon_\alpha$ takes the form

$$\begin{aligned} E_\alpha + \varepsilon_\alpha &= \hbar^2(\mathbf{k}_0 + \mathbf{q})^2(2\mu_\alpha)^{-1} + \varepsilon_\alpha, \\ &\approx E + \hbar v_\alpha(\hat{k}_0 \cdot \mathbf{q}), \end{aligned} \quad (1.18)$$

where the quadratic term has been dropped, as mentioned above, and where

$$E \equiv \frac{\hbar^2 k_0^{\ 2}}{2\mu_\alpha} + \varepsilon_\alpha$$

is the average total incident energy. The **q**-dependence of k_β is obtained from the equation of energy conservation,

$$\frac{\hbar^2 k_\alpha^{\ 2}}{2\mu_\alpha} + \varepsilon_\alpha = \frac{\hbar^2 k_\beta^{\ 2}}{2\mu_\beta} + \varepsilon_\beta, \tag{1.19}$$

by expanding k_β about $\bar{k}_\beta$. It follows that

$$k_\beta \approx \bar{k}_\beta + \frac{v_\alpha}{v_\beta}(\hat{k}_0 \cdot \mathbf{q}), \tag{1.20}$$

where again terms of quadratic or higher order in **q** have been dropped.

On insertion of (1.18), (1.20), and (1.4) into (1.15), that equation reduces to

$$\Phi(t) = e^{-(iE/\hbar)(t-t_0)}\Big\{\psi_\alpha e^{i(\mathbf{k}_0\cdot\mathbf{r}_\alpha)}\int a(\mathbf{q})e^{i(\mathbf{q}\cdot\mathbf{r}_\alpha)-i(\mathbf{q}\cdot\mathbf{v}_\alpha)t}\, d^3q$$
$$+ \sum_\beta \psi_\beta r_\beta^{\ -1} e^{ik_\beta r_\beta}\int a(\mathbf{q}) f(\mathbf{k}_\alpha, \hat{r}_\beta) e^{i(\mathbf{k}_0\cdot\mathbf{q})(v_\alpha/v_\beta)[r_\beta - v_\beta t]}\, d^3q\Big\}. \tag{1.21}$$

Of course, the integral in the first line of (1.21) is trivial and simply restores the incident wave packet at a displaced location. The integrals in the second line are more complicated; to perform them it is necessary to discuss the **q**-dependence of the scattered amplitudes, the $f(\mathbf{k}_\alpha, \hat{r}_\beta)$. It is interesting to remark first that under asymptotic conditions the integrals that stand in (1.21) tend to oscillate rapidly with respect to **q**, hence they tend to average to zero. For example, within the range $|\mathbf{q}| \lesssim w^{-1}$ in which the magnitude of the integrand is large the phase of the integrand can vary by as much as z_0/w, a very large number. It is such rapid variations of phase that isolate wave-packet properties in (1.21). Only for certain combinations of the parameters in (1.21) do the rapid phase variations balance so that the net phase variation is slow and the result of the integration, large.

The important range of values of **q** in (1.21) is so small that to lowest approximation the amplitudes may be considered independent of the incident momentum and removed from the integrals. By this step the scattered amplitudes of the stationary-state wavefunction are made to be the amplitudes of the outgoing wave packets, and the cross-section formulas of the stationary-state theory are recovered. Some thought must be given to this step because rapid variations of scattered amplitudes with respect to both energy and angle are well known; for example, resonances are known that have widths of the order of a few electron volts and thus lifetimes of the order 10^{-15} sec.

However, the durations of packets formed by accelerators are likely to be of the order of rf periods, or collision times in ion sources, and to be, say, 10^{-10} sec. For such packets the spread of incident momenta is indeed small enough to be disregarded. Although other results might be obtained if much briefer packets were to scatter from much narrower resonances, it is clear that the approximation of factoring out $f(\mathbf{k}_\alpha, \hat{r}_\beta)$ from the integrals of (1.21) will normally be valid. By introducing this approximation and performing the integrals that remain, (1.21) reduces to

$$\Phi(t) = e^{-(iE/\hbar)(t-t_0)}\Big\{\psi_\alpha e^{i(\mathbf{k}_0\cdot\mathbf{r}_\alpha)}G(\mathbf{r}_\alpha - \mathbf{v}_\alpha t) + \sum_\beta \psi_\beta r_\beta^{-1} e^{ik_\beta r_\beta} f(\mathbf{k}_0, \hat{r}_\beta) G\Big(\hat{k}_0\Big[\frac{r_\beta - v_\beta t}{v_\beta/v_\alpha}\Big]\Big)\Big\}. \quad (1.22)$$

The physical interpretation of (1.22) is straightforward. The first term is the ongoing incident wave packet; its center moves classically with the velocity $\mathbf{v}_\alpha$. Each of the scattered wave terms is a packet in the form of a spherical shell of flux that moves outward radially with the velocity v_β. The shape of one of these outgoing packets as a function of radius is the same as the shape of the incoming packet as a function of z_α, with $x_\alpha = y_\alpha = 0$. However, the radial thickness of the outgoing packet generally is not equal to the length of the incoming packet. The ratio of the outgoing thickness to the incoming length is v_β/v_α. This ratio expresses the fact that during the time the packet interacts with the scattering center the front of the packet moves with speed v_β, whereas the rear moves with speed v_α. It may also be noted that the radial variable r_β is positive; therefore the radially outgoing packets exist with appreciable probability only for $t \geqslant 0$. Thus these packets are formed at times $t \approx 0$ during the scattering interaction.

The spherical wave packet in the channel $\beta = \alpha$ is the elastically scattered outgoing packet. Embedded in this packet and moving with the same speed is the ongoing incident wave packet. Interference between these two parts of the wavefunction is related to depletion of the incident beam by the scattering center. (Unfortunately, a complete analysis of this interference requires study of the spreading of the wave packets, which we omitted above.)

An improved approximate integration of (1.21) may be obtained by carrying to first order the influence of the scattered amplitudes on the phases of the integrands. We express the amplitudes in the form

$$\begin{aligned} f(\mathbf{k}_\alpha, \hat{r}_\beta) &= \exp[\ln f(\mathbf{k}_\alpha, \hat{r}_\beta)], \\ &\approx f(\mathbf{k}_0, \hat{r}_\beta)\exp[\mathbf{q}\cdot\nabla_\mathbf{k} \ln f(\mathbf{k}_0, \hat{r}_\beta)]. \end{aligned} \quad (1.23)$$

Here $\nabla_\mathbf{k} \ln f(\mathbf{k}_0, \hat{r}_\beta)$ represents the derivatives of $\ln f$ with respect to the components of $\mathbf{k}_\alpha$, evaluated at $\mathbf{k}_\alpha = \mathbf{k}_0$. Introduction of (1.23) in (1.21) yields the improved result

$$\Phi(t) = e^{-(iE/\hbar)(t-t_0)}\Big\{\psi_\alpha e^{i(\mathbf{k}_0\cdot\mathbf{r}_\alpha)}G(\mathbf{r}_\alpha - \mathbf{v}_\alpha t)$$

$$+ \sum_\beta \psi_\beta r_\beta^{-1} e^{ik_\beta r_\beta} f(\mathbf{k}_0, \hat{r}_\beta) G\left(\hat{k}_0\left[\left(\frac{v_\alpha}{v_\beta}\right)r_\beta - v_\alpha t\right] - i\,\nabla_\mathbf{k} \ln f\right)\Big\}. \quad (1.24)$$

The gradient term in the argument of G expresses a displacement of the envelope function of the outgoing wave packet. The two components of this gradient that are perpendicular to $\mathbf{k}_0$ are small; they are derivatives of f with respect to the scattering angle, multiplied by the wavelength k_0^{-1}. It is only the component parallel to $\mathbf{k}_0$ that can become appreciable and that we will retain. If in this component the derivative with respect to momentum is replaced by a derivative with respect to energy, the envelope function in channel β reduces to

$$G\left(\hat{k}_0 v_\alpha\left[\frac{r_\beta}{v_\beta} - t + Q_{\alpha\beta}\right]\right), \quad (1.25)$$

where

$$Q_{\alpha\beta} = -i\hbar\frac{\partial}{\partial E_\alpha} \ln f(\mathbf{k}_0, \hat{r}_\beta) \quad (1.26)$$

is the *time delay* of the reaction. Because of this delay the time when the center of the outgoing packet appears at some given radius r_β is not r_β/v_β, but rather $(r_\beta/v_\beta) + Q_{\alpha\beta}$.

It is especially easy to understand the meaning of $Q_{\alpha\beta}$ if the entire energy dependence of f can be expressed by a single phase shift, δ, so that f is proportional to $\exp(i2\delta)$. In this case

$$Q_{\alpha\beta} = 2\hbar\frac{\partial\delta}{\partial E_\alpha}. \quad (1.27)$$

Near an isolated resonance the phase shift δ is given by the Breit-Wigner formula, and (1.27) then yields a time delay at resonance that exactly equals the lifetime of the resonant state.* This example shows that narrow resonances can cause large time delays. In association with these delays there are large radial displacements of the scattered wave packets; these can be as large, say, as 10^4 times the wavelength k_0^{-1}.

The significance of the above time delay, however, is better appreciated if we recall the earlier estimate that wave packets produced by accelerators have durations of the order of 10^{-10} sec. In comparison with these durations even

* It is possible to turn this discussion around and to base the derivation of the Breit-Wigner formula on the idea of time delay [1].

the longest nuclear time delays are very small. For this reason the energy dependence of the scattered amplitude and the consequent time delay cause only minor end effects, such that the ends of the incident and scattered wave packets have slightly different shapes. Parts of the scattered amplitude that vary slowly with energy contribute little to these end effects.

1.4 Cross Sections

All dimensions of the incident wave packet are much larger than the dimensions of the physical region in which the scattering interaction takes place. The incident packet therefore presents to the scattering center a uniform density of probability flux per unit area. The net incident flux per unit area is obtained by integrating over the entire length of the packet, along the line of impact $x_\alpha = y_\alpha = 0$. The net incident flux per unit area therefore is

$$\mathscr{F} = \int_{-\infty}^{\infty} |G(\hat{k}_0 z_\alpha)|^2 \, dz_\alpha. \tag{1.28}$$

The net outgoing flux in channel β per unit solid angle is obtained in a corresponding fashion by integrating with respect to r_β across the full thickness of the outgoing wave packet. Equations 1.22 and 1.24 both give for the outgoing flux per unit solid angle

$$\int_{-\infty}^{\infty} \left| f(\mathbf{k}_0, \hat{r}_\beta) G\left(\hat{k}_0 \left[\frac{v_\alpha}{v_\beta}\right] r_\beta\right) \right|^2 dr_\beta = \left(\frac{v_\beta}{v_\alpha}\right) |f(\mathbf{k}_0, \hat{r}_\beta)|^2 \, \mathscr{F}. \tag{1.29}$$

The differential cross section for the reaction $\alpha \to \beta$ is the ratio of the two fluxes,

$$\frac{d\sigma_{\alpha\beta}}{d\Omega_\beta} = \left(\frac{v_\beta}{v_\alpha}\right) |f(\mathbf{k}_0, \hat{r}_\beta)|^2. \tag{1.30}$$

Averages of the differential cross section taken with respect to orientations of angular momenta will be required later. These are computed from (1.30) by summing and averaging on the indices α and β, which describe all the properties of the incoming and outgoing states of motion.

1.5 Transition Amplitude

The transition matrix gives the amplitudes of outgoing waves of momenta $\mathbf{k}_\beta$ in channels β in terms of the amplitude of a plane wave of momentum $\mathbf{k}_\alpha$ in channel α. Under physical conditions the total energies in all channels are the same, and the transition amplitudes reduce to simple multiples of $f(\mathbf{k}_\alpha, \hat{r}_\beta)$:

$$T_{\alpha\beta} = -\frac{2\pi\hbar^2}{\mu_\beta} f(\mathbf{k}_\alpha, \hat{r}_\beta). \tag{1.31}$$

When written in terms of transition amplitudes, the expression for the cross section becomes

$$\frac{d\sigma_{\alpha\beta}}{d\Omega_\beta} = \frac{\mu_\alpha \mu_\beta}{(2\pi\hbar^2)^2}\left(\frac{k_\beta}{k_\alpha}\right)|T_{\alpha\beta}|^2. \tag{1.32}$$

These equations are used later.

Insertion of the mass factor in (1.31) causes $T_{\alpha\beta}$ to have more convenient symmetry properties than $f(\mathbf{k}_\alpha, \hat{r}_\beta)$. These symmetry properties are best summarized in terms of the S-matrix,

$$S = 1 - 2\pi i T.$$

The S-matrix is unitary [1, 2].

1.6 Recapitulation

Typical wave packets emitted by an accelerator probably are formed in the ion source with dimensions of the order of the de Broglie wavelength at thermal energies. During the process of acceleration a packet becomes stretched out along its direction of motion until its length may become a noticeable fraction of a centimeter. During the same time its transverse dimensions broaden until they become of the order of 10^{-3} cm. All of these dimensions are much greater than the dimensions of the region of strong interaction with a target nucleus and much greater than the de Broglie wavelength of the rapidly moving projectile. At the same time these typical packet properties display lateral confinement much better than the beam of a typical experiment, and they display energy definition many orders of magnitude better than such a beam. Although beam-forming devices select streams of packets that lie within small ranges of energies and directions, evidently the range of properties of individual packets is much smaller than the range in a beam.

The individual packets scatter in a fashion that is approximated closely by stationary-state eigenfunctions. Thus, because they are large compared with k_0^{-1} and initially far from the scattering center, the packets incident upon the scattering center may be expanded in terms of stationary-state eigenfunctions whose range of incident momenta has the narrow distribution of an ordinary Fourier expansion of the packets. This superposition of energy eigenfunctions is preserved as time proceeds and eventually yields the properties of the radially outgoing scattered wave packets. Provided the momentum distribution of an incident packet is narrow enough, it does not matter at all. Either in (1.22) or in the more accurate (1.24) it is seen that the outgoing wave packet *has the same shape* as the incoming packet; therefore the detailed packet properties cancel when cross sections are evaluated from ratios of fluxes. In the ratios only the scattered amplitudes $f(\mathbf{k}_0, \hat{r}_\beta)$ survive. The stationary-state theory immediately gives just these amplitudes.

The wave-packet description of a scattering experiment need not always reduce so simply to the use of stationary-state scattered amplitudes. There are two steps at which the picture outlined above may fail: (a) the expansion of the incident packet may not be approximated sufficiently well by the limiting expression of (1.13); (b) it may not be possible to factor out the scattered amplitudes from the integrals of (1.21), either in the fashion that yields (1.22) or in the fashion that yields (1.24). Consideration of the parameters of typical experiments, as discussed above, makes it clear that neither of these causes of failure of the stationary-state picture is ordinarily very likely.

Rapid energy variation of the scattered amplitudes in (1.21) leads to outgoing packets whose shapes, as functions of r_β, are not the same as the longitudinal shape of the incident packet. Such alterations of shape occur chiefly at the two ends of the packets and may be quite interesting if the scattered amplitudes should be mixtures of rapidly varying and slowly varying parts. However, over most of the length of a packet its shape is not altered by such effects. Experiments that measure bremsstrahlung emitted in coincidence with nuclear reactions [3] may be considered as attempts to measure these changes of shape in wave packets.

1.7 Time-reversed Motion

Let us imagine a wave packet that is localized about a point $\mathbf{z}_0$ on the positive z_α-axis and has an average momentum directed *away from the scattering center*. Such a packet may be produced as the outcome of a reaction. To study the development of the packet as a function of time its wavefunction must be expanded in terms of stationary eigenfunctions, as done for the incoming packet in (1.13). Further insight into the derivation of (1.13) is gained by studying this example.

Equation 1.13 is an expansion of an incoming packet in terms of stationary-state wavefunctions that obey outgoing-wave boundary conditions. Because the scattered waves are outgoing, they have poor overlap with the incoming packet and do not affect the Fourier coefficients. This same argument indicates that the scattered-wave terms of $\Psi'^{(+)}(\mathbf{k}_\alpha)$ must have excellent overlap with an outgoing wave packet; therefore (1.13) cannot be a correct expansion for such a packet.

The expansion of an outgoing packet is best done in terms of stationary-state wavefunctions that obey *ingoing-wave boundary conditions*. These wavefunctions are obtained from the $\Psi'^{(+)}(\mathbf{k}_\alpha)$ by simple operations derived from the theory of time reversal [4]. It is sufficient to carry through the present discussion for particles that are spinless, and in this case the time-reversed wavefunction is

$$\Psi^{(-)}(\mathbf{k}_\alpha) = \Psi^{(+)}(-\mathbf{k}_\alpha)^*. \tag{1.33}$$

The wavefunction $\Psi^{(-)}$ is a solution of the same time-independent Schrödinger equation that governs $\Psi^{(+)}$; however, it is chosen to have the asymptotic form

$$\Psi^{(-)}(\mathbf{k}_\alpha) = \psi_\alpha e^{i(\mathbf{k}_\alpha \cdot \mathbf{r}_\alpha)} + \sum_\beta \psi_\beta f^*(-\mathbf{k}_\alpha, \hat{r}_\beta) r_\beta^{-1} e^{-ik_\beta r_\beta}. \tag{1.34}$$

Because the asymptotic radial waves of the functions $\Psi^{(-)}(\mathbf{k}_\alpha)$ are ingoing, they have negligible overlap with an outgoing wave packet. Therefore the expansion of the packet in terms of the $\Psi^{(-)}(\mathbf{k}_\alpha)$ is given by the expression

$$\Phi_{\text{out}}(t_0) = \int a(\mathbf{k}_\alpha + \mathbf{k}_0) e^{-i(\mathbf{k}_\alpha \cdot \mathbf{z}_0) - i(\mathbf{k}_0 \cdot \mathbf{z}_0)} \Psi^{(-)}(\mathbf{k}_\alpha)\, d^3k_\alpha, \tag{1.35}$$

in which the coefficients of the $\Psi^{(-)}$ are the Fourier coefficients for the expansion of Φ_{out} in terms of plane waves.

Goldberger and Watson [1] show that the cross section for any reaction can be computed in terms of an overlap between a precollision packet formed from the $\Psi^{(+)}$ and a postcollision packet formed from the $\Psi^{(-)}$. The functions $\Psi^{(-)}$ are used in the present book as final-state wavefunctions in reaction calculations.

1.8 Three-body Final States

Our wave-packet formalism has been designed to describe reactions initiated by the collision of two particles. So far only those final states in which two particles also emerge from the reaction have been discussed. However, no essential changes are required to discuss final states that have three or more particles emerging. It is necessary only to insert in the wave-packet formalism the limiting expression to which the stationary-state wavefunction reduces as the separations among the product particles become asymptotic. This calculation is carried through now for the case in which there are three product particles, both for use later in the book and to illustrate the straightforward nature of the wave-packet approach.

The asymptotic limiting form of the stationary-state wavefunction in three-body channels is not so well known as that in two-body channels. Equation 1.4 is the asymptotic wavefunction for the two-body channels and serves to define the scattered amplitudes. It was presented without discussion. To obtain the corresponding expression for the three-body channels it is best to go back to fundamentals and to construct an appropriate Green's function solution of the time-independent Schrödinger equation (1.3). To do this we project (1.3) onto a channel function ψ_β of the sort introduced in (1.1):

$$\psi_\beta = \psi_{\beta 1}\psi_{\beta 2}\psi_{\beta 3},$$

where $\psi_{\beta 1}$, $\psi_{\beta 2}$, and $\psi_{\beta 3}$ are the internal wavefunctions of the three separating fragments. The resulting differential equation is

$$\left(-\frac{\hbar^2}{2M_1}\nabla_1{}^2 - \frac{\hbar^2}{2M_2}\nabla_2{}^2 - E_\beta\right)F_\beta(\mathbf{r}_1, \mathbf{r}_2) = -(\psi_\beta, V_\beta\Psi^{(+)}), \qquad (1.36)$$

where $\mathbf{r}_1$ and $\mathbf{r}_2$ are the displacements of the centers of mass of particles 1 and 2 from the center of mass of particle 3, and F_β is the wavefunction for the relative motion of the centers of mass of the three particles. For the sake of simplicity in defining the coordinates $\mathbf{r}_1$ and $\mathbf{r}_2$ it has been assumed that particle 3 of this example has infinite mass. Also

$$E_\beta = E_\alpha + \varepsilon_\alpha - \varepsilon_\beta,$$

where ε_β is the sum of the internal energies of the three fragments. The potential V_β includes all interactions that are not internal to any of the fragments. The scalar-product notation in the right-hand side of (1.36) naturally refers to integration only over the internal coordinates of ψ_β; therefore the scalar product remains a function of $\mathbf{r}_1$ and $\mathbf{r}_2$.

The solution of (1.36) is

$$F_\beta(\mathbf{r}_1, \mathbf{r}_2) = -\int G(\mathbf{r}_1, \mathbf{r}_2 \mid \mathbf{r}_1', \mathbf{r}_2')(\psi_\beta, V_\beta\Psi^{(+)})\, d^3r_1'\, d^3r_2', \qquad (1.37)$$

where $G(\mathbf{r}_1, \mathbf{r}_2 \mid \mathbf{r}_1', \mathbf{r}_2')$ is the outgoing-waves Green's function for three free particles. An explicit analytic form for this Green's function is known to be [5, 6]

$$G(\mathbf{r}_1, \mathbf{r}_2 \mid \mathbf{r}_1', \mathbf{r}_2') = \frac{iE_\beta(M_1M_2)^{1/2}}{(2\pi\hbar^2)^2\mathscr{R}^2}\, H_2^{(1)}(\xi\mathscr{R}), \qquad (1.38)$$

where $H_2^{(1)}$ is the outgoing-waves Hankel function of second order and where

$$\xi^2 = 2\hbar^{-2}E_\beta(M_1M_2)^{1/2}, \qquad (1.39)$$

$$\mathscr{R}^2 = (M_1M_2)^{-1/2}[M_1(\mathbf{r}_1 - \mathbf{r}_1')^2 + M_2(\mathbf{r}_2 - \mathbf{r}_2')^2]. \qquad (1.40)$$

We see that $\mathscr{R}$ is the mass-weighted, six-dimensional displacement from the source point $\mathbf{r}_1'$, $\mathbf{r}_2'$ to the field point $\mathbf{r}_1$, $\mathbf{r}_2$. It may be noted that (1.37) is a *particular solution* of the differential equation (1.36). No additional *homogeneous solution* appears in (1.37) because there are no three-body incident waves in channel β.

The asymptotic limit of $F_\beta(1, 2)$ is obtained by introducing the asymptotic Green's function into (1.37). This is derived from (1.38) by expanding $\mathscr{R}$ in powers of $(r_1'/\mathscr{R}_\beta)$ and $(r_2'/\mathscr{R}_\beta)$, where

$$\mathscr{R}_\beta{}^2 = (M_1M_2)^{-1/2}(M_1r_1{}^2 + M_2r_2{}^2),$$

and by using the limiting form of the Hankel function:

$$H_2^{(1)} \to -\left(\frac{2}{\pi\xi\mathscr{R}}\right)^{1/2} \exp i\left(\xi\mathscr{R} - \frac{\pi}{4}\right). \tag{1.41}$$

Then the asymptotic limit of $F_\beta(1, 2)$ is found to be

$$F_\beta(\mathbf{r}_1, \mathbf{r}_2) \to i\left(\frac{2}{\pi\xi}\right)^{1/2} \frac{E_\beta(M_1M_2)^{1/2}}{(2\pi\hbar^2)^2} \frac{e^{i\xi\mathscr{R}_\beta}}{\mathscr{R}_\beta^{5/2}}$$
$$\times \int e^{-i(\mathbf{K}_1\cdot\mathbf{r}_1')-i(\mathbf{K}_2\cdot\mathbf{r}_2')}(\psi_\beta, V_\beta\Psi^{(+)})\, d^3r_1'\, d^3r_2', \tag{1.42}$$

where the momenta $\mathbf{K}_1$ and $\mathbf{K}_2$ have the directions of $\mathbf{r}_1$ and $\mathbf{r}_2$, respectively, and the magnitudes

$$K_1 = \xi\left(\frac{M_1}{M_2}\right)^{1/2}\left(\frac{r_1}{\mathscr{R}_\beta}\right), \qquad K_2 = \xi\left(\frac{M_2}{M_1}\right)^{1/2}\left(\frac{r_2}{\mathscr{R}_\beta}\right), \tag{1.43}$$

so that

$$\frac{\hbar^2K_1{}^2}{2M_1} + \frac{\hbar^2K_2{}^2}{2M_2} = E_\beta. \tag{1.44}$$

Because $\xi\mathscr{R}_\beta = K_1r_1 + K_2r_2$, we see that (1.42) has the classically expected property that particles 1 and 2 both move in outgoing waves, with velocities $v_1 = K_1/M_1$, $v_2 = K_2/M_2$ in the ratio of their distances from particle 3. It also has the interesting property that the magnitude of F_β falls as $\mathscr{R}_\beta^{-5/2}$. This fall off is faster by one power of $\mathscr{R}_\beta^{-1/2}$ than would be expected for a product of two independent fluxes that originate from a common center. The extra $\mathscr{R}_\beta^{-1/2}$ is a result of the *correlation* between the energies of particles 1 and 2 [see (1.44)]; this correlation causes the correlation between given *radial intervals* Δr_1 and Δr_2 to sharpen up as measurements are made at larger radii. Fortunately the implied correlation between intervals of K_1 and K_2, the quantities of experimental interest, does not depend on $\mathscr{R}_\beta$.

We derive the outgoing wave packet in channel β by inserting the part F_β of $\Psi^{(+)}$ into the wave-packet formalism of (1.21). For the sake of conciseness (1.42) is first rewritten as

$$F_\beta(\mathbf{r}_1, \mathbf{r}_2) = \mathscr{R}_\beta^{-5/2} e^{i\xi\mathscr{R}_\beta} f_\beta(\mathbf{K}_1, \mathbf{K}_2; \mathbf{k}_\alpha), \tag{1.45}$$

whereby we have defined $f_\beta(\mathbf{K}_1, \mathbf{K}_2; \mathbf{k}_\alpha)$, the scattered amplitude for three-body breakup. Because the wave packet is made from a very narrow distribution of incident momenta we may reasonably disregard the energy variation of this amplitude, as for two-body breakup. The principal dependence of F_β

on the incident energy then comes through the energy dependence of ξ and is found by methods similar to those of (1.19) and (1.20). If we define an average *effective, weighted outgoing velocity*

$$v_\beta = \frac{\hbar\xi}{\sqrt{M_1 M_2}}, \tag{1.46}$$

the formal expression for the three-body wave packet is almost identical with that of (1.22). It is found to be

$$e^{-(iE/\hbar)(t-t_0)}\psi_\beta \mathscr{R}_\beta^{-5/2} e^{i\xi\mathscr{R}_\beta} f_\beta(\overline{\mathbf{K}}_1, \overline{\mathbf{K}}_2; \mathbf{k}_0) G\left(\hat{k}_0\left[\frac{\mathscr{R}_\beta - v_\beta t}{v_\beta/v_\alpha}\right]\right) \tag{1.47}$$

where $\overline{\mathbf{K}}_1$ and $\overline{\mathbf{K}}_2$ are the values of $\mathbf{K}_1$ and $\mathbf{K}_2$ at the central energy of the incident wave packet. Because the motions of particles 1 and 2 are correlated it is not surprising that the packet function depends on $\mathscr{R}_\beta$, the mass-weighted root-mean-square radial position of the two particles.

The above analysis deduces the asymptotic form of the stationary-state three-body wavefunction by constructing a Green's function solution of the reduced Schrödinger equation (1.36). This analysis, however, is somewhat defective, and it is helpful to remark that Gerjuoy [5] proved in general that the asymptotic three-body wavefunction is proportional to the Green's function of (1.38), just as found above. Therefore our basic wave-packet analysis is correct.

The defect of the Green's function analysis is seen most clearly in the expression for the scattered amplitude (1.42). In this expression one term of V_β is the interaction $V_{12}(\mathbf{r}_1' - \mathbf{r}_2')$, the interaction *between* particles 1 and 2. In the integrations over $\mathbf{r}_1'$ and $\mathbf{r}_2'$ the interaction V_{12} enforces convergence only with respect to $\mathbf{r}_1' - \mathbf{r}_2'$ and not with respect to $\mathbf{r}_1' + \mathbf{r}_2'$. Therefore the convergence of the integrations is in doubt, and the amplitude f_β is poorly defined. To solve this problem Gerjuoy [5] suggested that it would be preferable to define f_β by projecting the time reverse of $\Psi^{(+)}$ onto channel α. Under this definition the expression for f_β depends on the interaction V_α, and because V_α is an interaction in a two-body breakup channel the convergence of all integrals is assured.

However, the wave-packet analysis suggests an alternative well-defined expression for the three-body amplitude, which is related more closely to (1.42). We observe that the time-dependent propagation of a wave packet is inherently well behaved and presents no convergence difficulties. This propagation is described, for example, by (1.7) or by (1.15). Therefore a well-defined three-body breakup amplitude is obtained by left multiplying either of these two equations by the wavefunction ψ_β that projects onto the three-body channel. Difficulties arise only when ψ_β is taken under the integral sign in (1.15) and the time-independent Schrödinger equation is applied to the analysis of the scalar product $(\psi_\beta, \Psi^{(+)}(\mathbf{k}_\alpha))$. It is this step that introduces infinite integrals over the relative coordinates in channel β.

To exploit these observations we note further that the contribution V_{12} introduces into (1.42) is not really very badly behaved. For very large values of $|\mathbf{r}_1' + \mathbf{r}_2'|$ the product $V_{12}\Phi^{(+)}$ in the integrand is dominated by the particular two-body open-channel parts of $\Psi^{(+)}$ in which particle 3 moves in relation to a *bound state* of particles 1 and 2. However, because of energy conservation, the momentum of the center of mass of particles 1 and 2 does not have the same value in such two-body channels that it has in the three-body channel in which we are interested. Therefore the integration over $|\mathbf{r}_1' + \mathbf{r}_2'|$ oscillates indefinitely as the variable becomes large, but does not diverge. A well-behaved amplitude can be obtained by introducing a convergence factor, with the result, for example, that the amplitude of (1.42) goes over to

$$f_\beta = i\left(\frac{2}{\pi\xi}\right)^{1/2} E_\beta(M_1 M_2)^{1/2}(2\pi\hbar^2)^{-2} \lim_{\lambda\to 0} \int e^{-\lambda|\mathbf{r}_1'+\mathbf{r}_2'|} \times e^{-i(\mathbf{K}_1\cdot\mathbf{r}_1')-i(\mathbf{K}_2\cdot\mathbf{r}_2')} (\psi_\beta, V_\beta \Psi^{(+)})\, d^3r_1'\, d^3r_2'. \quad (1.48)$$

The immediate introduction of this convergence factor into (1.15) would not have altered our basic description of wave-packet propagation, however it would have allowed us to interchange orders of integration and still retain well-defined expressions. This use of a convergence factor is equivalent to the retention of a small positive imaginary part in the energy when calculating the stationary eigenfunctions to use in the wave-packet formalism [1]. Huby and Mines [7] were the first to suggest that such a convergence factor could be used in practical calculations to quench the oscillations of f_β. Further discussion of their work is given in Section 5.8.

C. M. Vincent [8] has presented a formal derivation of (1.48) in the context of stationary-state scattering theory, using a procedure that parallels the above qualitative discussion.

References

[1] M. L. Goldberger and K. M. Watson, *Collision Theory*, Wiley, New York, 1964.
[2] Albert Messiah, *Quantum Mechanics*, Wiley, New York, 1963.
[3] R. M. Eisberg, D. R. Yennie, and D. H. Wilkinson, *Nucl. Phys.* **18,** 338 (1960); T. Ericson, *Phys. Rev. Letters* **5,** 430 (1960); H. Feshbach and D. R. Yennie, *Nucl. Phys.* **37,** 150 (1962); R. M. Eisberg, *Rev. Mod. Phys.* **36,** 1100 (1964).
[4] E. P. Wigner, *Göttingen Nachr.* **31,** 546 (1932).
[5] E. Gerjuoy, *Ann. Phys.* **5,** 58 (1958).

[6] Arnold Sommerfeld, *Partial Differential Equations in Physics*, Academic, New York, 1949, p. 231; P. M. Morse and H. Feshbach, *Methods of Theoretical Physics*, McGraw-Hill, New York, 1953, p. 1732; H. Chew, *Bull. Am. Phys. Soc.* **12,** 49 (1967).
[7] R. Huby and J. R. Mines, *Rev. Mod. Phys.* **37,** 406 (1965).
[8] C. M. Vincent, *Phys. Rev.*, **175,** 1309 (1968); also see F. S. Levin, *Ann. Phys.* **46,** 41 (1968).

CHAPTER 2

Direct Reactions

Nuclear reactions concern continuum states of many-body systems. Because these states have high energies, reactions tend to involve complicated excitations of many degrees of freedom. Reactions that do possess such complications are said to proceed by the *compound-nucleus* reaction mechanism (CN). However, it may sometimes happen that only a few degrees of freedom are excited, whereas the other degrees of freedom of the many-body system remain effectively passive. Reactions that possess this simplicity are said to proceed by the *direct-reaction* mechanism (DI). Both the DI and CN reaction mechanisms lead to characteristic properties of predicted cross sections, of types that are discussed subsequently. Because cross sections of both types are found experimentally, it is necessary to build theories that are adapted to both the DI and CN reaction mechanisms. Often the reaction mechanisms tend to be analyzed by quite different theoretical techniques, with DI analyses stressing detailed solutions of the Schrödinger equation and CN analyses stressing statistical ideas. However, it is clear that the CN and DI mechanisms normally are present simultaneously and are two aspects of any given nuclear reaction.

This chapter presents qualitative discussions of theoretical and experimental aspects of the CN and DI reaction mechanisms. Some consideration is given to the rather speculative question of how to separate these two mechanisms when both are present simultaneously. It is argued that this separation is often somewhat arbitrary and that it tends to be governed by questions of taste and convenience. A more quantitative unified treatment of the CN and DI reaction mechanisms is given in Chapter 11.

Section 2.1 considers the possibility of distinguishing among reaction mechanisms on the basis of a model in which the reaction amplitude is separated into parts that have different energy dependences. This separation is related to wave-packet ideas; these ideas suggest that direct reactions should be fast and compound nucleus reactions should be slow. Although the wave-packet ideas do not quite correspond to actual experiments, they suggest a mathematical model that has definite experimental consequences and yields formulas that may be associated with energy-averaged experiments.

In Section 2.2 it is pointed out that theoretical models of the dynamics of a reaction can be classified as being of DI or CN character, largely according to how complicated they are, and that to a *limited* extent this classification of dynamical models can be placed in one-to-one correspondence with the classification of the amplitudes. As a corollary a DI process is defined as one in which no spatially confined excitations are formed.

Section 2.3 is a brief review of the properties of statistical CN theories, for comparison with DI results.

In Section 2.4 an attempt is made to sketch the physical conditions under which reactions ought to show CN or DI character, on the basis of the definitions and discussions of Sections 2.1 and 2.2.

Section 2.5 is a brief discussion of the *boundary-matching theories* used in most CN discussions. In these theories the explicit treatment of the nuclear dynamics is transformed away in favor of an orderly set of empirical parameters.

In Section 2.6 some remarks are given about the use of angular distributions and angular correlations as guides to reaction mechanisms.

Section 2.7 is a mathematical appendix to this chapter and gives a wave-packet treatment of the relation between CN and DI in three-body breakup.

2.1 Time Delay: Energy Averaging

Time delay is a semiclassical concept that appears prominently in heuristic thinking about nuclear reactions and is related closely to experimental observables. Because a direct reaction is simple, it probably takes place quickly (hence its name) by some kind of glancing interaction between the target nucleus and the incident projectile. The duration of such an event should be of the order of 10^{-22} sec, corresponding to one transit across a nucleus by a medium-energy projectile. By contrast the CN reaction mechanism involves complicated intermediate states whose lifetimes may be 10^4 or more times greater than the basic transit time.

The formula for the time delay of a wave packet (1.26),

$$Q_{\alpha\beta} = -i\hbar \frac{\partial}{\partial E_\alpha} \ln f(\mathbf{k}_\alpha, \hat{r}_\beta), \tag{2.1}$$

shows that the above ideas are related to the observed energy dependence of cross sections. Except near thresholds, the amplitude for the very rapid DI process can show appreciable variations only over energy intervals of several million electron volts. The amplitude for the slow CN process, however, should show variations localized within 1 keV or less.

Cross sections that show the above two types of energy variation are well known. The excitation functions for some reactions are observed to have rapid, narrow resonance or fluctuation structure. Other excitation functions, however, even those for reactions leading to discrete states of product nuclei, are observed to change in a gradual, stable fashion over intervals of several million electron volts. It is clear that these two types of energy dependence are to be associated with CN and DI reaction mechanisms respectively.

If CN and DI mechanisms are present simultaneously, the above ideas suggest that a *mathematical model* suitable for the separation of the two mechanisms might be

$$f(\mathbf{k}_\alpha, \hat{r}_\beta) = f_{\mathrm{DI}}(\mathbf{k}_\alpha, \hat{r}_\beta) + f_{\mathrm{CN}}(\mathbf{k}_\alpha, \hat{r}_\beta), \tag{2.2}$$

where over an energy interval of the order of 1 MeV the direct amplitude f_{DI} is *defined* to have negligible energy variation, and the compound nucleus amplitude f_{CN} is *defined* to fluctuate rapidly with average value zero. Insertion of (2.2) in the equation that generates the outgoing wave packet (1.21) seems to yield two separated packets, the packet due to f_{CN} having a long time delay and the packet due to f_{DI} having a short time delay.*

The model expressed by (2.2) may be extended; for example, analyses of *intermediate structure* suggest that the amplitude may possess three qualitatively different rates of energy variation [1, also see Chapter 11]. However, this extended model is not discussed.

Unfortunately, a too casual wave-packet interpretation of (2.2) is misleading. It is true that the packet generated from f_{DI} is delayed by about 10^{-22} sec and the packet generated from f_{CN} is delayed by a time that is orders of magnitude greater. However, it has already been pointed out that practical packets produced by accelerators probably have durations in the range of 10^{-12} to 10^{-10} sec. Under these circumstances the minute difference in time delay of the packets generated from f_{DI} and f_{CN} is without meaning; because of their great durations the two scattered packets overlap almost perfectly. They are effectively one packet, and when the cross section is computed the f_{DI} and f_{CN} amplitudes interfere. Because the two terms of (2.2) do not lead to distinct packets, it is not clear that they refer to distinct physical processes.

Friedman and Weisskopf [2] attempted to clarify the meaning of the time delay. They imagined initiating an experiment with a very brief wave packet, perhaps about 10^{-21} sec long. If the momentum distribution for this hypothetical incident packet is inserted in (1.21), the scattered packets yielded from the two terms of (2.2) are found to have very different shapes. The term

* Of course, simple addition of independent amplitudes risks violation of flux conservation (unitarity). In most practical applications of (2.2) this is not a problem. Further discussion is given in Chapter 11.

$f_{\rm DI}$ yields a scattered packet that is a brief pulse of the same duration as the incident pulse. However, the term $f_{\rm CN}$ yields a scattered packet whose intensity decays exponentially, with a mean life that is the average of the lifetimes of CN states whose energies lie in the packet. Two packets of such different shapes have negligible overlap and yield independent (incoherent) contributions to the cross section.

Thus use of a brief incident wave packet eliminates interference between the DI and CN amplitudes. Such a brief wave packet spans a broad range of incident energies. This observation suggests that DI-CN interference can be eliminated in a practical fashion by performing experiments with poor energy resolution, by measuring energy-averaged cross sections. This is not to say that an experiment performed with a resolution interval I is identical to an experiment initiated by packets of duration $(\hbar/I)$. In a practical experiment each individual packet that comes from the accelerator has fine energy resolution; the fact that I is large merely means that the experiment spans a large ensemble of these packets. Each packet of the ensemble effectively initiates an independent high resolution experiment. Therefore the average cross sections that are measured in low resolution are effectively averaged over independent experiments.

Brief wave packets and poor-resolution experiments, however, both use broad distributions of incident energies; their difference lies only in the phase relations among the stationary eigenstates that make up the distributions. These phase relations play no role when the average cross section is computed. The cross section is proportional to the square of (2.2) and contains a DI-CN interference term that is linear in $f_{\rm CN}$. Therefore the fluctuations of $f_{\rm CN}$ cause DI-CN interference to average to zero, no matter which incident phase relations appear in the averaging process. The averaged cross section is found to be

$$\left\langle \frac{d\sigma_{\alpha\beta}}{d\Omega_\beta} \right\rangle_I = \left(\frac{v_\beta}{v_\alpha}\right)\{|f_{\rm DI}|^2 + \langle |f_{\rm CN}|^2 \rangle_I\}, \tag{2.3}$$

where the angular brackets indicate the energy averaging and v_β/v_α has been treated as constant. Equation 2.3 has been the basic expression used to separate experimental cross sections into DI and CN parts. For elastic scattering the two terms of (2.3) are known as the *shape elastic* and *compound elastic* parts of the cross section, respectively. Independent theories are used to compute these two parts of the cross section.

Because (2.3) is used frequently, it is worthwhile to remark that correct elimination of DI-CN interference requires averaging over an interval I that is large compared with the width and spacing of CN states, so that the average may include many rapid fluctuations of $f_{\rm CN}$. Experiments that average over smaller energy intervals are influenced by effects of DI-CN interference.

2.2 Alternative Definition of DI

The above separation of the amplitude into two parts labeled DI and CN is mathematically simple and is well suited for application to the analysis of experiments. These parts are separated on the basis of their energy variation, a measurable property. However, our object in studying the mechanism of a reaction is primarily to determine which theoretical models are important *in the dynamics* of this reaction. Theoretical models are introduced by dividing either the Hamiltonian or the wavefunction into DI and CN parts; only if such parts can be assigned clearly can a reaction be said to be understood. It is therefore necessary to ask, especially if DI and CN are in strong competition, whether the DI and CN parts of the amplitude, as already defined, can be placed in one-to-one correspondence with DI and CN dynamical models.

1. To formulate DI and CN dynamical models for a system the stationary-state wavefunction at energy E is expressed as

$$\Psi(E) = \Psi(E)_{\text{closed}} + \Psi(E)_{\text{open}}. \tag{2.4}$$

Qualitative discussion does not require that we define the two terms of (2.4) very precisely. However, it is understood that Ψ_{closed} is confined to a limited volume of configuration space; thus it is a normalizable wavefunction and does not exhibit flux at large distances. By contrast Ψ_{open} is understood to include all those parts of $\Psi(E)$ that are not confined, that do not vanish in the channel region of configuration space, and that do exhibit flux at large distances. These definitions of Ψ_{closed} and Ψ_{open} are not complete; however the further steps required to give a full specification of these functions are best regarded as aspects of the construction of models for particular applications. (It may be presumed that the functions Ψ_{closed} and Ψ_{open} are antisymmetrized, but this property is of little importance in the present qualitative discussion.)

One aspect of the incompleteness of the above definitions is that Ψ_{open} has been defined only in the channel region of configuration space. Most dynamical models make the definition more explicit by requiring Ψ_{open} to be a sum over products of internal (channel) wavefunctions and relative wavefunctions and using this expression for Ψ_{open} throughout all configuration space. We adopt this more explicit definition and hence have

$$\Psi_{\text{open}} = \sum_{\alpha} \psi_{1\alpha}\psi_{2\alpha}\xi_{\alpha}(\mathbf{r}_{1\alpha} - \mathbf{r}_{2\alpha}) + \text{three-body channels} + \text{etc.} \tag{2.5}$$

Equation 2.5 is still an incomplete mathematical specification of Ψ_{open} because the relative wavefunctions ξ_α are still known only asymptotically. More precise definitions of the ξ_α are left for particular applications. Normally only the two-body terms in (2.5) are retained.

It is clear that in general the various terms of (2.5) cannot be expected to be orthogonal to one another. It is also clear that the set of basis functions used in (2.5) and in any reasonable Ψ_{closed} probably is overcomplete. Neither of these failings causes difficulties in the present qualitative discussion. Improvement of these failings may be left for particular applications.

> Dynamical models of DI type are now understood to be models that discuss the coupling among the various terms of Ψ_{open} without any consideration of Ψ_{closed}. By contrast dynamical models of CN type are understood to require the various terms of Ψ_{open} to be coupled by an intermediate coupling to some part of Ψ_{closed}.

Although nuclear interactions normally cause couplings of both kinds, particular reactions may emphasize one or the other. It is clear that the vagueness in the definition of Ψ_{open} and Ψ_{closed} leads to a vagueness of distinction between CN and DI dynamical models. This vagueness allows the free invention of DI dynamical models for particular reactions. It is interesting that any one definite choice of basis functions for the expansion of $\Psi(E)$ tends to eliminate this vagueness [3].

2. It is next important to observe that the function Ψ_{open} is probably much *simpler* than Ψ_{closed}. To give this remark meaning we note in (2.5) that the number of basis functions used to construct Ψ_{open} is equal to the number of distinct products $\psi_{1\alpha}\psi_{2\alpha}$, that is, to the number of channels open at the energy in question. At moderate energies the number of open channels is not large and therefore the complications of Ψ_{open} are limited. Further limitation of complications appears because at moderate energies radical differences among the internal wavefunctions of the open channels are not likely. In summary it may be said that Ψ_{open} does not depend on very many *active* degrees of freedom.

By contrast, although Ψ_{closed} must be localized, it is not limited in any other way and it may be expanded in essentially any basis that is complete for the overall reacting system. As an example, let us consider the set of states of the independent-particle model (IPM). These states are antisymmetrized products of single-particle orbitals and their properties are known by examining those orbitals. We note that if all the orbitals in a given IPM state are bound then this state is of the type Ψ_{closed}. We shall speak of such a state as belonging to a *closed configuration* of the IPM. Because the energy of an

IPM state is the sum of the energies of the single-particle orbitals from which it is constructed, we see that closed configurations can have positive energies and can serve as *bound states in the continuum*. Such bound-state wavefunctions belong to a discrete energy spectrum. Presumably Ψ_{closed} is a linear combination of IPM states that belong to closed configurations whose energies are near the energy E of the reacting system. At most energies of interest in nuclear reaction experiments such closed configurations must contain two or more single-particle orbitals that have energies greater than in the IPM ground state. Only in this manner can the energy of excitation be distributed sufficiently widely so that no one orbital lies in the single-particle continuum. Of course, other nearby IPM states may contain one or more orbitals whose energies are in the single-particle continuum. Such states are parts of Ψ_{open}. They may be said to belong to *open configurations* of the IPM. We now observe that in the region of high level density, in which the excitation energy of a system has become large compared with the single-particle level spacing, *the closed IPM states must be much more numerous than the open ones*. This fact follows because the closed configurations contain more excited orbitals than the open ones, and because the number of states in any configuration increases rapidly with the number of excited orbitals it contains. Because the closed IPM states are so numerous, the number of basis states that mix together to form Ψ_{closed} is likely to be much larger than the number of open channels; therefore Ψ_{closed} is likely to be more complicated than Ψ_{open}.

3. Another aspect of the relation between Ψ_{closed} and Ψ_{open} is seen by considering the energy dependence of $\Psi(E)$ of (2.4). The energy E appears in Ψ_{closed} in the combination coefficients with which the discrete basis functions are added together and it appears in Ψ_{open} in the relative wavefunctions ξ_α. Presumably, if only the direct couplings among the open channels were taken into account, the ξ_α would have a fairly smooth and weak energy dependence. On the other hand, because the number of basis states in Ψ_{closed} is very large, the coefficients that describe the makeup of Ψ_{closed} at each energy may be expected to vary erratically as we go from energy to energy. (Often these coefficients are treated as random variables [4].) Therefore the fact that Ψ_{closed} is part of the wavefunction $\Psi(E)$ must cause the ξ_α and the scattering amplitudes to have a rapidly fluctuating dependence on energy. This rapid energy dependence would be absent if only DI couplings among the open channels were carried.

Thus the CN and DI couplings induce in the reaction amplitude just the two kinds of energy dependence that were considered in Section 2.1. It appears that the role of Ψ_{closed} is to extend the lifetime of whatever compound nucleus a reacting system may form, whereas the role of Ψ_{open} is to shorten this lifetime. Of course, it is not necessary to study the energy dependence of the scattered amplitude to recognize that Ψ_{closed} extends the time during which strong interactions are effective. Because Ψ_{closed} exhibits no flux asymptotically, it does not contribute to the decay of the system, and for this reason alone it extends lifetimes. Only Ψ_{open} allows decay of the compound system, and all excitations of Ψ_{closed} must decay through Ψ_{open}. Inversion of this argument emphasizes that excitation of many degrees of freedom extends the lifetime by causing the nuclear wavefunction to be more confined.

4. We return now to the question whether DI and CN dynamical models can be in one-to-one correspondence with DI and CN parts of the transition amplitude, with the one part of the amplitude smooth and the other part fluctuating about a zero mean. We have seen that a classification of dynamical models is based on a classification of the matrix elements of the nuclear interaction; a CN model uses matrix elements that link open parts of $\Psi(E)$ with closed parts, whereas a DI model uses matrix elements that link together the open parts of $\Psi(E)$. Because the relation between the interaction and the transition amplitude tends to be strongly nonlinear, a one-to-one correspondence between additive dynamical models and additive parts of the amplitude must normally be impossible. As a graphic illustration that such a correspondence is false [5] we note that the idea that independent dynamical models produce independent parts of the amplitude implies that a sum of two independent unitary matrices is a unitary matrix. Such a result is not possible.

Thus the calculation of the ξ_α and the amplitudes is affected in a nonlinear fashion by the combined DI and CN couplings. When both sets of couplings are strong, it seems hopeless to try to identify simple dynamical models with simple parts of the amplitude. If the direct couplings could be treated as perturbations, it would at least be possible to identify a part of the amplitude that would be linear in the DI couplings. However, this DI part of the amplitude would still show fluctuations because of the accompanying CN couplings [3].

Only two reasonable bases for a meaningful separation of DI and CN effects seem to exist: (a) under special circumstances one or the other type of coupling may be strongly dominant (examples are given in Section 2.3); (b) under more general circumstances it may be shown that the *energy-averaged transition amplitude* is governed primarily by the DI couplings, with only minor modifications (optical potentials) that express in a simplified way the average influence of the CN couplings. This energy-averaging approach [3] seems to be the major basis for the separation of DI and CN effects. It is discussed at length in Chapter 4 and in Chapter 11.

The average amplitude is governed by the DI couplings and by the optical potentials. However, the average amplitude does not exhaust the influence of the DI couplings. Subtraction of the average amplitude leaves a fluctuating remainder amplitude whose mean is zero. *The presence of the DI couplings correlates the fluctuations* of this remainder amplitude [3, 5, 6]. Unfortunately, most theories of cross section fluctuations ignore [7] this type of correlation, therefore the significance of these theories is doubtful.*

Some sophisticated *unified theories* omit part of the CN couplings from the process of energy averaging in order that a small number of closed states may be included in the explicit calculations performed with Ψ_{open}. Further remarks about these theories appear in Chapters 4, 6, and 11. Such theories yield amplitudes that possess a resonance structure. Each discrete closed state carried in these theories gives rise to a distinct resonance.

5. DI theories that only incorporate couplings among the open channels also sometimes yield amplitudes that show rapid variation with energy as if narrow isolated resonances were present [8]. In this manner these theories mimic the effects of CN theories. Because few degrees of freedom are involved, the number of such resonances is extremely small; nevertheless their mere existence can be a source of confusion. Resonances of this type often seem to be consequences of resonances in the spectrum of single-particle states that are used as a basis for the reaction theory. It is therefore helpful to recognize that single-particle resonances are not frequent in practical calculations based on *optical potentials*. Exploratory studies of the single-particle spectrum often show resonances; however, these resonances tend to be quenched when practical optical model parameters are used systematically [9, 10]. This quenching occurs because use of a smooth nuclear surface and of normal optical potential absorption strengths greatly attenuates multiple reflection effects. Figure 2.1 illustrates what turning on the absorption does to an otherwise prominent resonance.

Single-particle resonances are of practical significance only at energies very close to thresholds, where appreciable mismatch of wavelengths inside and outside the nucleus can occur; for example, such resonances affect calculations of the *strength function* for slow neutron scattering [11].

2.3 Consequences of CN Models

We now present a brief survey of statistical consequences of CN models, to provide background for later discussions of mixed situations. It is assumed here that DI couplings are absent.

At each energy the function Ψ_{closed} is a linear combination of the states of a dense, discrete set, defined by some model Hamiltonian (such as the IPM). It is readily seen (e.g., see Chapter 11 or Ref. 3, 5, 12) that coupling to this Ψ_{closed} induces a resonance structure in the ξ_α of Ψ_{open}, which causes poles to appear in the reaction amplitudes. The density of these resonances (or

* We note now that (2.2) must be regarded as defective because it does not recognize either DI effects in the fluctuating amplitude or CN effects in the average amplitude.

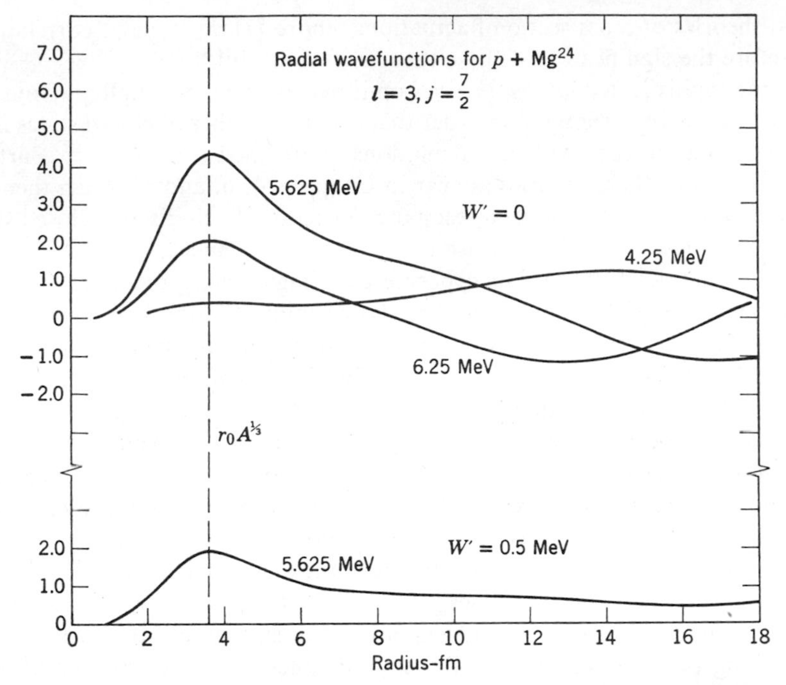

Fig. 2.1 Even a weak imaginary potential (10 MeV is realistic) quenches the 5.625 MeV resonance. All wavefunction magnitudes are normalized to unit amplitude at large radius. (From [9].)

poles) per unit energy is likely to be much the same as that of the basis states from which Ψ_{closed} is assembled. An occasional pole may involve only one basis state. Coupling to Ψ_{open}, however, broadens the region of energy over which a given resonance participates in $\Psi(E)$. Therefore it is only for very light nuclei and very low energies that the density of resonances is so low that individual ones are of concern.

Because the nuclear interaction can link the closed basis states with each other and the CN part of this interaction can link the closed states with Ψ_{open}, we may expect each resonance in $\Psi_{\text{closed}}(E)$ to be a vast linear combination of the nearby closed basis states. At any particular energy E the particular basis states that appear in $\Psi_{\text{closed}}(E)$ determine with which open channels it is linked. It is understandable that if the amplitudes are averaged over energy the special effects due to particular basis states are averaged out, with the result that *on the average each open channel is coupled to* Ψ_{closed} *with the same strength.* This idea is the starting point of statistical theories of CN-dominated reactions. Under conditions of high level density the overlapping of resonances causes a similar averaging over basis states, even without any energy averaging.

If the properties of individual basis states are averaged away, the determination of the reaction amplitudes is dominated by properties of the ξ_α that arise in the region external to the overlap with Ψ_{closed}. The major properties to consider are the *penetrability coefficients* P_α that measure the ratio of $|\xi_\alpha(\mathbf{r}_\alpha)|^2$ at small r_α to $|\xi_\alpha(\mathbf{r}_\alpha)|^2$ at large r_α. Each P_α is determined by the Coulomb and centrifugal barriers in channel α. It is clear that any cross section involving channel α must be proportional to the penetrability P_α.

Channels that differ in their angular momenta or in the sharing of the total charge may have strikingly different values for the P_α. Therefore there is a *competition among the open channels*, such that the total cross section for CN formation is shared among open exit channels in proportion to their penetrabilities. Particles with large penetrabilities, such as neutrons, normally are emitted with overwhelmingly greater probability than particles with small penetrabilities, such as massive, composite charged particles. A dramatic experimental illustration of this fact is the quenching of the CN contribution to the elastic scattering of charged particles as the bombarding energy is raised high enough to allow neutron emission [13]. CN decay by the emission of S-wave neutrons tends to be so probable that it suppresses CN decay by the emission of particles with much lower penetrabilities. (Exit channels with low P_α tend to be populated by one or another sort of DI process.) The spectrum of the emitted S-wave neutrons tends to have a characteristic *evaporation shape* that is determined by the number of neutron exit channels per unit energy multiplied by the (energy-dependent) penetrabilities for these channels.

Quantitative theories of statistical CN-dominated reactions are obtained if it is observed that the competition among channels cannot mix values of conserved quantities, such as total J and parity [14]. Explicit insertion of flux conservation (unitarity) further improves these statistical theories [15].

The angular distributions of CN-dominated reaction cross sections have qualitative properties that depend on the extent of averaging over resonances. Thus an isolated resonance incorporates basis states of only one parity, therefore in each exit channel it couples only to partial waves that are all odd or all even. Therefore the ensuing differential cross sections are symmetric about 90° scattering angle. Any overlap of resonances of opposite parity allows simultaneous coupling to both odd and even partial waves and therefore allows fore-aft asymmetry to appear. The angular distribution may peak in either the forward or backward hemisphere. However, this interference of resonances of opposite parity may be eliminated by averaging over many resonances, either by averaging over E or by going to circumstances in which many resonances overlap. Such averaging again yields a cross section that is symmetric about 90°. *Under special conditions* [14] this cross section may be isotropic.

2.4 Conditions for Direct Reactions

The question whether the conditions of a given reaction more nearly favor direct or compound processes may now be studied. We do this by considering how these conditions affect the coupling between Ψ_{open} and Ψ_{closed}, hence how they affect the extent to which confined wavefunctions can be formed. The coupling between closed and open wavefunctions is seen to depend on particle types and energies, on relative angular momenta, and on which exit channels are considered. Figure 2.4 presents a partial summary of this discussion.

1. There is a competition for the incident flux, as illustrated in Fig. 2.2. Flux enters the reaction through the entrance channel term of Ψ_{open} as shown. Much of this flux emerges again in this same channel after only suffering elastic scattering by the DI interaction that is diagonal in the channel index. In addition, flux may be coupled to other terms of Ψ_{open}, again with the possibility of immediate departure from the scene of the reaction, or it may be coupled to various terms of Ψ_{closed}. One may now ask which terms of Ψ_{open} and Ψ_{closed} are best coupled to the entrance channel. Good coupling to other terms of Ψ_{open} causes a direct reaction. Good coupling to any of the terms of Ψ_{closed} causes confinement of the flux and initiates compound nucleus formation. Both sorts of couplings can exist simultaneously so that the two reaction processes can compete (and interfere) with each other.

The largest of the matrix elements that couple to the entrance channel are the ones that link it with wavefunctions as much like it as possible. Probably the strongest is the diagonal interaction that is responsible for shape-elastic scattering. Nondiagonal interactions of a comparable nature, however, appear for inelastic scattering caused by collective excitations or by isobaric-analog excitations. Such processes may be thought of as *quasi-elastic scattering*, on the grounds that the excited states formed under these circumstances belong to the same *intrinsic wavefunction* as the ground state. It is understandable that the entrance channel is coupled strongly to such states and that large DI cross sections are found for these particular open channels.

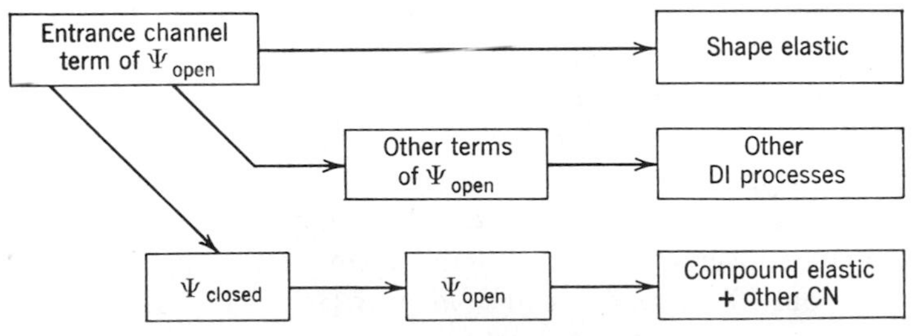

Fig. 2.2 Competition for the incident flux.

Most matrix elements that link with the entrance channel are much smaller than the ones just mentioned. This particularly must be true for the matrix elements that connect with the basis states of Ψ_{closed}, because these states differ from the ground state of the target nucleus, as explained in Section 2.2, by the introduction of one or more single-particle excitations. However, even though the matrix elements that connect with these complicated states are small, the density of these states is so high, at all except the lowest bombarding energies, that most of the flux may well go into their formation. Unless there are special reasons to the contrary CN is likely to be favored whenever the level density of the compound system becomes very high.

One immediate special effect that inhibits the easy excitation of complicated CN states is seen in reactions initiated by single incident nucleons. Because the basic nuclear force is (mainly) a two-body interaction, it couples with the entrance channel only those few nearby closed IPM states that have just *one additional single-particle excitation*. To link in closed basis states that are more complicated the nuclear interaction must act repeatedly. Very complicated states become mixed into the wavefunction only gradually, in higher and higher stages of perturbation theory. Because the various terms of Ψ_{closed} are coupled in by stages and the nuclear force that couples them is comparatively weak, we see that the relative weights with which the various basis states appear in Ψ_{closed} may be very different from their relative statistical weights. Weisskopf has especially emphasized this picture: the complicated parts of the overall nuclear wavefunction are coupled in gradually; therefore the DI effects associated with the simple parts of the wavefunction are enhanced [16]. Griffin showed how semiclassical analyses of the step-by-step coupling to the closed IPM basis states could predict departures from the neutron evaporation spectrum of the statistical CN theory [17].

Angular momentum plays an important part in the inhibition of CN excitation and is as effective for composite incident projectiles as for nucleons:

a. One effect of angular momentum is based on the fact that Ψ_{closed} is localized and tends to be confined within some *nuclear radius* R. Therefore incident partial waves that have high angular momentum ($l > kR$) have poor overlap with Ψ_{closed}. Because such partial waves of high angular momentum can have good overlap with corresponding high partial waves in other open channels, their poor overlap with Ψ_{closed} inhibits CN more than DI.

b. Another effect appears because most basis states available for Ψ_{closed} have rather low J. This well-known property of nuclear level densities follows theoretically from statistical analyses of the angular momentum coupling of IPM states. Because of this property the number of CN states to which partial waves of high angular momentum can couple is limited, and CN excitation is further inhibited. All in all, incident partial waves that have high angular momentum preferentially excite DI reactions. Because the angular momentum in the entrance channel is unlimited and partial waves with high angular momentum have high statistical weight (this goes as $2l + 1$), we see that a major role for DI processes is indicated.

The above discussion about angular momenta is paralleled by the semiclassical observation that projectiles incident on a nucleus with such large impact parameters as to miss the main body of the nucleus are not likely to cause CN formation. They have little overlap with the confined wavefunctions that are characteristic for CN formation. A further aspect of this observation is that large impact parameters present a large effective target area; therefore their DI contributions to the reaction may be significant.

2. Besides asking how CN and DI excitations compete for the incident flux, we must ask which process more effectively feeds flux into *given final states*. We observed in Section 2.3 that competition with neutron emission tends to suppress CN emission of particles that have low penetrabilities, such as massive charged particles. Because DI processes are simple (i.e., because they are *fast*), DI emission into unfavored exit channels is not quenched by competition with favored exit channels. Therefore exit channels that have low penetrabilities tend to be formed preferentially by DI processes. Much early emphasis on DI reaction mechanisms was initiated by the observation of appreciable cross sections for such exit channels [18].

3. As a corollary of the arguments given above it may be noted that collisions that have low impact parameters and are initiated by composite projectiles (projectiles other than neutrons or protons) can be expected to cause strong CN excitation. Under such circumstances nuclei tend to be *black* and give complete absorption of the projectile out of the entrance channel (see Sections 5.2, 5.3.)

4. Bombarding energy has a complicated and varied influence on the balance between reaction mechanisms. This shows up in competition for the incident flux and in the selection of final states that are more or less probable for formation by CN or DI.

The selection among final states is the easier question to discuss. It is clear that both CN and DI cross sections rise rapidly just above threshold, as penetration factors increase. However, as the energy is raised further, there is a rapid increase in the number of open exit channels. Since the CN process populates all exit channels with comparable likelihood, there is a consequent rapid decrease of the CN contribution to any one exit channel. The DI process, on the other hand, does not form an intermediate dynamical system that has a choice of decay modes, therefore it is not affected by competition among the exit channels. Therefore the DI contribution to a given final state simply falls slowly as the energy rises and the increasing momenta gradually reduce values of overlap integrals. Thus competition among final states eliminates CN contributions to the favored DI exit channels and allows DI to be studied in isolation from CN.

A dramatic illustration of the effect of competition among exit channels is found in experiments by Cranberg and Zafiratos [19], who studied the inelastic scattering of neutrons by lead nuclei. At bombarding energies of 8 and 6 MeV strong DI excitation of the 2.7 MeV collective 3^- state of these nuclei is found. However, at 4 MeV the DI contribution to this state is swamped by the CN contribution. This comes about because a 2-MeV decrease in bombarding energy causes a tenfold decrease in the number of competing open exit channels.

To discuss competition for the incident flux we must distinguish three energy regions: At low energy, barrier penetration effects may be seen to favor CN over DI [20]. This comes about because DI amplitudes are proportional to the overlap of entrance and exit channel wavefunctions; therefore DI cross sections are proportional to the products of the penetration factors for the two channels and vanish strongly as the energy goes to zero. By contrast, to excite the CN it is necessary only that the entrance channel have overlap with Ψ_{closed}. The CN excitation function is proportional only to the one penetration factor for the entrance channel, the so-called *sticking probability* for CN formation. Therefore CN excitation falls off more slowly than that of DI as the energy goes to zero. (We note that penetration factors for exit channels affect the CN lifetime and competition among CN exit channels, but do not affect competition for the incident flux.) The difference between the threshold behaviors of DI and CN excitation functions may be seen in the reaction $U^{238}(n, n')$, which was studied by Chase, Wilets and Edmonds [11]. Near threshold the DI cross section for the $0^+ \to 2^+$ rotational excitation, computed by the methods discussed in Chapter 5, is much smaller than the cross section predicted by the WHF statistical theory [14]. The ratio of DI and CN cross sections is given in Fig. 2.3.

It is not always possible, however, to establish the relative importance of DI and CN just by counting barrier penetration factors. DI particle transfer reactions often take place at large impact parameters and involve much larger penetration factors than are possible for CN formation. This effect is common in (d, p) reactions. In (d, p) the DI process may dominate even at very low bombarding energies ([21, 22] and Section 5.9).

As the bombarding energy gradually rises, two effects compete in modifying the balance between CN and DI. One is that the density of nearby closed states increases; this favors an increased rate of transition into the CN. The other effect, however, is that a large fraction of the states that are linked to the entrance channel to first order in the nuclear interaction become open for

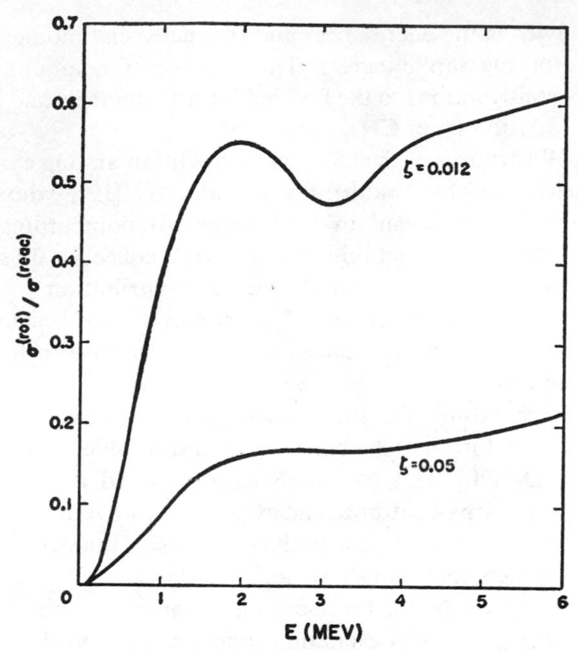

Fig. 2.3 Fraction of reaction cross section contributed by the DI process, as a function of bombarding energy, for the reaction $U^{238}(n, n')$. Results are shown for two different values of the absorptivity, ζ. (From [11], Fig. 4.)

particle emission; this favors an increased rate into DI. We may define *medium energy* as the condition in which both effects are strongly developed but in which multiple-particle emission is not yet likely. The medium energy range for nucleons may lie, very roughly, at about 10–50 MeV. It is difficult to find any general rules with which to decide how the flux divides between DI and CN under medium-energy conditions. This division apparently must be discussed in a detailed manner, as done earlier, on the basis of particle types, angular momentum, and the spectrum of available states. Undoubtedly most of the flux that arrives in low partial waves goes into CN excitations.

For rather high bombarding energy, say for nucleon-induced reactions at greater than about 100 MeV, it again becomes possible to make a number of simple and general statements about the balance between CN and DI. Because of the high energy, nearly all the IPM configurations coupled to the entrance channel to first or second order in the two-nucleon interaction must be open and must have large penetration factors for particle emission. For light nuclei the first one or two collisions between the incident projectile and the particles of the target nucleus yield immediate DI emission of a small number of particles that carry off most of the energy. For heavier target nuclei the initial collisions are not so decisive. The matrix elements of the two-nucleon force strongly favor such linear combinations of open IPM configurations as are excited by collisions with small momentum transfers. The initial collisions, therefore, produce particles that move into the nucleus in the direction of the incident particle; therefore they tend to be followed by further collisions that lead on to CN formation. For this reason only initial collisions near the periphery of a heavy nucleus lead to DI emission. We note that the above high-energy discussion pictures nuclear reactions as proceeding by a succession of simple nucleon-nucleon collisions. Experimental evidence that cross sections for inelastic scattering of 160 MeV protons conform to this picture is presented by Wall and Roos [23].

In the above picture of high-energy reactions the emission of energetic nucleons is almost purely a single-collision, DI process. This agrees with the picture that any CN part of the wavefunction is confined and may not contain highly-excited single-particle states. It is interesting that the wavefunction that describes the DI process may nevertheless contain an important CN part. Such a CN part of the wavefunction would describe secondary boil off of low-energy nucleons, following a DI first stage. Although secondary boil off does not involve a large fraction of the incident energy, it may cause emission of a *greater number of particles* than emerge in the DI first stage.

One way high-energy DI processes can cause CN excitations is by the creation of hole states; for example, the incident projectile may collide with a nucleon in the target nucleus and excite it from a deep inner shell to a high-lying continuum state. In all probability the nucleon thus excited emerges from the nucleus without further collisions and is a reaction product. Removal of this first nucleon, however, leaves a hole in the inner shell and therefore leaves an intermediate nucleus whose wavefunction is closed but highly excited. Hence the reaction can have a second stage in which a nucleon from an outer shell drops into the hole and gives up sufficient energy to eject one or more other nucleons. We thus get typical CN boil off. Considerable energy can be left in deep hole states; for example, it has been shown experimentally [24] that 65 MeV is required to remove a $1s$ proton from Al^{27}. Therefore the Mg^{26} system left by removing a $1s$ proton from Al^{27} is excited by 65 MeV and can easily boil off several nucleons. It is reasonable that some of the best verifications of CN evaporation formulas have been found in high-energy reactions initiated by DI first stages [25].

When a reaction has a DI first step that causes CN formation, the two processes are not in competition. If the several steps of such a reaction are independent except for energetics, the reaction amplitude breaks up into factors. One factor would describe the particle(s) emitted in the DI first step, another factor would describe the emission of the first boil-off particle, etc.

Such a product of DI and CN factors contrasts with the conditions at low energy, where DI and CN are alternatives to each other. These remarks suggest a mathematical model that includes CN amplitudes of both types:

$$f \approx f_{\mathrm{CN}}^{(1)} + f_{\mathrm{CN}}^{(2)} f_{\mathrm{DI}}. \tag{2.6}$$

Equation 2.6 is a convenient summary of the ideas just discussed. Here $f_{\mathrm{CN}}^{(1)}$ is the CN amplitude that is *alternative* to DI and is important at low energy. On the other hand, $f_{\mathrm{CN}}^{(2)}$ is the CN amplitude that *accompanies* DI.

Further understanding of the product term in (2.6) can be gained by examining its consequences for the time development of a brief incident wave packet, along the lines suggested by Friedman and Weisskopf (see Ref. [2] and Section 2.1). Section 2.7 presents details of the wave-packet analysis, in which it is assumed that f_{DI} has negligible variation over the energy range spanned by the incident packet and that $f_{\mathrm{CN}}^{(2)}$ is a sum over independent Breit-Wigner resonances. It is found that the outgoing wave packet has a sharp onset, marked by the motion of the particle described by f_{DI}. This onset is followed by a set of exponential tails, one for each resonance of $f_{\mathrm{CN}}^{(2)}$; the decay length of each exponential is derived from the lifetime of the corresponding resonance. Of course, f_{DI} is a factor common to all terms of the outgoing wave packet, so that integration over the exponential tails (this is tantamount to nonobservation of the second-emitted particle) leaves f_{DI} by itself as the packet amplitude. This analysis emphasizes that although the wavefunction for a high-energy reaction has its CN aspects the experiment can be arranged so that only the DI first step is studied.

Thus at high bombarding energy the CN aspects of the wavefunction show up primarily in the channels for multiple-particle emission, and even these channels may have clearly defined DI aspects. Exit channels for such simple processes as inelastic scattering or single-nucleon transfer are entirely free from CN contamination.

5. Figure 2.4 presents a partial summary of our lengthy discussion of conditions for direct reactions. It illustrates the competition for the incident flux and emphasizes the important control that angular momentum has over this competition. High partial waves form the CN with difficulty; however, they easily give DI transitions to exit channels with which they have good overlap (the *parentage* is close). This rule applies both for nucleons and for composite projectiles. For low partial waves it is necessary to distinguish among several cases:

a. If both the colliding particles are composite the low partial waves always give strong CN formation merely because coupling to the more complicated terms of Ψ_{closed} goes easily (*mean free paths* are small). Even under these conditions small DI admixture terms may yield dominating

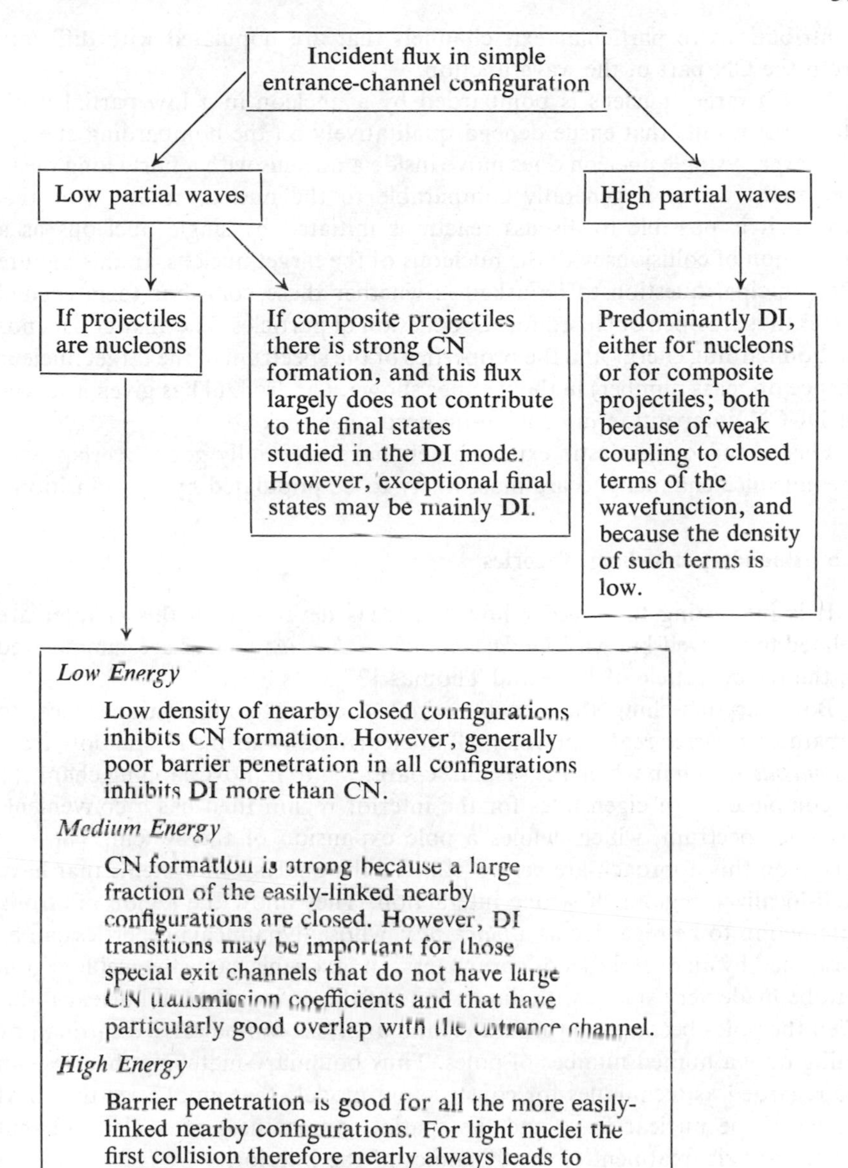

Fig. 2.4 Conditions for direct reactions.

contributions to particular exit channels that are populated with difficulty from the CN part of the wavefunction.

b. If a target nucleus is bombarded by a nucleon in a low partial wave state, the results that ensue depend qualitatively on the bombarding energy. However, a single nucleon does move inside a nucleus with a fairly long mean free path, which is generally comparable to the nuclear radius. For this reason it is possible to discuss reactions initiated by single nucleons as a succession of collisions with the nucleons of the target nucleus. In this picture the principal question to be asked is whether these collisions form excited states that are open or closed for the emission of particles. The answer depends on bombarding energy and the properties of the spectrum of the target nucleus (hence on mass number) in the manner shown. Cindro [26] has given a survey of DI-CN competition in neutron-induced reactions.

Elastic and quasi-elastic exit channels have especially good overlap with the entrance channel and are most likely to be populated by DI transitions.

2.5 Boundary-matching Theories

It is interesting to consider how the ideas developed in this chapter are related to the well-known *boundary-matching theories* such as are summarized in the review article of Lane and Thomas [27].

Boundary-matching theories introduce a cut in configuration space, to separate an *inside region* in which all nucleons are in strong interaction, from an *outside region* in which the system separates into nonoverlapping channels. A complete set of eigenstates for the interior region then has a convenient, discrete spectrum, which defines a pole expansion of the system. Theories based on this approach are very useful for the analysis of systems that have well-localized regions of strong interaction. They allow the region of strong interaction to be regarded as a *black box*, whose dynamical properties can be described by an orderly set of parameters. In favorable cases the inside region can be made very small, with dimensions of the order of usual nuclear radii; then the poles become well separated and analyses can be carried through by using only a limited number of poles. Thus boundary-matching theories can be regarded as techniques for constructing models that emphasize the short range of the nuclear force and the spatial compactness of nuclei and that avoid explicit treatment of the dynamics of the interior.

In contrast, DI theories freely use explicit models of nuclear dynamics, as in analyses of bound states, and do not make much use of nuclear compactness. Therefore it is appropriate that DI theories be developed *de novo* as simplified solutions of the Schrödinger equation. Demonstrations [28, 29] that certain DI theories can be exhibited as infinite sums over correlated poles of the boundary-matching theories are not so much derivations of these DI theories as they are demonstrations of the consistency of alternative methods for solving the Schrödinger equation.

2.6 Angular Distributions, Angular Correlations

The angular distribution is of great help in identifying the mechanism of a reaction:

1. Rapidly oscillatory angular distributions involve states of high angular momentum [30]; therefore, by arguments given earlier, they probably involve a DI mechanism.

2. Angular distributions that show a fore-aft asymmetry that *persists over a large energy range* also must involve a DI reaction mechanism. Persistent asymmetries indicate persistent phase relations among overlapping CN levels of opposite parity and these indicate the presence of underlying simple reaction mechanisms. Although a pure CN mechanism (Section 2.3) also allows asymmetric angular distributions, it causes backward peaking as often as forward.

The persistent DI angular distributions often show strong forward peaking. *Forward peaking* in this case means the lighter weight reaction product tends to be emitted (in the center-of-mass system) in the direction of motion of the lighter weight incident projectile. To see why a DI reaction mechanism should lead to forward peaking we first note that in the cm coordinate system the light incident projectile and the heavy target nucleus have momenta oppositely directed and equal in magnitude. Under DI circumstances, however, the projectile interacts with only a few particles from the substructure of the target nucleus, and the net backward momentum carried by these particles is smaller in magnitude than the forward momentum of the light projectile. Therefore the net momentum of the strongly interacting particles lies in the forward direction and forward peaking may be expected. Of course, this argument has been oversimplified. The light particles interact in the field of the heavy core, an effect that is described by using *distorted waves* for the light particles. The light particles thereby exchange some momentum with the core. We see that strong distortion effects can spoil the forward peaking of DI angular distributions and can even cause peaking in the backward hemisphere. Such failures of forward peaking are especially likely at low bombarding energy.

It was noted in Section 2.4 that flux incident in low partial waves (small impact parameters) is often entirely removed into CN excitations. For this reason the partial waves that are most strongly distorted by the field of the core tend not to initiate DI processes. This may be the reason why forward peaking survives as a major aspect of DI reactions.

The angular distribution of coincidence gamma radiation is potentially a sensitive indicator of the mechanism of a nuclear reaction. Gamma-ray angular distributions depend only on the gamma-ray multipolarities and the distribution of population in the magnetic substates of the radiating system. Because only one or two multipoles participate in any given gamma decay, careful theoretical fits of observed angular correlations can determine from experiment both the gamma-ray multipolarity and the *population parameters* of the intermediate magnetic substates [31]. It has been pointed out that for reactions such as $(d, p\gamma)$ the initial (d, p) stage yields very different substate populations, depending on whether it proceeds by CN or DI reaction mechanisms [32, 33]. The energy dependence of the substate population might be an especially sensitive indicator of the reaction mechanism [33].

2.7 Wave Packet Study of Three-body Breakup

This section gives a wave-packet development that illustrates further the kind of three-body breakup implied by the second term of (2.6). To obtain interesting results it will be necessary to consider artificially-brief incident packets, of the kind discussed by Friedman and Weisskopf [2]. For such packets the function $a(\mathbf{q})$, the momentum transform of the incident packet, has a large half breadth and the energy dependence of the amplitude assumes importance.

We get the appropriate asymptotic wavefunction by inserting the second term of (2.6) for f_β in (1.45). This asymptotic wavefunction is inserted into (1.15), the basic formula for the time development of the packet $\Phi(t)$. Then partial linearization* of phase factors with respect to $\mathbf{q}$ yields for the three-body term of $\Phi(t)$ the expression

$$\Phi_3(t) = \mathscr{R}_\beta^{-5/2} e^{-(iE/\hbar)(t-t_0)} \int a(\mathbf{q}) e^{-i(\mathbf{q}\cdot\mathbf{v}_\alpha)t} \times e^{i(K_1 r_1 + K_2 r_2)} f_{\mathrm{DI}}(\mathbf{k}_\alpha, K_1, \hat{r}_1) f_{\mathrm{CN}}^{(2)}(\mathbf{k}_\alpha, K_2, \hat{r}_2)\, d^3q, \tag{2.7}$$

where the identity $\xi \mathscr{R}_\beta = K_1 r_1 + K_2 r_2$ has been introduced. The amplitudes f_{DI} and $f_{\mathrm{CN}}^{(2)}$ have been indicated as depending on the incident momentum, the magnitudes of the momenta K_1 and K_2, and the directions of emission of particles 1 and 2. However, for simplicity it is assumed henceforth that the CN state emits only S-wave particles; therefore $f_{\mathrm{CN}}^{(2)}$ does not depend on $\hat{r}_2$.

For explicit integration of (2.7) it is necessary to employ simple analytic

* It was noted in Chapter I that linearization of phase factors with respect to $\mathbf{q}$ eliminates wave packet spreading. The brief packets now being considered do give appreciable spreading, however this does not cause qualitative changes of the discussion and we continue to ignore it.

models that give reasonable momentum dependences to the amplitudes f_{DI} and $f_{\mathrm{CN}}^{(2)}$. Of course, the amplitude for the particle produced by a direct process, the first-emitted particle, is treated as being momentum independent and is removed from the integral as a constant factor $\bar{f}_{\mathrm{DI}}$. The amplitude $f_{\mathrm{CN}}^{(2)}$ is more complicated. To give this amplitude the rapidly fluctuating energy dependence caused by wavefunctions of CN type we choose it to be a linear combination of Breit-Wigner resonances

$$f_{\mathrm{CN}}^{(2)} = \sum_s \frac{i\Gamma_s^{\beta}}{E_s - E_2 - (i\Gamma_s/2)}, \tag{2.8}$$

where E_2 is the energy associated with particle 2,

$$E_2 = \frac{\hbar^2 K_2^{\,2}}{2M_2} = E_{\beta} - \frac{\hbar^2 K_1^{\,2}}{2M_1}. \tag{2.9}$$

Also, E_s is the energy of a resonance, Γ_s is the total width, and Γ_s^{β} is the width for breakup in channel β.

In terms of the above models of the amplitudes, (2.7) goes over to

$$\begin{aligned} \Phi_3(t) &= i\mathscr{R}_{\beta}^{-5/2} e^{-(iE/\hbar)(t-t_0)} \bar{f}_{\mathrm{DI}} \\ &\times \sum_s \Gamma_s^{\beta} \int a(\mathbf{q}) \left(E_s - E_2 - \frac{i\Gamma_s}{2}\right)^{-1} \exp\{i[K_1 r_1 + K_2 r_2 - (\mathbf{q}\cdot\mathbf{v}_{\alpha})t]\}\, d^3q. \end{aligned} \tag{2.10}$$

The properties of the outgoing wave packets can now be obtained by examining this integral.

However, because of the richness of the motions described by the three-body wavefunctions, the packet integration in (2.10) does not necessarily yield results that are easy to interpret; for example, the single integration with respect to $\mathbf{q}$ does not lead to individual wave packets for $\mathbf{r}_1$ and $\mathbf{r}_2$. The difficulty of interpretation can be overcome by careful choice of the variables in terms of which the outgoing packet is to be described. Let us choose K_2 and r_1, because the amplitude is sensitive to K_2 and because we expect to see a narrow packet dependence on r_1. Choice of these variables suggests that the linearization of the exponent in (2.10) be completed in the form

$$\begin{aligned} K_2 &\equiv K_{2s} + \sigma, \\ K_1 &= K_{1s} + v_1^{-1}[(\mathbf{v}_{\alpha}\cdot\mathbf{q}) - v_2\sigma], \end{aligned} \tag{2.11}$$

where

$$\begin{aligned} \frac{\hbar^2 K_{2s}^{\,2}}{2M_2} &= E_s, \\ \frac{\hbar^2 K_{1s}^{\,2}}{2M_1} &= E - E_s. \end{aligned} \tag{2.12}$$

Thus the entire **q**-dependence of the outgoing wavefunction has been put into K_1, a momentum that is not restricted by the resonance amplitude. Integration with respect to **q** is now trivial and yields

$$\Phi_3(t) = i\mathscr{R}_\beta^{-5/2}\bar{f}_{\rm DI}e^{-(iE/\hbar)(t-t_0)}G\left(\mathbf{v}_\alpha\left[t - \frac{r_1}{v_1}\right]\right)$$
$$\times \sum_s \Gamma_s^{\ \beta}\left(E_s - E_2 - \frac{i\Gamma_s}{2}\right)^{-1} e^{i(K_{1s}r_1+K_{2s}r_2)}e^{i\sigma[r_2-(v_2/v_1)r_1]}, \quad (2.13)$$

where G is the envelope function of the (narrow) incoming wave packet. Evidently $\Phi_3(t)$ constrains r_1 to a narrow packet that has the classically-expected motion. The envelope of the outgoing packet is a rapidly varying function of K_2, that is, of σ.

As a last step of manipulation let us integrate over a small range of σ to average out the rapid energy dependence introduced by the resonance denominators. This integration may be regarded as a representation of the limited energy resolution of the system that detects particle 2. We obtain

$$\overline{\Phi_3(t)} = -\left(\frac{2\pi\bar{f}_{\rm DI}}{\hbar v_2\mathscr{R}_\beta^{5/2}}\right)e^{-(iE/\hbar)(t-t_0)}G\left(\mathbf{v}_\alpha\left[t - \frac{r_1}{v_1}\right]\right)$$
$$\times \sum_s \Gamma_s^{\ \beta} \exp i(K_{1s}r_1 + K_{2s}r_2) \exp \frac{\Gamma_s}{2\hbar}\left(\frac{r_2}{v_2} - \frac{r_1}{v_1}\right) \quad (2.14a)$$

if $(r_2/v_2) < (r_1/v_1)$ and

$$\overline{\Phi_3(t)} = 0 \quad (2.14b)$$

if $(r_2/v_2) > (r_1/v_1)$. Thus under the given conditions of observation particle 1 emerges in a sharp pulse and particle 2 follows after particle 1 in a set of exponential tails whose decay rate is determined by the resonance lifetimes.

The above calculation shows that (2.6), the assumed product form for the reaction amplitude at high energy, leads to physical results that are a priori reasonable. It should further be noted that nonobservation of particle 2 is expressed mathematically by integrating $|\overline{\Phi_3(t)}|^2$ over all values of r_2. This step yields a narrow outgoing pulse for particle 1 alone, as if DI emission of particle 1 were the only physical event that took place. This result is also reasonable.

References

[1] A. K. Kerman, L. S. Rodberg, and J. E. Young, *Phys. Rev. Letters* **11,** 422 (1963); H. Feshbach, in *Nuclear Structure Study with Neutrons*, M. N. de Mévergnies, P. Van Aasche, and J. Verview, Ed., North-Holland, Amsterdam, 1966. Also see [13] in Chapter 11.

[2] F. L. Friedman and V. F. Weisskopf, in *Niels Bohr and the Development of Physics*, W. Pauli, Ed., Pergamon, London 1955.
[3] J. Hüfner, C. Mahaux, and H. A. Weidenmüller, *Nucl. Phys.* **A105,** 489 (1967).
[4] C. E. Porter, *Statistical Theories of Spectra*, Academic, New York, 1965; N. Rosenzweig, in *Statistical Physics*, W. A. Benjamin, New York, 1963; C. E. Porter, *Rev. Mod. Phys.* **36,** 1094 (1964).
[5] K. F. Ratcliff, Ph.D. thesis, University of Pittsburgh, 1965, unpublished; K. F. Ratcliff and N. Austern, *Ann. Phys.* **42,** 185 (1967).
[6] P. A. Moldauer, *Phys. Rev.* **157,** 907 (1967); *Phys. Rev. Letters* **18,** 249 (1967); in *Proceedings of International Conference on Nuclear Structure*, Tokyo (1967); *Phys. Rev.* **171,** 1164 (1968); N. Ullah, *Nucl. Phys.* **A111,** 335 (1968).
[7] T. Ericson, *Ann. Phys.* **23,** 390 (1963); D. M. Brink and R. O. Stephen, *Phys. Letters* **5,** 77 (1963); T. Ericson and T. Mayer-Kuckuk, in *Ann. Rev. Nuclear Sci.*, Vol. 16, E. Segré, G. Friedlander, and H. P. Noyes, Ed., Annual Reviews, Palo Alto, 1966.
[8] See Chap. 6, Ref. 17–20.
[9] R. M. Drisko and N. Austern, (unpublished).
[10] K. W. McVoy, L. Heller, and M. Bolsterli, *Rev. Mod. Phys.* **39,** 245 (1967).
[11] D. M. Chase, L. Wilets, and A. R. Edmonds, *Phys. Rev.* **110,** 1080 (1958).
[12] N. Austern, *Ann. Phys.* **45,** 113 (1967).
[13] W. F. Waldorf and N. S. Wall, *Phys. Rev.* **107,** 1602 (1957); G. W. Greenlees and P. M. Rolph, *Proc. Phys. Soc.* (London) **75,** 201 (1960).
[14] L. Wolfenstein, *Phys. Rev.* **82,** 690 (1951); W. Hauser and H. Feshbach, *Phys. Rev.* **87,** 366 (1952). An excellent review of these theories is given by E. Vogt, in *Advances in Nuclear Physics*, Vol. 1, M. Baranger and E. Vogt, Ed., Plenum, New York, 1968.
[15] P. A. Moldauer, *Phys. Rev.* **123,** 968 (1961); *Rev. Mod. Phys.* **36,** 1079 (1964); G. R. Satchler, *Phys. Letters* **7,** 55 (1963).
[16] V. F. Weisskopf, *Phys. Today* **14,** 18 (1961); *Rev. Mod. Phys.* **29,** 174 (1957); *Physica* **22,** 952 (1958).
[17] J. J. Griffin, *Phys. Rev. Letters* **17,** 478 (1966); *Phys. Letters* **24B,** 5 (1967).
[18] E. B. Paul and R. L. Clarke, *Can. J. Phys.* **31,** 267 (1953); R. M. Eisberg, *Phys. Rev.* **94,** 739 (1954); F. L. Ribe, in *Fast Neutron Phys.*, Vol. II, J. B. Marion and J. L. Fowler, Ed., Interscience, New York, 1963.
[19] L. Cranberg and C. D. Zafiratos, *Phys. Rev.* **142,** 775 (1966).
[20] H. Taketani and W. P. Alford, *Nucl. Phys.* **32,** 430 (1962).
[21] B. A. Robson, *Nucl. Phys.* **86,** 649 (1966).
[22] J. R. Oppenheimer and M. Phillips, *Phys. Rev.* **48,** 500 (1935).
[23] N. S. Wall and P. R. Roos, *Phys. Rev.* **150,** 811 (1966).
[24] U. Amaldi, Jr., et al., *Phys. Rev. Letters* **13,** 341 (1964).
[25] K. J. Le Couteur, in *Nuclear Reactions*, P. M. Endt and M. Demeur, Ed., North-Holland, 1959; K. J. Le Couteur and D. W. Lang, *Nucl. Phys.* **13,** 32 (1959); G. D. Harp, J. M. Miller, and B. J. Berne, *Phys. Rev.* **165,** 1166 (1968).
[26] N. Cindro, *Rev. Mod. Phys.* **38,** 391 (1966).
[27] A. M. Lane and R. G. Thomas, *Rev. Mod. Phys.* **30,** 257 (1958).
[28] C. Bloch, *J. phys. radium* **17,** 510 (1956); *Nucl. Phys.* **3,** 137 (1957).

[29] G. E. Brown and C. T. de Dominicis, *Proc. Phys. Soc.* (London) **A70,** 668 (1957).
[30] R. G. Sachs, *Nuclear Theory*, Addison-Wesley, Cambridge, 1953. p. 317.
[31] A. E. Litherland and A. J. Ferguson, *Can. J. Phys.* **39,** 788 (1961).
[32] L. J. B. Goldfarb, *Nucl. Phys.* **57,** 4 (1964); L. J. B. Goldfarb and K. K. Wong, *Phys. Letters* **22,** 310 (1966).
[33] Jean M. O'Dell, Ph.D. thesis, University of Kansas, 1966, (unpublished); J. M. O'Dell, R. W. Krone, and F. W. Prosser, Jr., *Nucl. Phys.* **82,** 574 (1966).

CHAPTER 3

Some Useful Formulas

The scattered amplitude $f_{\alpha\beta}(\mathbf{k}_\alpha, \hat{r}_\beta)$ describes the breakup into channel β of a reaction initiated through channel α. It is the normalization factor [see (1.4)] of the asymptotically outgoing-wave term of the stationary-state wavefunction that lies in channel β. In principle the set of scattered amplitudes of a given system may be obtained straightforwardly by computing the stationary-state wavefunction and picking the amplitudes of the various outgoing waves from it. Such a procedure is used most notably in the method of coupled channels (Chapters 4, 6). However, use of this straightforward method for obtaining the amplitudes implies a complete and accurate knowledge of the stationary-state eigenfunction. It is helpful to derive expressions for the scattered amplitude that are better suited for the introduction of approximate methods of calculation.

In this chapter Green's theorem is used to derive several expressions for the scattered amplitude that are integrals over products of various wavefunctions and potentials associated with the reacting system. In these expressions the stationary-state eigenfunction enters only as a *source term* for the particular outgoing amplitude of interest. The expressions obtained are convenient for the introduction of approximations. It may be interesting to refer to the textbook of Messiah [1] for a more thorough discussion of some of the material in this chapter.

3.1 Projection of $\Psi^{(+)}$ on Channel β

To calculate the asymptotic wavefunction in channel β we start with the Schrödinger equation for the full stationary-state wavefunction,

$$H\Psi^{(+)} = E\Psi^{(+)}. \tag{3.1}$$

This equation is projected onto the internal wavefunction in channel β, yielding

$$(\psi_\beta, H\Psi^{(+)}) = E(\psi_\beta, \Psi^{(+)}). \tag{3.2}$$

Here the scalar product notation indicates integration over only the internal variables of ψ_β. It is (3.2) that is solved by use of Green's theorem.

We limit discussion to the important case of two-body breakup; hence for ψ_β we use the product form

$$\psi_\beta = \psi_b \psi_B. \tag{3.3}$$

The functions ψ_b and ψ_B refer to definite energy states and spin orientations of each of the two product nuclei. (Naturally, b, B, or both may be single nucleons.) It is implicit in the product form of (3.3) that the nucleon labels of ψ_b are distinct from those of ψ_B, since otherwise these two wavefunctions would not refer to independent nuclei. This means we may introduce a running index i that labels all nucleons of the system, so that

$$\begin{array}{lll} \psi_b & \text{contains nucleons labelled} & i = 1, 2, \ldots, n_b, \\ \psi_B & \text{contains nucleons labelled} & i = n_b + 1, \ldots, n_b + n_B. \end{array} \tag{3.4}$$

Here n_b is the total number of nucleons in b, and n_B is the total number of nucleons in B. The use of (3.3) evidently implies that full consideration of the identity of particles is put aside for the time being. This creates no problems. The present discussion merely concerns a mathematical procedure by which solutions of the Schrödinger equation are obtained. The problem of solving this linear equation is quite independent of the task of constructing antisymmetric linear combinations of its solutions. However, it may as well be assumed immediately that ψ_b and ψ_B are individually antisymmetrized with respect to their internal variables.

Because the variables in ψ_b and ψ_B are distinct the matrix element of H that appears in (3.2) can be broken up in a useful manner, enabling us to obtain considerable simplification of that equation. The channel-radius variable $\mathbf{r}_\beta$ is introduced. It is the displacement between the centers of mass of b and B:

$$\mathbf{r}_\beta = \frac{1}{n_b} \sum_{i=1}^{n_b} \mathbf{r}_i - \frac{1}{n_B} \sum_{i=n_b+1}^{n_b+n_B} \mathbf{r}_i. \tag{3.5}$$

Here all nucleons are assumed to have the same mass. We identify the kinetic energy operator associated with the variable $\mathbf{r}_\beta$. It is

$$T_\beta = -\frac{\hbar^2}{2\mu_\beta} \nabla_{\mathbf{r}_\beta}^{\;2}, \tag{3.6}$$

where

$$\mu_\beta = \frac{M_b M_B}{M_b + M_B} \tag{3.7}$$

is the reduced mass in channel β. (Because $\mathbf{r}_\beta$ is the displacement between centers of mass the momentum $\nabla_{\mathbf{r}_\beta}$ only appears in the Hamiltonian in the kinetic energy operator T_β. There are no cross terms between $\nabla_{\mathbf{r}_\beta}$ and any momenta internal to b or B.) We also identify the potential energy operator V_β which is the sum of all interactions not internal to either b or B. If the potential energy is a sum of two-body interactions, as is usually the case, we have

$$V_\beta = \sum_{\substack{i=1,2,\ldots,n_b \\ j=n_b+1,\ldots,n_b+n_B}} V(i,j). \tag{3.8}$$

We now can define $H_\beta = H - T_\beta - V_\beta$, the portion of the Hamiltonian that operates only on the variables internal to b and B. Because ψ_b and ψ_B are bound, the operator H_β is reliably Hermitean and may be permitted to operate to the left in (3.2) and to bring out as an eigenvalue the channel eigenenergy ε_β, a c-number. Further simplification of (3.2) follows because T_β commutes with ψ_β. As a result of these steps (3.2) reduces to

$$\{T_\beta + \varepsilon_\beta - E\}(\psi_\beta, \Psi^{(+)}) = -(\psi_\beta, V_\beta \Psi^{(+)}). \tag{3.9}$$

Finally, because the scalar product $(\psi_\beta, \Psi^{(+)})$ signifies integration over only the internal variables of ψ_β it defines a function of $\mathbf{r}_\beta$, which is denoted

$$(\psi_\beta, \Psi^{(+)}) \equiv \xi_\beta(\mathbf{r}_\beta). \tag{3.10}$$

(If the model wavefunction of (2.5) were inserted for $\Psi^{(+)}$ in (3.10), for example, the function ξ_β defined above would be the relative wavefunction in (2.5) that is denoted by the same symbol.) Insertion of (3.10) into (3.9) yields

$$\{T_\beta + \varepsilon_\beta - E\}\xi_\beta = -(\psi_\beta, V_\beta \Psi^{(+)}). \tag{3.11}$$

Equation 3.11 is easy to solve.

Equation 3.11 is an inhomogeneous differential equation for the function ξ_β. It is solved by using the familiar Green's function [1, 2] for the left-hand side of (3.11),

$$K_\beta^{0}(\mathbf{r}_\beta, \mathbf{r}_\beta') = \left(\frac{2\mu_\beta}{\hbar^2}\right) \frac{e^{ik_\beta|\mathbf{r}_\beta - \mathbf{r}_\beta'|}}{4\pi\,|\mathbf{r}_\beta - \mathbf{r}_\beta'|}, \tag{3.12}$$

where

$$\{T_\beta + \varepsilon_\beta - E\}K_\beta^{0}(\mathbf{r}_\beta, \mathbf{r}_\beta') = \delta(\mathbf{r}_\beta - \mathbf{r}_\beta'), \tag{3.13}$$

and

$$\frac{\hbar^2 k_\beta^{2}}{2\mu_\beta} = E - \varepsilon_\beta. \tag{3.14}$$

As a result

$$\xi_\beta = \left(\frac{-\mu_\beta}{2\pi\hbar^2}\right) \int \frac{e^{ik_\beta|\mathbf{r}_\beta - \mathbf{r}_\beta'|}}{|\mathbf{r}_\beta - \mathbf{r}_\beta'|} (\psi_\beta, V_\beta \Psi^{(+)})\, d^3 r_\beta'. \tag{3.15}$$

Equation 3.15 is a particular solution of (3.11). To obtain the most general solution of the differential equation an arbitrary regular eigenfunction of the left hand side of (3.11) must be added to (3.15), so that the scattering boundary conditions are satisfied. However, the Green's function $K_\beta{}^0$ yields flux that is asymptotically only radially outgoing, therefore (3.15) already satisfies the boundary conditions in channels other than the incident channel. We obtain an expression that is valid for all channels by adding the appropriate regular eigenfunction of the left-hand side of (3.11), namely, the incoming plane wave in channel α, to (3.15). Therefore the full solution of (3.11) is

$$\xi_\beta = e^{i(\mathbf{k}_\alpha\cdot\mathbf{r}_\alpha)}\,\delta_{\alpha\beta} - \left(\frac{\mu_\beta}{2\pi\hbar^2}\right)\int \frac{e^{ik_\beta|\mathbf{r}_\beta-\mathbf{r}_\beta'|}}{|\mathbf{r}_\beta - \mathbf{r}_\beta'|}(\psi_\beta, V_\beta\Psi^{(+)})\,d^3r_\beta'. \tag{3.16}$$

This expression may be regarded as a convenient device for extracting the properties of one particular channel from $\Psi^{(+)}$.

The asymptotic form of ξ_β is obtained from (3.16) by passing to the limit $r_\beta \gg r_\beta'$. It is

$$\xi_\beta \to e^{i(\mathbf{k}_\alpha\cdot\mathbf{r}_\alpha)}\,\delta_{\alpha\beta} - \left(\frac{\mu_\beta}{2\pi\hbar^2}\right)\frac{e^{ik_\beta r_\beta}}{r_\beta}(\psi_\beta e^{i(\mathbf{k}_\beta\cdot\mathbf{r}_\beta')}, V_\beta\Psi^{(+)}), \tag{3.17}$$

where

$$\mathbf{k}_\beta = k_\beta\hat{r}_\beta, \tag{3.18}$$

and the scalar-product notation has been extended to include the integration over $\mathbf{r}_\beta'$. Comparison between (3.17) and (1.4), which defined the scattering amplitudes, yields

$$f_{\alpha\beta}(\mathbf{k}_\alpha, \hat{r}_\beta) = \frac{-\mu_\beta}{2\pi\hbar^2}(\psi_\beta e^{i(\mathbf{k}_\beta\cdot\mathbf{r}_\beta')}, V_\beta\Psi^{(+)}). \tag{3.19}$$

In place of the scattered amplitude it is convenient to go one step further and to introduce the transition amplitude of (1.31). We obtain

$$T_{\alpha\beta} = (\psi_\beta e^{i(\mathbf{k}_\beta\cdot\mathbf{r}_\beta')}, V_\beta\Psi^{(+)}). \tag{3.20}$$

In terms of the present expressions, $T_{\alpha\beta}$ is neater algebraically than $f_{\alpha\beta}$. Most of the calculations in this book are organized in terms of expressions related to (3.20).

3.2 Spherical Distorted Waves

It is useful to modify (3.11) by inserting on both sides an arbitrary potential $U_\beta(r_\beta)$ that is a function only of the magnitude of the channel radius variable $\mathbf{r}_\beta$. The modified equation is

$$\{T_\beta + U_\beta + \varepsilon_\beta - E\}\xi_\beta = -(\psi_\beta, [V_\beta - U_\beta]\Psi^{(+)}). \tag{3.21}$$

To verify that (3.21) is equivalent to (3.11) it is sufficient to observe that U_β on the RHS of (3.21) does not involve the variables of ψ_β, which are the integration variables of the RHS, and to recall the definition of ξ_β, (3.10). Considerable adjustment of the procedure for computing ξ_β can be achieved by suitable choice of the potential U_β. The potential U_β is known as a *distorting potential.* In most of the work in this book U_β is considered to include both the Coulomb interaction between the two nuclei in channel β and a short-range interaction that is regarded as some kind of average nuclear interaction between them.

Equation 3.21 may be solved in the same manner as (3.11), that is, by explicit construction of the Green's function for the LHS. Although this Green's function generally cannot be obtained in closed form, as was done for (3.11), it can be expanded in partial waves and expressed as a sum over closed-form radial Green's functions for each of the partial waves. (A three-dimensional closed-form Green's function is available for the purely-Coulomb problem [3].) The partial-wave expansion of the Green's function is reduced to an easily interpretable closed form when the asymptotic ξ_β is computed.

Before constructing the Green's function for (3.21) it is helpful to introduce additional notation. This is done most conveniently by discussing the stationary-state eigenfunctions of the LHS of (3.21). Let $\chi_\beta^{(+)}(\mathbf{k}_\beta, \mathbf{r}_\beta)$ be the scattering wavefunction governed by the potential U_β, so that

$$\{T_\beta + U_\beta + \varepsilon_\beta - E\}\chi_\beta^{(+)}(\mathbf{k}_\beta, \mathbf{r}_\beta) = 0, \tag{3.22}$$

and so that, asymptotically,

$$\chi_\beta^{(+)} \to e^{i(\mathbf{k}_\beta \cdot \mathbf{r}_\beta)} + \text{outgoing scattered waves.} \tag{3.23}$$

The magnitude of the momentum $\mathbf{k}_\beta$ is given by (3.14). In partial-wave expansion $\chi_\beta^{(+)}$ becomes

$$\chi_\beta^{(+)} = \frac{4\pi}{k_\beta r_\beta} \sum_{l,m} i^l e^{i\sigma_{\beta l}} f_{\beta l}(k_\beta, r_\beta) Y_l^{\,m}(\hat{r}_\beta) Y_l^{\,m*}(\Theta, \Phi). \tag{3.24}$$

The angles Θ, Φ indicate the direction of $\mathbf{k}_\beta$ with respect to the coordinate axes. The radial functions $f_{\beta l}$ are required to vanish at $r_\beta = 0$ in order that $\chi_\beta^{(+)}$ be regular. At radii so large that the nuclear part of U_β is negligible these functions go over to the form

$$f_{\beta l} \to \frac{i}{2}\,[H_l^*(k_\beta r_\beta) - \eta_{\beta l} H_l(k_\beta r_\beta)], \tag{3.25}$$

where $\eta_{\beta l}$ is the reflection coefficient for the lth partial wave. The function $H_l(k_\beta r_\beta)$ is defined by Hull and Breit [4, 5] to be the Coulomb analogue of $ik_\beta r_\beta h_l^{(1)}$, where $h_l^{(1)}$ is the outgoing spherical Hankel function [2]. In terms of the regular and irregular radial Coulomb functions this function is

$$H_l(k_\beta r_\beta) = G_l(k_\beta r_\beta) + iF_l(k_\beta r_\beta), \tag{3.26}$$

where asymptotically

$$F_l \to \sin \theta_{\beta l},$$
$$G_l \to \cos \theta_{\beta l}, \tag{3.27}$$

and

$$\theta_{\beta l} = k_\beta r_\beta - n_\beta[\ln (2k_\beta r_\beta)] - \frac{l\pi}{2} + \sigma_{\beta l},$$
$$\sigma_{\beta l} = \arg \Gamma (l + 1 + in_\beta), \tag{3.28}$$
$$n_\beta = \frac{Z_b Z_B \mu_\beta e^2}{\hbar^2 k_\beta}.$$

The spherical harmonic expansion in (3.24) is orderly and easy to use in practical numerical calculations.

To build a Green's function for (3.21) it is necessary to supplement the regular radial wavefunctions $f_{\beta l}$ of (3.22), which vanish at the origin, with other solutions of (3.22) that do not vanish at $r_\beta = 0$. It is convenient to consider radial solution functions of (3.22) that go over asymptotically to purely outgoing functions

$$h_{\beta l}(k_\beta, r_\beta) \to H_l(k_\beta r_\beta), \qquad \text{at large } r_\beta. \tag{3.29}$$

Then a suitable Green's function for the radial differential equation is

$$k_\beta^{-1} f_{\beta l}(k_\beta, r_<) h_{\beta l}(k_\beta, r_>), \tag{3.30}$$

where $r_<$, $r_>$ are the lesser or greater of r_β, r'_β. The Green's function of (3.30) satisfies the equation

$$\left\{-\frac{d^2}{dr_\beta^{\,2}} + \frac{l(l+1)}{r_\beta^{\,2}} + U_\beta - k_\beta^{\,2}\right\} k_\beta^{-1} f_{\beta l} h_{\beta l} = \delta(r_\beta - r'_\beta). \tag{3.31}$$

With the use of (3.30) it is easy to write in partial-wave expansion

$$K_\beta(\mathbf{r}_\beta, \mathbf{r}'_\beta) = \frac{2\mu_\beta}{\hbar^2} \sum_{l,m} \frac{f_{\beta l}(k_\beta, r_<) h_{\beta l}(k_\beta, r_>) Y_l^m(\hat{r}_\beta) Y_l^{m*}(\hat{r}'_\beta)}{k_\beta r_\beta r'_\beta}, \tag{3.32}$$

the Green's function for the full differential equation of interest, (3.21). This Green's function is regular at $r_\beta = 0$ and is purely radially outgoing at asymptotic values of r_β.

Using the outgoing-waves Green's function of (3.32) we immediately write a particular solution of (3.21),

$$\xi_\beta = -\int K_\beta(\mathbf{r}_\beta, \mathbf{r}'_\beta)(\psi_\beta, [V_\beta - U_\beta]\Psi^{(+)})\, d^3 r'_\beta. \tag{3.33}$$

The only regular homogeneous solutions of (3.21) that can be added to (3.33) and still maintain consistency with the boundary conditions of the scattering problem are those that have incoming flux in channel α, the incident channel, and no incoming flux in other channels, $\beta \neq \alpha$. Only one eigenfunction of that type exists, yielding

$$\xi_\beta = \chi_\alpha^{(+)} \delta_{\alpha\beta} - \int K_\beta(\mathbf{r}_\beta, \mathbf{r}'_\beta)(\psi_\beta, [V_\beta - U_\beta]\Psi^{(+)}) \, d^3r'_\beta. \tag{3.34}$$

for the desired solution of (3.21).

The scattering amplitude is obtained from (3.34) by letting the value of r_β become asymptotically large. In this limit the Green's function of (3.34) simplifies and becomes

$$K_\beta(\mathbf{r}_\beta, \mathbf{r}'_\beta) \to \frac{\mu_\beta}{2\pi\hbar^2} \frac{e^{i\{k_\beta r_\beta - n_\beta[\ln(2k_\beta r_\beta)]\}}}{r_\beta} \chi_\beta^{(-)*}(\mathbf{k}_\beta, \mathbf{r}'_\beta), \tag{3.35}$$

where

$$\chi_\beta^{(-)*}(\mathbf{k}_\beta, \mathbf{r}'_\beta) \equiv \frac{4\pi}{k_\beta r_\beta} \sum_{l,m} i^{-l} e^{i\sigma_{\beta l}} f_{\beta l}(k_\beta, r'_\beta) \, Y_l^m(\hat{k}_\beta) \, Y_l^{m*}(\hat{r}'_\beta), \tag{3.36}$$

and $\mathbf{k}_\beta$ has the direction of $\mathbf{r}_\beta$ as in (3.10). We then insert (3.35) into (3.34), and pick off the scattering amplitude in conformity with the definition in (1.4). (The Coulomb phase factor in (3.35) introduces nothing new.) The amplitude is

$$f_{\alpha\beta}(\mathbf{k}_\alpha, \hat{r}_\beta) = f_{\alpha\alpha}^{(0)} \delta_{\alpha\beta} - \frac{\mu_\beta}{2\pi h^2} (\chi_\beta^{(-)}\psi_\beta, [V_\beta - U_\beta]\Psi^{(+)}), \tag{3.37}$$

where the scalar product now includes the integration over $\mathbf{r}'_\beta$ and $f_{\alpha\alpha}^{(0)}$ is the elastic-scattered amplitude yielded from $\chi_\alpha^{(+)}$. The transition amplitude follows from (3.37) in the usual manner, and is

$$T_{\alpha\beta} = T_{\alpha\alpha}^{(0)} \delta_{\alpha\beta} + (\chi_\beta^{(-)}\psi_\beta, [V_\beta - U_\beta]\Psi^{(+)}). \tag{3.38}$$

It is also interesting to use (3.20) to express the amplitude $T_{\alpha\alpha}^{(0)}$ as a matrix element of U_β. We thereby obtain the alternative expression

$$T_{\alpha\beta} = (e^{i(\mathbf{k}_\alpha \cdot \mathbf{r}_\beta)}\psi_\beta, U_\beta \psi_\alpha \chi_\alpha^{(+)})\delta_{\alpha\beta} + (\chi_\beta^{(-)}\psi_\beta, [V_\beta - U_\beta]\Psi^{(+)}) \tag{3.39}$$

In this form it is clear that the introduction of the distorting potential yields a rather general transformation, whereby $T_{\alpha\beta}$ is converted from (3.20) to (3.39). This is known as the *Gell-Mann, Goldberger transformation* [6]. It is discussed again in the next section.

It is not customary in the literature to carry a coefficient $\delta_{\alpha\beta}$ in the first terms of equations such as (3.39) and those discussed earlier. It is considered obvious that terms like these make zero contributions to $T_{\alpha\beta}$ if $\alpha \neq \beta$. Indeed, if $T_{\alpha\beta}$ should be an amplitude for inelastic scattering the first term of (3.39) would vanish because of the orthogonality of ψ_α and ψ_β. However, if the channels α and β should differ by rearrangement, so that bB and aA are different modes of sharing the nucleons that are present, then the first term of (3.39) would not vanish for any simple reasons of orthogonality. The channel functions ψ_α and ψ_β are governed by H_α and H_β which are different fragments of the full Hamiltonian. Careful discussions of this question [7] point out that two channels that differ by rearrangement can be regarded as *orthogonal* in the sense that wave packets located at asymptotically large values of the channel radii, $\mathbf{r}_\alpha$ and $\mathbf{r}_\beta$, are necessarily nonoverlapping. The method that led to the $\delta_{\alpha\beta}$ that appears above made use of the asymptotic boundary conditions and is related closely to this wave-packet argument.

The function $\chi_\beta^{(-)}$ derived above is known as the *time-reversed scattering wavefunction* for the potential U_β. It is expressed in terms of the scattering wavefunction of (3.24) by the Wigner time-reversal relation [1, 7, 8]

$$\chi^{(-)*}(\mathbf{k}, \mathbf{r}) = \chi^{(+)}(-\mathbf{k}, \mathbf{r}), \tag{3.40}$$

which takes this simple form because particle spins are not yet being considered. The time-reversed wavefunction is asymptotically of the form of a plane wave with momentum $\mathbf{k}_\beta$, together with *ingoing radial waves*. The importance of this function is seen in Chapter 1, where it is shown that only in terms of this function is it possible to build outgoing wave packets that are not confused by the scattered waves generated by U_β. Thus $T_{\alpha\beta}$ connects an ingoing packet of average momentum $\mathbf{k}_\alpha$ with an outgoing packet of average momentum $\mathbf{k}_\beta$. However, none of these considerations is necessary in the present calculation, which merely concerns the evaluation of the component of $\Psi^{(+)}$ in the channel β.

The present derivation of (3.39) is so straightforward mathematically that it is obvious it would go through without alteration if the distorting potential U_β were complex. This fact is valuable, because complex distorting potentials are of importance later. More formal derivations of (3.39) tend to introduce Hermitian conjugation as one step in calculation. This sometimes leads to confusion when U_β is complex.

The functions $\chi_\beta^{(+)}$ and $\chi_\beta^{(-)}$ are known as *distorted waves*. (The adjective is intended to suggest that they are not plane waves!) It is shown later that by suitable choice of the potential U_β these distorted waves can be made excellent zero-order representations of part of the physics of the scattering problem.

The calculation given in this section would not be altered significantly if the distorting potential were taken to depend both on $\mathbf{r}_\beta$ and on the spins of the nuclei b and B; for example, spin-orbit interactions might be added to the distorting potential. Such terms do not raise new problems because they do not operate on the internal coordinates of b and B. Calculations that use spin-dependent distorting potentials are referred to later.

3.3 Generalized Distorting Potential

The distorting potentials considered in the preceding section are introduced arbitrarily in each open channel β. Hence the set of distorting potentials introduced in this manner is a potential matrix that is diagonal in the space spanned by the open channels. It is also possible to consider distorting-potential matrices that are not diagonal, which couple a number of channels or which couple all the members of some class of channels. Such non-diagonal distorting-potential matrices are referred to as *generalized distorting potentials*.

It is seen in this section that the Gell-Mann, Goldberger transformation referred to in (3.39) remains valid even if the distorting potentials couple different channels.

Messiah [1] shows how the Gell-Mann, Goldberger transformation can be proved in general, using a method that generalizes the Green's function method of the preceding section. In each case the method used implies the comparison of two Hamiltonians, $H = T + V$ and $\hat{H} = T + \hat{V}$, that contain the same kinetic energy operator but somewhat different short-ranged potentials, V and $\hat{V}$. We consider two stationary eigenfunctions of H and $\hat{H}$ that will be denoted by $\Psi'^{(+)}(\mathbf{k}_\alpha)$ and $\hat{\Psi}'^{(-)}(\mathbf{k}_\beta)$, respectively. Here $\Psi'^{(+)}(\mathbf{k}_\alpha)$ has a plane wave in channel α and only outgoing scattered waves; $\hat{\Psi}'^{(-)}(\mathbf{k}_\beta)$ has a plane wave in channel β (which evidently is required to be an open channel of $\hat{H}$ at the energy E) and only ingoing scattered waves.

The above two eigenfunctions obey the two Schrödinger equations

$$H\Psi'^{(+)}(\mathbf{k}_\alpha) = E\Psi'^{(+)}(\mathbf{k}_\alpha), \tag{3.1}$$

$$\hat{H}\hat{\Psi}'^{(-)}(\mathbf{k}_\beta) = E\hat{\Psi}'^{(-)}(\mathbf{k}_\beta). \tag{3.41}$$

If we multiply (3.1) by $\hat{\Psi}'^{(-)*}(\mathbf{k}_\beta)$ and the complex conjugate of (3.41) by $\Psi'^{(+)}(\mathbf{k}_\alpha)$ and subtract, we obtain

$$0 = [\hat{\Psi}'^{(-)*}(T\Psi'^{(+)}) - (T\hat{\Psi}'^{(-)})^*\Psi'^{(+)}] + [\hat{\Psi}'^{(-)*}(V\Psi'^{(+)}) - (\hat{V}\hat{\Psi}'^{(-)})^*\Psi'^{(+)}]. \tag{3.42}$$

This equation is now summed over the spin variables and integrated over the whole of configuration space. The integration of the second term of (3.42) is straightforward because the potentials are short ranged. For this reason, the operator $\hat{V}$, whether or not it is entirely Hermitian, may be transferred from one wavefunction to the other without detailed consideration of the properties of the wavefunctions. We exploit this property and rewrite (3.42) in the form

$$0 = [\hat{\Psi}'^{(-)*}(T\Psi'^{(+)}) - (T\hat{\Psi}'^{(-)})^*\Psi'^{(+)}] + \hat{\Psi}'^{(-)*}(V - \hat{V}^\dagger)\Psi'^{(+)}. \tag{3.43}$$

The integration of the first term of (3.42) or (3.43) requires more care. Because there are open channels the kinetic energy operator is not Hermitian and the two contributions to the first term do not cancel. It is necessary to calculate the integral over a finite volume of configuration space and then take the limit when this volume is extended to infinity. The volume integral can be transformed into a surface integral by use of Green's theorem. The surface integral then breaks down into a sum of contributions from the open channels that are common to $\Psi'^{(+)}$ and $\hat{\Psi}'^{(-)}$; the contributions from channels α and β yield the required scattered amplitudes. Assuming that only two-body breakup occurs, we finally obtain

$$T_{\alpha\beta} = \hat{T}_{\alpha\beta} + (\hat{\Psi}'^{(-)}(\mathbf{k}_\beta), (V - \hat{V}^\dagger)\Psi'^{(+)}(\mathbf{k}_\alpha)), \tag{3.44}$$

the generalized Gell-Mann, Goldberger transformation. Here $T_{\alpha\beta}$ is the amplitude for the transition $\alpha \to \beta$ as governed by H, and $\hat{T}_{\alpha\beta}$ is the amplitude as governed by $\hat{H}$.

In the difference $V - \hat{V}^\dagger$ the interactions internal to nuclei b and B are Hermitian and are common to V and $\hat{V}^\dagger$. Therefore they cancel. For this reason, $V - \hat{V}^\dagger$ reduces to

$$V - \hat{V}^\dagger = V_\beta - U, \tag{3.45}$$

where V_β is the channel potential of (3.8), and U is the generalized distorting potential.

In the most interesting applications of the above generalized expressions the potential U is chosen to couple with channel β only those channels that differ from β by *inelastic scattering*. Such channels involve different states of excitation of b and B, but do not involve any rearrangements of those nuclei. All these channels utilize the same radius variable $\mathbf{r}_\beta$, therefore U depends on $\mathbf{r}_\beta$ and some internal variable(s) h that couple the various states of excitation of b and B. Accordingly, the generalized potential may be written $U(h, \mathbf{r}_\beta)$. This notation is used later. The calculation of the amplitude $\hat{T}_{\alpha\beta}$ requires a coupled-channels calculation (Chapters 4 and 6) with the potential $U(h, \mathbf{r}_\beta)$.

The above derivation of the Gell-Mann, Goldberger transformation proceeds in the same manner if β is a three-body exit channel. The asymptotic Green's function for a three-body channel, derived in Chapter 1, is used in this calculation. However, the timc-reversed wavefunction in channel β then has the awkward property of describing a reaction initiated by the collision of three free particles.

3.4 Some Integral Equations

The Green's functions of the preceding sections all were discussed explicitly in coordinate representation. However, it is often more convenient and more flexible to work directly with the equivalent operators, avoiding the introduction of any explicit representation. A brief and rather heuristic description of operator methods is now given. Numerous sophisticated discussions of this material are available [1, 7].

We consider first a formal trick for solving the differential equation of (3.11) to obtain (3.16) for the channel function $\xi_\beta(\mathbf{r}_\beta)$. In this trick an operator solution of (3.11) is obtained by dividing both sides by the differential operator that is its left-hand side (LHS). The solution obtained this way has the great advantage that its relation with the original differential equation is unmistakeable. On the other hand, the operator by which we divide is capable of having zero eigenvalues and the division can be a rich source of error. For reliable results the reciprocal operator must be applied only to vectors of finite norm [1]. Because V_β has short range the RHS of (3.11) does have finite norm and the method may be used. The reciprocal operator often is expanded in eigenfunctions and replaced by the reciprocals of its eigenvalues.

If we examine the eigenfunction expansion of the above-described formal solution of (3.11) we find that this solution is ambiguous as to the boundary conditions it obeys. A simple technique to resolve this ambiguity and select the outgoing-wave solution is to add $i\varepsilon$ to the energy E and select the solution obtained by letting ε go to zero through positive values. This $i\varepsilon$ technique is convenient because it can be used in either the eigenfunction representation or the operator expression. With the use of this technique it is possible to write an unambiguous formal, operator solution of (3.11),

$$\xi_\beta = e^{i(\mathbf{k}_\alpha \cdot \mathbf{r}_\alpha)}\,\delta_{\alpha\beta} + \lim_{\varepsilon\to 0}\,(E + i\varepsilon - \varepsilon_\beta - T_\beta)^{-1}(\psi_\beta, V_\beta \Psi^{(+)}). \tag{3.46}$$

Equation 3.46 is identical in content with (3.16). Normally the limiting operation in such expressions as (3.46) is not indicated explicitly, but is understood.

An expression similar to (3.46) may be obtained from the Schrödinger equation for the full stationary-state wavefunction $\Psi^{(+)}$. When the Schrödinger equation is rearranged to

$$(E - H + V_\beta)\Psi^{(+)} = V_\beta \Psi^{(+)}, \tag{3.47}$$

division results in

$$\Psi^{(+)} = \delta_{\alpha\beta} e^{i(\mathbf{k}_\alpha \cdot \mathbf{r}_\alpha)}\psi_\alpha + (E + i\varepsilon - H + V_\beta)^{-1} V_\beta \Psi^{(+)}, \tag{3.48}$$

where the first term is that eigenfunction of $(E - H + V_\beta)$ that has the correct incoming wave if $\beta = \alpha$. Equation 3.46 is recovered if (3.48) is projected onto the channel state ψ_β. However, eigenfunction expansions may be used to verify that the second term of (3.48) yields correct outgoing waves in all channels. It appears, therefore, that (3.48) is a generalized Green's function solution of the Schrödinger equation, and that this solution yields the full wavefunction $\Psi^{(+)}$ in all channels, no matter which channel-potential V_β is used in (3.48). Of course, our procedure of solution actually has yielded in (3.48) an integral equation for $\Psi^{(+)}$. This formal conversion of the Schrödinger equation to an integral equation often is helpful for theoretical manipulation.

The special form of (3.48) for the case $\beta = \alpha$ is

$$\Psi^{(+)} = e^{i(\mathbf{k}_\alpha \cdot \mathbf{r}_\alpha)}\psi_\alpha + (E + i\varepsilon - H + V_\alpha)^{-1}V_\alpha\Psi^{(+)}. \tag{3.49}$$

Equation 3.49 is known as the *Lippmann-Schwinger equation* [7, 9]. It is more inclusive than the Schrödinger equation in that it incorporates the boundary conditions of the scattering problem. The inhomogeneous term of this equation represents the incoming plane wave in channel α and forces the solution of the equation to be unique [10].

The function $\Psi^{(+)}$ of (3.49) obeys a differential equation that differs from the Schrödinger equation by terms of order ε. This differential equation is obtained by multiplying both sides of (3.49) by $(E + i\varepsilon - H + V_\alpha)$, with the result that

$$(E + i\varepsilon - H)\Psi^{(+)} = i\varepsilon e^{i(\mathbf{k}_\alpha \cdot \mathbf{r}_\alpha)}\psi_\alpha. \tag{3.50}$$

Equation 3.50 is inhomogeneous and much more symmetrical than, say, (3.47). The $i\varepsilon$ terms carry the effects resulting from the boundary conditions. By dividing (3.50) by its LHS we obtain

$$\begin{aligned}\Psi^{(+)} &= i\varepsilon(E + i\varepsilon - H)^{-1}e^{i(\mathbf{k}_\alpha \cdot \mathbf{r}_\alpha)}\psi_\alpha, \\ &= \{1 + (E + i\varepsilon - H)^{-1}(H - E)\}e^{i(\mathbf{k}_\alpha \cdot \mathbf{r}_\alpha)}\psi_\alpha, \\ &= \{1 + (E + i\varepsilon - H)^{-1}V_\alpha\}e^{i(\mathbf{k}_\alpha \cdot \mathbf{r}_\alpha)}\psi_\alpha.\end{aligned} \tag{3.51}$$

Equation 3.51, which exhibits several different forms, is known as the *formal solution* of the Schrödinger equation. It relates the stationary state eigenfunction to the incoming plane wave. It uses a Green's function derived from the full Hamiltonian of the scattering problem. (This Green's function is normally not available in explicit form.)

The Green's functions discussed above all yield asymptotically *radially-outgoing* scattered waves. Green's functions that yield asymptotically *radially-ingoing* scattered waves are derived from the ones above by reversing the sign of ε. By use of these Green's functions we obtain the *time-reversed stationary-state scattering wavefunction*,

$$\Psi^{(-)} = \delta_{\alpha\beta}e^{i(\mathbf{k}_\alpha \cdot \mathbf{r}_\alpha)}\psi_\alpha + (E - i\varepsilon - H + V_\beta)^{-1}V_\beta\Psi^{(-)}, \tag{3.52}$$

or

$$\begin{aligned}\Psi^{(-)} &= -i\varepsilon(E - i\varepsilon - H)^{-1}e^{i(\mathbf{k}_\alpha \cdot \mathbf{r}_\alpha)}\psi_\alpha, \\ &= \{1 + (E - i\varepsilon - H)^{-1}V_\alpha\}e^{i(\mathbf{k}_\alpha \cdot \mathbf{r}_\alpha)}\psi_\alpha.\end{aligned} \tag{3.53}$$

It is pointed out in Chapter 1 that the time-reversed stationary-state wavefunction is appropriate for building wave packets that move away from a scattering center. It is no surprise that $\Psi^{(-)}$ comes up naturally in the course of many calculations.

As one elementary application of the time-reversed wavefunctions, it is interesting to convert the transition amplitude of (3.20) to an alternative form. Thus (3.20) may be rewritten

$$T_{\alpha\beta} = (\psi_\beta e^{i(\mathbf{k}_\beta\cdot\mathbf{r}_\beta)}, V_\beta\Psi_\alpha^{(+)}), \tag{3.54}$$

$$= (\psi_\beta e^{i(\mathbf{k}_\beta\cdot\mathbf{r}_\beta)}, V_\beta[1 + (E + i\varepsilon - H)^{-1}V_\alpha]\psi_\alpha e^{i(\mathbf{k}_\alpha\cdot\mathbf{r}_\alpha)}), \tag{3.55}$$

where the stationary-state wavefunction is given a subscript α to indicate in which channel there is an incident wave, and the formal solution is used. By use of the Hermiticity properties of the operators in (3.55) this equation becomes

$$T_{\alpha\beta} = ([1 + (E - i\varepsilon - H)^{-1}V_\beta]\psi_\beta e^{i(\mathbf{k}_\beta\cdot\mathbf{r}_\beta)}, V_\alpha\psi_\alpha e^{i(\mathbf{k}_\alpha\cdot\mathbf{r}_\alpha)}), \tag{3.56}$$

$$= (\Psi_\beta^{(-)}, V_\alpha\psi_\alpha e^{i(\mathbf{k}_\alpha\cdot\mathbf{r}_\alpha)}). \tag{3.57}$$

It is well known that (3.54) and (3.57) are equivalent expressions for the transition amplitude. More careful demonstrations of the equivalence of (3.55) and (3.56) are based on Green's theorem, or on integration by parts, and include the demonstration that because V_α and V_β are short ranged the open channels do not yield any surface terms.

Finally, we present an operator identity that is useful for deriving relations among different Green's functions. The identity is

$$\frac{1}{A} - \frac{1}{B} = \frac{1}{A}(B - A)\frac{1}{B} = \frac{1}{B}(B - A)\frac{1}{A}. \tag{3.58}$$

Therefore, for example,

$$\frac{1}{A} = \left\{1 + \frac{1}{A}(B - A)\right\}\frac{1}{B}. \tag{3.59}$$

In the case that

$$\begin{aligned} A &= (E + i\varepsilon - H), \\ B &= (E + i\varepsilon - H + V_\alpha - U(r_\alpha)), \end{aligned} \tag{3.60}$$

it is seen that (3.59) gives a new expression for the *formal solution* of the Schrödinger equation, (3.51):

$$\begin{aligned} \Psi^{(+)} &= i\varepsilon(E + i\varepsilon - H)^{-1}e^{i(\mathbf{k}_\alpha\cdot\mathbf{r}_\alpha)}\psi_\alpha, \\ &= \{1 + (E + i\varepsilon - H)^{-1}(V_\alpha - U)\}\{i\varepsilon(E + i\varepsilon - H + V_\alpha - U)^{-1} \\ &\quad \times e^{i(\mathbf{k}_\alpha\cdot\mathbf{r}_\alpha)}\psi_\alpha\}. \end{aligned} \tag{3.61}$$

In this expression the calculation of $\Psi'^{(+)}$ is separated into two steps. In the first step the formal solution of the scattering problem governed by the Hamiltonian $(H - V_\alpha + U)$ is calculated. Thus in this step V_α is removed from H and replaced by the presumably simpler potential U. In the second step the term $(V_\alpha - U)$ is put back into the problem, very much as the full V_α would have been put in the original (3.51). Such a stepwise division of a scattering calculation obviously is valuable in many applications. This division may be used as a basis for the derivation of the Gell-Mann, Goldberger transformation.

3.5 Antisymmetrization

To avoid writing lengthy equations the discussion in this section proceeds on the basis that *all* nucleons are indistinguishable, that there is only one kind of nucleon. Accordingly, to apply the equations given here it is necessary that they either be used with the *i*-spin formalism or generalized in the manner shown later.

In principle antisymmetrization raises no problems of any difficulty. It is merely necessary to represent all observables with operators that are symmetric under exchange of particles, and to select from the set of all solutions of the Schrödinger equation those that are antisymmetric under exchange of particles. However, some questions of mathematical technique do arise, because methods for equation solving do not tend to be formulated in a symmetrical manner. Moreover, it is necessary to examine how the use of antisymmetric wavefunctions fits in with the definition of "cross section".

The properties of $\Psi'^{(+)}$, the stationary-state eigenfunction, are governed both by H, which is a symmetrical operator, and by the scattering boundary conditions. No particular symmetry properties have yet been imposed on these boundary conditions. However, both the Schrödinger equation and the boundary conditions are linear, therefore it is easy to introduce additional symmetry requirements by constructing linear combinations of the solutions already obtained. This may be done in conjunction with one or another of the Green's function expressions derived earlier in this chapter. Perhaps the most convenient of these is the formal solution of (3.51), in which $\Psi'^{(+)}$ is generated from the incoming plane wave in channel α by the application of the Green's function based on H, the full Hamiltonian of the problem. Because this Green's function is a symmetrical operator the entire task of antisymmetrizing $\Psi'^{(+)}$ can be accomplished simply by antisymmetrizing the incoming plane wave. Thus far this plane wave has been written in factored form:

$$e^{i(\mathbf{k}_\alpha\cdot\mathbf{r}_\alpha)}\psi_\alpha = e^{i(\mathbf{k}_\alpha\cdot\mathbf{r}_\alpha)}\psi_a(1, 2, \ldots, n_a)\, \psi_A(n_a + 1, \ldots, n_a + n_A), \quad (3.62)$$

where

$$\mathbf{r}_\alpha = \frac{1}{n_a}\sum_{i=1}^{n_a}\mathbf{r}_i - \frac{1}{n_A}\sum_{i=n_a+1}^{n_a+n_A}\mathbf{r}_i \quad (3.63)$$

and both ψ_a and ψ_A are antisymmetrized with respect to their internal variables. The variables of ψ_a and ψ_A are the position and spin variables of each of the nucleons whose labels are shown. Evidently, the linear operation $\mathscr{A}$ that antisymmetrizes (3.62) automatically antisymmetrizes the entire $\Psi'^{(+)}$ generated from (3.62), namely,

$$\Psi'^{(+)}_A = \mathscr{A}\Psi'^{(+)}. \tag{3.64}$$

We now need to discuss the form of $\mathscr{A}$.

By assumption the functions ψ_a and ψ_A already are antisymmetrized individually, and are normalized to unity. Therefore, the only remaining permutations of the particle labels in (3.62) are the ones that exchange labels *between* the two sets

$$\{1, 2, \ldots, n_a\}, \{n_a + 1, n_a + 2, \ldots, n_a + n_A\}.$$

Let us denote a permutation of this type by the symbol P. Then an antisymmetrized version of (3.62) is

$$\sum_P (-)^P P e^{i(\mathbf{k}_\alpha \cdot \mathbf{r}_\alpha)} \psi_\alpha, \tag{3.65}$$

where $(-)^P$ is understood to be positive if P exchanges an even number of particles between the two nuclei, and to be negative if P exchanges an odd number of particles. It is clear that the total number of permutations included in (3.65) is

$$N_\alpha = \frac{(n_a + n_A)!}{n_a!\, n_A!}; \tag{3.66}$$

this is the number of permutations of all kinds, divided by the number internal to a and the number internal to A.

To achieve a full definition of $\mathscr{A}$, the antisymmetrizer introduced in (3.64), it is necessary to find a suitable normalization for (3.65). We choose this normalization so that the incident flux in channel α has the same value after antisymmetrization that it had beforehand.

At first sight the calculation of the incident flux might appear rather obscure. Because the various terms of (3.65) differ only in the labelling of the particles, and because a plane wave pervades all space with equal probability, it appears that all the terms of (3.65) must overlap and that they must interfere when the flux is computed. However, this view is mistaken. It must be borne in mind that cross section is a classical concept, and that in an actual experiment the incident flux of particles in channel α is measured when the two nuclei a and A are well separated from each other. Hence, to avoid confusion we should not antisymmetrize an incident plane wave, but rather should start with an incident packet state and antisymmetrize this. In fact, we only use plane waves in (3.65) by virtue of the rules established in Chapter 1. In a more careful description of the scattering, nuclei a and A are localized initially in wave packets that do not overlap, and that occupy very small regions in configuration space. Then each new permutation of the particle labels produces two new nonoverlapping wave packets, which lie in new regions of configuration space and have no overlap with the packets produced by any other permutation. As a result the various permutations in (3.65) make independent, noninterfering contributions to the incident flux. We therefore define the normalized antisymmetrizer for channel α to be

$$\mathscr{A}_\alpha = N_\alpha^{-1/2} \sum_P (-)^P P, \tag{3.67}$$

where N_α is given in (3.66). The antisymmetrized stationary-state wavefunction is obtained, as indicated in (3.64), by applying the operator $\mathscr{A}_\alpha$ to the wavefunction $\Psi^{(+)}$ that is computed from only the first term of (3.65).

To get the cross section for the transition $\alpha \rightarrow \beta$ we use the Green's function formula of (3.20) (or one of the equivalent formulas) and obtain from $\Psi_A^{(+)}$ the amplitude for breakup into channel β with the particular set of labels

$$\psi_\beta = \psi_b(1, 2, \ldots, n_b)\, \psi_B(n_b + 1, n_b + 2, \ldots, n_b + n_B). \tag{3.68}$$

In this application (3.20) yields the projection of $\Psi_A^{(+)}$ on channel β in exactly the same manner in which it yielded the projection of $\Psi^{(+)}$ on channel β. However, ψ_β of (3.68) displays only one particular labelling of channel β. Identical amplitudes are yielded by each of the various labellings that can be assigned to ψ_β. Again, because actual experiments are conducted under classical circumstances it is clear that the various labellings of ψ_β refer to nonoverlapping wave packets. Therefore each of these different labellings refers to an independent and equal contribution to the outgoing flux, hence to an independent and equal contribution to the cross section. We therefore obtain the cross section for the physical transition $\alpha \rightarrow \beta$ by using (3.20) to calculate the cross section for a single labelled breakup mode, and then multiplying by the number of different labellings of ψ_β. Because ψ_b and ψ_B are antisymmetrized individually the number of different labellings of ψ_β is found, as in (3.66), to be

$$N_\beta = \frac{(n_b + n_B)!}{n_b!\, n_B!} = \frac{(n_a + n_A)!}{n_b!\, n_B!}, \tag{3.69}$$

where it is noted that $n_b + n_B = n_a + n_A$. Therefore the cross section we seek [see (1.32)] is

$$\frac{d\sigma_{\alpha\beta}}{d\Omega_\beta} = \frac{\mu_\alpha \mu_\beta}{(2\pi\hbar^2)^2}\left(\frac{k_\beta}{k_\alpha}\right) N_\beta\, |(\psi_\beta e^{i(\mathbf{k}_\beta \cdot \mathbf{r}_\beta)}, V_\beta \Psi_A^{(+)})|^2. \tag{3.70}$$

Of course, any one of the mathematically equivalent formulas for the transition amplitude may be used in (3.70).

By introducing the explicit antisymmetrization operator $\mathscr{A}_\alpha$ of (3.67), we find that (3.70) reduces to

$$\frac{d\sigma_{\alpha\beta}}{d\Omega_\beta} = \frac{\mu_\alpha\mu_\beta}{(2\pi\hbar^2)^2}\left(\frac{k_\beta}{k_\alpha}\right)\left(\frac{N_\beta}{N_\alpha}\right)|\mathscr{T}_{\alpha\beta}|^2, \tag{3.71}$$

where

$$\mathscr{T}_{\alpha\beta} = \sum_P (-)^P(\psi_\beta e^{i(\mathbf{k}_\beta\cdot\mathbf{r}_\beta)}, V_\beta P\Psi^{(+)}), \tag{3.72}$$

and

$$\left(\frac{N_\beta}{N_\alpha}\right) = \frac{n_a!\, n_A!}{n_b!\, n_B!}. \tag{3.73}$$

Equations 3.71–3.73 express the consequences of antisymmetrization.

If neutrons and protons are treated as distinguishable particles the above discussion is changed only in that the permutations of nucleons P are replaced by products of independent permutations of neutrons P_ν and protons P_π. Accordingly, (3.71–3.73) are changed to

$$\frac{d\sigma_{\alpha\beta}}{d\Omega_\beta} = \frac{\mu_\alpha\mu_\beta}{(2\pi\hbar^2)^2}\left(\frac{k_\beta}{k_\alpha}\right)\left(\frac{N_\beta}{N_\alpha}\right)|\mathscr{T}_{\alpha\beta}|^2, \tag{3.71'}$$

$$\mathscr{T}_{\alpha\beta} = \sum_{P_\nu}\sum_{P_\pi} (-)^{P_\nu+P_\pi}(\psi_\beta e^{i(\mathbf{k}_\beta\cdot\mathbf{r}_\beta)}, V_\beta P_\pi P_\nu\Psi^{(+)}), \tag{3.72'}$$

where now

$$\left(\frac{N_\beta}{N_\alpha}\right) = \left[\frac{Z_a!\, Z_A!}{Z_b!\, Z_B!}\right]\left[\frac{N_a!\, N_A!}{N_b!\, N_B!}\right]. \tag{3.73'}$$

Often only one of the permutations in (3.72) or (3.72′) is carried. However, the normalizing coefficients N_β/N_α must be carried in all calculations. For rearrangement collisions these coefficients can be quite large.

Further discussion of antisymmetrization is given in Section 4.8 and Chapter 10.

Many authors assert that it is more satisfactory to introduce antisymmetrization by use of the techniques of particle-number quantization (second quantization) than by explicit consideration of particle labels. However, it may be noted that both procedures are only methods for the derivation of (3.71) and (3.72) and it is far easier to derive those equations than to apply them. Furthermore, the method of particle-number quantization only avoids explicit consideration of particle labels because it introduces instead a rather sizeable formal apparatus. Reaction calculations then make only limited use of this apparatus. With these reasons in mind, personal taste led the present author to keep the discussion in terms of explicitly-labelled particles.

References

[1] Albert Messiah, *loc. cit.*, Chap. 1, Ref. 2.
[2] L. I. Schiff, *Quantum Mechanics*, McGraw-Hill, New York, 1949.
[3] L. Hostler and R. H. Pratt, *Phys. Rev. Letters* **10,** 469 (1963).
[4] M. H. Hull, Jr., and G. Breit, in *Encyclopedia of Physics*, Vol. 41, S. Flügge, Ed., Springer-Verlag, Berlin, 1959, Part 1, p. 410.
[5] E. S. Rost and N. Austern, *Phys. Rev.* **120,** 1375 (1960).
[6] M. Gell-Mann and M. L. Goldberger, *Phys. Rev.* **91,** 398 (1953).
[7] M. L. Goldberger and K. M. Watson, *loc. cit.*, Chapter 1, Ref. 1.
[8] E. P. Wigner, *loc. cit.*, Chapter 1, Ref. 4.
[9] B. Lippmann and J. Schwinger, *Phys. Rev.* **79,** 469 (1950).
[10] L. Foldy and W. Tobocman, *Phys. Rev.* **105,** 1099 (1957).

CHAPTER 4

Basic DI Theories

4.1 Introduction of Simple Models

In this chapter we set up practical theories for direct nuclear reactions. This is done by introducing simple models that seem appropriate for the reactions that are discussed, and then using the methods of Chapter 3 to work out the consequences of these models. It is understandable that the DI models for the various reactions are introduced somewhat arbitrarily, inasmuch as (see Chapter 2) there is no precise theoretical distinction between CN and DI processes of nuclei. However, the appropriateness of each particular DI model usually is sufficiently clear that it is worthwhile to devote major efforts to working out its consequences and comparing them with experiment. Most current work with DI theories, and most of this book, is devoted to nothing except the development of accurate methods for working out the consequences of the simple DI dynamical models. The development of these methods is begun in the present chapter.

It is tempting to seek to derive and improve the DI models by exhibiting them as leading terms of appropriate series expansions of some basic theory of rearrangement collisions. However, such derivations do not yet seem to have much significance, perhaps because the association of a model with a series is a step that tends to be as arbitrary as the initial invention of the model. It seems best to consider the DI models to be devices for the generation of good trial wavefunctions, such as might be used in variational calculations, and to acknowledge that heuristic thinking is always the basis for the discovery of trial wavefunctions. This sort of justification for the DI models is similar in spirit to the justification for the shell model. In fact, many DI models may be regarded simply as extrapolations of the shell model into the continuum.

The formulation of DI models is guided by the discussion in Chapter 2, where *closed* and *open* parts of the stationary-state wavefunction are defined. In that discussion the wavefunction Ψ_{open} is a sum of products of internal and relative wavefunctions for the particles that separate from each other in each open channel, as in (2.5). We construct a practical DI model by limiting this sum and including in it only those *few open channels* that the nuclear force is able to couple strongly to the entrance channel. With this proviso in mind the wavefunction for the DI model is written as

$$\Psi_{\text{model}} = \sum_{\alpha} \psi_{1\alpha}\psi_{2\alpha}\xi_{\alpha}(\mathbf{r}_{1\alpha} - \mathbf{r}_{2\alpha}). \tag{4.1}$$

We then attempt to determine the ξ_α by diagonalizing the Hamiltonian in the portion of Hilbert space spanned by (4.1), subject to the limited set of internal functions $\psi_{1\alpha}$ and $\psi_{2\alpha}$ chosen for the model. The theory just described is related closely to the *resonating group* theory of Wheeler [1], as well as to the Tamm-Dancoff theory [2]. Because of the dynamical simplicity of this DI theory the ξ_α must be expected to be slowly-varying functions of bombarding energy.*

Some generalization of (4.1) may be of interest; for example, sometimes it may be useful to include into the sum over α one or two simple closed configurations that are coupled particularly strongly to the entrance channel. It also may be useful to include into the sum one or two terms that express breakup of the system into three or more parts. These terms would have the internal wave functions $\psi_{1\alpha}$, $\psi_{2\alpha}$, $\psi_{3\alpha}$, etc. For such breakup alternatives the ξ_α would have to depend on two or more vector variables that would express the displacements among the various parts.

To complete the formulation of the DI theories we still must consider one further aspect of Ψ_{model}. Namely, (4.1) is only a part of the physical wave function of the system, and it may be a rather small part. DI theories therefore take the important step of basing the calculation of the ξ_α on a Hamiltonian that contains *optical potentials*. These are potentials that operate within the Hilbert space spanned by (4.1) but that are constructed so they include into the calculation of Ψ_{model} the energy-average of the coupling between Ψ_{model} and the remainder of the physical wave function, the part we otherwise tacitly dismiss as CN. These optical potentials are complex and their imaginary parts describe the transfer of flux into processes not otherwise included in (4.1). The optical potentials introduce an element of vagueness in the DI theories, inasmuch as many of their properties must be determined phenomenologically. On the other hand, because these potentials represent energy-averaged properties of the system, they reduce neither the dynamical simplicity of the DI theories nor the slow energy dependence of the ξ_α. It is clear that the use of optical potentials limits the equivalence between DI theories and the resonating-group theory or the Tamm-Dancoff theory, mentioned earlier. It is also clear that the use of optical potentials introduces considerable

* A coupled-channels theory occasionally yields a few isolated resonances (see Chapter 6). These are exceptions to the otherwise general slow energy dependence of the ξ_α.

unification in the overall analysis of the physical system. A preliminary discussion of the use of complex potentials is given in the following two sections. Further discussion of this method of constructing *unified theories* is given in Chapter 11.

4.2 Elimination of the Compound Nucleus

We now discuss the coupling of Ψ_{model} with other parts of the physical wavefunction. To do this it is convenient to define a division of Hilbert space into two disjoint parts. In accordance with the notation introduced by Feshbach [3, 4, 5], P and Q will be defined to be projection operators that select each of these two parts of Hilbert space. Such operators must obey

$$P^2 = P, \qquad Q^2 = Q, \tag{4.2}$$

$$QP = PQ = 0, \tag{4.3}$$

and

$$P + Q = I, \tag{4.4}$$

where I is the identity operator. The particular wavefunctions selected by P are understood to be those constructed by introducing a complete set of all possible relative functions ξ_α into (4.1). The functions selected by Q are those that cannot be constructed in this manner. In terms of these operators we write for the physical wavefunction of the system:

$$\Psi^{(+)} = \Psi_P + \Psi_Q, \tag{4.5}$$

where

$$\Psi_P = P\Psi^{(+)}, \tag{4.6}$$

$$\Psi_Q = Q\Psi^{(+)}. \tag{4.7}$$

It is clear that the above definitions select for inclusion in Ψ_P all the entrance-channel parts of $\Psi^{(+)}$, because Ψ_{model} includes the entrance channels. Accordingly the terms left for inclusion in Ψ_Q must either be closed or obey outgoing-wave boundary conditions. The division of $\Psi^{(+)}$ thus treats the boundary conditions in an unsymmetrical manner.

We note [3] that the above division of $\Psi^{(+)}$ does not require that all open-channel parts of $\Psi^{(+)}$ be included in Ψ_P. Some open-channel terms that carry outgoing flux are omitted from Ψ_{model} for reasons of convenience or simplicity; these therefore are included in Ψ_Q.

Although the function Ψ_P of (4.6) has the mathematical structure of (4.1), it is not identical with Ψ_{model} because the coupling to Ψ_Q has not yet been smoothed in any manner. Later we will consider the identification

$$\Psi_{\text{model}} = \overline{\Psi}_P, \tag{4.8}$$

that is, it will be considered that Ψ_{model} is obtained from Ψ_P by some suitable average with respect to energy.

The Schrödinger equation for $\Psi^{(+)}$,

$$(E - H)\Psi^{(+)} = 0, \tag{4.9}$$

takes the form of a pair of coupled equations for the two functions Ψ_P and Ψ_Q. Thus through left-multiplication by P and by Q, (4.9) reduces to

$$(E - H_{PP})\Psi_P = H_{PQ}\Psi_Q, \tag{4.10}$$

$$(E - H_{QQ})\Psi_Q = H_{QP}\Psi_P. \tag{4.11}$$

Here

$$\begin{aligned} H_{PP} &\equiv PHP, & H_{QQ} &\equiv QHQ, \\ H_{PQ} &\equiv PHQ, & H_{QP} &\equiv QHP. \end{aligned} \tag{4.12}$$

By definition, the operator H_{PP} only has matrix elements that connect wave-functions of the form of the various terms of (4.1); the operator H_{QP} only has matrix elements that connect functions of the form of (4.1) with functions not included in that expression; etc. It is clear that CN effects are introduced into Ψ_P by the term $H_{PQ}\Psi_Q$, which is the right hand side of (4.10).

The right hand side of (4.10) is seldom small enough to disregard its effects upon Ψ_P. On the other hand, our original choice of the DI model, as expressed by the selection of the terms in (4.1), is motivated by the wish to avoid working with any larger space than that spanned by Ψ_P. To accomplish this wish, we derive an operator that is defined in the space of Ψ_P and that eliminates the explicit appearance of Ψ_Q in the RHS of (4.10). Although this operator carries all the complications of Ψ_Q, its introduction enables us to develop suitable approximations of the RHS of (4.10). The most interesting general property of the new operator is that one part of it is anti-Hermitian (a.h.), a consequence of the unsymmetrical treatment of the boundary conditions that is mentioned earlier. The new operator has an a.h. part both because Ψ_Q may have some open channels that carry outgoing flux, and because Ψ_Q is coupled to an outgoing-wave part of Ψ_P that we later exclude from Ψ_{model}.

The coupling to Ψ_P influences the calculation of Ψ_Q and must be treated first. To display this effect we construct a Green's function solution of (4.10), using the methods discussed in Chapter 3:

$$\Psi_P = \mathring{\Psi}_P + (E + i\varepsilon - H_{PP})^{-1}H_{PQ}\Psi_Q. \tag{4.13}$$

The second term of (4.13) uses the outgoing-waves Green's function; therefore it carries no ingoing flux. The first term of (4.13) is a solution of the homogeneous part of (4.10),

$$(E - H_{PP})\mathring{\Psi}_P = 0; \tag{4.14}$$

it carries the appropriate ingoing-wave parts of Ψ_P as required by the scattering boundary conditions. Insertion of (4.13) in (4.11) yields a complicated uncoupled differential equation for Ψ_Q,

$$[E - H_{QQ} - H_{QP}(E + i\varepsilon - H_{PP})^{-1}H_{PQ}]\Psi_Q = H_{QP}\mathring{\Psi}_P. \quad (4.15)$$

It is the third term of the left-hand side of (4.15) that carries into the calculation of Ψ_Q the effects caused by coupling to outgoing-wave parts of Ψ_P.

The Green's function in (4.15) may be expanded in bilinear form by introducing a complete set of regular eigenstates of H_{PP},

$$H_{PP}\,|P(\nu, E')\rangle = E'\,|P(\nu, E')\rangle, \quad (4.16)$$

where ν distinguishes among degenerate states. Then the complicated operator in (4.15) becomes

$$H_{QP}(E + i\varepsilon - H_{PP})^{-1}H_{PQ} = \sum_\nu \int dE' \frac{H_{QP}\,|P(\nu, E')\rangle\langle P(\nu, E')|\, H_{PQ}}{(E - E') + i\varepsilon}. \quad (4.17)$$

The numerator of the right-hand side of (4.17) quite clearly is Hermitian. Furthermore, it has only a slow dependence on the energy E' because the operators H_{QP} and H_{PQ} are localized and the $|P(\nu, E')\rangle$ are governed by H_{PP}, the part of H whose matrix elements connect simple wavefunctions. Thus, although the operator of (4.17) is not Hermitian it may be split into Hermitian and anti-Hermitian parts merely by splitting the denominator into its real and imaginary parts. This procedure gives

$$H_{QP}(E + i\varepsilon - H_{PP})^{-1}H_{PQ} = \mathscr{K}_1 - i\mathscr{K}_2, \quad (4.18)$$

$$\mathscr{K}_1 = \sum_\nu \int dE' \frac{H_{QP}\,|P(\nu, E')\rangle(E - E')\langle P(\nu, E')|\, H_{PQ}}{(E - E')^2 + \varepsilon^2}, \quad (4.19)$$

$$\mathscr{K}_2 = \sum_\nu H_{QP}\,|P(\nu, E)\rangle\rho(E)\langle P(\nu, E)|\, H_{PQ} \int \frac{\varepsilon\, dE'}{(E - E')^2 + \varepsilon^2}. \quad (4.20)$$

In (4.20) the slowly-varying numerator has been removed from the integral.* The function $\rho(E)$ is the density of eigenstates of H_{PP} at the energy E. The integral that remains in (4.20) is easily evaluated, and $\mathscr{K}_2$ reduces to

$$\mathscr{K}_2 = \pi \sum_\nu H_{QP}\,|P(\nu, E)\rangle\rho(E)\langle P(\nu, E)|\, H_{PQ}. \quad (4.21)$$

* Of course, in the limit $\varepsilon \to 0$ the function $\mathscr{K}_2$ only can be nonvanishing if the range of integration over E' includes the value $E' = E$. In the present case this condition is fulfilled automatically.

This $\mathscr{K}_2$ is positive definite, and is a sum over matrix elements that connect Ψ_Q to the set of regular eigenfunctions of H_{PP} at the energy E. The presence of $\mathscr{K}_2$ in (4.15) introduces a decay width in the calculation of Ψ_Q. The operator $\mathscr{K}_1$ is less interesting. It shifts the energies of the various eigenstates of (4.15).

A simple example illustrates how $\mathscr{K}_2$ may be traced to the non-Hermitian nature of the outgoing-waves Green's function: For a wavefunction Ψ_P that includes only one open channel, and therefore only represents elastic scattering, the Green's function may be written out explicitly in partial-wave expansion as in (3.32). Thus

$$\begin{aligned}(E + i\varepsilon - H_{PP})^{-1} &= -K(\mathbf{r}_\alpha, \mathbf{r}_\alpha') \\ &= -\frac{2\mu_\alpha}{\hbar^2} \sum_{l,m} \frac{f_l(r_<)h_l(r_>)}{kr_\alpha r_\alpha'} Y_l^m(\hat{r}_\alpha) Y_l^{m*}(\hat{r}_\alpha'), \end{aligned} \tag{4.22}$$

where f_l is the regular solution of the radial Schrödinger equation for partial wave l and is real, and h_l is the outgoing-wave solution of the radial Schrödinger equation and is complex. If $h_l(r_\alpha)$ is written as a linear combination of regular and irregular solution functions f_l and g_l, both of which are real, then

$$h_l = g_l + if_l, \tag{4.23}$$

and the operators of (4.19) and (4.20) become

$$\mathscr{K}_1 = -\frac{2\mu_\alpha}{\hbar^2 k} \sum_{l,m} H_{QP} Y_l^m(\hat{r}_\alpha) \frac{f_l(r_<)g_l(r_>)}{r_\alpha r_\alpha'} Y_l^{m*}(\hat{r}_\alpha') H_{PQ}, \tag{4.24}$$

$$\mathscr{K}_2 = \frac{2\mu_\alpha}{\hbar^2 k} \sum_{l,m} \{H_{QP} r_\alpha^{-1} f_l(r_\alpha) Y_l^m(\hat{r}_\alpha)\}\{h.c.\}. \tag{4.25}$$

In (4.25) the first factor under the summation sign is multiplied by its Hermitian conjugate. It is thus clear both that $\mathscr{K}_2$ is positive and that the sign of $\mathscr{K}_2$ is a consequence of the sign of the imaginary term of (4.23); the latter sign, in turn, is a consequence of the use of the outgoing-wave solution function h_l in (4.22). Evidently, Green's functions that satisfy the outgoing-waves boundary condition are not Hermitian operators.

We now obtain an exact differential equation for Ψ_P by solving (4.15) for Ψ_Q and using that solution to eliminate Ψ_Q from (4.10). The solution of (4.15) yielded by the insertion of (4.18) is the expression

$$\Psi_Q = (E + i\mathscr{K}_2 - H_{QQ} - \mathscr{K}_1)^{-1} H_{QP} \mathring{\Psi}_P. \tag{4.26}$$

No homogeneous term is needed in this expression because, by definition, all parts of Ψ_Q obey outgoing-waves boundary conditions. (Of course, the term $i\mathscr{K}_2$ automatically selects outgoing-waves boundary conditions in

(4.26), and no additional $i\varepsilon$ term is needed.) With the use of (4.26), (4.10) becomes

$$(E - H_{PP})\Psi_P = \mathcal{O}\mathring{\Psi}_P, \tag{4.27}$$

in which we at last see that the operator* that eliminates the explicit appearance of Ψ_Q is

$$\mathcal{O} \equiv H_{PQ}(E + i\mathcal{K}_2 - H_{QQ} - \mathcal{K}_1)^{-1}H_{QP}. \tag{4.28}$$

The operator $\mathcal{O}$ carries into the calculation of Ψ_P all the consequences of coupling between Ψ_P and Ψ_Q.

The structure of the operator $\mathcal{O}$ is much like that discussed in connection with (4.17). If the term $i\mathcal{K}_2$ were omitted (an $i\varepsilon$ term would have to be introduced), and if the Green's function of (4.28) were expanded in a complete set of eigenstates of $H_{QQ} + \mathcal{K}_1$, then, if some of these eigenstates were to carry nonvanishing outgoing flux $\mathcal{O}$ would have an anti-Hermitian part like that in (4.21). The presence of the decay-width operator $\mathcal{K}_2$ causes little alteration of these results. Because $\mathcal{K}_2$ is present, the regular eigenfunctions of the operator $(H_{QQ} + \mathcal{K}_1 - i\mathcal{K}_2)$ are not orthogonal, and they must be combined with conjugate functions to form a biorthogonal set. The eigenvalues of this operator possess small negative imaginary parts. Thus,

$$(H_{QQ} + \mathcal{K}_1 - i\mathcal{K}_2)\,|Q(\mu, E'')\rangle = E''\,|Q(\mu, E'')\rangle, \tag{4.29}$$

where μ distinguishes among degenerate states, and where

$$E'' = E' - \frac{i}{2}\Gamma(E', \mu), \tag{4.30}$$

with E' and Γ real. We give the states $|Q(\mu, E'')\rangle$ the brief name *compound-nucleus states*. Because $\mathcal{K}_1$ and $\mathcal{K}_2$ depend on E, the bombarding energy of the scattering problem, it is clear that our compound states must depend on E parametrically.† However, if the DI model from which Ψ_P was developed is fairly complete, so that Ψ_P includes all the open channels coupled strongly to the entrance channel, then the $|Q(\mu, E'')\rangle$ tend to be closed states ("bound states in the continuum") with only small widths for decay through Ψ_P. Such states conform closely to usual ideas about compound states. The spectrum of their eigenenergies E'' tends to be discrete.

* The operator $\mathcal{O}$ is related to the optical potential, as is seen shortly. However, this relationship is not an identity.

† Similar compound states arise naturally in discussions of the Kapur-Peierls theory. A relevant discussion of these states is given by Feshbach [4], section IVC. A clear discussion of biorthogonal expansions is given by W. MacDonald, *Nucl. Phys.* **54**, 393 (1964).

If the compound states are long lived so that the Γ are small, then up to first order the Γ need only be carried in the energy E'', and may be omitted from the compound-state wavefunctions themselves. The wavefunctions thereby become orthogonal. This simplification is now assumed, with the result

$$\mathcal{O} \approx \sum_{\mu} \int dE' \frac{H_{PQ}\,|Q(\mu, E')\rangle\langle Q(\mu, E')|\,H_{QP}}{E - E' + (i/2)\Gamma(E', \mu)}. \tag{4.31}$$

Once again, as with (4.17), the above expression is split into Hermitian and anti-Hermitian parts by splitting the denominator into its real and imaginary parts. The two parts of $\mathcal{O}$ are

$$\tfrac{1}{2}(\mathcal{O} + \mathcal{O}^{\dagger}) = \sum_{\mu} \int dE' \frac{H_{PQ}\,|Q(\mu, E')\rangle(E - E')\langle Q(\mu, E')|\,H_{QP}}{(E - E')^2 + (\Gamma^2/4)}, \tag{4.32}$$

and

$$\tfrac{1}{2}(\mathcal{O} - \mathcal{O}^{\dagger}) = \sum_{\mu} \int dE' \frac{H_{PQ}\,|Q(\mu, E')\rangle(-i\Gamma/2)\langle Q(\mu, E')|\,H_{QP}}{(E - E')^2 + (\Gamma^2/4)}. \tag{4.33}$$

Because the widths Γ are not infinitesimals it is not possible at the present stage to carry through any further reduction of these equations that would parallel, for example, the reduction of (4.20) to (4.21). However, we do see from (4.33) that for the anti-Hermitian part of $\mathcal{O}$ to be nonvanishing there must be compound states that lie at energies E' within an interval Γ of the energy E. We also see that this result obtains whether or not the compound states themselves carry any outgoing flux.

The major new property of $\mathcal{O}$ is that the strength of this operator fluctuates rapidly according to whether or not E lies near one of the E'. This effect should be most prominent at low energies, where the E' tend to be widely spaced. Because the anti-Hermitian part of $\mathcal{O}$ is dominated by the few nearby compound states that nearly conserve energy, the fluctuations of this part of $\mathcal{O}$ must be especially strong. By contrast, the Hermitian part of $\mathcal{O}$ receives contributions from distant levels and these contributions may be large. They provide a background on which the fluctuations sit. It is interesting that the Hermitian contributions to $\mathcal{O}$ indicate a point at which some judgment regarding the calculation of Ψ_P can be exercised. The Hermitian contributions from distant levels can be carried as such, or they can be minimized by redefining the DI model for Ψ_P to minimize the matrix elements that couple Ψ_P to distant compound states [6].

The complications of the operator $\mathcal{O}$ are no surprise. Equation 4.27 is an exact equivalent of the original coupled equations (4.10, 4.11). The operator $\mathcal{O}$ carries the complications of Ψ_Q into the calculation of Ψ_P, and thereby introduces CN effects in Ψ_P. However, it is a great convenience to have isolated $\mathcal{O}$ as the source of the complications in (4.27), because these complications can now be reduced very easily by substituting a smoothed operator $\bar{\mathcal{O}}$ in (4.27) in place of $\mathcal{O}$. The operator $\bar{\mathcal{O}}$ is obtained by averaging $\mathcal{O}$ with respect to energy, to eliminate the fluctuations. Further discussion of the smoothed operator $\bar{\mathcal{O}}$ is given in Section 4.3. At the present stage we only recognize that substitution of $\bar{\mathcal{O}}$ for $\mathcal{O}$ in (4.27) allows the previously discussed DI model wavefunction to be defined quantitatively as the solution of the equation

$$(E - H_{PP})\Psi_{\text{model}} = \bar{\mathcal{O}}\mathring{\Psi}_P. \tag{4.34}$$

Equation 4.34 retains the energy-averaged influence of Ψ_Q upon Ψ_P.

The reaction amplitudes yielded by (4.34) do not contain CN fluctuations, therefore they tend to be slowly-varying functions of E. They are very nearly equal to the energy-averaged parts of the exact reaction amplitude. Hence we have achieved a division of Ψ_P into two parts,

$$\Psi_P = \Psi_{\text{model}} + (\Psi_P - \Psi_{\text{model}}), \tag{4.35}$$

that yield noninterfering contributions to energy-averaged cross sections. The second term of (4.35) gives the CN contributions to the reaction amplitudes in the various open channels; it generally is calculated by the methods of Wolfenstein, and Hauser and Feshbach [7]. It gives, for example, the amplitude for *compound-elastic* scattering. Throughout most of this book we concentrate on the calculation of the first term of (4.35).

Equation 4.34 is not homogeneous, and therefore it cannot be regarded as a reduced Schrödinger equation for the model wavefunction. Algebraic methods can be used (see Chapter 11, also [6, 3, 4]) to transform (4.34) into an exact homogeneous equation for Ψ_{model}. However, the equation so derived contains an effective interaction that is considerably more complicated than $\bar{\mathcal{O}}$, and that must be handled by perturbative methods [6, 3, 4]. For the present discussion, rather than derive this complicated homogeneous equation that must then be simplified, it is more straightforward to introduce perturbative approximations immediately. We therefore recognize that Ψ_P and Ψ_{model} are approximately equal and use this fact to arrive immediately at the homogeneous equation

$$(E - H_{PP} - \bar{\mathcal{O}})\Psi_{\text{model}} \approx 0. \tag{4.36}$$

The relative wavefunctions ξ_α of the DI model are computed from (4.36). The operator $\bar{\mathcal{O}}$ is treated phenomenologically, subject to the qualitative discussions already given. Equation 4.36 is an *optical model* for Ψ_{model} and the various DI reactions it describes.

A wave-packet discussion further illustrates the content of the above analysis: Suppose that at $t = 0$ the Schrödinger wavefunction of the system is by some means made to be identical with Ψ_{model}. How does such a wavefunction evolve as a function of time? Of course, the evolution of the wavefunction is dominated by the coupling to the compound states $|Q(\mu, E')\rangle$. As time proceeds these states are coupled in with larger and larger amplitudes. This effect is accompanied by a decrease in the amplitude of the original simple part of the wavefunction, so that the matrix element

$$(\Psi_{\text{model}}, e^{-iHt/\hbar}\Psi_{\text{model}}) \tag{4.37}$$

is a decreasing function of time. It is easily seen that the rate of decrease of the above matrix element is measured by the anti-Hermitian part of the operator $\bar{\mathcal{O}}$. It is also seen to be plausible that the rate of decay of the original packet is governed only by the rate at which flux goes over to the compound states, and is not affected by the manner in which these states subsequently decay. This observation is in accord with the detailed structure of the anti-Hermitian part of $\bar{\mathcal{O}}$, as is seen in the next section. The expressions derived there make no reference to the manner in which the compound states decay, whether by the emission of flux into the continuum of Ψ_Q or by coupling to Ψ_P. (This unsymmetrical description of the damping of Ψ_{model} resembles the treatments of such other phenomena as mechanical friction or radiation damping.)

It is also possible to apply the above wave-packet discussion if Ψ_{model} is a bound state, that is, if neither Ψ_P nor the compound states are open for the emission of flux. In this case (4.37) initially decreases with time in much the same manner as in the continuum, because as time proceeds the complicated states $|Q(\mu, E')\rangle$ gradually become mixed in with the simple state Ψ_{model}. Once again the rate of decrease of (4.37) is measured by the anti-Hermitian part of $\bar{\mathcal{O}}$. What is new in the bound-state case is that (4.37) does not decay indefinitely. Because there is no emission of flux the states $|Q(\mu, E')\rangle$ do not decay; their phases maintain coherence and eventually the original simple state Ψ_{model} recurs. Thus only in the continuum does a non-Hermitian operator have meaning for the calculation of a stationary-state wavefunction.

Wave packets change their structures [as in the decay of (4.37)] because the states from which they are assembled have different energies; therefore these states evolve at different rates and gradually drop out of phase. Thus the rate of decay of (4.37) measures the range of energies from which eigenstates of H must be assembled to build Ψ_{model}. Therefore the anti-Hermitian part of $\bar{\mathcal{O}}$ also measures this range of energies, the range over which a simple state is spread by being mixed with complicated states.

4.3 Further Discussion of the Optical Model: Complex Potentials

It is necessary to specify more carefully what is meant by the averaging procedure introduced in the preceding section. Let us first consider the simple rectangular average

$$\bar{\mathcal{O}}(E) \equiv \frac{1}{I}\int_{E-I/2}^{E+I/2} \mathcal{O}(E'')\,dE'', \tag{4.38}$$

where I is the averaging interval. If Γ is small compared with the spacing of E', then, aside from endpoint effects, (4.38) yields for the anti-Hermitian part of $\bar{\mathcal{O}}$ the result

$$\tfrac{1}{2}(\bar{\mathcal{O}} - \bar{\mathcal{O}}^{\dagger}) = -i\pi \sum_{\mu,E'} H_{PQ}\,|Q(\mu, E')\rangle\langle Q(\mu, E')|\, H_{QP}, \tag{4.39}$$

just as in (4.21). Here the sum over E' includes only those E' that lie in the interval I centered at the energy E. However, endpoint effects become troublesome in (4.39) if Γ is comparable to, or larger than, the spacing of the E'. In this case the popular procedure [6] of averaging $\mathcal{O}$ with a Lorentzian weighting factor,

$$\bar{\mathcal{O}}(E) \equiv \frac{I}{\pi}\int \frac{\mathcal{O}(E'')\,dE''}{(E''-E)^2 + I^2}, \tag{4.40}$$

yields the better-behaved expression

$$\bar{\mathcal{O}}(E) = \mathcal{O}(E + iI), \tag{4.41}$$

so that

$$\tfrac{1}{2}(\bar{\mathcal{O}} - \bar{\mathcal{O}}^{\dagger}) = -iI \sum_{\mu} \int dE' \,\frac{H_{PQ}\,|Q(\mu, E')\rangle\langle Q(\mu, E')|\, H_{QP}}{(E-E')^2 + I^2}, \tag{4.42}$$

where Γ has been neglected because $\Gamma \ll I$. Although the Lorentzian weight factor overemphasizes the importance of distant compound states (Feshbach, private communication), we see that (4.42) is not very different in content from (4.39) and either serves our present purpose. Equations 4.39 and 4.42 make no explicit reference to the manner in which the compound states decay, now that the averaging has caused Γ to disappear from these expressions.

Is there an optimum value for the magnitude of the averaging interval, I? We observe immediately that a minimum magnitude for I is determined by the fact that I must be at least large enough to smooth the fluctuations of $\mathcal{O}$ to a satisfactory extent. A maximum magnitude for I is not determined by the properties of $\mathcal{O}$, but rather by the fact that the averaging of $\mathcal{O}$ is accompanied by an averaging of the scattering amplitude. If the latter average were taken over too broad an energy range it could obscure energy-dependent structures that are characteristic of the DI model itself. Thus we see there are limiting conditions that tend to define an optimum value of I. The optical model can have meaning only if the two limiting conditions can be fulfilled simultaneously.

At high bombarding energies there is a wide range of values of I that meet both the above conditions, because the density of compound states is so large that $\mathcal{O}$ has very little fluctuation structure. Discussions of the optical model for high energies [8] seldom mention the question of averaging.

On the other hand, the averaging is essential for applications of the optical model at low energies, and for these applications it is necessary to inquire with great care whether the physics of real nuclei admits the determination of a value of I that meets the two limiting conditions. On the basis of perturbative studies of $\bar{\mathcal{O}}$, Brown [6] concluded that low-energy nucleon-induced reactions do tend to meet the above two conditions. This conclusion is supported by the generally successful use of the optical model to analyze low and medium-energy experiments. Brown's analysis suggests that typical values for I are of the order of 0.5 MeV. However, it must be noted that Brown's analysis provided only marginal justification of the averaging procedure at low energy, and frequent failures in this application of the optical model may be expected. (Recent discussions [9, 10] of *intermediate structure* relax the requirements on I by redefining the energy average so that some fluctuation structure of $\mathcal{O}$ is retained.)

Averaging $\mathcal{O}$ with respect to bombarding energy has further theoretical importance in that it greatly increases the number of compound states contributing to the effective interaction. As a result the calculation of $\bar{\mathcal{O}}$ tends to be independent of details of the compound states, and we can hope this calculation would not change too much if the compound states were replaced by uncoupled basis states.

We now discuss the basic optical model equation of (4.36) in more practical terms. In that equation the operator $\bar{\mathcal{O}}$ introduces the corrections required because Ψ_{model} is an incomplete description of the system. The most interesting part of $\bar{\mathcal{O}}$ is an anti-Hermitian term that expresses loss of the flux transferred from Ψ_{model} into the *compound nucleus* parts of the wavefunction. The flux lost from Ψ_{model} may be emitted in open channels not included in Ψ_{model}, or it may be emitted in the same open channels that make up Ψ_{model} but in the form of contributions that fluctuate strongly with respect to energy. The averaging procedure that selects Ψ_{model} excludes the latter sort of contribution. There is also a part of $\bar{\mathcal{O}}$ that is Hermitian, but that hopefully is only a small correction to the Hermitian operator H_{PP}.

As noted earlier, numerical details of $\bar{\mathcal{O}}$ are to be determined by fitting to experiment. Before this can be done, it is necessary to develop qualitative theoretical understanding of the structure of $\bar{\mathcal{O}}$ as a function of the variables in Ψ_{model}. There has been very little work of this kind. Clearly, since $\bar{\mathcal{O}}$ is a correction to H_{PP}, its nature and purpose should be expected to change as the model that selects H_{PP} is changed. In fact, $\bar{\mathcal{O}}$ seems to have been investigated theoretically only for the circumstance that Ψ_{model} describes nonexchange elastic scattering! In this case (4.1) consists of the one term,

$$\Psi_{\text{model}} = \psi_{1\alpha}\psi_{2\alpha}\xi_\alpha(\mathbf{r}_\alpha), \tag{4.43}$$

and (4.36) reduces to a single second-order differential equation for the relative wavefunction ξ_α. The operator $\bar{\mathcal{O}}$ reduces to a function of the one variable $\mathbf{r}_\alpha$, and (4.36) takes the form

$$\left[E + \frac{\hbar^2}{2\mu_\alpha}\nabla_{\mathbf{r}_\alpha}{}^2 - U_\alpha(\mathbf{r}_\alpha)\right]\xi_\alpha(\mathbf{r}_\alpha) = 0, \tag{4.44}$$

in which $U_\alpha(\mathbf{r}_\alpha)$ is known as the *optical potential*. This potential is complex. Its imaginary term is identical with the anti-Hermitian part of $\bar{\mathcal{O}}$; its real term consists mainly of the potential term of H_{PP} (the Hartree potential) but with a small correction from the Hermitian part of $\bar{\mathcal{O}}$. It is clear from inspection of (4.39) and (4.42) that the imaginary term of U_α probably is quite nonlocal.*

With minor exceptions (see [11, 12, 13] and Section 5.2) $U_\alpha(\mathbf{r}_\alpha)$ of (4.44) is assumed to be a *local* function of $\mathbf{r}_\alpha$, with a radial shape that is either proportional to the nuclear density distribution or closely related to it. This assumption originated in derivations of the optical potential for high-energy scattering [8, 14], in which case the number of compound states that contribute to $\mathcal{O}$ is so great that it is possible to replace these states with the states of a Fermi gas model, with the result that the various expressions for $\mathcal{O}$ can be evaluated explicitly. Similar schematic investigations at lower energy serve chiefly to show that the imaginary term of U_α must be concentrated near, or even somewhat outside, the nuclear surface [15]. Figure 4.1 shows one such set of results. The surface peaking apparently arises because in the nuclear interior the Pauli exclusion principle limits the compound-state wavefunctions that participate in the determination of U_α. Unfortunately, it is a little risky to use schematic models to calculate the imaginary potential for low-energy elastic scattering. We see in (4.39) and (4.42) that this potential is determined by the compound states in a fairly narrow band of energies. It may be that the compound states in this band are not typical of the overall nuclear spectrum and are represented poorly by a schematic model. Some recent work tends to emphasize the importance of compound states based on collective excitations.

* Indeed, those equations express the imaginary term of U_α as a sum of *separable interactions*, one for each compound state. This fact has seldom been exploited. However, see R. Lipperheide, *Z. Physik* **202**, 58 (1967).

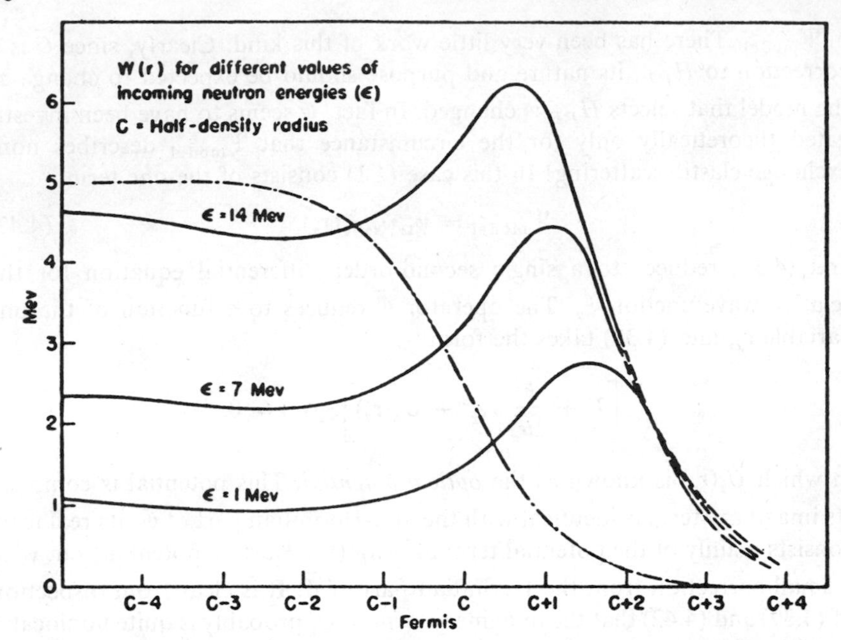

Fig. 4.1 The imaginary part W of the optical potential at the nuclear surface, for different values of incoming neutron energies ε. The vertical scale is in MeV and the horizontal scale in fm. The dot-dash curve is the nuclear density function $\rho(r)$ in arbitrary vertical scale; C is the half-density radius. (From [15], Fig. 2.)

If the DI model includes other processes there is much less guidance for the treatment of $\bar{\mathcal{O}}$ than in the discussion of elastic scattering that was just given. Typical direct reaction studies make (4.36) explicit by separately inserting in each open channel an optical potential of the sort introduced in (4.44). Although it is followed in this book, this procedure is rather arbitrary. It may be hoped that future work will be based more firmly on multi-channel theoretical expressions for $\bar{\mathcal{O}}$.

The detailed parameters of the optical potentials in (4.36) are determined by fitting them to energy-averaged elastic scattering experiments. One can ask: What effects would be introduced if the range of averaging in these elastic scattering experiments were not sufficiently broad to smooth out all the fluctuations of $\mathcal{O}$? Is it meaningful to fit potentials that follow to some extent the energy-dependent effects caused by coupling to the compound states? In discussing these questions it is necessary to recognize that the averaging enters the calculation in two independent steps. One step consists of the use of Ψ_{model} in the calculation of cross sections, the other consists of the calculation of Ψ_{model} from (4.36). The use of Ψ_{model} is affected by the fact that this wavefunction is only the DI part of the full open-channel wavefunction Ψ_P. With incomplete averaging the interference between the DI and CN parts of Ψ_P would not vanish, and would have to be treated with considerable detail. Incomplete averaging would have a more subtle influence on the calculation of Ψ_{model}. It would affect the parameter values of the optical potentials used to calculate Ψ_{model} and the interpretations of these potentials. Sufficient averaging brings into the determination of the optical potentials such a broad range of compound states that the details of these states are rendered unimportant, so that simple, universal forms for the optical potentials are rendered appropriate. With incomplete averaging the optical potentials would be forced to follow the detailed energy dependence of the scattering cross sections, and as a result they might need to have both energy-dependent parameters and energy-dependent functional forms. Energy dependence of the latter kind would be difficult to handle.*

4.4 The DI Transition Amplitude

We compute Ψ_{model} by solving (4.36) for the relative wave functions ξ_α, using some suitable mixture of experimental and theoretical information to achieve a full definition of the operator $\bar{\mathcal{O}}$. This calculation is subject to the usual scattering boundary conditions, with a plane wave in the entrance channel and outgoing scattered waves.

The amplitude for breakup into channel β is obtained from Ψ_{model} by the methods discussed in Chapter 3. It is obtained by either picking from Ψ_{model} the asymptotic amplitude of the term that lies in channel β, or using Green's theorem to project Ψ_{model} onto channel β in the region of strong nuclear interactions. In coupled-channels calculations the amplitudes are picked out asymptotically. This method is discussed in Section 4.6. Only the results obtained by use of Green's theorem are given in this section. Thus, for a two-body breakup channel, substitution of Ψ_{model} in (3.20) yields

$$T_{\alpha\beta}{}^{DI} = \left(\psi_{1\beta}\psi_{2\beta}e^{i(\mathbf{k}_\beta\,\mathbf{r}_\beta)},\ V_\beta \sum_\gamma \psi_{1\gamma}\psi_{2\gamma}\xi_\gamma(\mathbf{r}_\gamma)\right), \tag{4.45}$$

where V_β is the interaction between the (labelled) nucleons of $\psi_{1\beta}$ and the (labelled) nucleons of $\psi_{2\beta}$ and the incident wave is taken in the channel $\gamma = \alpha$. Equation 4.45 is based on the use of the plane-wave Green's function. The result that follows from the distorted-waves Green's function is obtained by

* On the other hand, the latter effect would have its principal importance in the nuclear interior, and might be of little significance for certain kinds of practical calculations. See Section 5.3.

substituting Ψ_{model} in (3.38):

$$T_{\alpha\beta}{}^{DI} = \left(\psi_{1\beta}\psi_{2\beta}\chi_\beta^{(-)}(\mathbf{r}_\beta), [V_\beta - U_\beta] \sum_\gamma \psi_{1\gamma}\psi_{2\gamma}\xi_\gamma(\mathbf{r})\right). \tag{4.46}$$

Here U_β is the distorting potential. The term that expresses elastic scattering by U_β is omitted from (4.46). It should be noted that (4.45) and (4.46) are exact expressions for the transition amplitude predicted from Ψ_{model}, provided the ξ_α on the right hand sides of these expressions are computed without approximation from (4.36). It also should be recalled that the use of a complex potential U_β to generate the distorted waves does not affect the validity of (4.46) as an exact expression for the amplitude.

Equations 4.45 and 4.46 are known as *post* forms of the reaction amplitude, because they are based on the interactions in the exit channel. Alternative expressions are obtained if the wavefunction of the time-reversed reaction is projected on the entrance channel:

$$T_{\alpha\beta}{}^{DI} = \left(\sum_\gamma \psi_{1\gamma}\psi_{2\gamma}\xi_\gamma^{(-)}(\mathbf{r}_\gamma), V_\alpha\psi_{1\alpha}\psi_{2\alpha}e^{i(\mathbf{k}_\alpha\cdot\mathbf{r}_\alpha)}\right), \tag{4.47}$$

and

$$T_{\alpha\beta}{}^{DI} = \left(\sum_\gamma \psi_{1\gamma}\psi_{2\gamma}\xi_\gamma^{(-)}(\mathbf{r}_\gamma), [V_\alpha - U_\alpha]\psi_{1\alpha}\psi_{2\alpha}\chi_\alpha^{(+)}(\mathbf{r}_\alpha)\right). \tag{4.48}$$

Here the model wavefunction on the left-hand side is subjected to the conditions that it have a plane wave in channel β and *ingoing* scattered waves. The superscript $(-)$ on the radial wavefunctions indicates use of these boundary conditions. Because (4.47, 4.48) are based on the interactions in the entrance channel they are known as *prior* forms of the reaction amplitude.

From now until Chapter 11 all our expressions concern the DI amplitude. We therefore omit the explicit superscript DI.

4.5 Distorted-waves Approximation

Physically-useful results are obtained by developing approximate methods for the computation of Ψ_{model}, namely, by developing suitable approximate methods by which to obtain the relative wavefunctions ξ_γ of (4.45, 4.46). The simplest and most general of these is the method of distorted waves (DW). It emphasizes the role played by the entrance channel, which has $\gamma = \alpha$. To develop this method it is necessary to use (4.46) for the transition amplitude. Early applications of this method in nuclear physics are listed in Chapter 5, [1–4].

Let us recopy (4.46), splitting the sum over γ into two parts, so that special attention may be given to the term $\gamma = \alpha$. The amplitude becomes

$$T_{\alpha\beta} = (\psi_{1\beta}\psi_{2\beta}\chi_\beta^{(-)}(\mathbf{r}_\beta), [V_\beta - U_\beta]\psi_{1\alpha}\psi_{2\alpha}\xi_\alpha(\mathbf{r}_\alpha))$$
$$+ \left(\psi_{1\beta}\psi_{2\beta}\chi_\beta^{(-)}(\mathbf{r}_\beta), [V_\beta - U_\beta]\sum_\gamma{}' \psi_{1\gamma}\psi_{2\gamma}\xi_\gamma(\mathbf{r}_\gamma)\right), \tag{4.49}$$

where the prime on the summation in the second line indicates that the term $\gamma = \alpha$ is not included. It should be recalled that (4.49) is invariant to the choice of the distorting potential U_β. Then the distorted-waves method drops the second line of (4.49) and in the first line uses the approximation

$$\xi_\alpha(\mathbf{r}_\alpha) \approx \chi_\alpha^{(+)}(\mathbf{r}_\alpha), \tag{4.50}$$

where the wavefunction $\chi_\alpha^{(+)}$ is calculated from an optical potential that fits the energy-averaged elastic scattering in channel α. Under these approximations $T_{\alpha\beta}$ reduces to the form

$$T_{\alpha\beta}{}^{DW} = (\psi_{1\beta}\psi_{2\beta}\chi_\beta^{(-)}(\mathbf{r}_\beta), [V_\beta - U_\beta]\psi_{1\alpha}\psi_{2\alpha}\chi_\alpha^{(+)}(\mathbf{r}_\alpha)), \tag{4.51}$$

in which product wavefunctions now appear on both sides of the matrix element. This approximate form of the DW amplitude is not now invariant to the choice of U_β; therefore it only yields useful results if U_β is chosen correctly.

The second of the above approximations [expressed by (4.50)] is probably rather accurate. Because the functions $\chi_\alpha^{(+)}$ and ξ_α predict the same elastic scattering in channel α they must be identical everywhere outside the nucleus. They may differ inside the nucleus, either because we have neglected (small) exchange contributions to elastic scattering, or because of uncertainties about details of the optical potential from which $\chi_\alpha^{(+)}$ is computed. However, neither of these effects is likely to be large and they do not provide much difficulty for the justification of (4.50).

It is more difficult to justify the omission of the second line of (4.49). The terms of Ψ_{model} lying in the excited channels are seldom so much weaker than the term in the incident channel that they can be omitted on this basis alone. In general, the contributions of the excited channels can be ignored only if we first minimize the error by judicious choice of U_β. Suitable choice of U_β minimizes the contributions of the discarded terms and adjusts the contribution of the retained term, so that the latter may be a good approximation to the exact $T_{\alpha\beta}$. We almost always choose for U_β the optical potential that fits elastic scattering in channel β.

As an example of the technique for selection of U_β, we note that one of the largest contributors to the second line of (4.49) probably is the term of the primed summation that has $\gamma = \beta$. The contribution that this term makes to $T_{\alpha\beta}$ is

$$(\psi_{1\beta}\psi_{2\beta}\chi_\beta^{(-)}(\mathbf{r}_\beta), [V_\beta - U_\beta]\psi_{1\beta}\psi_{2\beta}\xi_\beta(\mathbf{r}_\beta))$$

$$= \int d^3r_\beta \chi_\beta^{(-)*}(\mathbf{r}_\beta)\{\langle\psi_{1\beta}\psi_{2\beta}| V_\beta |\psi_{1\beta}\psi_{2\beta}\rangle - U_\beta\}\xi_\beta(\mathbf{r}_\beta), \quad (4.52)$$

where the bra-ket notation indicates integration over only the internal coordinates of $\psi_{1\beta}$ and $\psi_{2\beta}$. Evidently we can reduce the magnitude of (4.52) by choosing a potential U_β that is approximately equal to the quantity $\langle\psi_{1\beta}\psi_{2\beta}| V_\beta |\psi_{1\beta}\psi_{2\beta}\rangle$. The optical potential that fits elastic scattering in channel β possesses this property. Hence this potential is generally used for U_β in distorted waves calculations; this potential not only helps to minimize such matrix elements as are shown in (4.52) but it possesses the important additional property that it can be determined from experiment.

However, use of the final state optical potential may not provide an ideal minimization of the errors of the DW method. The optical potential for channel β is affected by the coupling to other channels, therefore it is not quite identical with the diagonal matrix element $\langle\psi_{1\beta}\psi_{2\beta}| V_\beta |\psi_{1\beta}\psi_{2\beta}\rangle$. Besides, it is not clear how the use of this particular U_β would affect the smaller terms of (4.49) that have $\gamma \neq \beta$. One is therefore tempted to use the diagonal matrix element itself, $\langle\psi_{1\beta}\psi_{2\beta}| V_\beta |\psi_{1\beta}\psi_{2\beta}\rangle$, for U_β. Unfortunately, it is difficult to obtain a reliable estimate of this alternative potential because it is not very closely related to experiment. Furthermore, this new choice of U_β would still leave in doubt the terms that have $\gamma \neq \beta$. Evidently neither of the simple potentials associated with the DW method provides a best minimization of the errors of the approximation. It would be idle, however, to attempt to define a "best potential" for the reduction from (4.49) to (4.51), because a definition introduced in this way would provide little assistance for the explicit determination of a potential for practical calculation of $T_{\alpha\beta}{}^{DW}$. *The distorted waves method is approximate.* Not only must the results of DW calculations be sensitive to the choice of U_β, but it also appears that the range of error of the DW method is closely correlated with the range of uncertainty in the choice of U_β. The sensitivity to the choice of U_β would disappear only if the terms of Ψ_{model} that lie in excited channels were very small, that is, if Born approximation were valid. It seems clear that the DW method is better than Born approximation, even if the error terms of this method are not under very obvious control. The DW method appears to be a convenient and valuable first approximation for the DI amplitude, provided its limitations are borne in mind. To improve on the DW results it appears necessary to go over to much more complete treatments of Ψ_{model}, based upon careful coupled-channels solutions of (4.36), like those described in the next section.

It should be noted that like any other optical potential the final-state optical potential is complex. It is appropriate that such a complex potential be used for U_β in DW calculations. Elimination of the CN parts of the wave-function means that the V_β of a DI theory must in part be anti-Hermitian, as described in Sections 4.2 and 4.3. The imaginary part of the optical potential tends to cancel the anti-Hermitian part of V_β.

When we consider channels α, β that concern only nucleon-nucleus relative motion the DW amplitude of (4.51) takes on an obvious interpretation as an extension of ordinary shell-model calculations. The products $\psi_{1\alpha}\psi_{2\alpha}\chi_\alpha^{(+)}$ and $\psi_{1\beta}\psi_{2\beta}\chi_\beta^{(-)}$ are then zero-order independent particle model eigenfunctions in the shell-model potential well, and $T_{\alpha\beta}{}^{DW}$ is a matrix element of the residual interaction. Presumably $T_{\alpha\beta}{}^{DW}$ should give the magnitudes of DI cross sections about as well as corresponding bound-state calculations give level splittings.

The above derivation of the DW amplitude is based on (4.46), the *post* form of the exact transition amplitude, derived by projecting Ψ_{model} on the exit channel. A similar DW amplitude may be derived from (4.48), the *prior* form of the exact amplitude. The prior form of the DW amplitude then differs from (4.51) only in that $(V_\beta - U_\beta)$ is replaced by $(V_\alpha - U_\alpha)$. However, we may use the Schrödinger equations for channels α and β to show that the two forms of the DW amplitude are equal on the energy shell. Therefore we are free to calculate with either the post or prior form of the DW amplitude, whichever is more convenient.

We now begin the process of practical evaluation of the distorted-waves amplitude. For conciseness we change the labelling of the breakup particles in channels α and β, so that instead of

$$1\alpha + 2\alpha \rightarrow 1\beta + 2\beta,$$

we use

$$a + A \rightarrow b + B.$$

With the introduction of this notation in (4.51) and the explicit display of the integrations over the relative coordinates $\mathbf{r}_\alpha = \mathbf{r}_a - \mathbf{r}_A$, $\mathbf{r}_\beta = \mathbf{r}_b - \mathbf{r}_B$, the distorted waves amplitude becomes

$$T_{\alpha\beta}{}^{DW} = \mathscr{J}\int d^3r_\alpha \int d^3r_\beta \chi_\beta^{(-)*}(\mathbf{k}_\beta, \mathbf{r}_\beta)\langle B, b|\, V\,|A, a\rangle \chi_\alpha^{(+)}(\mathbf{k}_\alpha, \mathbf{r}_\alpha), \qquad (4.53)$$

where $\mathscr{J}$ is the Jacobian of the transformation to the relative coordinates. The momenta on which the distorted waves depend are indicated in (4.53).

The matrix element that remains in (4.53) is taken between the internal states of the colliding pairs,

$$\langle B, b|\, V\,|A, a\rangle \equiv \langle \psi_B\psi_b|\, V_\beta - U_\beta\,|\psi_A\psi_a\rangle. \qquad (4.54)$$

It is written in terms of a bra-ket notation that signifies integration over all coordinates independent of $\mathbf{r}_\alpha$ and $\mathbf{r}_\beta$. This nuclear matrix element is a function of $\mathbf{r}_\alpha$ and $\mathbf{r}_\beta$, and plays the role of an effective (and, in general, nonlocal) interaction for the transition between the distorted waves $\chi_\alpha^{(+)}$ and $\chi_\beta^{(-)}$. It contains all the information on nuclear structure, on angular momentum selection rules, and even on the type of reaction being considered. The factorization of the integrand of (4.53) has great advantages in practical calculations. It enables us to treat the dynamics of the distorted waves while making use of only the rotational properties of the nuclear matrix element and rough descriptions of its radial shape. By having a number of typical radial shapes (form factors) as options in a computer code we may study a variety of nuclear models and reaction modes without having to organize a new calculation for each case. The "physics" of any given reaction affects only the magnitude of the nuclear matrix element, which factors out of the calculation, and it selects which form factor must be used.

To proceed further it is necessary to give attention to angular momenta. For the sake of standardization the notation of Satchler and his collaborators is adopted as far as possible. Particular use is made of [16]. Much of [16], however, is devoted to the introduction of spin-orbit interactions in the distorting potentials. This complication is omitted from the present analysis, and the results obtained using spin-orbit interactions are only quoted from time to time. We also do not make reference here to *i*-spin.*

To handle the angular momenta, the nuclear matrix element of (4.54) is expanded into a series of *multipoles* each of which corresponds to the transfer to the target nucleus of a definite angular momentum j composed of an orbital part l and a spin part s. If the projectiles a and b have definite spins s_a and s_b, and if the target and residual nuclear spins are J_A and J_B, respectively, we define the transfer angular momenta to be

$$\mathbf{j} = \mathbf{J}_B - \mathbf{J}_A, \quad \mathbf{s} = \mathbf{s}_a - \mathbf{s}_b, \quad \mathbf{l} = \mathbf{j} - \mathbf{s}. \tag{4.55}$$

Then the multipole series may be written with Clebsch-Gordan coefficients, corresponding to the vector coupling in (4.55), so that the nuclear matrix

* The multipoles of (4.56) express definite angular momentum transfer. Each of these can be expanded into a sum of multipoles in charge space, that express definite *i*-spin transfer. Each $G_{lsj,m}$ then becomes a sum of terms weighted by

$$\langle T_A t; M_{T_A} m_t \mid T_B M_{T_B} \rangle \langle t_b t; m_{t_b} m_t \mid t_a m_{t_a} \rangle,$$

corresponding to the transfer of *i*-spin t in analogy with (4.55),

$$\mathbf{t} = \mathbf{T}_B - \mathbf{T}_A = \mathbf{t}_a - \mathbf{t}_b.$$

element becomes†

$$\mathscr{J}\langle J_B M_B, s_b m_b | V | J_A M_A, s_a m_a \rangle$$
$$= \sum_{lsj} i^{-l} G_{lsj,m}(\mathbf{r}_\beta, \mathbf{r}_\alpha; bB, aA) \tag{4.56}$$
$$\times (-)^{s_b - m_b} \langle J_A j; M_A, M_B - M_A | J_B M_B \rangle \langle s_a s_b; m_a, -m_b | s, m_a - m_b \rangle$$
$$\times \langle ls; m, m_a - m_b | j, M_B - M_A \rangle,$$

where $m = M_B + m_b - M_A - m_a$. The symbols bB, aA as arguments of $G_{lsj,m}$ denote its dependence on the various nuclear quantum numbers (other than z-components of spin). The multipole operator $G_{lsj,m}$ may be explicitly displayed by the inverted form of the above expansion,

$$G_{lsj,m} = i^l \left(\frac{2l+1}{2J_B+1} \right) \sum_{M_B, M_A, m_b, m_a} \mathscr{J}\langle J_B M_B, s_b m_b | V | J_A M_A, s_a m_a \rangle$$
$$\times (-)^{s_b - m_b} \langle J_A j; M_A, M_B - M_A | J_B M_B \rangle$$
$$\times \langle s_a s_b; m_a, -m_b | s, m_a - m_b \rangle$$
$$\times \langle ls; m, m_a - m_b | j, M_B - M_A \rangle. \tag{4.57}$$

The factor i^l is included to ensure convenient time-reversal properties. By its construction, $G_{lsj,m}$ is seen to transform under rotation of the coordinate system like the spherical harmonic Y_l^{m*} (to which it often reduces). Another property carried by $G_{lsj,m}$ is the parity change in the nuclear transition, $\pi(a)\pi(b)\pi(A)\pi(B)$, where $\pi(i)$ is the parity of the internal state of the ith particle. Since $G_{lsj,m}$ is a function of two position vectors this parity is not uniquely related to l. Only in the *zero-range approximation* (see Chapter 5) is the parity of $G_{lsj,m}$ necessarily given by $(-)^l$; this is sometimes referred to as "normal parity." Finally, we note that it is often helpful to write $G_{lsj,m}$ as a product of two factors,

$$G_{lsj,m}(\mathbf{r}_\beta, \mathbf{r}_\alpha) \equiv A_{lsj} f_{lsj,m}(\mathbf{r}_\beta, \mathbf{r}_\alpha). \tag{4.58}$$

This separation into a *spectroscopic coefficient* A_{lsj} and a *form factor* $f_{lsj,m}$ is one of convenience, so that standard types of form factors with simple normalization may be used in computation. It is natural to choose A_{lsj} to include such quantities as fractional parentage coefficients for the initial or final nuclear states and the interaction strength.

One benefit of introducing the above multipole series is that very often only one value each of l, s and j is allowed, or is important, in a given transition.

† The Jacobian $\mathscr{J}$ of (4.53) is incorporated in the definition of $G_{lsj,m}$ in (4.56), following the practice of [16]. An earlier reference [17] that used the same notation does not incorporate this Jacobian into $G_{lsj,m}$.

Which multipoles are important is determined by the selection rules and the dynamics of the transition, as expressed by the A_{lsj}.

With the introduction of (4.56) into (4.53) the DW amplitude reduces to a sum over multipole contributions that we adopt as our standard form:

$$\begin{aligned} T_{\alpha\beta}{}^{DW} &= \sum_{l,s,j}(2l+1)^{1/2}A_{lsj}(-)^{s_b-m_b}\langle J_A j; M_A, M_B - M_A \mid J_B M_B\rangle \\ &\quad \times \langle ls; m, m_a - m_b \mid j, m - m_b + m_a\rangle \\ &\quad \times \langle s_a s_b; m_a, -m_b \mid s, m_a - m_b\rangle \beta_{sj}{}^{lm}. \end{aligned} \tag{4.59}$$

Here the *reduced amplitude* $\beta_{sj}{}^{lm}$ is defined so that

$$(2l+1)^{1/2}\, i^l \beta_{sj}{}^{lm} = \int d^3r_\alpha \int d^3r_\beta \chi_\beta^{(-)*}(\mathbf{k}_\beta, \mathbf{r}_\beta) f_{lsj,m}(\mathbf{r}_\beta, \mathbf{r}_\alpha)\chi_\alpha^{(+)}(\mathbf{k}_\alpha, \mathbf{r}_\alpha). \tag{4.60}$$

To calculate the (unpolarized) differential cross section we insert the amplitude into (1.32), sum over final spin projections and average over initial spin projections. Therefore

$$\frac{d\sigma}{d\Omega} = \frac{\mu_\alpha\mu_\beta}{(2\pi\hbar^2)^2}\left(\frac{k_\beta}{k_\alpha}\right)\left(\frac{N_\beta}{N_\alpha}\right)(2J_A+1)^{-1}(2s_a+1)^{-1}\sum_{\substack{M_A, M_B,\\ m_a, m_b}} |T_{\alpha\beta}|^2. \tag{4.61}$$

The factor N_β/N_α of (3.71–3.73) is inserted into (4.61) to account for the numbers of equivalent arrangements of nucleons in channels α and β. Because the spin projections are displayed explicitly in (4.59) the sums in (4.61) are easily performed, with the result that

$$\frac{d\sigma}{d\Omega} = \frac{\mu_\alpha\mu_\beta}{(2\pi\hbar^2)^2}\left(\frac{k_\beta}{k_\alpha}\right)\left(\frac{N_\beta}{N_\alpha}\right)\frac{(2J_B+1)}{(2J_A+1)(2s_a+1)}\sum_{lsj}|A_{lsj}|^2\sum_m|\beta_{sj}{}^{lm}|^2. \tag{4.62}$$

The sum over l, s, j is incoherent, as already mentioned, as if (4.62) were a description of a total cross section for the "capture" of the l, s, j multipole. *Thus, different multipoles make noninterfering contributions to the differential cross section. This property is a very basic result of the distorted-waves approximation.* (It is clear that if spin-orbit coupling were introduced in the computation of the distorted waves only the sum over j would be incoherent.)

Detailed procedures for the evaluation of (4.60) are discussed in Chapter 5.

4.6 The Method of Coupled Channels

For sufficiently simple DI models, which incorporate only a small number of open channels, we may directly substitute (4.1) for Ψ_{model} into the optical-model Schrödinger equation of (4.36) and we may attempt to obtain the relative wavefunctions ξ_γ by solving the equation that results. Thus the equation to be solved is

$$(E - H_{PP} - \bar{\mathcal{O}}) \sum_\gamma \xi_\gamma(\mathbf{r}_\gamma)\psi_{1\gamma}\psi_{2\gamma} = 0. \tag{4.63}$$

The operator $\bar{\mathcal{O}}$ in this equation accounts for the effects caused by the parts of the full physical wavefunction omitted from Ψ_{model}.

A set of coupled equations for the relative functions ξ_γ is obtained from (4.63) by left multiplying this equation by each of the function pairs $\psi_{1\gamma}\psi_{2\gamma}$ in turn. The number of coupled equations (see below) is equal to the number of channels carried in Ψ_{model}. A typical one of the set of equations is

$$\sum_\delta \langle \psi_{1\gamma}\psi_{2\gamma}|\, E - H - \bar{\mathcal{O}}\, |\xi_\delta(\mathbf{r}_\delta)\, \psi_{1\delta}\psi_{2\delta}\rangle = 0, \tag{4.64}$$

where in each term the bra-ket symbol indicates integration over all *internal* coordinates of the function pair $\psi_{1\gamma}\psi_{2\gamma}$, namely, over all coordinates other than $\mathbf{r}_\gamma$. The full Hamiltonian H, rather than H_{PP}, is used in (4.64). These two operators are equivalent in (4.64) inasmuch as H_{PP} is obtained originally by projecting H onto the space spanned by Ψ_{model}.

For each function pair in (4.64) we split H into a term that operates only on the internal coordinates of the separated nuclei, plus a term that also involves the relative coordinate. Thus

$$H = H_\delta + T_\delta + V_\delta, \tag{4.65}$$

where

$$T_\delta = -\frac{\hbar^2}{2\mu_\delta}\nabla_{\mathbf{r}_\delta}{}^2, \tag{4.66}$$

and

$$H_\delta\psi_{1\delta}\psi_{2\delta} = (\varepsilon_{1\delta} + \varepsilon_{2\delta})\, \psi_{1\delta}\psi_{2\delta}. \tag{4.67}$$

Then if the term with $\delta = \gamma$ is split off from the summation, (4.64) becomes

$$\begin{aligned}\{E - \varepsilon_{1\gamma} - \varepsilon_{2\gamma} - T_\gamma - \langle \psi_{1\gamma}\psi_{2\gamma}|\, V_\gamma + \bar{\mathcal{O}}\, |\psi_{1\gamma}\psi_{2\gamma}\rangle\}\xi_\gamma(\mathbf{r}_\gamma) \\ = -\sum_{\delta\neq\gamma} \langle \psi_{1\gamma}\psi_{2\gamma}|\, E - \varepsilon_{1\delta} - \varepsilon_{2\delta} - T_\delta - V_\delta - \bar{\mathcal{O}}\, |\psi_{1\delta}\psi_{2\delta}\xi_\delta(\mathbf{r}_\delta)\rangle.\end{aligned} \tag{4.68}$$

The right hand side of (4.68) expresses the influence exercised upon ξ_γ by the functions ξ_δ that have $\delta \neq \gamma$. If the reaction under discussion involves rearrangement, different function pairs $\psi_{1\gamma}\psi_{2\gamma}$ and $\psi_{1\delta}\psi_{2\delta}$ involve different sets of variables, and the integrals of (4.68) are very difficult to perform [18].

Nonexchange inelastic scattering is the one case in which the above equations simplify sufficiently to be used for practical calculations. In this case the projectile wavefunction is common to all terms of Ψ_{model} and has the same variables in each term. Hence we may take

$$\psi_{1\gamma} = \psi_{1\delta} = \phi, \qquad \text{for all } \gamma, \delta. \tag{4.69}$$

The internal Hamiltonian H_δ is the same for all terms of Ψ_{model}; it breaks into the two terms

$$H_\delta = H_{\text{projectile}} + H_{\text{target}}, \tag{4.70}$$

with

$$H_p\phi = \varepsilon\phi. \tag{4.71}$$

The target nucleus wavefunctions are $\psi_{2\gamma}$, $\psi_{2\delta}$, etc. For convenience these are denoted ψ_j, so that now

$$\Psi_{\text{model}} = \phi \sum_j \psi_j \xi_j(\mathbf{r}), \tag{4.72}$$

and

$$H_t\psi_j = \varepsilon_j\psi_j. \tag{4.73}$$

It is also convenient to define

$$E_j = E - \varepsilon - \varepsilon_j. \tag{4.74}$$

When we substitute (4.69)–(4.74) into (4.68) that equation becomes

$$\{E_k - T - \langle\phi\psi_k|\, V + \bar{\mathcal{O}}\, |\phi\psi_k\rangle\}\xi_k(\mathbf{r}) = \sum_{j\neq k} \langle\phi\psi_k|\, V + \bar{\mathcal{O}}\, |\phi\psi_j\rangle\xi_j(\mathbf{r}), \tag{4.75}$$

in which the variable $\mathbf{r}$, the displacement between the projectile and the target nucleus, does not appear in any of the wavefunctions ϕ, ψ_k, ψ_j. The various matrix elements of $(V + \bar{\mathcal{O}})$ in (4.75) are functions of $\mathbf{r}$ and play the roles of *effective interactions* that govern the relative wavefunctions ξ_j. The determination of these effective interactions from suitable models of nuclear structure precedes the calculation of the $\xi_j(\mathbf{r})$ and is independent of that calculation.

It is straightforward to solve the system of equations (4.75) by numerical means. The details of solution are discussed in Chapter 6. Now let us only ask, *how many equations are there*? Evidently, the number of equations equals the number of channels included in Ψ_{model}, that is, it equals the number of different, completely labelled modes of breakup carried in Ψ_{model}. Thus the number of equations equals *the number of linearly independent functions* $\psi_{1\gamma}\psi_{2\gamma}$. Unfortunately, it is clear that in any practical calculation of inelastic scattering this number is much greater than the number of states of excitation of the target nucleus that are of interest. Typical nuclear states have non-zero angular momentum, and each distinct projection of angular momentum defines a distinct channel. As a result, practical calculations are likely to require large numbers of coupled equations (see Chapter 6).

Because the method of coupled channels automatically requires the calculation of accurate relative wavefunctions $\xi_\gamma(\mathbf{r})$ in every open channel of interest, it is easy to obtain the transition amplitude for any particular exit channel by picking off the asymptotic amplitude of the relative wavefunction for that channel. Accordingly, in all coupled channels calculations the transition amplitude is obtained in this manner. The amplitude thus obtained is identical with the one obtained from Ψ_{model} by the use of (4.45) or (4.46), expressions based on the use of Green's theorem. There is no advantage gained by inserting Ψ_{model} into (4.46), the distorted waves expression for the amplitude. In particular, the use of the distorted waves expression can in no way compensate for the higher excited states omitted from the coupled-channels calculation.

Green's theorem, however, may be used to obtain formal solutions of the coupled-channels equations for inelastic scattering, and with the aid of these formal solutions we gain some insight into the errors of the distorted-waves approximation. If we solve for ξ_k of (4.75) in terms of the quantities on the RHS of that equation the amplitude for channel k is found to be

$$T_{0k} = (\phi\psi_k\chi_k^{(-)}(\mathbf{r}), (V + \bar{\mathcal{O}}) \sum_{j \neq k} \phi\psi_j\xi_j(\mathbf{r})), \tag{4.76}$$

where $j = 0$ labels the entrance channel and it is understood that $k \neq 0$. The distorted wave $\chi_k^{(-)}$ obeys

$$\{E_k - T - \langle\phi\psi_k|\, V + \bar{\mathcal{O}}\, |\phi\psi_k\rangle\}\chi_k^{(-)}(\mathbf{r}) = 0. \tag{4.77}$$

It is interesting that channel k is excluded from the RHS of (4.76). Three aspects of (4.76) distinguish it from the distorted waves approximation: (a) The function $\chi_k^{(-)}$ is not affected by coupling to the other channels, therefore it is not the wavefunction that fits elastic scattering in channel k. (b) The right hand side of (4.76) may include significant admixtures of the terms lying in excited channels, channels other than the entrance channel. (c) The relative wave-function in the entrance channel $\xi_0(\mathbf{r})$ may not be describable as the solution of some simple optical scattering problem that fits elastic scattering in channel 0.

Of the three differences between (4.76) and the DW approximation the last probably is the least important. Certainly, the function $\xi_0(\mathbf{r})$ exactly fits the cross section for elastic scattering, just as the optical wavefunction $\chi_0^{(+)}(\mathbf{r})$ does. Therefore, $\xi_0(\mathbf{r})$ and $\chi_0^{(+)}(\mathbf{r})$ must be identical outside the nucleus. It is not implausible, for now, to make the simplifying assumption that these two functions are identical throughout all space.

Let us limit (4.76) to a case in which only two channels, 0 and k, are open. Then, with the use of the assumed equality of $\xi_0(\mathbf{r})$ and $\chi_0^{(+)}(\mathbf{r})$, (4.76) becomes

$$T_{0k} = (\phi\psi_k\chi_k^{(-)}(\mathbf{r}), (V + \bar{\mathcal{O}})\phi\psi_0\chi_0^{(+)}(\mathbf{r})). \tag{4.78}$$

At this stage the only remaining difference between (4.78) and the result of the DW approximation is the fact that $\chi_k^{(-)}$ in (4.78) does not fit elastic scattering in channel k. Modifying $\chi_k^{(-)}$ to fit elastic scattering would necessitate introducing into (4.77) (which governs $\chi_k^{(-)}$) a term that expresses the influence of channel 0 upon channel k. Because such a term would be *second order* in the coupling between the two channels it would tend to be weak and would not tend to cause important changes. For this reason, it probably is quite satisfactory to adopt the usual DW approximation and to allow $\chi_k^{(-)}$ to fit elastic scattering in channel k. Figure 4.2, taken from work by Perey and Satchler [19], illustrates the errors caused by introducing DW approximation into (4.78). The full curves are the coupled-equations predictions. The dashed curves are DW predictions; in each case the optical potentials used in the DW calculations were fitted to the elastic scattering predicted by the coupled equations calculations, just as if they were being fitted to experimental elastic scattering cross sections. The deformation parameter β characterizes the particular deformed optical model of inelastic scattering that was studied in these calculations. On the whole the two sets of calculations are seen to agree quite well, with the discrepancies becoming larger as the bombarding energy is lowered. Calculations by Stamp and by Edwards led to the same conclusions [20].

Similar results obtain if three or more channels are coupled strongly. Let us consider three coupled channels, labelled 0, 1, k. Equation (4.76) becomes

$$T_{0k} = (\phi\psi_k\chi_k^{(-)}(\mathbf{r}), (V + \bar{\mathcal{O}})\{\phi\psi_0\chi_0^{(+)}(\mathbf{r}) + \phi\psi_1\xi_1(\mathbf{r})\}), \qquad (4.79)$$

where the equality of $\xi_0(\mathbf{r})$ and $\chi_0^{(+)}(\mathbf{r})$ has again been assumed. Let us introduce for $\xi_1(\mathbf{r})$ the approximate expression generated from $\chi_0^{(+)}(\mathbf{r})$ in first order,

$$\xi_1 \approx K_1\langle\phi\psi_1|\ V + \bar{\mathcal{O}}\ |\phi\psi_0\rangle\chi_0^{(+)}, \qquad (4.80)$$

where K_1 is the Green's function for channel 1,

$$K_1 = (E_1 - T - \langle\phi\psi_1|\ V + \bar{\mathcal{O}}\ |\phi\psi_1\rangle)^{-1}. \qquad (4.81)$$

With the insertion of (4.80) in (4.79) there results

$$T_{0k} \approx (\phi\psi_k\chi_k^{(-)}, (V + \bar{\mathcal{O}})\{1 + |\phi\psi_1\rangle K_1\langle\phi\psi_1|\ (V + \bar{\mathcal{O}})\}\phi\psi_0\chi_0^{(+)}), \quad (4.82)$$

an expression equivalent to

$$T_{0k} \approx (\{1 + |\phi\psi_1\rangle K_1\langle\phi\psi_1|\ (V + \bar{\mathcal{O}})\}^\dagger\phi\psi_k\chi_k^{(-)}, (V + \bar{\mathcal{O}})\phi\psi_0\chi_0^{(+)}). \quad (4.83)$$

The wave operator whose Hermitian conjugate now stands on the left-hand side of (4.83) has many of the properties required to transform $\chi_k^{(-)}$ into the final-state elastic scattering wavefunction, as used in the distorted-waves approximation. However, even this altered version of $\chi_k^{(-)}$ does not include the second-order effects caused by coupling to $\chi_0^{(+)}$. In summary, the above perturbative discussion of the coupled-channels treatment of inelastic scattering would seem to provide fairly strong support for the version of the distorted-waves approximation that uses elastic-scattering eigenfunctions in

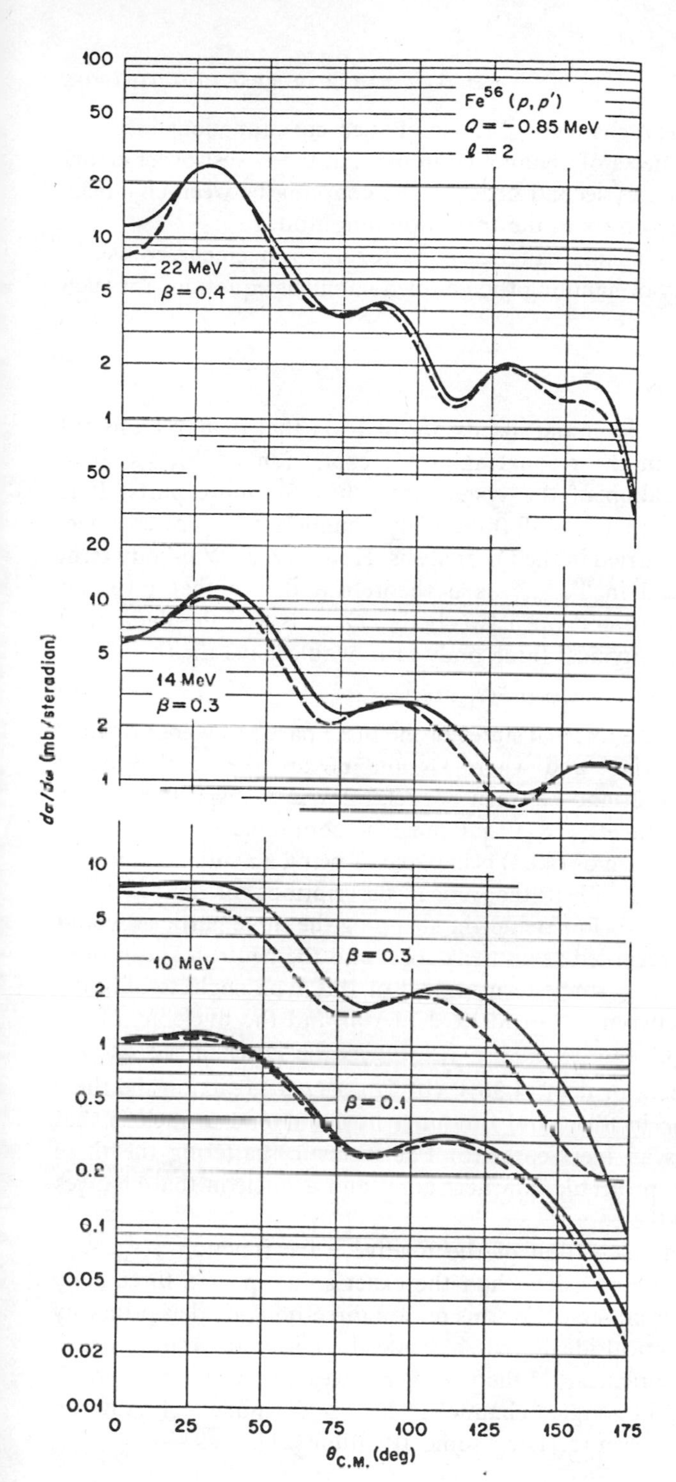

Fig. 4.2 Inelastic scattering cross sections calculated for the $l = 2$ excitation of Fe^{56} by protons, at 22 MeV, 14 MeV, 10 MeV bombarding energies. The full curves are coupled equations predictions, and the dashed curves are self-consistent DW predictions based on potentials that fit the elastic scattering predicted by the coupled equations. The DW curves are normalized with the same deformation parameters as the coupled equations curve. (From [19], Fig. 2.)

both the entrance and exit channels. The use of such eigenfunctions incorporates so much of the effects of channel coupling that the lowest-order errors in the wavefunctions are of second order in the coupling between channels. These imply third-order errors in the transition amplitude.

No such definite statements can be made for rearrangement collisions because of the difficulty of handling the coupled-channels equations for such cases.

4.7 Three-body Models

It is noted as part of the initial discussion of (4.1) that it may be useful sometimes to include in the resonating-group expansion of Ψ_{model} some terms that express breakup of the system into three or more parts. It is necessary to include such terms if three-body channels are allowed energetically and are to be carried in the DI analysis. However, a three-body term also may be introduced in Ψ_{model} as a theoretical device that gives an inclusive description of several open two-body channels and the dynamical relations among them. A typical three-body term would have the form

$$\psi_1\psi_2\psi_3\xi(\mathbf{r}_{12}, \mathbf{r}_{23}), \tag{4.84}$$

where ψ_1, ψ_2 and ψ_3 are the internal states of the three particles whose relative motion we are considering, and where $\mathbf{r}_{12}$ and $\mathbf{r}_{23}$ are the displacements between particles 1 and 2, and between particles 2 and 3, respectively. It is the relative wavefunction $\xi(\mathbf{r}_{12}, \mathbf{r}_{23})$ that must be computed by solution of (4.36). One term of the form of (4.84) is capable of being an entire DI model.

Three-body models are ideal theoretical descriptions of many direct reactions; for example: (a) For deuteron stripping the three particles would be the target nucleus, regarded as an inert core, and the neutron and proton of the incident deutron. A single-term model of this type neglects only the effects of exchange between the incident deuteron and the nucleons of the core, and the effects of core excitation [except as the latter effects are accounted for by the operator $\bar{\mathcal{O}}$ of (4.36)]. (b) For a (p, n) reaction the three particles would be the incident and outgoing nucleons and a nucleus that would be regarded as an inert core. (c) For inelastic scattering the three particles would be the projectile, an inert core, and a nucleon that changes its state of binding to the core.

Of course the exact calculation of the relative wave function $\xi(\mathbf{r}_{12}, \mathbf{r}_{23})$, using (4.36), requires nothing less than the exact solution of a three-body problem. This seldom is possible. For this reason three-body models primarily play the role of theoretical ideals that indicate directions in which to seek improvements to such practical DI theories as are based on the DW approximation or the method of coupled channels. Three-body models of deuteron stripping have recently been receiving some attention [21].

4.8 Antisymmetrization

In principle, antisymmetrization raises no questions of any difficulty, and we simply apply the methods described in Section 3.5. Given a function $\Psi^{(+)}$ that is an exact solution of some Schrödinger equation and describes a reaction among definite labelled particles, we go over to the corresponding function $\Psi_A^{(+)}$ that is a solution of the same Schrödinger equation and is antisymmetric, by applying the operator $\mathscr{A}_\alpha$ of (3.67) to $\Psi^{(+)}$. The operator $\mathscr{A}_\alpha$ antisymmetrizes the incident plane wave part of $\Psi^{(+)}$ and thereby makes the entire function antisymmetric. On the other hand, because practical calculations in nuclear physics do not produce exact solutions of the Schrödinger equation we are compelled to inquire how the technique of antisymmetrization fits in with the usual methods of approximation.

Our standard approximate wavefunction is Ψ_{model} of (4.1), a series of products that is summed over the various open channels of interest. Antisymmetrization of this series converts it into another series of the same form. Accordingly, the antisymmetric wavefunction is subject to the same set of interpretations given for Ψ_{model}. Antisymmetrization only imposes certain relationships among the relative wavefunctions ξ_γ that appear in Ψ_{model}, and compels the use of a certain minimum number of terms in the sum over channels.

By compelling the use of many terms for the construction of Ψ_{model}, antisymmetrization by implication changes the techniques of calculation of the ξ_γ. However, before considering how the relative wavefunctions ξ_γ are to be calculated it is convenient to inquire what kinds of terms are associated by antisymmetrization, and what kinds of contributions these terms make to the transition amplitude.

The simplest version of Ψ_{model} in the earlier discussions is the single-term expression used in the distorted waves approximation of (4.51). In this case we carry only the one term that lies in the entrance channel,

$$\Psi_{\text{model}} = \psi_a \psi_A \chi_\alpha^{(+)}(\mathbf{r}_\alpha).$$

To produce an equivalent antisymmetrized wavefunction we now must also carry those additional terms obtained by applying the antisymmetrization operator of (3.67) to Ψ_{model}. The antisymmetrized wavefunction is

$$\begin{aligned} \Psi_A &= \mathscr{A}_\alpha \Psi_{\text{model}}, \\ &= N_\alpha^{-1/2} \sum_P (-)^P P \psi_a \psi_A \chi_\alpha^{(+)}(\mathbf{r}_\alpha). \end{aligned} \tag{4.85}$$

The amplitude for the transition to channel β is obtained by inserting (4.85) on the RHS of the DW matrix element of (4.51). To accord with the normalization used for the calculation of cross sections [see (4.61), (4.62)], the factor

$N_\alpha^{-1/2}$ of (4.85) is omitted from this amplitude. Therefore,

$$T_{\alpha\beta}{}^{DW} = \sum_P (-)^P \left(\psi_b \psi_B \chi_\beta^{(-)}(\mathbf{r}_\beta), [V_\beta - U_\beta] P \psi_a \psi_A \chi_\alpha^{(+)}(\mathbf{r}_\alpha)\right). \tag{4.86}$$

It is seen that the various permutations on the RHS of (4.86) play different roles, inasmuch as only one definite set of particle labels appears on the LHS. The term of (4.86) in which the ordering of labels of the RHS is as much as possible like the LHS will be called the *direct term.* All other terms of (4.86) are *exchange terms* and tend to be smaller than the direct term.

The distinction between the direct and exchange terms is clearest in the case of inelastic scattering of nucleons. If we ignore spin in the present discussion $\psi_a = \psi_b = 1$ and the coordinates of the projectile particle appear only in the two distorted waves. Let us also ignore recoil, and measure all coordinates from the center of mass of the target (or residual) nucleus. Then $\mathbf{r}_\alpha$ and $\mathbf{r}_\beta$ become simply the coordinates of the incident and outgoing nucleons. For the LHS of (4.86) we introduce the labelling

$$\psi_b \psi_B \chi_\beta^{(-)} \rightarrow \psi_B(1, 2, \ldots, n) \chi_\beta^{(-)}(0). \tag{4.87}$$

Then the interactions are

$$[V_\beta - U_\beta] = \left[\sum_{i=1}^{n} V(i0) - U_\beta(0)\right], \tag{4.88}$$

and the distorted-waves amplitude becomes

$$T_{\alpha\beta}{}^{DW} = \left(\chi_\beta^{(-)}(0)\psi_B(1, 2, \ldots, n), \left[\sum_{i=1}^{n} V(i0) - U_\beta(0)\right]\left[1 - \sum_{j=1}^{n} P_{j0}\right] \times \chi_\alpha^{(+)}(0)\psi_A(1, 2, \ldots, n)\right). \tag{4.89}$$

Here the direct term has nucleon 0 as the incident projectile and has nucleons $1 - n$ in the target nucleus. The exchange terms are obtained by interchanging, in turn, the incident nucleon with each of the nucleons of the target nucleus. The direct term is given a positive sign by convention.

In the sum over interchanges in (4.89) the index j refers to the set of coordinate variables of the target nucleus. Both the product $\psi_B^* \psi_A$ and the interaction $\sum_{i=1}^{n} V(i0)$ are symmetric in this set of variables. Therefore the exchange terms of (4.89) are all identical, and that equation reduces, say, to

$$\begin{aligned} T_{\alpha\beta}{}^{DW} = &\left(\chi_\beta^{(-)}(0)\psi_B(1, 2, \ldots, n), \right. \\ &\left. \times \left[\sum_{i=1}^{n} V(i0) - U_\beta(0)\right] \chi_\alpha^{(+)}(0)\psi_A(1, 2, \ldots, n)\right) \\ &- n\left(\chi_\beta^{(-)}(0)\psi_B(1, 2, \ldots, n), \right. \\ &\left. \times \left[\sum_{i=1}^{n} V(i0) - U_\beta(0)\right] \chi_\alpha^{(+)}(1)\psi_A(0, 2, 3, \ldots, n)\right). \end{aligned} \tag{4.90}$$

The coefficient n should not be taken as an indication of the relative importance of the exchange term in (4.90). In all explicit calculations this coefficient is cancelled by the normalization coefficients of ψ_A and ψ_B. The exchange term actually is likely to be smaller than the direct term because its structure requires the two distorted waves, $\chi_\alpha^{(+)}$ and $\chi_\beta^{(-)}$, to simultaneously have good overlap with ψ_A and ψ_B. Such complicated overlaps tend to be small. Furthermore, absorption reduces the magnitudes of the distorted waves in the required region of overlap [22]. Unfortunately, estimates of the importance of the exchange term have not been backed up very often with careful numerical calculations because the simplifications that reduce the direct term to manageable proportions (especially the *zero-range approximation*, see Chapter 5) cannot be used for the exchange term.

To facilitate further study of the exchange term in (4.90) we distinguish three different parts of this term on the basis of the type of interaction engaged in by nucleon 0:

(a) The *knockon term* is based on the interaction $V(10)$. It is

$$T_{\alpha\beta}{}^{DW}(\text{knockon}) = -n(\chi_\beta^{(-)}(0)\psi_B(1,\ldots,n),\ V(10)\chi_\alpha^{(+)}(1)\psi_A(0,2,3,\ldots,n)). \tag{4.91}$$

This matrix element has the character of a collision between two nucleons, labelled 0 and 1, that move in the field of a passive core. We see that if $V(10)$ were of zero range this term would be identical with the direct term.

(b) The *heavy particle stripping term* (HPS) is based on the interactions with the core. It is

$$T_{\alpha\beta}{}^{DW}(\text{HPS}) = -n\left(\chi_\beta^{(-)}(0)\psi_B(1,\ldots,n,\ \sum_{i=2}^{n} V(i0)\chi_\alpha^{(+)}(1)\psi_A(0,2,3,\ldots,n)\right). \tag{4.92}$$

Here nucleon 0 is ejected by interaction with core excitations or by recoil from the center-of-mass motion of the core. (We did omit some of the effects of recoil. These are restored later.)

(c) The *distorting-potential exchange term* is

$$T_{\alpha\beta}{}^{DW}(\text{distort}) = +n(\chi_\beta^{(-)}(0)\psi_B(1,\ldots,n),\ U_\beta(0)\chi_\alpha^{(+)}(1)\psi_A(0,2,3,\ldots,n)). \tag{4.93}$$

It is clear that there may be considerable cancellation among the three parts of the exchange term. It is also clear that terms of types (b) and (c) would vanish if the continuum function $\chi_\alpha^{(+)}$ were orthogonal to the bound single-particle orbitals that make up ψ_B. To discuss questions such as these it is necessary to organize very accurate numerical calculations of the various matrix elements.

We now consider antisymmetrization of a simple rearrangement collision, taking as an example the reaction $A(d, p)B$. Spin and recoil are ignored as in the discussion of inelastic scattering. We antisymmetrize separately on neutrons and protons. Then (4.86) becomes

$$T_{\alpha\beta}{}^{DW} = \Bigg(\chi_\beta^{(-)}(0_p)\psi_B(1_p, \ldots, n_p; 0_n, \ldots, n_n), \Bigg[\sum_{i=1_p}^{n_p} V(i0_p) + \sum_{i=0_n}^{n_n} V(i0_p) - U_\beta(0_p)\Bigg]\Bigg[1 - \sum_{j=1_p}^{n_p} P_{j0_p}\Bigg] \times \Bigg[1 - \sum_{j=1_n}^{n_n} P_{j0_n}\Bigg]\psi_A(1_p, \ldots, n_p; 1_n, \ldots, n_n)\phi_d(0_p, 0_n)\chi_\alpha^{(+)}(0_p, 0_n)\Bigg), \tag{4.94}$$

where ϕ_d is the internal wavefunction of the deuteron. Once again the proton exchange terms in (4.94) are identical, as are the neutron exchange terms. Therefore the antisymmetrizers in (4.94) may be replaced by the expressions

$$\begin{aligned} 1 - \sum_{j=1_p}^{n_p} P_{j0_p} &= 1 - n_p P_{0_p 1_p}, \\ 1 - \sum_{j=1_n}^{n_n} P_{j0_n} &= 1 - n_n P_{0_n 1_n}. \end{aligned} \tag{4.95}$$

Further reduction may be obtained by noting that ψ_B is antisymmetric in the interchange of 0_n with 1_n, and that the interactions are symmetric in this interchange. Therefore the neutron direct and exchange terms combine to yield a factor $n_n + 1$, and (4.94) becomes

$$T_{\alpha\beta}{}^{DW} = (n_n + 1)\Bigg(\chi_\beta^{(-)}(0_p)\, \psi_B\, (1_p, \ldots, n_p; 0_n, \ldots, n_n), \times \Bigg[\sum_{i=1_p}^{n_p} V(i0_p) + \sum_{i=0_n}^{n_n} V(i0_p) - U_\beta(0_p)\Bigg] \times [1 - n_p P_{0_p 1_p}]\psi_A(1_p, \ldots, n_p; 1_n, \ldots, n_n)\phi_d(0_p, 0_n)\chi_\alpha^{(+)}(0_p, 0_n)\Bigg). \tag{4.96}$$

Once again we classify the various parts of the exchange term, in analogy with the discussion of inelastic scattering:

(a) The *knockon term* is

$$T_{\alpha\beta}{}^{DW}(\text{knockon}) = -n_p(n_n + 1)(\chi_\beta^{(-)}(0_p)\psi_B(1_p, \ldots, n_p; 0_n, \ldots, n_n), \times [V(1_p, 0_p) + V(0_n, 0_p)]\psi_A(0_p, 2_p, \ldots, n_p; 1_n, \ldots, n_n)\phi_d(1_p, 0_n)\chi_\alpha^{(+)}(1_p, 0_n)). \tag{4.97}$$

This term has the character of a collision between the incident deuteron and the ejected proton. We note that this term is not like the knockon term of inelastic scattering, because the introduction of zero-range two-nucleon interactions does not make it identical with the direct stripping term.

(b) The *heavy-particle stripping term* is

$$T_{\alpha\beta}{}^{DW}(\text{HPS}) = -n_p(n_n+1)\Big(\chi_\beta^{(-)}(0_p)\psi_B(1_p,\ldots,n_p;0_n,\ldots,n_n),$$

$$\times\left[\sum_{i=2_p}^{n_p}V(i0_p)+\sum_{i=1_n}^{n_n}V(i0_p)\right]\psi_A(0_p,2_p,\ldots,n_p;1_n,\ldots,n_n)\phi_d(1_p,0_n)\chi_\alpha^{(+)}(1_p,0_n)\Big). \tag{4.98}$$

Here proton 0_p is ejected by its interaction with the core.

(c) The *distorting potential exchange term* is

$$T_{\alpha\beta}{}^{DW}(\text{distort}) = n_p(n_n+1)(\chi_\beta^{(-)}(0_p)\psi_B(1_p,\ldots,n_p;0_n,\ldots,n_n),$$
$$\times\ U_\beta(0_p)\psi_A(0_p,2_p,\ldots,n_p;1_n,\ldots,n_n)\phi_d(1_p,0_n)\chi_\alpha^{(+)}(1_p,0_n)). \tag{4.99}$$

These exchange terms probably are small, for the same reasons mentioned for the exchange terms in inelastic scattering.

None of the above exchange terms factorize in the simple fashion described in (4.53), for what we now recognize to have been the direct term of the DW approximation. To calculate the exchange terms it is necessary to use microscopic descriptions of the structures of nuclei A and B. Collective descriptions of these nuclei would obscure the coordinates of the individual nucleons.

4.9 Influence of Antisymmetrization on the Calculation of the Relative Wavefunctions

The wavefunction Ψ_{model} is a truncated version of the full physical wavefunction of the system. The errors of truncation are in part compensated by carrying the operator $\bar{\mathcal{O}}$ in the calculation of Ψ_{model}. Because antisymmetrization now compels us to introduce many more terms in Ψ_{model} it alters the point of truncation of the problem, it alters the calculation of the individual product terms in Ψ_{model}, and it alters the optical operator $\bar{\mathcal{O}}$. We are presented with interesting questions concerning the consistency of carrying the exchange terms in Ψ_{model} while omitting other, similar terms that do not happen to be brought in by the antisymmetrization operator. It is pertinent to consider some of these questions in a little detail.

We note first that the exchange terms which must be carried because nucleons are identical are related very closely to rearrangement terms that would arise if nucleons were not identical. To clarify this remark we consider as an example the elastic collision of a nucleon with a nucleus. The antisymmetric model wavefunction for this problem is

$$\psi_A(1, \ldots, n)\xi(0) - \sum_{i=1}^{n} \psi_A(1, \ldots, i-1, 0, i+1, \ldots, n)\xi(i). \quad (4.100)$$

However, if nucleons were not identical, the permutations of nucleon 0 with the nucleons of the target would label distinct channels that would correspond to rearrangements of the nonidentical particles. The wavefunction that includes all possible rearrangement channels of this sort would be

$$\psi_A(1, \ldots, n)\bar{\xi}(0) - \sum_{i=1}^{n} \psi_A(1, \ldots, i-1, 0, i+1, \ldots, n)\eta_i(i), \quad (4.101)$$

where $\bar{\xi}$ contains the incident wave of the collision and the η_i are asymptotically radially outgoing waves. The similarity of the above wavefunctions is apparent. Provided that the two systems are governed by the same Hamiltonian the relative wavefunction ξ of (4.100) is a simple linear combination of $\bar{\xi}$ and the η_i of (4.101).

Obviously, although the model wavefunction for an antisymmetrized system of identical particles contains the same kinds of terms that arise in the study of rearrangement scattering of nonidentical particles, the problem with identical particles uses many fewer relative wavefunctions and is much the simpler of the two. To calculate $\bar{\xi}$ and the η_i it is necessary to solve a system of $(n + 1)$ coupled differential equations, whereas to calculate ξ it is only necessary to solve one equation. However, this simplicity is deceptive. Because the single relative wavefunction that appears in the study of identical particles is a linear combination of the many relative wavefunctions that appear in the study of nonidentical particles it must share much of the complication of these other functions.

The study of the rearrangement scattering of nonidentical particles gives some understanding of the importance of the exchange terms that appear with identical particles. It is clear that the exchange terms of the antisymmetrized problem can have an appreciable influence on the determination of the relative wavefunction ξ only if in the problem with nonidentical particles there is fairly strong coupling between the incident channel and the rearrangement channels. We now see the nature of the truncation question raised at the beginning of this section: Namely, is the coupling to the rearrangement channels suggested by antisymmetrization necessarily more significant than the coupling to rearrangement channels of other types? This question is not pursued.

To obtain an explicit example of the calculation of the relative wavefunctions in antisymmetrized problems, we now substitute the antisymmetrized wavefunction for one open channel into the Schrödinger equation. We obtain

$$(E - H_{PP} - \bar{\mathcal{O}})\left[1 - \sum_{j=1}^{n} P_{j0}\right]\xi(0)\psi_A(1, \ldots, n) = 0. \qquad (4.102)$$

Left multiplication by $\psi_A^*(1, \ldots, n)$, followed by integration, yields

$$\langle\psi_A(1, \ldots, n)|\, E - H - \bar{\mathcal{O}}\, |\psi_A(1, \ldots, n)\rangle\xi(0)$$
$$- n\langle\psi_A(1, \ldots, n)|\, E - H - \bar{\mathcal{O}}\, |\xi(1)\psi_A(0, 2, \ldots, n)\rangle = 0 \qquad (4.103)$$

because the exchange terms are all identical to each other. Insertion of (4.65) then yields

$$\{E - \varepsilon_A - T_0 - \langle\psi_A(1, \ldots, n)|\, V_{0A} + \bar{\mathcal{O}}\, |\psi_A(1, \ldots, n)\rangle\}\xi(0)$$
$$- n\langle\psi_A(1, \ldots, n)|\, E - \varepsilon_A - T_0 - V_{0A} - \bar{\mathcal{O}}\, |\xi(1)\psi_A(0, 2, \ldots, n)\rangle$$
$$= 0, \qquad (4.104)$$

where

$$V_{0A} = \sum_{i=1}^{n} V(0i) \qquad (4.105)$$

is the interaction between nucleon 0 and the labelled nucleons of the target nucleus. The optical operator $\bar{\mathcal{O}}$ naturally is symmetric in the coordinates of all $(n + 1)$ nucleons.

Equation 4.104 is an integrodifferential equation for the function ξ. The exchange term, in which ξ appears in an integral, is the complication that carries over from the problem of nonidentical particles. In that problem [see (4.68)] terms of similar structure express the coupling to the rearrangement channels. We note in (4.104) that the reaction of the exchange term on the direct term influences the calculation of the relative wavefunction ξ, much as inelastic channels influence the relative wavefunction in the elastic channel.

The complicated exchange term in (4.104) is primarily of interest at low and medium bombarding energies. Toward high bombarding energy the magnitude of this term is reduced because of poor overlap between $\xi(1)$ and $\psi_A(1, \ldots, n)$. Another complication in (4.104) lies in the optical operator $\bar{\mathcal{O}}$ which may need to be given a special form appropriate to the exclusion of an antisymmetrized function space. Further discussion of these topics occurs in Section 10.1.

References

[1] J. A. Wheeler, *Phys. Rev.* **52,** 1107 (1937).
[2] I. Tamm, *J. Phys.* (USSR) **9,** 449 (1945).
[3] S. M. Dancoff, *Phys. Rev.* **78,** 382 (1950).

[3] H. Feshbach, *Ann. Phys.* **5,** 357 (1958).
[4] H. Feshbach, *Ann. Phys.* **19,** 287 (1962).
[5] Y. Hahn, *Phys. Rev.* **142,** 603 (1966).
[6] G. E. Brown, *Rev. Mod. Phys.* **31,** 893 (1959).
[7] L. Wolfenstein; W. Hauser and H. Feshbach, *loc. cit.*, Chapter 2, Ref. 14.
[8] W. B. Riesenfeld and K. M. Watson, *Phys. Rev.* **102,** 1157 (1956).
R. J. Glauber, in *Lectures in Theoretical Physics*, Vol. 1, Interscience, New York, 1959.
[9] B. Block and H. Feshbach, *Ann. Phys.* **23,** 47 (1963).
C. M. Shakin, *Ann. Phys.* **22,** 373 (1963).
A. Landé and B. Block, *Phys. Rev. Letters* **12,** 334 (1964); also see Chapter 11, Ref. 13.
[10] A. Landé and G. E. Brown, *Nucl. Phys.* **75,** 344 (1966).
[11] W. E. Frahn and R. H. Lemmer, *Nuovo Cimento* **5,** 523, 1564 (1957).
R. H. Lemmer and A. E. S. Green, *Phys. Rev.* **119,** 1043 (1960).
P. J. Wyatt, J. G. Wills, and A. E. S. Green, *Phys. Rev.* **119,** 1031 (1960).
[12] F. G. Perey and B. Buck, *Nucl. Phys.* **32,** 353 (1962).
[13] G. Ripka, *Nucl. Phys.* **42,** 75 (1963).
[14] A. M. Lane and C. F. Wandel, *Phys. Rev.* **98,** 1524L (1955).
[15] L. C. Gomes, *Phys. Rev.* **116,** 1226 (1959).
[16] G. R. Satchler, *Nucl. Phys.* **55,** 1 (1964).
[17] N. Austern, R. M. Drisko, E. C. Halbert and G. R. Satchler, *Phys. Rev.* **133,** B3 (1964).
[18] A. P. Stamp, *Nucl. Phys.* **83,** 232 (1966).
[19] F. G. Perey and G. R. Satchler, *Phys. Letters* **5,** 212 (1963).
[20] A. P. Stamp and J. R. Rook, *Nucl. Phys.* **53,** 657 (1964); V. R. W. Edwards, *Nucl. Phys.* **A101,** 17 (1967).
[21] A. N. Mitra, *Phys. Rev.* **139,** B1472 (1965); R. Aaron and P. E. Shanley, *Phys. Rev.* **142,** 608 (1966); S. T. Butler, et al., *Ann. Phys.* **23,** 282 (1967); *Phys. Rev.* **162,** 1061(1967); J. M. Bang and C. A. Pearson, *Nucl. Phys.* **A100,** 1, 24 (1967), and references given there; A. S. Reiner and A. I. Jaffe, *Phys. Rev.* **161,** 935 (1967); S. T. Butler, R. G. L. Hewitt and J. S. Truelove, *Phys. Letters* **26B**, 264 (1968); C. F. Clement, *Phys. Rev. Letters* **20,** 22 (1968).
[22] G. L. Strobel and B. L. Scott, *Phys. Rev.* **140,** B311 (1965).

CHAPTER 5

Applications of the DW Method

In this chapter we discuss typical properties of the distorted waves amplitude of Section 4.5, we consider practical methods for the calculation of this amplitude, and we apply these methods to the study of a number of reactions of physical interest.

Only the direct term is treated in this chapter. Further discussion of the exchange terms described in Section 4.8 is given in Chapter 10. On the other hand, it is sometimes useful to reinterpret the exchange term of one reaction mechanism as the direct term of a different mechanism. On this basis some remarks about "knockout" and "heavy-particle stripping" are included in Section 5.12.

5.1 Zero-range Approximation

In Section 4.5 the procedure for carrying out a distorted waves calculation of a nuclear reaction is divided into two steps. In the first step careful consideration is given to the formulation of a model for the reaction, and the internal structure of the nuclei A, a, B, b is taken into account. As a result the spectroscopic coefficient A_{lsj} and the form factor $f_{lsj,\,m}(\mathbf{r}_\beta, \mathbf{r}_\alpha)$ defined in (4.58) are computed. In the second step the form factor is inserted into the calculation of the reduced amplitude $\beta_{sj}{}^{lm}$ of (4.60). We now discuss the second of these two steps.

Equation 4.60 is of the form of a six-dimensional integral over the two channel-displacement variables $\mathbf{r}_\alpha$ and $\mathbf{r}_\beta$:

$$(2l+1)^{1/2} i^l \beta_{sj}{}^{lm} = \int d^3 r_\alpha \int d^3 r_\beta \chi_\beta^{(-)*}(\mathbf{k}_\beta, \mathbf{r}_\beta) f_{lsj,m}(\mathbf{r}_\beta, \mathbf{r}_\alpha) \chi_\alpha^{(+)}(\mathbf{k}_\alpha, \mathbf{r}_\alpha). \quad (4.60)$$

For inelastic scattering reactions the two channel-displacement variables are identical, and the six-dimensional integral reduces automatically to a three-dimensional integral. However, for reactions that involve any sort of rearrangement (this includes exchange) the presence of different variables in the two channels prevents easy reduction of the integral, and therefore (4.60) is difficult to evaluate. It is particularly unpleasant that the two distorted waves in (4.60) normally depend on different vector variables. Because these distorted waves must be handled numerically, it is seen that in general (4.60) needs to be handled by six-dimensional numerical integration.

In most current distorted waves calculations the above difficulty is removed by the assumption that the form factor is of very small range, perhaps because it is proportional to an interaction that has a small range, or perhaps because the internal wavefunctions of one or more of the projectiles have small ranges. The appropriateness of this assumption becomes clear as we come to consider various individual reactions (see also Section 5.13). When applicable, this zero-range assumption has the physical meaning that the light particle in channel β, particle b, is assumed to be emitted at the same point at which the light particle in channel α, particle a, is absorbed. If the reaction should, say, involve the particle transfer

$$B - A = x = a - b,$$

then the zero-range assumption implies

$$\mathbf{r}_a = \mathbf{r}_b = \mathbf{r}_x,$$

and therefore

$$\begin{aligned} \mathbf{r}_\beta &= \mathbf{r}_b - \mathbf{r}_B, \\ &= \mathbf{r}_a - \frac{M_A\mathbf{r}_A + M_x\mathbf{r}_a}{M_A + M_x}, \\ &= \frac{M_A}{M_B}\mathbf{r}_\alpha, \end{aligned} \tag{5.1}$$

where $M_B = M_A + M_x$. The form factor can then be written, say,

$$\begin{aligned} f^{(\text{zero})}_{lsj,m}(\mathbf{r}_\beta, \mathbf{r}_\alpha) &= \delta\left(\mathbf{r}_\beta - \frac{M_A}{M_B}\mathbf{r}_\alpha\right)\int f_{lsj,m}\left(\mathbf{s} + \frac{M_A}{M_B}\mathbf{r}_\alpha, \mathbf{r}_\alpha\right) d^3s, \\ &\equiv F_{lsj}(r_\alpha)Y_l^{m*}(\hat{r}_\alpha)\,\delta\left(\mathbf{r}_\beta - \frac{M_A}{M_B}\mathbf{r}_\alpha\right). \end{aligned} \tag{5.2}$$

As a result (4.60) reduces to the zero-range expression

$$(2l + 1)^{1/2}i^l\beta_{sj}{}^{lm} \approx \int d^3r\chi_\beta^{(-)*}\left(\mathbf{k}_\beta, \frac{M_A}{M_B}\mathbf{r}\right)F_{lsj}(r)Y_l{}^{m*}(\hat{r})\chi_\alpha^{(+)}(\mathbf{k}_\alpha, \mathbf{r}), \tag{5.3}$$

which is considerably easier to evaluate. In zero-range approximation *all DW calculations reduce to the form of* (5.3).

It is seen in Section 5.13 that a more accurate approximate treatment of the finite-range problem yields an expression that still retains the form of (5.3). It is therefore reasonable to organize the analysis of $\beta_{sj}{}^{lm}$ around the analysis of (5.3). The integrand of the improved expression contains only one additional slowly varying factor.

Many computer codes have been constructed with which (5.3) may be evaluated numerically without further approximation. All these codes are organized in the same manner: The distorted waves are inserted into (5.3) in the form of the partial-wave expansions of (3.24) and (3.36). The angular integrations of (5.3) are done analytically, yielding

$$\beta_{sj}{}^{lm} = \sum_{L_\alpha L_\beta M_\alpha M_\beta} i^{L_\alpha - L_\beta - l} e^{i\sigma_\alpha L_\alpha + i\sigma_\beta L_\beta} \left[\frac{4\pi(2L_\beta + 1)}{(2L_\alpha + 1)}\right]^{1/2} I^{lsj}_{L_\beta L_\alpha}$$
$$\times \langle L_\beta l; 00 \mid L_\alpha 0\rangle \langle L_\beta l; M_\beta m \mid L_\alpha M_\alpha\rangle Y_{L_\beta}{}^{M_\beta}(\hat{k}_\beta) Y^{M_\alpha *}_{L_\alpha}(\hat{k}_\alpha), \quad (5.4)$$

where

$$I^{lsj}_{L_\beta L_\alpha} = \frac{M_B}{M_A} \frac{4\pi}{k_\alpha k_\beta} \int_0^\infty F_{lsj}(r)\, f_{\beta L_\beta}\left(k_\beta, \frac{M_A}{M_B} r\right) f_{\alpha L_\alpha}(k_\alpha, r)\, dr. \quad (5.5)$$

The radial integrals $I^{lsj}_{L_\beta L_\alpha}$ then are obtained by numerical integration of (5.5), using radial wavefunctions $f_{\beta L_\beta}$ and $f_{\alpha L_\alpha}$ that have been computed by numerical integration of the (optical model) radial Schrödinger equations in channels α and β. This procedure* for the calculation of $\beta_{sj}{}^{lm}$ is extremely orderly, and is well adapted for use by a computing machine. Because F_{lsj} is derived from the overlap of bound wavefunctions the integral of (5.5) and the partial wave expansions of (5.4) have good convergence. Indeed, the summation over partial waves really contains only one free index, say, L_α. The other indices are tied to this one by the requirement that the angular momenta of the two partial waves add vectorially to l, m.

It is often of interest to choose the coordinate axes so that the z-axis lies along the direction of the incident momentum $\mathbf{k}_\alpha$ and the y-axis lies along $\mathbf{k}_\alpha \times \mathbf{k}_\beta$. In this case (5.4) reduces to the more compact form

$$\beta_{sj}{}^{lm} = \sum_{L_\alpha L_\beta} i^{L_\alpha - L_\beta - l} e^{i\sigma_\alpha L_\alpha + i\sigma_\beta L_\beta} \sqrt{2L_\beta + 1}\, I^{lsj}_{L_\beta L_\alpha}$$
$$\times \langle L_\beta l; 00 \mid L_\alpha 0\rangle \langle L_\beta l; -m, m \mid L_\alpha 0\rangle Y_{L_\beta}{}^{-m}(\Theta, 0), \quad (5.6)$$

where Θ is the *scattering angle*, the angle between $\mathbf{k}_\alpha$ and $\mathbf{k}_\beta$.

We observe in (5.4) and (5.6) the presence of the special Clebsch-Gordan coefficient $\langle L_\beta l; 00 \mid L_\alpha 0\rangle$. This coefficient vanishes unless $(l + L_\alpha + L_\beta)$ is an even number. As a result the parity transfer of the nuclear reaction is given uniquely in terms of the multipole index l, and has the value $(-)^l$. This is the

* The first applications of the above method in nuclear reaction work were made in 1952 by Daitch and French and in 1953 by Horowitz and Messiah [1]. Extensive calculations were later performed by Tobocman and collaborators [2, 3] and by Levinson and Banerjee [4]. The method has now become widespread and is standard in many laboratories.

condition known as *normal parity* (Section 4.5). Evidently the use of a zero-range form factor always leads to normal parity. With finite-range form factors we also sometimes find normal parity, however, it cannot generally be expected (Section 5.13). (The parity of an excitation in inelastic scattering is said to be *normal* if it is given by $(-)^j$. This rule can be broken either if the form factor has finite range or if it is spin dependent.)

Semi-classical estimates are good guides to the important values of angular momenta in (5.4–6). To perform such estimates, we note that the factor $F_{lsj}(r)$ in (5.5) limits the range of impact parameters that contribute to the integration. This factor causes the range of impact parameters to have an upper limit of the order of, or somewhat greater than, the radius of nucleus A or of nucleus B. A rough guess for this upper limit is

$$R_0 \approx (1.4A^{1/3} + 2.0)\text{fm},$$

where A is now the mass number of the target nucleus or of the residual nucleus. For a medium-weight nucleus this yields $R_0 \approx 8$fm. Then for channel α, say, the angular momentum that corresponds to this limit is

$$l_\alpha(\text{max}) \approx k_\alpha R_0 \approx \frac{R_0\sqrt{\mu_\alpha E_\alpha}}{4.553},$$

where μ_α is the reduced mass number for channel α, and E_α is the relative kinetic energy, measured in MeV. For a 20 MeV proton the above estimate gives $l_\alpha(\text{max}) \approx 8$. For a 40 MeV alpha particle it gives $l_\alpha(\text{max}) \approx 16$. For a 20 MeV deuteron it gives $l_\alpha(\text{max}) \approx 12$, etc. Evidently the number of partial waves that must be considered is substantial but not unwieldy.

Often only a few terms in the sum over partial waves contain radial integrals that are at all large, and these few terms dominate the properties of β_{sj}^{lm}. Whenever this effect occurs we obtain useful insights into the structure of the DW calculation.

One quantity that affects the magnitudes of the radial integrals in (5.5) is the change of linear momentum, $k_\alpha \to k_\beta$. To see this we note that the zero-range form factor $F_{lsj}(r)$ is derived from integrals over bound wavefunctions. Because these tend to have small momenta this form factor tends to be a slowly-varying function of r. Therefore the values of the radial integrals of (5.5) are dominated by the overlap of the radial distorted waves $f_{\beta L_\beta}$ and $f_{\alpha L_\alpha}$. This overlap is reduced if the momenta k_α and k_β are very different, for then the distorted waves oscillate at very different rates and the overlap averages to a small value. *To yield large cross sections, reactions must provide good momentum matching* $k_\alpha \approx k_\beta$.

Angular-momentum matching also plays a role in the radial integrals. The wavefunction of each partial wave tends to have its largest maximum near the classical turning point. This is understood classically from the fact that near the turning point a projectile has very little kinetic energy and moves very slowly. Therefore it is advantageous for good overlap for the first maxima of the two radial waves in (5.5) to coincide. This occurs if

$$\frac{L_\alpha}{k_\alpha} \approx \frac{L_\beta}{k_\beta}.$$

We see that if the conditions of the reaction yield $k_\alpha \approx k_\beta$ then $L_\alpha \approx L_\beta$ is favored; therefore small values of l are favored. However, if $k_\alpha \neq k_\beta$ then best overlap favors $L_\alpha \neq L_\beta$, and this might favor $l > 0$. Many experiments are known that show such *preferential population* of higher l values [5]. Further discussion of this effect is given in Sections 5.10 and 5.11.

Finally, it is interesting to remark that the very use of the partial wave expansion for the calculation of the reduced amplitude automatically and exactly incorporates the basic geometrical facts that the nucleus tends to be spherically symmetric, that it is of finite extent, and that angular momentum is conserved. Approximations of the radial integrals, like those introduced later, only approximate a fairly limited part of the calculation of the reduced amplitude, and have a good chance of being reliable. Methods of approximation that do not use explicit quantum mechanical treatments of angular momentum are less likely to be reliable.

5.2 The Optical Potentials

To calculate the relative radial wavefunctions in channels α and β it is necessary to know the optical potentials that describe the relative motion of a and A in channel α and of b and B in channel β. It is already agreed (Section 4.8) that these optical potentials are fitted to the observed elastic scattering in these two channels.

Let us consider the optical potential in channel α. It is expressed in coordinate representation as a function of $\mathbf{r}_\alpha$ and the spins of nuclei a and A. It may well be a matrix function of $\mathbf{r}_\alpha$, that is, its dependence on $\mathbf{r}_\alpha$ may be nonlocal. Reasons for such nonlocality are already discussed (Sections 4.2, 4.3). Because a nonlocal dependence on $\mathbf{r}_\alpha$ is equivalent to a dependence on $\mathbf{p}_\alpha$, the momentum conjugate to $\mathbf{r}_\alpha$, there is no need to include any additional explicit dependence on $\mathbf{p}_\alpha$. On the other hand, because the optical potential is not a Hermitian operator it may depend on the energy E_α in some fashion that cannot be transformed into a nonlocal dependence on $\mathbf{r}_\alpha$ and that must be carried explicitly (see Section 4.2). In other words, besides being nonlocal with respect to $\mathbf{r}_\alpha$ the optical potential may be nonlocal with respect to time. Practical applications of optical potentials always have used expressions that are explicitly energy dependent. Only recently [6, 7] has there been much application of expressions that are nonlocal functions of $\mathbf{r}_\alpha$. For these reasons it is suitable to begin the present discussion by regarding the optical potential as a local function of $\mathbf{r}_\alpha$, as a function of the spins of nuclei a and A, and as explicitly dependent on E_α. Further details about nonlocality with respect to $\mathbf{r}_\alpha$ are given at the end of this section.

We see from (4.38) that the optical potential in channel α is obtained theoretically as a sum composed of the operator $\bar{\mathcal{O}}$, which compensates for the elimination of other channels, plus the ground-state expectation of the potential terms of H_{PP}. Thus

$$U_\alpha(\mathbf{r}_\alpha, \text{spins}) = \bar{\mathcal{O}} + \langle \psi_a \psi_A | \sum_{i=1}^{n_a} \sum_{j=1}^{n_A} V(ij) \, | \psi_a \psi_A \rangle. \tag{5.7}$$

The second term of (5.7) probably is the larger part of the optical potential. At least for this term it is clear that the interaction between a large nucleus composed of n_A nucleons and a small nucleus composed of n_a nucleons probably is roughly proportional to n_a. Therefore the potential for a deuteron-nucleus interaction probably is about twice as deep as that for a nucleon-nucleus interaction, and so forth. The presence of $\bar{\mathcal{O}}$ keeps this rule from being strictly true. Nevertheless it is useful as a guide for further analysis.

The second term of (5.7) is calculated as an expectation over the internal states of the two nuclei, a and A. Therefore the $\mathbf{r}_\alpha$-dependence of this term must resemble the distribution of nucleon density in nucleus A. Because the real, spin-independent part of U_α derives mainly from the second term of (5.7), it too may be expected to resemble the nuclear density distribution;* it may be expected to be constant inside the nucleus, and to fall gradually to zero in the vicinity of some rather definite surface radius. These properties are found conveniently in the Woods-Saxon function [9]

$$f(x) = (1 + e^x)^{-1},$$
$$x \equiv \frac{r_\alpha - R_0}{a}, \tag{5.8}$$

and therefore this function is almost always used for the shape of the real, spin-independent part of U_α. The contributions yielded by $\bar{\mathcal{O}}$, the first term of (5.7), do not change the shape of the real potential much. Because $\bar{\mathcal{O}}$ is averaged over a broad range of intermediate states its contributions can have only a simple dependence on $\mathbf{r}_\alpha$. Although the $\mathbf{r}_\alpha$-dependence of these contributions tends to be surface peaked (Section 4.3) no drastic changes of the real potential need ensue. On the other hand, the contributions of $\bar{\mathcal{O}}$ dominate

* Greenlees and collaborators [8] have emphasized methods of analysis that exploit this relationship.

the imaginary, spin-independent term of U_α, therefore this term does tend to be surface peaked. To represent this behavior it is customary to use for the shape of the imaginary potential a mixture composed of a volume term plus a surface term. The volume imaginary term is taken of Woods-Saxon shape as in (5.8). The surface imaginary term generally is taken of the form of a derivative of the Woods-Saxon shape $df(x)/dx$. The surface term tends to dominate at low bombarding energies and the volume term tends to dominate at high bombarding energies.

The spin-dependent parts of U_α also are expected to be concentrated near the nuclear surface. The most important spin-dependent part of U_α is the spin-orbit term that is proportional to the operator $(\mathbf{L}_\alpha \cdot \mathbf{s}_a)$, where $\mathbf{L}_\alpha$ is the relative angular momentum operator and $\mathbf{s}_a$ is the spin operator for nucleus a. For collisions between a nucleon and a spin-zero nucleus this spin-orbit term is the only one that can be constructed. (Expressions that involve higher powers of the Pauli spin operator can be linearized.) This term is expected to be strongest near the surface of nucleus A because it is only near the surface that the interaction between a and A is likely to depend on the direction of their relative momentum. Terms that involve the spin of nucleus A can also be constructed if this spin is nonzero. However, experience suggests that terms of this kind make only small contributions to the optical potential. Still other spin-dependent terms can be constructed if nucleus a has $s_a > \frac{1}{2}$, for then there is no trivial linearization of powers of $\mathbf{s}_a$. The most notable case of this type is the deuteron-nucleus interaction, for which $s_a = 1$. With spin $s_a = 1$, tensor-type terms appear in the optical potential. These terms involve $\mathbf{s}_a$ quadratically. Many authors have discussed the terms of this type and suggested experiments by which they can be measured [10]. No certain results have yet been obtained. For deuteron-nucleus scattering the tensor terms may well be as important as the spin-orbit term. It may be best to obtain these terms theoretically from other known properties of the deuteron-nucleus system. These terms are not discussed further.

The above remarks describe rather fully the qualitative properties of the optical potential. It is important that this potential is described in coordinate space, and that its various terms are either of the shape of the nuclear density distribution or peaked near the surface of this density distribution. These geometrical properties of the optical potential are related to nuclear saturation. They establish relations between the scattering of the various different partial waves. Thus the optical potential is not a general quantum-mechanical operator that must be investigated partial wave by partial wave. The characteristic geometry of nuclei almost suffices to determine its form. Nonlocality (see below) makes only minor modifications of these observations. We note, in contrast, that separable interactions do not exploit the important geometrical properties of nuclei.

The analytic expression that authors attempt to fit to experimentally observed elastic-scattering angular distributions and polarizations then is of the form

$$U_\alpha = -Vf(x) - iW'g(x') - iW''f(x'') - (V_{so} + iW_{so})\, h(r)(\boldsymbol{\sigma}\cdot\mathbf{l}_\alpha), \quad (5.9)$$

where $f(x)$ is the Woods-Saxon function of (5.8), and $g(x')$ is the derivative of $f(x')$

$$g(x') \equiv 4\left|\frac{df(x')}{dx'}\right|. \quad (5.10)$$

The function $g(x')$ is surface peaked, and is normalized so that $g(0) = 1$. The arguments x, x', x'' are defined to be

$$x = \frac{r_\alpha - R_0}{a}, \qquad x' = \frac{r_\alpha - R_0'}{a'}, \qquad x'' = \frac{r_\alpha - R_0''}{a''}.$$

The spin-orbit term of (5.9) is customarily taken to be related to the central terms in the manner of a Thomas-type spin-orbit interaction [11]

$$h(r) \equiv (2.000\text{fm}^2)r^{-1}\,|df(x''')/dr|,$$
$$x''' = \frac{r_\alpha - R_0'''}{a'''}. \quad (5.11)$$

The coefficient in (5.11) is introduced arbitrarily to make $h(r)$ dimensionless. (This coefficient is approximately equal to the square of the pion Compton wave-length.) The operator $\mathbf{l}_\alpha$ in (5.9) is the dimensionless orbital angular momentum operator $\mathbf{l}_\alpha = -i[\mathbf{r}_\alpha \times \nabla_{\mathbf{r}_\alpha}]$. The operator $\boldsymbol{\sigma}$ is a dimensionless multiple of $\mathbf{s}_a$, the spin angular momentum of particle a, such that

$$\mathbf{s}_a = \begin{cases} \dfrac{\hbar}{2}\boldsymbol{\sigma}, & \text{for nucleons and for } H^3, He^3, \\ \hbar\boldsymbol{\sigma}, & \text{for deuterons.} \end{cases}$$

Thus for nucleons $\boldsymbol{\sigma}$ is the Pauli spin operator.

The optical potential should include a Coulomb term if particles a and A are charged. For this term authors generally use the expression

$$\begin{aligned} U_c(r_\alpha) &= \left(\frac{Z_a Z_A e^2}{2R_c}\right)\left(3 - \frac{r_\alpha^2}{R_c^2}\right), \qquad & r_\alpha \leqslant R_c, \\ &= \frac{Z_a Z_A e^2}{r_\alpha}, & r_\alpha \geqslant R_c, \end{aligned} \quad (5.13)$$

the Coulomb potential of a uniform sphere of charge of radius R_c. This Coulomb term was omitted from U_α in the present discussion, in order that it may later be displayed explicitly in the Schrödinger equation. Thus the net optical potential in channel α is $U_\alpha + U_c$.

The net optical potential contains fourteen adjustable parameters, which must be determined by fitting the cross sections predicted by the optical potential to the cross sections measured in elastic scattering experiments. These fourteen parameters are the nine geometrical parameters, R_0, a, R_0', a', R_0'', a'', R_0''', a''', R_c, and the five well depths V, W', W'', V_{so} and W_{so}. Authors generally attempt a least-squares fit between the experimental and theoretical cross sections. Details of the fitting procedure are described in many references [10, 13, 15, 17, etc.]. It is only of interest here to cite a few of the characteristic difficulties and to quote some typical result.

Although (5.9) is sufficiently limited in form that sometimes no set of parameters gives an acceptable fit to experiment [10], a more typical difficulty is the discovery of a large range of parameters that all give equally acceptable fits to experiment. Such ambiguities in parameter values may be either *discrete* or *continuous*. When there are discrete ambiguities it is found that acceptable fits are given by a series of real well depths, $V_1, V_2 \cdots$ that are roughly integer multiples of some least well depth. Examination of the wavefunction [12, 13] discloses that this discrete series of well depths corresponds to the discrete number of half waves that lie inside the potential. Scattering experiments do not test this property of the wavefunction. Discrete ambiguities often can be resolved by noting (see earlier) that the potential for composite projectiles should be given very nearly by multiplying the nucleon optical potential by the number of nucleons in the composite projectiles.

Three continuous ambiguities are well known. These are, (a) an equivalence between volume and surface absorption, so that equally good fits to experiment can be obtained by adjustment of W' or W''; (b) an equivalence between range and depth of the real potential, so that values of V and R_0 that satisfy $VR_0^n =$ constant, where $n \approx 2$, give equally good fits to experiment; (c) an equivalence between absorption and surface diffuseness, so that adjustment of a, W' or W'' give equally good fits to experiment. These ambiguities do not appear consistently in all analyses, and often seem to disappear as the quality of the experimental data is improved.

Igo [14] noted that an important continuous ambiguity may arise for projectiles that are absorbed very strongly. For such cases the optical model wavefunction tends to be small in the nuclear interior, and the elastic scattering tends to be affected only by the potential in the region of the nuclear surface. Therefore, potentials that are equivalent in the surface region tend to give equivalent cross sections.

For any given collision experiment a best representation of the wavefunction in the elastic channel is obtained by making an optical model fit to experiments for the system in question and at the energy in question. However, because the optical potential is based on an average over details of intermediate states (Section 4.3) the properties of this potential are likely to be divorced from those details of nuclear structure that distinguish one system from another. Therefore the parameters of the potential are likely to vary smoothly as a function of mass number, in much the same way that the nuclear density and binding energy vary smoothly with A. In particular, because the nuclear radius varies as $A^{1/3}$ and the nuclear surface thickness tends to be constant, it may be supposed that the radii should vary as $R_0 = r_0 A^{1/3}$, $R_0' = r_0' A^{1/3}$, $R_0'' = r_0'' A^{1/3}$, $R_0''' = r_0''' A^{1/3}$, $R_c = r_c A^{1/3}$, and that the diffuseness parameters a, a', a'', a''' all should be independent of A. Likewise, the well depth parameters should be independent of A. A number of authors have attempted to identify average potentials whose parameters show the expected simple variation with respect to A. Such parameter systematics may help to resolve ambiguities, or may provide approximate potentials for cases for which experiments have not yet been done, or may serve as a context against which the behavior of one or another case may be seen to be unusual.

Several average potentials are listed in Table 5.1. On the whole these potentials are rough averages of best-fit potentials obtained for a large range of target nuclei. The proton potential for 40 MeV is obtained from the work of Fricke, Gross, Morton, and Zucker [18]. The proton potential for 29.6 MeV is obtained from the work of Greenlees and Pyle [17]. (A similar potential was found by Satchler [17].) It is noteworthy that at both these energies an accurate fit to polarization data requires $r_0''' < r_0$. The proton potential for 9–22 MeV is obtained, on the whole, from the work of Perey [15]. His study did not include polarization data. A study of the polarization at 17.8 MeV bombarding energy, given by Baugh, Griffith and Roman [16] shows the necessity for reducing the Perey value of r_0''' to the value given in Table 5.1. Similar conclusions were reached by Kossanyi-Demay, et al. [16]. An independent study of 10.5 and 14.5 MeV polarization data was given by Rosen, Beery, Goldhaber and Auerbach [16]. Unfortunately these authors did not consider the possibility $r_0''' \neq r_0$, and therefore their results are not included in the table. (Rosen, et al., found for a', the imaginary diffuseness, a value equal to that found at higher energy and in disagreement with the value 0.47 given in Table 5.1. In fact, Rosen, et al. were unable to resolve an ambiguity between a' and W'. Consideration of reaction cross section data (see Greenlees and Pyle [17]) resolves this ambiguity in favor of the small value of a' given in the table.) The neutron potential of Table 5.1 is obtained from Bjorklund and Fernbach [19], except that their (large) value for V_{so} is reduced to conform with proton polarization data. The deuteron potential is taken from reference [13], guided by the results of a theoretical analysis in which the neutron and proton optical potentials are folded in with the

Table 5.1. **Average potentials for various simple projectiles. Energies are given in MeV. Lengths are given in 10^{-13} cm. Other useful average potentials for nucleons are discussed by F. D. Becchetti, Jr. and G. W. Greenlees, *Phys. Rev.* 182, 1190 (1969).**

	Protons			Neutrons	Deuterons	H^3	He^3	Alphas
	7–22 MeV	29.[illegible] MeV	40 MeV	4–14 MeV	15–27 MeV	15,20 MeV	40 MeV	24.7 MeV
	[15, 16]	[17]	[18]	[19]	[13, 20]	[22]	[23]	[24]
r_0	1.25	1.2	1.16	1.25	1.15	1.24	1.14	1.4
a	0.65	0.7	0.75	0.65	0.81	0.678	0.723	0.52
V_0	$53.3 - 0.55E$	[illegible]3.0	41.1	$52.4 - 0.6E$	$81 - 0.22E + 2.0ZA^{-1/3}$	150	175	185
V_1	27	[illegible]0	26.4	—	—	—	—	—
r_0'	1.25	1.25	1.37	1.25	1.34	—	—	—
a'	0.47	0.7	0.63	$b = 0.98$*	0.68	—	—	—
W'	13.5	6	~ 5	$5.4 + 0.4E$	$14.4 + 0.24E$	0	0	0
r_0''	—	1.25	1.37	—	—	1.45	1.60	1.4
a''	—	0.7	1.37	—	—	0.841	0.81	0.52
W''	0	2	~ 5	0	0	20	16	25
r_0	1.04	1.10	1.064	1.25	—	—	—	—
a'''	0.65	0.7	0.738	0.65	—	—	—	—
V_{so}	5.5	5.5	6	5.5	0	0	0	—
W_{so}	0	0	0	0	0	0	0	—
r_c	1.25	1.2	1.25	—	1.15	1.25	1.4	1.3

* Here b is the diffuseness parameter for a potential of Gaussian shape.

deuteron wavefunction [20]. (An extensive study of the optical potential for 52 MeV deuterons has been published by Hinterberger, et al. [21].) The H^3 potential is that given by Hafele, Flynn and Blair [22]. Similar H^3 potentials were given by Glover and Jones [22]. The He^3 potential is taken from a study by Gibson, et al. [23]. Work by Baugh, et al. [23] is in fair agreement with the average potential of Gibson, et al.

The studies of parameter systematics for neutrons and protons disclose that the real central well depth does not have quite the same value for all nuclei. In particular, it is necessary to recognize a dependence on symmetry, namely, a dependence on the neutron excess of the target nucleus [10, 15, 17]. It is also necessary to include for protons a term that expresses a Coulomb shift of the bombarding energy [15, 25] at which the potential is evaluated. The complete expressions for the real central well depths therefore are

$$V_p = V_0 + V_1\left(\frac{N - Z}{A}\right) + \gamma Z A^{-1/3}, \tag{5.14}$$

$$V_n = V_0 - V_1\left(\frac{N - Z}{A}\right). \tag{5.15}$$

(The imaginary potential also should depend on $(N - Z)/A$. This is ignored.) Values for V_0 and V_1 of (5.14) and (5.15) are given in Table 5.1. Naturally, V_1 is less accurately determined than V_0. It should be noted that the well-depth parameters for protons are taken from proton scattering experiments and those for neutrons from neutron scattering experiments. It is not clear whether these two independent determinations lead to the same values for the parameters, in the manner suggested by (5.14) and (5.15). A satisfactory value for γ of (5.14) appears to be $\gamma = 0.4$ MeV.

Attempts to determine parameter systematics for alpha particles have not achieved much success. The one very rough average potential for 24.7 MeV bombarding energy, given in Table 5.1, is taken from work by McFadden and Satchler [24]. Strong absorption allows the use of Woods-Saxon shapes for both the real and imaginary potential, a usage that is customary with alpha particles. Strong absorption also tends to cause the outer regions of the nuclear surface to dominate the scattering, as remarked by Igo [14]. For such regions of r the Woods-Saxon potential reduces to

$$U \approx -(Ve^{R_0/a})e^{-r/a} - i(W''e^{R_0''/a''})e^{-r/a''}. \tag{5.16}$$

If (5.16) indeed described the part of U that dominates the scattering there would be a continuous ambiguity in the determination of the parameters that fitted the scattering measurements: any two potentials that had the same a and a'' and the same values for the two products $(Ve^{R_0/a})$ and $(W''e^{R_0''/a''})$ would be equivalent. A change of R_0 would be compensated by a change of V, and a change of R_0'' would be compensated by a change of W''. Just such a continuous ambiguity was found with remarkable clarity by Bingham, Halbert and Bassel [26], in an analysis of the scattering of 65 MeV alpha particles by zirconium isotopes. This ambiguity shows up less reliably near 40 MeV bombarding energy, even though it was at this energy that Igo first discovered it. At 24.7 MeV there is almost no remaining trace of the Igo ambiguity [24].

Accurate optical potentials for individual cases must be obtained by consulting the latest literature. Most of the references mentioned above provide compilations of potentials for individual cases.

Only local optical potentials are discussed above. However, the theoretical expressions for the optical potentials (Section 4.2, 4.3) show clearly that nonlocality is expected, therefore the Schrödinger equation for single-particle motion should properly be

$$-\left(\frac{\hbar^2}{2\mu}\right)\nabla^2\chi(\mathbf{r}) + \int U(\mathbf{r}, \mathbf{r}')\chi(\mathbf{r}')\, d^3r' = E\chi(\mathbf{r}). \tag{5.17}$$

So far the only nonlocal interaction whose consequences have been studied in detail has the factored form [6, 7]

$$U(\mathbf{r}, \mathbf{r}') = U_0(\tfrac{1}{2}|\mathbf{r} + \mathbf{r}'|)H(|\mathbf{r} - \mathbf{r}'|). \tag{5.18}$$

The nonlocality function H is assumed normalized to unity with respect to integration over $\mathbf{r}$ or $\mathbf{r}'$. It is not clear [27] whether single-particle potentials derived from Hartree-Fock calculations really do have a structure that resembles that of (5.18). However, provided that the range of the nonlocality function H is reasonably short it is likely that (5.18) will lead to results indicative of those caused by other sorts of short-ranged nonlocalities.

Perey and Buck [7] solved (5.17) by straightforward numerical means, using the interaction of (5.18), and employed their analysis for a study of the elastic scattering of neutrons by nuclei in the energy range 0.4 MeV–24 MeV. They found the interesting result that the introduction of suitable nonlocality enabled them to fit the entire range of data with energy-independent optical model parameters. It is even more interesting that they took the differential cross sections computed from (5.17) and fitted these with *local optical potentials* as if these cross sections were experimental data. It was found that the equivalent local potential $U_L(r)$ obtained in this manner was consistently weaker than $U_0(r)$, and that these two potentials bore a simple relation to each other. It was also found [28] that within the range of U_0 the magnitude of the local eigenfunction $\chi_L(\mathbf{r})$ was consistently larger than that of $\chi(\mathbf{r})$. The latter result has been called the "Perey effect" [29].

It is natural for $\chi(\mathbf{r})$ and $\chi_L(\mathbf{r})$ to differ, inasmuch as these two wavefunctions are governed by different Schrödinger equations. What is unusual about the Perey effect is its almost complete independence from the details that distinguish individual cases.

It is possible to derive [29, 30] approximate analytic expressions that describe with good accuracy the relation between the nonlocal problem of (5.17), (5.18) and the "equivalent" local problem of Perey and Buck [7]. Let us introduce a function $F(\mathbf{r})$ in terms of which we formalize the study of the relation between $\chi(\mathbf{r})$ and $\chi_L(\mathbf{r})$. We define

$$\chi(\mathbf{r}) \equiv F(\mathbf{r})\chi_L(\mathbf{r}). \tag{5.19}$$

In order that $\chi(\mathbf{r})$ and $\chi_L(\mathbf{r})$ describe the same scattering they must be identical everywhere outside the nucleus. Therefore

$$F(\mathbf{r}) \to 1, \qquad \text{as} \qquad r \to \infty. \tag{5.20}$$

This is a boundary condition for $F(\mathbf{r})$ and is used in conjunction with a differential equation that is derived below.

The differential equation that governs $\chi_L(\mathbf{r})$ is determined by substituting (5.19) into (5.17). It is

$$-\left(\frac{\hbar^2}{2\mu}\right)\nabla^2\chi_L(\mathbf{r}) + U_L(r)\chi_L(\mathbf{r}) = E\chi_L(\mathbf{r}), \tag{5.21}$$

where

$$U_L(r) = \left\{-\frac{\hbar^2}{\mu}(\nabla F \cdot \nabla\chi_L) - \frac{\hbar^2}{2\mu}(\nabla^2 F)\chi_L + \int U(\mathbf{r}, \mathbf{r}')F(\mathbf{r}')\chi_L(\mathbf{r}')\, d^3r'\right\} \times [F(\mathbf{r})\chi_L(\mathbf{r})]^{-1} \tag{5.22}$$

is the equivalent local potential. We see that $U_L(r)$ depends on the properties of $\chi_L(\mathbf{r})$ and $F(\mathbf{r})$. It is now seen that by suitable choice of $F(\mathbf{r})$ it is possible to obtain from (5.22) an equivalent local potential with a simple and regular mathematical form, say, of the form of (5.9), as assumed by Perey and Buck [7]. Our choice of $F(\mathbf{r})$ minimizes the extent to which $U_L(r)$ may depend upon idiosyncracies of the wavefunction $\chi_L(\mathbf{r})$. In particular, our choice of $F(\mathbf{r})$ minimizes the dependence of $U_L(r)$ upon the phase of $\chi_L(\mathbf{r})$ so that $U_L(r)$ does not have singularities where $\chi_L(\mathbf{r})$ has zeroes [28].

It is also both desirable and possible to assume $F(\mathbf{r})$ to be a function that varies much more slowly than $\chi_L(\mathbf{r})$. By this assumption the potentials $U_0(r)$ and $U_L(r)$ are made to be as alike as possible.

Let us insert (5.18) into (5.22) and develop an explicit expression for that $F(\mathbf{r})$ that best meets the above conditions. It is perhaps easiest to do the integral in (5.22) by expanding the slowly varying factors of the integrand in powers of $\mathbf{s}$, where $\mathbf{s} = \mathbf{r} - \mathbf{r}'$. Thus

$$U_L(r) = -\left(\frac{\hbar^2}{2\mu}\right)\frac{\nabla^2 F}{F} + [F(\mathbf{r})\chi_L(\mathbf{r})]^{-1}$$
$$\times \left\{-\frac{\hbar^2}{\mu}(\nabla F \cdot \nabla\chi_L) + \int U_0(|\mathbf{r} - \tfrac{1}{2}\mathbf{s}|)H(s)F(\mathbf{r} - \mathbf{s})\chi_L(\mathbf{r} - \mathbf{s})\, d^3s\right\}, \quad (5.23)$$

and the first two terms of the power series expansion give

$$U_L(r) \approx -\left(\frac{\hbar^2}{2\mu}\right)\frac{\nabla^2 F}{F} + [F\chi_L]^{-1}\left\{-\frac{\hbar^2}{\mu}(\nabla F \cdot \nabla\chi_L)\right.$$
$$+ U_0(r)F(r)\int H(s)\chi_L(\mathbf{r} - \mathbf{s})\, d^3s$$
$$\left. - \tfrac{1}{2}F(r)\nabla U_0 \cdot \int \mathbf{s}H(s)\chi_L(\mathbf{r} - \mathbf{s})\, d^3s - U_0(r)\nabla F \cdot \int \mathbf{s}H(\mathbf{s})\chi_L(\mathbf{r} - \mathbf{s})\, d^3s\right\}. \quad (5.24)$$

If we consider the four terms in the braces in (5.24) we clearly see that the second term involves only a scalar average of χ_L at point $\mathbf{r}$, but the first, third, and fourth terms involve the average directed momentum of χ_L at the point $\mathbf{r}$. Not only is the latter quantity sensitive to details of the phase of $\chi_L(\mathbf{r})$, but it is not likely to have zeroes at the same values of $\mathbf{r}$ at which the denominator $F\chi_L$ has zeroes. We therefore determine F by requiring that those terms of (5.24) that depend on directed properties of χ_L must add up to zero. Equation 5.24 accordingly yields the two equations

$$0 = \frac{\hbar^2}{\mu}(\nabla F \cdot \nabla\chi_L) + [\tfrac{1}{2}F(\nabla U_0) + U_0(\nabla F)] \cdot \int \mathbf{s}H(s)\, \chi_L(\mathbf{r} - \mathbf{s})\, d^3s, \quad (5.25)$$

$$U_L = -\frac{\hbar^2}{2\mu}\left(\frac{\nabla^2 F}{F}\right) + U_0\left[\frac{\int H(s)\, \chi_L(\mathbf{r} - \mathbf{s})\, d^3s}{\chi_L(\mathbf{r})}\right]. \quad (5.26)$$

These equations must be solved for F and for U_L.

To reduce (5.25) and (5.26) to useable form we introduce "local WKB approximation" for the wavefunction $\chi_L(\mathbf{r})$:

$$\chi_L(\mathbf{r} - \mathbf{s}) \approx \chi_L(\mathbf{r})\int A(\boldsymbol{\varkappa})e^{-i(\mathbf{s}\cdot\boldsymbol{\varkappa}(\mathbf{r}))}\, d^2\,\hat{\kappa}, \quad (5.27)$$

with

$$\int A(\boldsymbol{\varkappa})\, d^2\hat{\kappa} = 1. \quad (5.28)$$

This approximation gives a good description of the dependence of χ_L upon *small displacements* $\mathbf{s}$ in the vicinity of a point $\mathbf{r}$. The momentum $\boldsymbol{\varkappa}(\mathbf{r})$ in

(5.27) is the "local momentum", whose magnitude has the value given by the local kinetic energy,

$$\frac{\hbar^2 \kappa^2}{2\mu} = E - U_L(r). \tag{5.29}$$

Equation 5.27 allows a distribution of directions of the local momentum.

The insertion of (5.27) in (5.26) immediately yields a simplified expression for the local potential,

$$U_L = -\left(\frac{\hbar^2}{2\mu}\right)\frac{\nabla^2 F}{F} + 4\pi U_0(r)\int_0^\infty s^2 j_0(\kappa s)\, H(s)\, ds, \tag{5.30}$$

in which j_0 is the spherical Bessel function of zero order. The insertion of (5.27) in (5.25) yields

$$0 = \frac{\hbar^2 \kappa}{\mu}\nabla F - 4\pi[\tfrac{1}{2}F(\nabla U_0) + U_0(\nabla F)]\int_0^\infty s^3 j_1(\kappa s)\, H(s)\, ds, \tag{5.31}$$

where we have at the same time inserted the reasonable assumption that F is a scalar function of $\mathbf{r}$, so that ∇F and ∇U_0 are parallel. Equation 5.31 then is an ordinary differential equation. It is readily solved for F, subject to the boundary condition of (5.20), with the result

$$F(r) = \left\{1 - \left(\frac{4\pi\mu}{\hbar^2 \kappa}\right)U_0(r)\int_0^\infty s^3 j_1(\kappa s)\, H(s)\, ds\right\}^{-1/2}. \tag{5.32}$$

In (5.32) we see the Perey effect: if $U_0(r)$ should be an attractive potential, then within the range of this potential $F(r) < 1$. Likewise, if we ignore the small first term of (5.30) we see that folding $H(s)$ with the oscillatory function $j_0(\kappa s)$ causes U_L to be weaker than U_0.

It now remains only to select an explicit normalized form for $H(s)$ and evaluate the integrals in (5.30) and (5.32). The Gaussian nonlocality function of Perey and Buck [7] is adopted,

$$H(s) = \pi^{-3/2}\beta^{-3}e^{-s^2/\beta^2}. \tag{5.33}$$

When we introduce (5.33) into (5.30) and (5.32) the expressions for U_L and $F(r)$ become

$$U_L = U_0 e^{-\kappa^2\beta^2/4} \tag{5.34}$$

and

$$F(r) = \left\{1 - \frac{\mu\beta^2}{2\hbar^2}U_0(r)e^{-\kappa^2\beta^2/4}\right\}^{-1/2}, \tag{5.35}$$

$$= \left\{1 - \frac{\mu\beta^2}{2\hbar^2}U_L(r)\right\}^{-1/2}, \tag{5.36}$$

where the small first term of (5.30) is omitted from U_L. The value of β that Perey and Buck found to give the best fit to the energy dependence of the neutron-nucleus scattering experiments was $\beta = 0.85$fm. For this value of β, and for $\mu = M$, the nucleon mass, we find

$$\frac{2\hbar^2}{\mu\beta^2} = 115 \text{ MeV}.$$

For $U_L(r) = -40$ MeV, say, (5.36) yields $F(r) = 0.74$.

It is clear from the above example that the Perey effect, a specific consequence of nonlocality in the optical potential, is capable of causing major modifications of DW matrix elements. These matrix elements often contain products of three or more single-particle wavefunctions. A twenty-five percent reduction of the magnitude of each single-particle wavefunction in such a matrix element can cause almost complete suppression of contributions from the nuclear interior.

It is likely that most practical calculations with optical interactions will continue to be based on the use of local potentials. However, it seems necessary to reinterpret these potentials as "equivalent local potentials", and to carry expressions of the form of (5.36) in the DW form factors as representations of the Perey effect. It hardly is appropriate to attempt a more elaborate approach.

5.3 Inelastic Scattering of Strongly-absorbed Projectiles

A projectile is absorbed strongly by a target nucleus, by definition, if the reflection coefficients $\eta_{\alpha L}$ [see (3.25)] are of small magnitude for all the low partial waves in the projectile-nucleus relative motion. In particular, this definition requires $|\eta_{\alpha L}| \ll 1$ for every partial wave whose classical impact parameter fulfills the relation $L/k_\alpha <$ nuclear radius. The nucleon-nucleus interaction is never truly strongly absorbing; the weak interaction evidenced by the existence of the shell model apparently persists toward higher energies. Fully developed strong absorption is found when composite projectiles (H^3, He^3, He^4, heavier ions) are scattered from nuclei.* With composite projectiles there is a large variety of reactions that can absorb any flux that manages to penetrate the centrifugal and Coulomb barrier; therefore reflection is inhibited. Figure 5.1 is a typical graph of $\eta_{\alpha L}$ for a case of strong absorption.

At one time it was believed that strong absorption would cause such extreme distortion of the wavefunctions as to invalidate the DW method.

* It is not yet clear whether the deuteron-nucleus interaction shows strong absorption. However, see [44, 48] in Chapter 6.

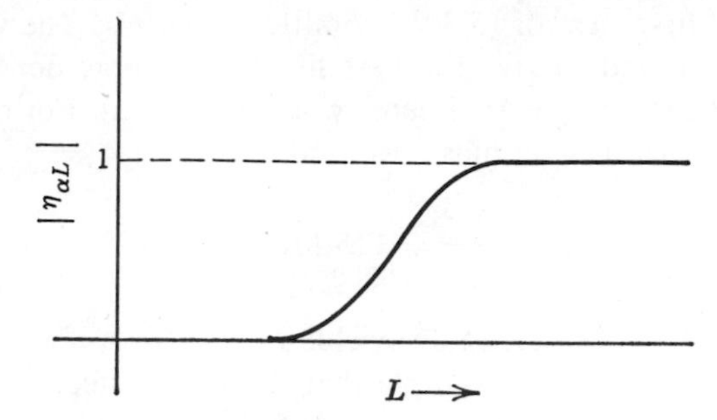

Fig. 5.1 Typical behaviour of $|\eta_{\alpha L}|$ in a case of strong absorption.

Therefore various diffraction models were developed (see Chapter 7). However, it later [31] became clear that the case of strong absorption actually is particularly suitable for DW analysis because it illustrates several simplifications that sometimes appear in the DW analysis. Accordingly, this is a good case with which to start discussions.

For inelastic scattering the distorted-waves discussion of Section 4.5 simplifies greatly: The channel displacement variables become identical, so that

$$\mathbf{r}_\alpha = \mathbf{r}_\beta = \mathbf{r}, \tag{5.38}$$

and the Jacobian reduces to $\mathscr{I} = 1$. The projectile wavefunctions become identical except for spin orientation, so that

$$\psi_a = \phi_{s_0\, m_a}, \qquad \psi_b = \phi_{s_0\, m_b}. \tag{5.39}$$

The distorting potential U_β drops out of the nuclear matrix element because ψ_A and ψ_B are orthogonal. On inserting all these simplifications (4.53) takes the form

$$T_{\alpha\beta}{}^{DW} = \int d^3r \chi_\beta^{(-)*}(\mathbf{k}_\beta, \mathbf{r}) \langle \psi_B \phi_{s_0,\, m_b} |\, V_\beta \,| \psi_A \phi_{s_0,\, m_a} \rangle \chi_\alpha^{(+)}(\mathbf{k}_\alpha, \mathbf{r}). \tag{5.40}$$

Here V_β is the net effective interaction between the (labelled) nucleons of ϕ and the (labelled) nucleons of ψ, as prescribed by the DI model that is being analyzed. Thus V_β consists of all those terms of $(H_{PP} + \bar{\mathcal{O}})$ that vanish as $r \to \infty$. The Coulomb interaction between the projectile and the target nucleus is one term of V_β.

The cross section is obtained by inserting (5.40) for $\mathscr{T}_{\alpha\beta}$ in (3.71). For inelastic scattering the number of equivalent arrangements of the identical particles is the same in channels α and β, therefore $N_\alpha = N_\beta$ in (3.71). A complete account of the identity of particles therefore would only require adding an exchange term to (5.40) as discussed in Section 4.8 and in Chapter 10.

The multipole expansion of (5.40) parallels the expansion of (4.53) and yields

$$T_{\alpha\beta}^{DW} = \sum_{lsj}(2l+1)^{1/2}A_{lsj}(-)^{s_0-m_b}\langle J_A j;\, M_A, M_B - M_A \mid J_B M_B\rangle$$
$$\times\, \langle ls;\, m, m_a - m_b \mid j, M_B - M_A\rangle\langle s_0 s_0;\, m_a, -m_b \mid s, m_a - m_b\rangle\beta_{sj}^{\;lm}, \quad (5.41)$$

where

$$(2l+1)^{1/2} i^l \beta_{sj}^{\;lm} = \int d^3r\, \chi_\beta^{(-)*}(\mathbf{k}_\beta, \mathbf{r}) F_{lsj}(r)\, Y_l^{m*}(\hat{r})\, \chi_\alpha^{(+)}(\mathbf{k}_\alpha, \mathbf{r}), \quad (5.42)$$

and where

$$\langle J_B M_B, s_0 m_b | V_\beta | J_A M_A, s_0 m_a\rangle = \sum_{lsj} i^{-l} A_{lsj} F_{lsj}(r)\, Y_l^{m*}(\hat{r})(-)^{s_0-m_b}$$
$$\times\, \langle J_A j;\, M_A, M_B - M_A \mid J_B M_B\rangle$$
$$\times\, \langle s_0 s_0;\, m_a, -m_b \mid s, m_a - m_b\rangle \quad (5.43)$$
$$\times\, \langle ls;\, m, m_a - m_b \mid j, M_B - M_A\rangle$$

The reduced amplitude of (5.42) is, of course, identical with (5.3), which was obtained by introducing zero-range approximation into (4.60). The notation in (5.42) emphasizes this identity.

The form factor $F_{lsj}(r)$ and the coefficient A_{lsj} depend on the detailed properties of the interaction V_β and the wavefunctions $\phi_{s_0, m}$ and ψ_A and ψ_B. Thus they depend on the details of the reaction model. A discussion of reaction models is given shortly. However, it is interesting to first recognize that angular distributions obtained under conditions of strong absorption tend to be model independent. They are determined primarily by the very limited overlap between $F_{lsj}(r)$ and the distorted waves $\chi_\alpha^{(+)}$ and $\chi_\beta^{(-)}$.

To see the extent to which the distorted waves determine the properties of the reduced amplitude, let us examine (5.6), which gives $\beta_{sj}^{\;lm}$ in partial-wave expansion. For inelastic scattering to low-lying states of the target nucleus we have $k_\beta \approx k_\alpha$. Then the properties of (5.6) are determined by the radial integrals

$$I_{L_\beta L_\alpha}^{lsj} \approx 4\pi k^{-2}\int_0^\infty F_{lsj}(r) f_{\beta L_\beta}(k, r) f_{\alpha L_\alpha}(k, r)\, dr, \quad (5.44)$$

with $k = \frac{1}{2}(k_\alpha + k_\beta)$, say. The properties of these integrals are determined in turn by the relation between L_α, L_β and the cutoff angular momentum L_0, where $L_0 k^{-1}$ = nuclear radius.*

The radial integrals necessarily are very small if L_α and L_β are *greater* than L_0, because centrifugal repulsion causes the high partial waves to have poor

* Here the term "nuclear radius" is used in a loose sense to indicate an interpretation of L_0. Other, related interpretations of L_0 are also used [32].

overlap with $F_{lsj}(r)$. We now see that with strong absorption the radial integrals also are very small if both L_α and L_β are *smaller* than L_0. However, the mechanism that quenches contributions from the low partial waves is related to the phases of the radial wavefunctions, more than to questions of spatial overlap. To see this we note, say, that for $L_\alpha < L_0$ the reflection coefficient $\eta_{\alpha L_\alpha}$ is very small; hence we see from (3.25) that outside the nucleus the associated radial wavefunction reduces to

$$f_{\alpha L_\alpha} \approx \left(\frac{i}{2}\right) H^*_{L_\alpha}(kr), \tag{5.45}$$

a purely ingoing travelling wave. Because the overall reflection is very small, there particularly must be negligible reflection at the nuclear surface; therefore $f_{\alpha L_\alpha}$ continues into the nuclear interior as an ingoing travelling wave of modified wavelength and modified amplitude. This travelling wave eventually is absorbed as it penetrates deep into the nucleus; however, the absorption causes only a gradual decrease of amplitude. In this situation the magnitudes of the radial integrals for the low partial waves are primarily controlled by the rapid oscillations that arise because $f_{\beta L_\beta} f_{\alpha L_\alpha}$ has reduced to a product of ingoing travelling waves. *The rapidly oscillating integrands tend to average to zero*. This effect has been called "phase averaging" [33–35].

Phase averaging excludes contributions to $\beta_{sj}{}^{lm}$ from low partial waves. It therefore tends to exclude contributions from the nuclear interior. However, it is by no means equivalent to a simple radial cutoff in the integrals of (5.44). Let us consider as an example that term of $F_{lsj}(r)$ that causes Coulomb excitation. This term falls off toward large r as $r^{-(l+1)}$, so that it is effective well outside the nucleus, where even the strongly-absorbed low partial waves have large amplitudes. Nevertheless, phase averaging eliminates any contributions of Coulomb excitation in the low partial waves. Strong absorption causes the product $f_{\beta L_\beta} f_{\alpha L_\alpha}$ to be oscillatory at all radii.

Evidently, the distinctive new phenomenon caused by strong absorption is that the contributions to the reaction amplitude are *localized in angular momentum space*. The amplitude is dominated by partial waves that have $L_\alpha \approx L_0$, $L_\beta \approx L_0$. Figure 5.2 depicts the localization of the radial integrals for 84 MeV alpha particles incident upon a neutral "Mg^{24} nucleus". The parameters chosen for the optical potential (see Section 5.2) are $V = 175.4$ MeV, $W'' = 107.24$ MeV, $R_0 = R_0'' = 3.841$fm, $a = a'' = 0.5$fm. The real and imaginary parts of $I^{lsj}_{L_\beta L_\alpha}$ are plotted versus the average angular momentum $\bar{L} = \frac{1}{2}(L_\alpha + L_\beta)$. It is seen that the localization does not depend very much on the value of $|L_\alpha - L_\beta|$.

Figure 5.3 shows a study of localization given by Rost [34]. He did not plot the radial integrals themselves, but rather the closely related quantity $\alpha_{L_\beta}{}^m$ which is found as the coefficient of $Y_{L_\beta}{}^m(\Theta, 0)$ in (5.6). The parameters of the

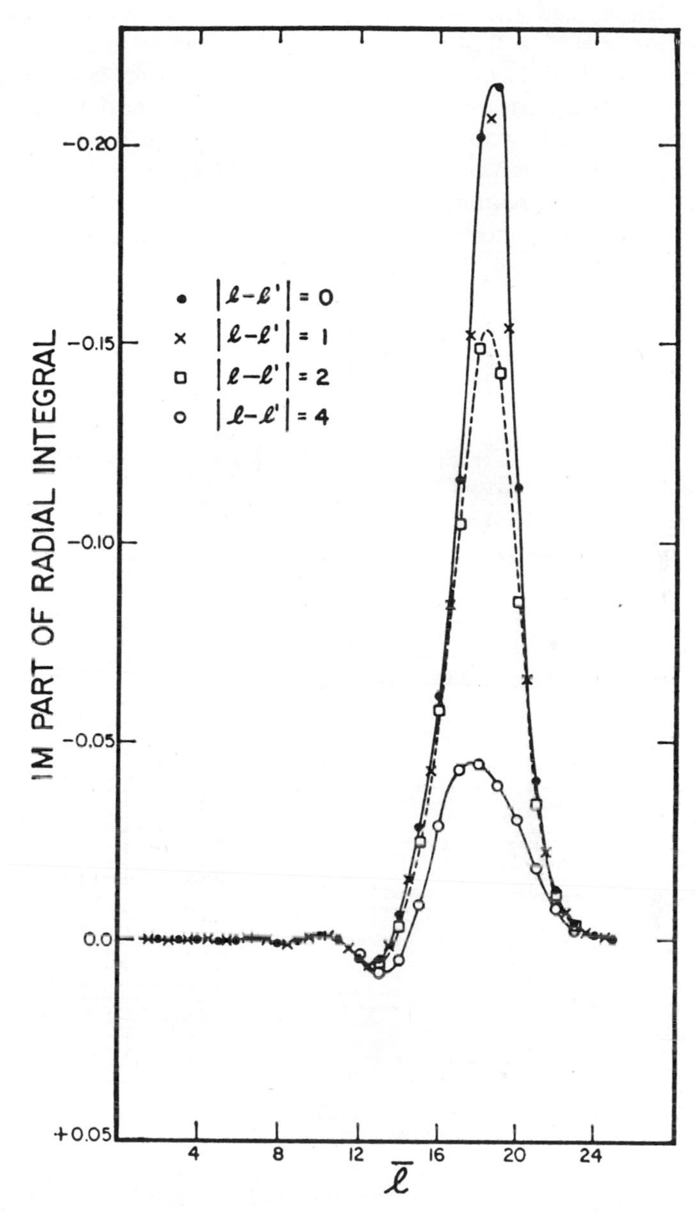

Fig. 5.2 Localization of the radial integrals for $Mg^{24}(\alpha, \alpha')$ at 84 MeV, as described in the text. (From [32], Fig. 1.)

optical potential had the values $V = 47.6$ MeV, $W'' = 13.8$ MeV, $r_0 = r_0'' = 1.585$fm, $a = a'' = 0.549$fm. The uppermost graph in Fig. 5.3 shows the localization obtained using a typical smoothly-varying form factor that has the shape of a derivative of the Woods-Saxon optical potential. The middle graph shows how the localization is destroyed if the form factor is taken to be a delta function at the nuclear surface. (A delta function suppresses phase averaging.) The bottom graph shows how localization is destroyed if the strongly-absorbed distorted waves are replaced by plane waves.

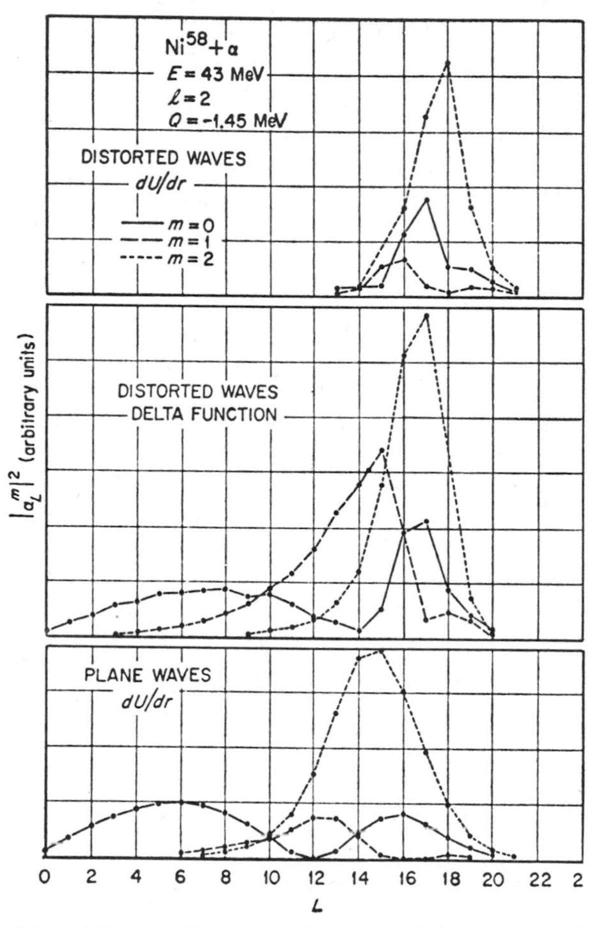

Fig. 5.3 Localization of the radial integrals for $Ni^{58}(\alpha, \alpha')$ at 43 MeV, as described in the text. Full DW calculation (top) is compared with delta-function radial localization (center) and with plane-wave calculation (bottom). Note the enhanced L-localization in the full calculation. (From [34], Fig. 2.)

The consequences of the L-space localization are seen most easily for monopole excitation, $l = 0$. In this case the partial wave expansion of (5.6) becomes

$$\beta_{sj}{}^{00} = \sum_L e^{i2\sigma_L}\sqrt{2L+1}\, I_{LL}^{0sj} Y_L{}^0(\Theta, 0). \tag{5.46}$$

This expression is of the form of an average of $Y_L{}^0$ [or of the Legendre function $P_L(\cos\Theta)$] over a few values of L near L_0. Thus

$$\beta_{sj}{}^{00} \propto \langle P_L(\cos\Theta)\rangle_{L\approx L_0}. \tag{5.47}$$

Because adjacent Legendre functions are in phase and have the value unity at $\Theta = 0$, and only gradually drop out of phase as Θ increases, it is clear that (5.47) predicts that $\beta_{sj}{}^{00}$ must oscillate with the period of the function $P_{L_0}(\cos\Theta)$, and that the amplitude of oscillation must decrease as Θ becomes large. Then the cross section is proportional to $|\beta_{sj}{}^{00}|^2$, and has the familiar form shown in Fig. 5.4.

The analysis of (5.6) is a little more complicated if $l \neq 0$. We have seen already that the radial integrals tend to be slowly-varying functions of $|L_\alpha - L_\beta|$. This fact may be used as the basis of an approximation that allows the double summation of (5.6) to be performed in two rather simple stages. In the first stage we sum over the rapidly varying Clebsch-Gordan coefficients to obtain

$$\sum_{L_\alpha} i^{L_\alpha - L_\beta}\langle L_\beta l; 00 \mid L_\alpha 0\rangle\langle L_\beta l; -mm \mid L_\alpha 0\rangle \approx [l\colon m], \tag{5.48}$$

where

$$\begin{aligned} [l\colon m] &\equiv i^l \frac{[(l-m)!\,(l+m)!]^{1/2}}{(l-m)!!\,(l+m)!!}, \qquad (l+m) \text{ even},\\ &\equiv 0, \qquad\qquad\qquad\qquad\qquad\quad (l+m) \text{ odd}. \end{aligned} \tag{5.49}$$

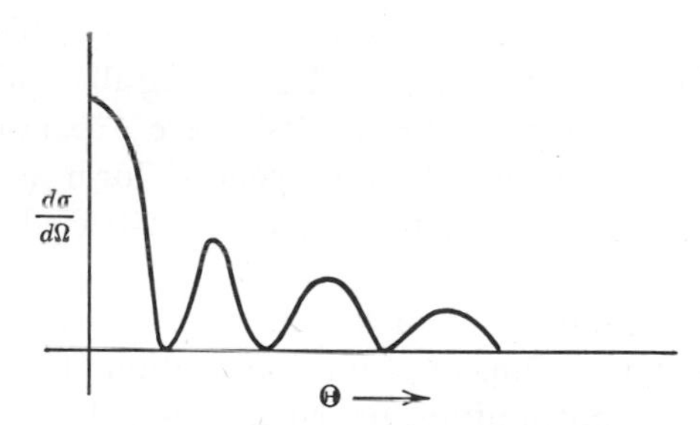

Fig. 5.4 Typical angular distribution for strongly-absorbing projectiles and small momentum change.

Here $N!! \equiv (N)(N-2)(N-4)\cdots(1 \text{ or } 2)$, as usual. The above approximation is valid [32] in the limit $l/L \to 0$, where $L = \frac{1}{2}(L_\alpha + L_\beta)$. In the second stage we substitute (5.48) in (5.6) and obtain the result

$$\beta_{sj}^{\ lm} \approx [l:m] \sum_L e^{i2\sigma_L} \sqrt{2L+1}\, I_{LL}^{lsj} Y_L^{-m}(\Theta, 0). \tag{5.50}$$

Thus the differential cross section becomes

$$\frac{d\sigma}{d\Omega} \propto \sum_{m,(l+m)\,\text{even}} [l:m]^2 \, |\langle Y_L^{-m}(\Theta, 0)\rangle_{L\approx L_0}|^2. \tag{5.51}$$

Now successive $Y_L^{\ m}$ are obtained from each other by differentiation. For $m \geqslant 0$ the rule is

$$Y_L^{\ m}(\Theta, 0) = \left[\left(\frac{2l+1}{4\pi}\right)\frac{(L-m)!}{(L+m)!}\right]^{1/2} (\sin \Theta)^m \frac{d^m P_L(\cos \Theta)}{d(\cos \Theta)^m}, \tag{5.52}$$

and from this rule we see that each successive differentiation reverses the phase of oscillation of $Y_L^{\ m}$. Therefore, because (5.51) uses only all odd or all even values of m, all terms of (5.51) must oscillate in phase. We thus verify one part of a "phase rule" discovered by Blair [36]: the oscillations of all cross sections with even l are in phase with each other and are out of phase with the oscillations of cross sections that have odd l. We also see in (5.52) that near small Θ the amplitudes are proportional to $(\sin \Theta)^m$. Therefore, the first peak of the angular distribution tends to be displaced to larger Θ as l increases. The Blair phase rule has been confirmed by numerous experiments. It is discussed again in Section 7.2.

The L-space localization is used as the basis for elaborate semiquantitative analyses of the transition amplitude (see Chapter 7). However, theoretical cross sections intended for practical comparison with experiment probably are best calculated by straightforward numerical evaluation of the distorted-waves expressions of (5.5) and (5.6). From this point of view the L-space localization is of interest because it provides some understanding of the structure of the partial wave expansion. For strong absorption, and for small scattering angles, the terms of the partial wave expansion combine constructively and each term has the same general form as the final answer. Therefore, DW numerical calculations may be expected to be stable with respect to approximations.

The L-space localization of the reaction amplitude is accompanied automatically by a certain amount of radial localization. The few partial waves that make important contributions to the amplitude have $L \approx L_0$, the cutoff angular momentum. These partial waves make only glancing collisions with the nuclear surface. In particular, these partial waves vanish inside the nucleus because of centrifugal repulsion, and have their first large peak in the surface region, about at the cutoff radius $R_0 = L_0/k$. (This radius usually is of the order of 2fm further out than the half point of the Woods-Saxon optical potential, apparently because partial waves that peak at smaller radii are not sufficiently shielded from the nuclear interior and are absorbed.) The radial localization of the important partial waves has the consequence that the amplitude is determined almost entirely by the part of the form factor $F_{lsj}(r)$ that lies at large radii, in the region of the nuclear surface. This part of the form factor is not very model dependent.

The manner in which the absorption mechanism operates to favor contributions from the surface region also is seen by considering the flux patterns associated with $\chi_\alpha^{(+)}$ and $\chi_\beta^{(-)}$. For short wavelengths, so that $kR_0 \gg 1$, these flux patterns are essentially classical (see Fig. 5.5). Because of absorption the flux inside the nucleus is either exclusively radially ingoing or exclusively radially outgoing. Accordingly, inside the nucleus the overlap integral of (5.42) always involves *large momentum transfer*. Because the form factor is smooth it cannot supply large momenta, therefore the contributions from the nuclear interior must be small. On the other hand, *small momentum transfer* can appear in overlaps in the region of the nuclear surface, provided that $k_\alpha \approx k_\beta$ and the scattering angle is small. Under these conditions the reaction amplitude is dominated by contributions from the surface region. Evidently the relative importance of the surface and volume contributions to the amplitude is affected both by particle type and by the magnitudes and directions of the momenta $\mathbf{k}_\alpha$ and $\mathbf{k}_\beta$.

We may now go back to (5.43) and discuss nuclear-structure models from which we determine the form factor $F_{lsj}(r)$. Because of strong absorption these models only need be accurate in the limited region of the nuclear surface in which the bound and continuum functions have good overlap.

The direct inelastic scattering of strongly-absorbed projectiles yields large cross sections whenever collective rotations or vibrations of the nuclear shape are excited [37, 38]. In the calculation of form factors such collective excitations may be described either *microscopically*, as correlated motions of individual nucleons, or *macroscopically*, as motions of collective variables that parameterize the nuclear shape. Of these two descriptions the microscopic

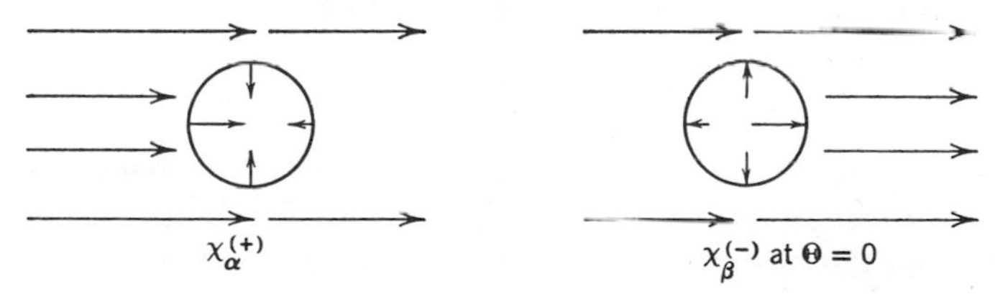

Fig. 5.5 Semiclassical flux patterns for the distorted waves $\chi_\alpha^{(+)}$ and $\chi_\beta^{(-)}$ showing the effects of absorption.

one is the more fundamental and may also be used for excitations that are not collective. A number of authors have published microscopic calculations of form factors [38–48]. Unfortunately, such microscopic calculations find it difficult to establish consistency between the nuclear-structure model that yields the form factor, and the optical model that yields the strongly-absorbed distorted waves [43]. Because of the limited overlap between the bound and continuum wavefunctions this lack of consistency easily causes large errors. Furthermore, it is difficult to have adequate knowledge of the interaction between a composite, strongly-absorbed projectile and the individual nucleons of the target nucleus. The microscopic calculations also find it difficult to represent the influence of the optical operator $\overline{\mathcal{O}}$ upon the interaction V_β; by ignoring this they omit any influence of the compound states upon the interaction that causes direct inelastic scattering. Because of these difficulties there is no further treatment of microscopic form factors in the present discussion of strongly-absorbed projectiles.

In the macroscopic description of collective excitations [49] the properties of the low-lying collective states are associated with deformations of the nuclear shape. This leads naturally to an extension of the optical model to include nonspherical potentials. The nonspherical parts of the potential then induce inelastic scattering to collective vibrational or rotational states. This model has the advantage that the form factor for inelastic scattering is determined from the optical model for elastic scattering, therefore consistency is maintained.

It is reasonable to suppose the strength of the optical potential depends only on the distance $(r - R)$ from some chosen "nuclear surface" and to introduce deformation by allowing this surface to be nonspherical:

$$R(\Theta', \Phi') = R_0\{1 - (4\pi)^{-1} \sum_{kq} |\alpha_{kq}|^2 + \sum_{kq} \alpha_{kq} Y_k^{q*}(\Theta', \Phi')\}. \tag{5.53}$$

Here the polar angles Θ', Φ' are referred to the body-fixed principal axes of the nucleus. The volume enclosed by this surface is constant to second order in the deformation parameters α_{kq}. The Hamiltonian that governs the collective excitations is a function of these parameters. Thus the low-lying states of the target nucleus, among which we make inelastic transitions, obey

$$H(\alpha)\,|i\rangle = \varepsilon_i\,|i\rangle. \tag{5.54}$$

The states $|i\rangle$ are members of rotational or vibrational bands. If $H(\alpha)$ is a quadratic function of the α_{kq} and the conjugate momenta, as is generally assumed, then the operators α_{kq} obey the usual selection rules for harmonic oscillators, and $\alpha_{kq}\,|i\rangle$ connects $|i\rangle$ only with the states $|i - 1\rangle$ and $|i + 1\rangle$ that lie immediately above or below it in the same band.

Our standard optical potential, given in (5.9), is of the form described above. It is a function of the displacement between r and a reference radius R_0. Some confusion arises because different reference radii, R_0, R_0', R_0'', R_0''' are used in the different terms of (5.9). It is not clear whether the same deformation parameters α_{kq} should be used for the nonspherical generalizations of these different radii, or whether some other relation among the deformations of these different radii must be found. As one solution to this question we may adopt the view that the various terms of (5.9) describe the structure of the surface of a droplet of nuclear matter; then it becomes appropriate to employ Blair's suggestion [32, 50] that for a given deformation of the nucleus all the different reference radii should undergo the same set of *displacements*,

$$\xi_{kq} = R_0 \alpha_{kq}. \tag{5.55}$$

Then R_0' of (5.9), for example, becomes

$$R_0' + \delta R, \tag{5.56}$$

where

$$\delta R = \sum_{kq} \xi_{kq} Y_k^{q*}(\Theta', \Phi') - (4\pi R_0)^{-1} \sum_{kq} |\xi_{kq}|^2. \tag{5.57}$$

This same nonspherical displacement δR is used for R_0'', R_0''', etc. Then $U(\mathbf{r}, \xi)$, the nonspherical generalization of the optical potential, is a function of $\mathbf{r}$ and of the ξ_{kq}. This generalized optical potential both scatters the projectile and excites deformations of the nuclear surface.

The macroscopic theory of collective excitations thus leads to a very simple expression for the nuclear matrix element of (5.40). It is

$$\langle \psi_B | \, U(\mathbf{r}, \xi) + \text{Coulomb term} \, | \psi_A \rangle, \tag{5.58}$$

where any consideration of the spin of the projectile is omitted. (Spin-dependent experiments with strongly-absorbed projectiles are rare.) The Coulomb term of (5.58) is treated in the next section. We now treat the nuclear term.

To display the operators ξ_{kq} that link ψ_A and ψ_B we make a Taylor expansion of the generalized optical potential in powers of δR:

$$U(\mathbf{r}, \xi) = U(r) - \delta R \frac{dU(r)}{dr} + \tfrac{1}{2}(\delta R)^2 \frac{d^2U(r)}{dr^2} + \cdots. \tag{5.59}$$

Only the one term of (5.59) that is linear in the ξ_{kq} contributes to transitions from the ground state of a collective nucleus to the lowest adjacent rotational or vibrational state of order k. The matrix element of this term is

$$\langle \psi_B | - \sum_{kq} \xi_{kq} Y_k^{q*}(\Theta', \Phi') \frac{dU(r)}{dr} | \psi_A \rangle. \tag{5.60}$$

Those terms of (5.59) that are of quadratic or higher order in the ξ_{kq} can connect the ground state to higher collective states, and can introduce small modifications of the ground-state elastic scattering. However, such terms cannot be carried consistently in a distorted wave calculation because terms of comparable order have been omitted from the wavefunctions. The higher-order terms of the wavefunctions and the higher-order terms of the interaction are of equal importance [51]. Transitions that require the use of these higher-order terms are most easily treated by the method of coupled channels (Chapter 6).

The r-dependence of the first-order nuclear matrix element factors out, yielding a complete separation of nuclear coordinates from projectile coordinates. Equation 5.60 becomes

$$-\frac{dU}{dr}\langle\psi_B|\sum_{kq}\xi_{kq}Y_k^{q*}(\Theta',\Phi')\,|\psi_A\rangle. \tag{5.61}$$

The form factor of (5.42), (5.43) then is obtained by selecting the interaction of order l and performing the integration over nuclear coordinates. This integration may be performed, for example, by rotating the spherical harmonic Y_k^{q*} to the space-fixed coordinate frame in which the collective wavefunctions ψ_A, ψ_B are defined. For rotational excitation of deformed, axially-symmetric even nuclei the result is

$$A_lF_l(r) = -i^l(2l+1)^{-1/2}(\beta_lR_0)\frac{dU}{dr}, \tag{5.62}$$

in which the deformation parameter β_l has been introduced. When expressed in terms of the deformation length $\delta_l = \beta_lR_0$ (5.62) becomes

$$A_lF_l(r) = -i^l(2l+1)^{-1/2}\,\delta_l\frac{dU}{dr}. \tag{5.63}$$

Exactly the same result is obtained for vibrational excitation of even nuclei provided we define the equivalent deformation parameter

$$\begin{aligned}\beta_l^2 &= \langle\sum_m|\alpha_{lm}|^2\rangle,\\ &= (2l+1)\frac{\hbar\omega_l}{2C_l},\end{aligned} \tag{5.64}$$

where ω_l is the frequency and C_l the "spring constant" of the vibration [49]. This β_l is the root-mean-square deformation in the ground state because of zero-point oscillations. The corresponding deformation length is

$$\delta_l^2 = \langle\sum_m|\xi_{lm}|^2\rangle. \tag{5.65}$$

The parameters β_l or δ_l determine the magnitude of the cross section for collective excitation.

After the insertion of (5.63) into (5.42) and (4.62) the cross section for collective excitation of an even nucleus by inelastic scattering is found to be

$$\frac{d\sigma}{d\Omega} = \left(\frac{\mu_\alpha}{2\pi\hbar^2}\right)^2 \left(\frac{k_\beta}{k_\alpha}\right) \delta_l^{\,2} \sum_m |\beta_{sj}{}^{lm}|^2, \tag{5.66}$$

where

$$(2l+1)^{1/2} i^l \beta_{sj}{}^{lm} = \int d^3r \chi_\beta^{(-)*}(\mathbf{k}_\beta, \mathbf{r}) \left(\frac{dU}{dr}\right) Y_l^{m*}(\hat{r})\, \chi_\alpha^{(+)}(\mathbf{k}_\alpha, \mathbf{r}). \tag{5.67}$$

The indices *sj* are redundant because the projectile is being treated as spinless.

The form factor in (5.67) is just the radial derivative of the spherical optical potential that fits elastic scattering of the projectile by the target nucleus. (Although slightly different optical potentials should in principle appear in the entrance and exit channels, their difference is undoubtedly very small and is being ignored.) It should be noted that (5.66) and (5.67) contain *no free parameters*. The optical potential is determined from elastic scattering, and the deformation length δ_l can be found from electromagnetic experiments.

The form factor dU/dr is complex, because U has both a real part and an imaginary part. The imaginary part of dU/dr expresses the influence upon the inelastic excitation caused by the coupling of the DI wavefunction to the compound nucleus states, states not explicitly included in the DI model. In the original formulation of DI models (Chapter 4) this sort of influence would be carried by the operator $\bar{\mathcal{O}}$. However, in the macroscopic theory this rather complicated effect is included automatically, merely by recognizing that the absorbing potential for a deformed nucleus must be deformed.

At one time calculations of inelastic scattering tended to omit the imaginary term of dU/dr. This omission led to both altered magnitudes of cross sections and altered angular distributions, inasmuch as the real and imaginary parts of U generally are of quite different shape; for example, the imaginary part of U frequently has much longer range than the real part, therefore it is the imaginary part of dU/dr that tends to have the better overlap with the distorted waves. Furthermore, with surface absorption the imaginary term of dU/dr changes sign at the nuclear surface, and this causes distinctive effects. Satchler and collaborators have verified repeatedly that fits with experiment are improved if both terms of the "complex form factor" are carried in the distorted-waves calculation [52–57, 60].

Figures 5.6–5.11 show several applications of the macroscopic theory to experimental results for the inelastic scattering of strongly-absorbed particles. Most typical are the results for (α, α') reactions. The cases in Figs. 5.6–5.8

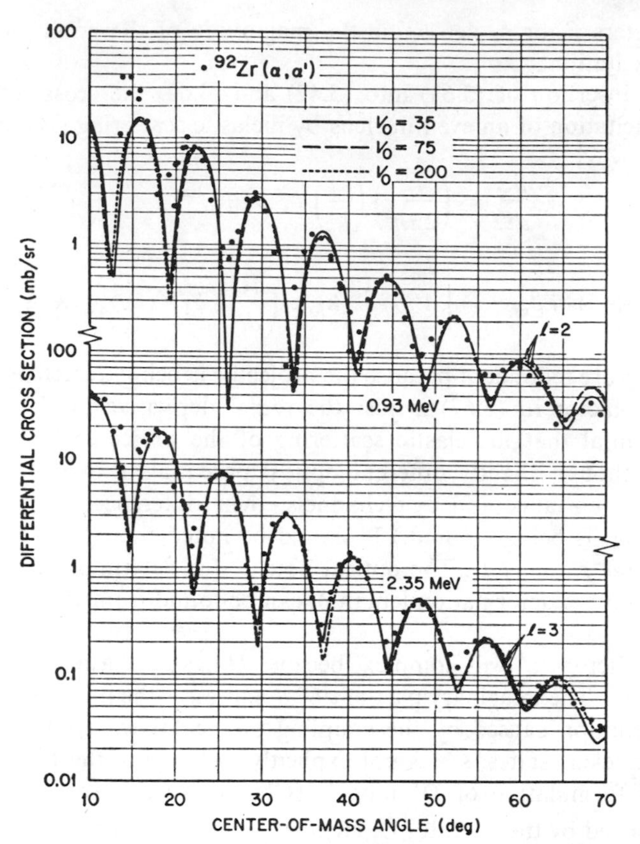

Fig. 5.6 Comparison of the DW theory with experimental data for $Zr^{92}(\alpha, \alpha')$ at 65 MeV bombarding energy. The energies and angular momenta of the Zr^{92} excited states are given on the graph. (From [58], Fig. 8.)

show (α, α') data that extends to particularly large scattering angles, at the three bombarding energies 65 MeV, 43 MeV, and 22 MeV. (The lowest of these energies may be marginal for successful application of the DW theory [62].) The very fine example of 43.7 MeV He^3 inelastic scattering in Fig. 5.9 extends over the angular range 15°–120°. Figures 5.10 and 5.11 show applications of the theory to the inelastic scattering of H^2 by medium-weight and heavy nuclei. In Fig. 5.10 the target is the heavy nucleus W^{182}, for 12 MeV bombarding energy. The DW theory is seen to give an acceptable fit to the data, provided that the complete complex form factor and Coulomb excitation (see Section 5.4) are carried.

In all the examples presented above the theoretical analysis follows the procedure described earlier: Optical potentials are obtained that give good

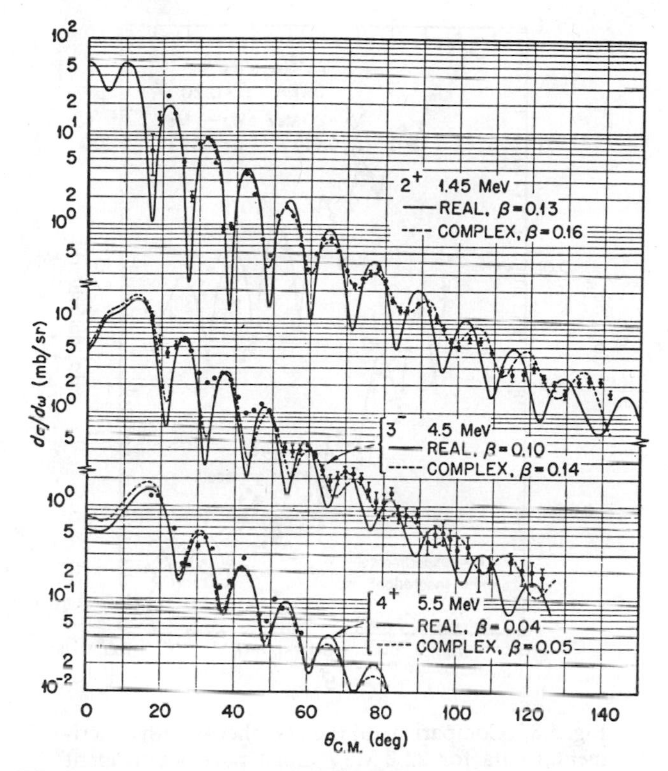

Fig. 5.7 Comparison of the DW theory with experimental data for $Ni^{58}(\alpha, \alpha')$ at 43 MeV bombarding energy, as in Fig. 5.6. Note the influence of "complex coupling." (From [55], Fig. 2.)

fits to the observed elastic scattering. Using these optical potentials, (5.66) and (5.67) are employed for the calculation of the cross section. We see that sufficiently good fits of the angular distributions for inelastic scattering are obtained to allow use of the magnitude of the cross section as a measure of the deformation parameter β_l, or of the deformation length δ_l. Values of β_l and δ_l extracted from experiment by this means tend to be independent of the nature of the projectile, and to agree well with values obtained from measurements of quadrupole moments or of $B(E2)$ [63]. However, it is not yet clear whether β_l or δ_l is the more acceptable measure of nuclear deformation.

5.4 Coulomb Excitation

The projectile-nucleus interaction includes a Coulomb term that is the sum of the Coulomb interactions between the projectile and the individual

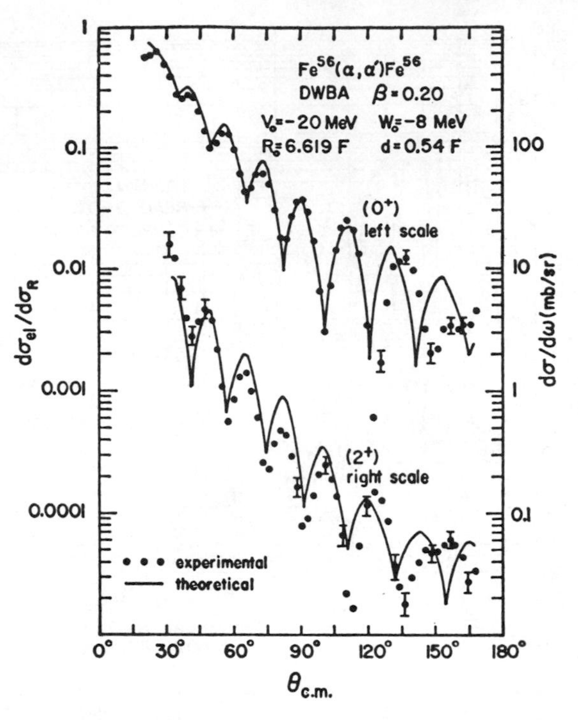

Fig. 5.8 Comparison of the DW theory with experimental data for 22.2 MeV alpha particles incident on Fe^{56}. Optical potential parameters used in the calculations are given on the graph. (From [59], Fig. 12.)

nucleons of the target nucleus:

$$V_c(\mathbf{r}_\alpha) = e^2 Z_a \sum_{i=1_p}^{n_p} |\mathbf{r}_\alpha - \mathbf{r}_i|^{-1}. \tag{5.68}$$

This term [see (5.58)] is not carried in the discussion in the preceding section. We now see that important effects are caused by this term. The effects caused by the Coulomb term are especially noticeable because this term is coherent with the nuclear term of (5.58), so that the Coulomb and nuclear terms interfere. This interference tends to be destructive.

In the macroscopic theory of collective excitations the summation in (5.68) becomes an integration over the continuous electric charge density of the target nucleus:

$$V_c(\mathbf{r}_\alpha) = e^2 Z_a \int d^3\rho \, |\mathbf{r}_\alpha - \mathbf{r}|^{-1}. \tag{5.69}$$

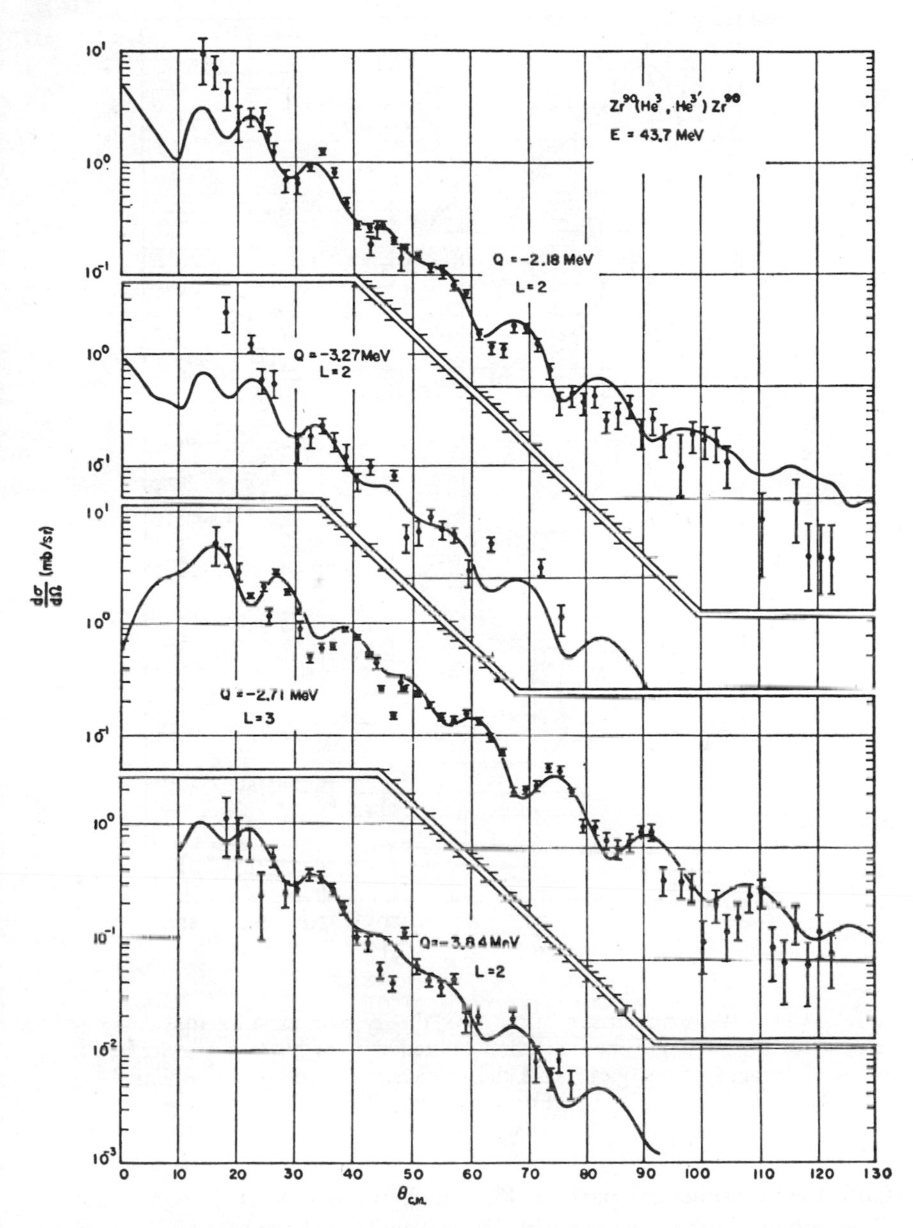

Fig. 5.9 Comparison of the DW theory with experimental data for $Zr^{90}(He^3, He^{3'})$ at 43.7 MeV bombarding energy, as in Fig. 5.6. (From [60], Fig. 14.)

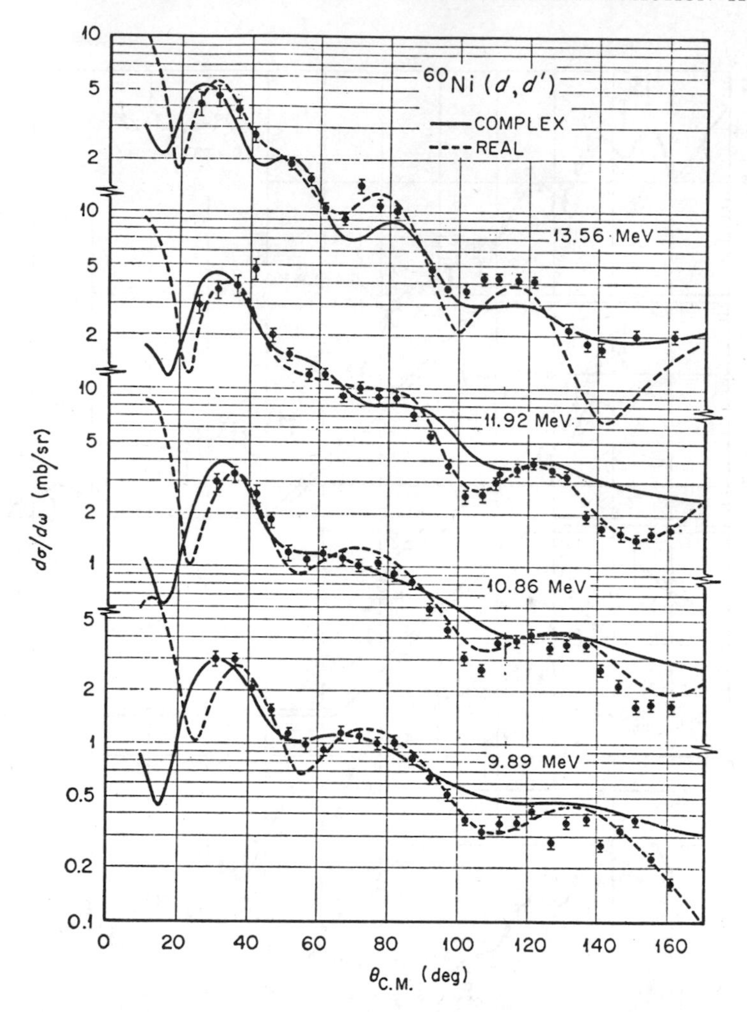

Fig. 5.10 Comparison of the DW theory with experimental data for $Ni^{60}(d, d')$ to the $l = 2$ first excited state of Ni^{60}, at a series of bombarding energies. Note the influence of "complex coupling." (From [56], Fig. 2.)

Only the nonspherical part of V_c, which we denote δV_c, contributes to Coulomb excitation. The quantity δV_c is usually evaluated under the assumption that the charge density is uniform within a nonspherical surface with a radius $R_c(\Theta', \Phi')$ in the body-fixed coordinate system, and that it is zero outside this surface. Under this assumption, and to first order in δR and in

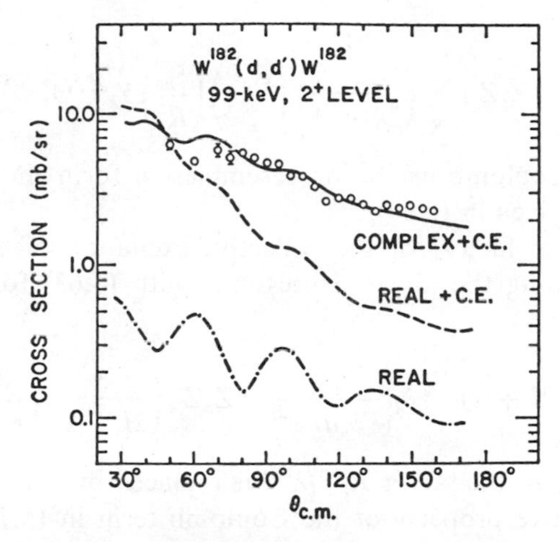

Fig. 5.11 Comparison of the DW theory with experimental data for $W^{182}(d, d')$ to the $l = 2$ first excited state of W^{182}, at 12 MeV bombarding energy. Calculations were performed with only the real nuclear optical potential deformed (REAL), with the deformed real potential plus Coulomb excitation (REAL + C.E.), and finally with the entire nuclear potential deformed and Coulomb excitation included (COMPLEX + C.E.). (From [61], Fig. 6.)

the ξ_{kq}, the volume integral of (5.69) becomes a surface integral,

$$\delta V_c(\mathbf{r}_\alpha) \approx e^2 Z_a \int \rho R_c^{\,2} \frac{\sum_{kq} \xi_{kq} Y_k^{q*}(\Theta', \Phi')}{|\mathbf{r}_\alpha - \mathbf{R}'|} \, d\Omega', \tag{5.70}$$

where ρ is the magnitude of the uniform charge density. With the insertion of the spherical harmonic expansion of $|\mathbf{r}_\alpha - \mathbf{R}'|^{-1}$, (5.70) yields

$$\delta V_c \approx e^2 Z_a \sum_{kq} \left(\frac{4\pi}{2k+1}\right)\left(\frac{R_c^{\,k}}{r_\alpha^{k+1}}\right)(R_c^{\,2}\rho\xi_{kq}) Y_k^{q*}(\Theta', \Phi') \tag{5.71a}$$

for $r_\alpha > R_c$, and

$$\delta V_c \approx e^2 Z_a \sum_{kq} \left(\frac{4\pi}{2k+1}\right)\left(\frac{r_\alpha^{\,k}}{R_c^{k+1}}\right)(R_c^{\,2}\rho\xi_{kq}) Y_k^{q*}(\Theta', \Phi') \tag{5.71b}$$

for $r_\alpha < R_c$. These expressions are simplified by inserting $\rho = (3Z_A/4\pi R_c^{\,3})$, the charge density of a uniform sphere of radius R_c. For $r_\alpha > R_c$, for example, we then obtain

$$\delta V_c = e^2 Z_a Z_A \sum_{kq} \left(\frac{3}{2k+1}\right)\left(\frac{R_c^{\,k}}{r_\alpha^{k+1}}\right)\left(\frac{\xi_{kq}}{R_c}\right) Y_k^{q*}(\Theta', \Phi'). \qquad (5.72)$$

This operator for Coulomb excitation resembles in form the operator for nuclear excitation, given in (5.60).

The complete form factor for the collective excitation of even nuclei is obtained by combining the above expressions with (5.63) for the nuclear form factor, giving

$$A_l F_l(r) = i^l (2l+1)^{-1/2}\, \delta_l \left\{-\frac{dU}{dr} + e^2 Z_a Z_A \left(\frac{3}{2l+1}\right)\left(\frac{R_c^{l-1}}{r^{l+1}}\right)\right\} \qquad (5.73)$$

for $r > R_c$. For $r < R_c$ the factor R_c^{l-1}/r^{l+1} is replaced by r^l/R_c^{l+2}.

The most distinctive property of the Coulomb term in (5.73) is its long range. As a result this term is effective for partial waves that do not overlap with the nuclear term. For this reason the Coulomb term is important at small scattering angles and at low bombarding energies.* Another result of the long range of the Coulomb term is that the convergence of the radial integrals of the DW calculation [see (5.44)] is controlled by the oscillations of the radial waves $f_{\beta L_\beta}$ and $f_{\alpha L_\alpha}$. Therefore the partial wave expansion for this term actually converges better at high bombarding energies than at low bombarding energies! To rough approximation, if $L_{\max}$ is the number of partial waves included in the DW calculation, then an accurate estimate of the effects of the Coulomb term is obtained [65] only for scattering angles greater than the classical angle of deflection for a projectile of angular momentum $L_{\max}$. This angle is $\theta_c \equiv 2n/L_{\max}$, where $n = Z_a Z_A e^2 \mu_\alpha/\hbar^2 k_\alpha$ is the usual Coulomb parameter. For $L_{\max} = 50$, for the case of 43-MeV alphas incident on Ni^{58}, we find $\theta_c = 6°$, and for 18-MeV alphas on Ni^{58} we find $\theta_c = 10°$. By contrast, for the treatment of the purely nuclear interaction no more than 30 partial waves are needed at 43 MeV, and no more than 20 at 18 MeV.

Figure 5.11 shows an example of the extreme importance of Coulomb excitation for inelastic scattering of deuterons by a heavy nucleus, even at the rather high bombarding energy of 12 MeV.

5.5 Inelastic Scattering of Nucleons

The distorted-waves formalism for the inelastic scattering of nucleons is identical with the formalism for the inelastic scattering of strongly-absorbed

* The classical Coulomb excitation experiments [64] are performed at such low bombarding energies that the nuclear interaction dU/dr makes negligible contributions in all partial waves. Naturally, small cross sections are obtained.

projectiles given in Sections 5.3 and 5.4. Thus, (5.40) is the appropriate expression for the DW amplitude. In that equation the projectile wavefunction $\phi_{s_0, m}$ now refers only to the orientation of the spin and i-spin of the nucleon. The multipole expansion of the amplitude is given by (5.41)–(5.43) as before.

The inelastic scattering of nucleons differs from the inelastic scattering of strongly-absorbed projectiles primarily in the greater clarity of physical interpretation that becomes possible because the projectiles are not composite. Microscopic derivations of the form factor become appropriate. Thus except for the CN influence in the operator $\overline{\mathcal{O}}$, which we do not know how to handle, the interaction V_β may be treated as a sum of nucleon–nucleon interactions, of either the sort familiar in nuclear structure studies or the sort of the free nucleon–nucleon t-matrix. The optical potentials with which the distorted waves are computed become related very closely to the self-consistent single-particle potentials of nuclear structure calculations. Therefore the calculation of inelastic scattering of nucleons tends to be a simple extension into the continuum of the kinds of calculations typical for bound states. Because the foundations of these calculations can be so definite, lengthy and detailed investigations of the inelastic scattering become worthwhile. Such investigations can go beyond the study of collective excitations of the nuclear surface and can treat excitations that have non-normal parity or a pronounced single-particle character, or those that involve other interesting approximate quantum numbers of the shell model.

The inelastic scattering of nucleons also tends to yield a richer body of experimental data than does the inelastic scattering of more complex projectiles. Polarizations often are measured. Angular distributions do not fall off rapidly with scattering angle, therefore measurements at large scattering angles are performed very frequently.

Because nucleons are not strongly absorbed the projectile wavefunctions can have good overlap with the nuclear interior. Experiments with nucleons therefore can test for possible localization of the interaction V_β and also can test those properties of the target-nucleus wavefunctions ψ_A and ψ_B that may be localized in the nuclear interior. Radial nodes and "interior mixing" of i-spin [66] are among properties of the latter kind. Unfortunately any sensitivity to the nuclear interior also implies a sensitivity to exchange effects. Good overlap between the projectile and the nucleons of the target nucleus may cause exchange terms to be large [67, 68]. These exchange terms are discussed in Chapter 10.

One further qualitative property of nucleon inelastic scattering that must not be overlooked is that a nucleon of given energy carries much less momentum than one of the composite, strongly-absorbed projectiles. Because of this, and because nucleons penetrate to smaller effective interaction radii than composite projectiles, the angular distributions obtained with nucleons tend to be less oscillatory than the ones described in Sections 5.3 and 5.4. The small momentum of nucleon projectiles is also a property that contributes to good overlap between these projectiles and the target nucleons, therefore it helps to cause large exchange terms.

We now proceed with the microscopic calculation of the form factors for inelastic scattering. We write for the interaction V_β the sum over two-body interactions

$$V_\beta = \sum_{i=1}^{A} v_{0i}(\mathbf{r}_{0i}, \boldsymbol{\sigma}_0, \boldsymbol{\sigma}_i, \boldsymbol{\tau}_0, \boldsymbol{\tau}_i), \tag{5.74}$$

in which 0 labels the projectile nucleon and i labels a nucleon of the target nucleus. The variables in v_{0i} are the displacement $\mathbf{r}_{0i} = \mathbf{r}_0 - \mathbf{r}_i$ and the spin and i-spin operators of the two nucleons. Bearing in mind that v_{0i} is an effective interaction, probably the "residual interaction" of shell-model studies, there is little reason to carry all the possible detail that might appear in this operator. It is customary to consider a simplified expression in which v_{0i} is a product composed of a local scalar function of $\mathbf{r}_{0i}$ multiplied by a linear combination of exchange operators:

$$v_{0i} = -[V_{00} + V_{01}(\boldsymbol{\tau}_0 \cdot \boldsymbol{\tau}_i) + V_{10}(\boldsymbol{\sigma}_0 \cdot \boldsymbol{\sigma}_i) + V_{11}(\boldsymbol{\sigma}_0 \cdot \boldsymbol{\sigma}_i)(\boldsymbol{\tau}_0 \cdot \boldsymbol{\tau}_i)]g(r_{0i}). \tag{5.75}$$

Here charge and spin exchange operators are employed, and a notation like that of Satchler and collaborators is adopted [45]. Then, to calculate the radial form factors defined in (5.43), the interaction V_β and the best available target-nucleus wavefunctions are inserted into the nuclear matrix element, and the multipole expansion of (5.43) is performed.

The exchange properties of the interaction v_{0i} are related to the exchange properties of the wavefunction. If the wavefunctions are fully antisymmetric in the space, spin, and charge coordinates of all pairs of nucleons, then an interaction written in terms of charge and spin exchange operators as in (5.75) is equivalent to one written in terms of space and spin exchange operators, because the operator identity $P_{0i}^{x}P_{0i}^{\sigma}P_{0i}^{\tau} = -1$ is valid. Here P_{0i}^{x} interchanges the space coordinates of nucleons 0 and i, P_{0i}^{σ} interchanges the spin coordinates, and P_{0i}^{τ} interchanges the charge coordinates. On the other hand, the exchange terms of the scattering wavefunction are being neglected in the present chapter (they are treated in Chapter 10), therefore the above-mentioned equivalence among alternative forms of the two-body interaction is broken. In this circumstance it is helpful to recognize that the expression chosen for (5.75) arranges the exchange operators of the problem in such a way that the terms we are omitting from the wavefunction are indeed the ones that would tend to give small contributions to the amplitude. Because charge and spin exchange operators are used in (5.75) the space-exchange operator

P_{0i}^x appears explicitly only in the wavefunction exchange terms, therefore it is only these exchange terms whose contributions to the amplitude might be reduced by poor spatial overlap of bound and continuum wavefunctions. For this reason it is at least *consistent* to use (5.75) for the interaction and to neglect wavefunction antisymmetrization. This conclusion obtains for both the ordinary inelastic scattering treated in the present section and the charge-exchange inelastic scattering treated in Section 5.7.

For nonexchange inelastic scattering it is not necessary to give much attention to the i-spin parts of ψ_A and ψ_B, and some simplification is available. Because particle 0 both enters and leaves as a proton, or both enters and leaves as a neutron, only the ζ-components of the $\boldsymbol{\tau}$ operators contribute to the nuclear matrix element. We find that

$$(\boldsymbol{\tau}_0 \cdot \boldsymbol{\tau}_i) = 1, \qquad \begin{cases} \text{if a proton interacts with a proton,} \\ \text{or if a neutron interacts with a neutron,} \end{cases}$$

$$(\boldsymbol{\tau}_0 \cdot \boldsymbol{\tau}_i) = -1, \qquad \text{if a proton interacts with a neutron.}$$

Therefore, (5.75) reduces to

$$v_{0i} = -[V_0 + V_1(\boldsymbol{\sigma}_0 \cdot \boldsymbol{\sigma}_i)]g(r_{0i}), \tag{5.76}$$

where $V_0 = V_{00} \pm V_{01}$ and $V_1 = V_{10} \pm V_{11}$ according to the rules discussed. The more complete expression of (5.75) is used in Section 5.7 in discussions of charge exchange.

The nuclear matrix element of (5.43) is then

$$\langle J_B M_B, \tfrac{1}{2}m_b| V_\beta |J_A M_A, \tfrac{1}{2}m_a\rangle$$
$$= -\langle \psi_{B,J_BM_B}\phi_{m_b}| \sum_{i=1}^{A} [V_0 + V_1(\boldsymbol{\sigma}_0 \cdot \boldsymbol{\sigma}_i)]g(r_{0i}) \,|\psi_{A,J_AM_A}\phi_{m_a}\rangle. \tag{5.77}$$

Certain selection rules are evident in (5.77) even before introducing the multipole expansion of the interaction. Thus, although V_β is a two-body operator, such that in first order V_β can change the orbits of two nucleons at a time, one of the two nucleons affected by V_β is always the projectile nucleon, labelled by the index 0. So far as the internal motion of the target nucleus is concerned V_β *is a one-body operator*, and its properties resemble those of the electromagnetic interaction or the beta decay interaction. If ψ_A and ψ_B are expanded in wavefunctions of the independent particle model, V_β will couple a given term of ψ_A only to those terms of ψ_B that differ from it in the motion of at most one nucleon. The properties of the form factors thus will be dominated by the properties of those few single-particle orbitals by which ψ_A and ψ_B differ; for example, because $g(r_{0i})$ tends to be short ranged, we see that if the active nucleons of ψ_A and ψ_B should all belong to the same single-particle orbital the form factors would tend to be of the shape of the square of the radial wavefunction of that orbital.

Interesting questions of coherence present themselves: Low-lying states of nuclei often are superpositions of many alternative single-nucleon excitations of the ground state. Each alternative excitation yields a contribution to the nuclear matrix element of (5.77). Large amplitudes for inelastic scattering can appear if these contributions should add constructively, because many nucleons of the target nucleus would then participate in the excitation process. A similar phenomenon occurs with the matrix elements of the usual electromagnetic multipole operators, especially with the operators $E1$, $E2$, $E3$, etc., that have normal parity. These operators have large matrix elements to the "collective" excited states of the nuclear spectrum, states whose wavefunctions can very nearly be described as products of the multipole operators times the wavefunction of the ground state. Because those multipoles of V_β that have normal parity weight the same radial regions of the nucleus* as do the electric l-pole operators, it may be expected that inelastic scattering and electromagnetic excitation will show collective enhancements of the same transitions [38]. Just such a correlation was found experimentally by Cohen and Rubin [37] in studies of the excitation of the lowest 2^+ states of nuclei by (p, p') reactions. In Fig. 5.12, taken from their paper, the inelastic cross section is seen to show a strong correlation with $B(E2)$.

There are some notable differences between the multipoles of V_β and the multipoles of the electromagnetic interaction. One difference lies in the i-spin dependence of the sum over nucleons. In (5.77) the weight factors in this sum are the interactions V_0 and V_1. In the corresponding sum over electromagnetic interactions, however, the weight factors are the charges and magnetic moments of the target nucleons, and these quantities have a much stronger dependence on i-spin.† This affects relative strengths of transitions. Another difference between the two types of interactions is that the spin-dependent term of V_β tends to be more important, relative to the spin-independent term, than in the case of electromagnetism.

To perform the multipole expansion of V_β we expand the scalar function $g(r_{0i})$ into multipoles:

$$g(r_{0i}) = 4\pi \sum_{lm} g_l(r_0, r_i)\, Y_l^m(\hat{r}_i)\, Y_l^{m*}(\hat{r}_0). \tag{5.78}$$

* The complete form of the operator that excites the nucleus is obtained by folding a multipole of V_β into the product of distorted waves $\chi_\beta^{(-)*}\chi_\alpha^{(+)}$. This operator gives greatest weight to the region of the nuclear surface, because the distorted waves are damped toward small r_0. The electric multipole operators also give great weight to the surface region because they are proportional to $r_0^{\,l}$.

† Clegg [69] noted that inelastic scattering and gamma ray absorption give particularly different probabilities for magnetic dipole excitations in which the i-spin does not change. (Such excitations correspond to the spherical tensor $T_{011,\nu}$ defined below.) There is almost complete cancellation among the terms that make up the $\Delta T = 0$ component of the magnetic dipole moment. The corresponding operator that governs inelastic scattering is not subject to these cancellations.

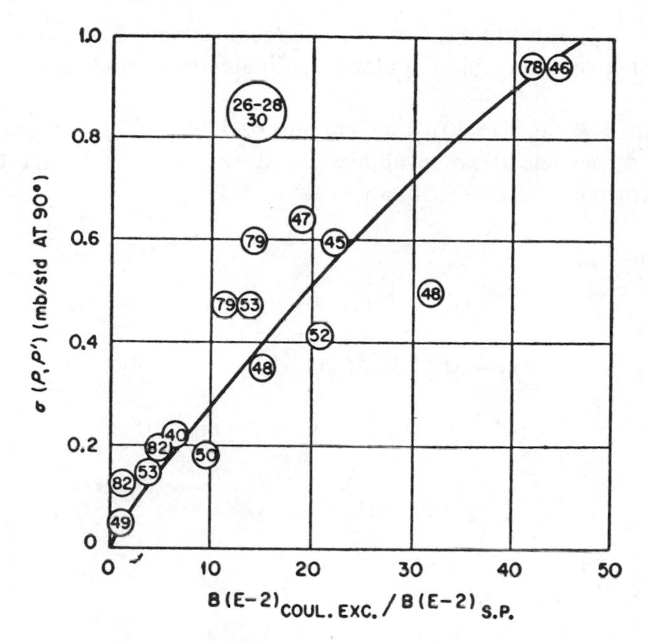

Fig. 5.12 Correlation between B(E2) and the cross section for proton inelastic scattering. (From [37], Fig. 9.)

Introduction of this expansion in the first term of (5.77) immediately yields the standard form of (5.43), with $s = 0$. We see, for example, that the total angular momentum transferred by the first term of (5.77) is $j = l$, and that the parity transfer is $(-)^j$, the condition known as "normal parity".

The second term of (5.77) uniquely couples up to spin transfer $s = 1$, and therefore is responsible for "spin-flip" excitations of the target nucleus. It is put into standard form by expressing the spin operators in terms of spherical components σ_μ, with

$$\sigma_1 = -\frac{\sigma_x + i\sigma_y}{\sqrt{2}}, \qquad \sigma_0 = \sigma_z, \qquad \sigma_{-1} = \frac{\sigma_x - i\sigma_y}{\sqrt{2}}.$$

These σ_μ transform under rotations like the $Y_1{}^\mu$. The scalar product becomes

$$(\boldsymbol{\sigma}_0 \cdot \boldsymbol{\sigma}_i) = \sum_\mu (-)^\mu \sigma_\mu(i)\sigma_{-\mu}(0). \tag{5.79}$$

With the aid of this expression, the multipole expansion becomes

$$g(r_{0i})(\boldsymbol{\sigma}_0 \cdot \boldsymbol{\sigma}_i) = 4\pi \sum_{l,\, j,\, \nu} g_l(r_0, r_i)(-)^{s+j+\nu} T_{lsj,\,\nu}(i) T_{lsj,\,-\nu}(0), \tag{5.80}$$

with

$$T_{lsj,\,\nu} \equiv \sum_{m,\, u} i^l Y_l{}^m \sigma^\mu \langle ls; m\, \mu \mid j\, \nu \rangle, \tag{5.81}$$

and with $s = 1$. Introduction of the spherical tensors $T_{lsj,\,\nu}$ has yielded a separation between the spin-angular coordinates of particle 0 and those of particle i.

Equations 5.78 and (5.80) now are inserted in (5.77), the spin matrix elements for particle 0 are evaluated, and the Wigner-Eckart theorem* is utilized for particles i. We obtain

$$
\begin{aligned}
&\langle J_B M_B, \tfrac{1}{2}m_b|\, V_\beta\, |J_A M_A, \tfrac{1}{2}m_a\rangle \\
&\quad = -4\pi\sqrt{2}\sum_{l,\,j} i^{-l} Y_l^{m*}(0)(-)^{1/2-m_b}(2J_B+1)^{-1/2} \\
&\qquad \times \langle J_A j; M_A, M_B - M_A \,|\, J_B M_B\rangle \\
&\qquad \times \left\{
\begin{aligned}
& V_0\langle B\| \sum_i g_l(r_0, r_i) i^l Y_l(i)\, \|A\rangle\langle \tfrac{1}{2}\tfrac{1}{2}; m_b, -m_b \,|\, 00\rangle \\
&\qquad\qquad \times \delta_{l,\,j}\,\delta_{m_b,\,m_a}\,\delta_{m,\,M_B-M_A} \\
& + V_1\langle B\| \sum_i g_l(r_0, r_i) T_{lsj}(i)\, \|A\rangle \\
& \times \langle \tfrac{1}{2}\tfrac{1}{2}; m_a, -m_b \,|\, 1, m_a - m_b\rangle \\
& \times \langle ls; m, m_a - m_b \,|\, j, M_B - M_A\rangle
\end{aligned}
\right\}, \qquad (5.82)
\end{aligned}
$$

Upon comparison with (5.43) the radial form factors for $s = 0$ and $s = 1$ are seen to be

$$
A_{l0j}F_{l0j}(r_0) = -4\pi\sqrt{2}V_0(2J_B+1)^{-1/2}\langle B\| \sum_i g_l(r_0, r_i) i^l Y_l(i)\, \|A\rangle\, \delta_{l,\,j}, \qquad (5.83)
$$

$$
A_{l1j}F_{l1j}(r_0) = -4\pi\sqrt{2}V_1(2J_B+1)^{-1/2}\langle B\| \sum_i g_l(r_0, r_i) T_{lsj}(i)\, \|A\rangle. \qquad (5.84)
$$

Often the nuclear states ψ_A and ψ_B are so simple that only one of the reduced matrix elements of (5.83) and (5.84) is nonzero. However, it also is often necessary to treat states that have appreciable configuration mixing, for which several multipoles have nonvanishing matrix elements. Therefore, it should be recalled (Section 4.5) that multipoles with different values of j make noninterfering contributions to the differential cross section of a distorted-waves calculation. If spin-orbit coupling should be omitted from the distorted waves the interference between different values of s and l also vanishes. Explicit calculations show† that the latter interferences sometimes are negligible even if spin-orbit optical potentials are used. Hence it is clear that the less important multipoles may safely be ignored.

* The Wigner-Eckart theorem is used in the symmetrical form

$$
\langle J_B, M_B|\, T_{j,\,\mu}\, |J_A, M_A\rangle = \langle J_A j; M_A \mu \,|\, J_B M_B\rangle(2J_B+1)^{-1/2}\langle J_B\|\, T_j\, \|J_A\rangle.
$$

† See footnote 17 of [45].

The reduced matrix elements of (5.83) and (5.84) are of a sort familiar from analyses of nuclear structure. While the evaluation of these particular matrix elements is not a routine aspect of structure analyses, there have been many specialized papers about them and about the inelastic-scattering cross sections they imply [44-54]. Three ingredients go into the discussion of these matrix elements: (a) the configuration mixtures that make up the states ψ_A and ψ_B; (b) the parameters of the effective two-body interaction; (c) the radial shapes of the single-particle wavefunctions that make up the states ψ_A and ψ_B. Topics (a) and (b) are related. Truncation of the configuration mixture often does not disturb the success of a structure analysis, but it does require modification of the two-body interaction that fits the structure. Analyses based on "realistic" two-body interactions [70] use fairly extensive mixtures of configurations. Topic (c) arises because of the presumed sensitivity of nuclear reaction calculations to overlaps at large radii. Insuring accuracy at large radii seems to require the use of single-particle wavefunctions governed by some realistic potential, like the Woods-Saxon potential, rather than that of wavefunctions of the harmonic oscillator potential. In fact, however, experience has shown [45, 46, 71] that calculations of the inelastic scattering of nucleons are not affected much by the introduction of wavefunctions that have realistic tails. In part this occurs because V_β has nonzero range, and in part because nucleons are weakly absorbed and the overlaps that contribute to nucleon inelastic scattering receive large contributions from the entire volume of the target nucleus. The use of realistic single-particle wavefunctions only seems to be important for particle-transfer reactions and for the inelastic scattering of strongly-absorbed particles. It is not discussed further in the present section.

The radial shapes of the single-particle wavefunctions also are affected by the Perey effect (Section 5.2). This effect, a consequence of nonlocality of the single-particle potential, causes a reduction of the magnitudes of single-particle wavefunctions in the nuclear interior. Because bound single-particle wavefunctions nevertheless remain normalized to unity the net result of the Perey effect is an enhancement of the tails of these wavefunctions. Although there has not been much experience with the introduction of this effect for bound wavefunctions it was carried by Johnson, Owen, and Satchler [45] in their derivation of the form factors for Zr^{90} (p, p'). In their work the Perey effect was found to give fifty percent reduction in the relative importance of contributions from the nuclear interior.

The article of Johnson, Owen, and Satchler [45] treats seventeen states of Zr^{90} that are excited by the inelastic scattering of 18.8 MeV protons. Five of these seventeen states receive special attention. These are states of spin 0^+, 2^+, 4^+, 6^+, and 8^+ that are believed to be formed from a single configuration composed of closed shells plus two protons in the open $g_{9/2}$ shell. These five states have the interesting property that only the spin-independent term of (5.77) can contribute to their excitation, and only the multipole with $l = J_B$ can contribute to the excitation of the level that has spin J_B. It therefore becomes possible to determine from experiment the range and shape of the interaction function $g(r_{0i})$. The authors favor a Yukawa interaction with a range of about 1 fm and a depth of about $V_0 = 200$ MeV. Thus they arrive at an effective interaction several times stronger than the free $^1S_0$ nucleon-nucleon interaction. Satchler has published a general study of the applicability of this type of effective interaction for nucleon inelastic scattering [72].

Several much earlier studies of nucleon inelastic scattering by C^{12} and Mg^{24} [4, 73] also needed to use unusually large strengths for the effective two-body interaction in order to fit the observed magnitudes of cross sections. In each case it became clear later that the wavefunctions ψ_A and ψ_B had been oversimplified [38, 39, 42]. Deformations of the "closed-shell" core had been ignored. These core deformations are induced by the nucleons in the open shells, and enhance the transition matrix elements associated with these shells. Because (5.83) and (5.84) sum over all the nucleons of the target nucleus, it is easily understood that the small contributions of a large number of core nucleons can add up coherently and cause major changes of the magnitudes of the matrix elements. It is just such enhancements that cause the effective charges often required for the very closely related matrix elements of the electromagnetic multipoles. It is likely that such coherent core contributions will affect *all studies of inelastic scattering*. It may be that the strong two-nucleon effective interaction found by Johnson et al. [45] is an indication of the omission of core contributions.* It also may be an indication of the omission of exchange contributions (see Section 10.3).

Core excitations also were omitted by Funsten, Roberson, and Rost [42] in their analysis of the scattering of 17.5 MeV protons by Ti^{50}, Cr^{52}, Fe^{54}, and V^{51}. The wavefunctions of these four nuclei were treated as having $N = 28$ closed neutron shells and active holes in the $f_{7/2}$ proton shell. Thus only one configuration was used in the description of each nucleus. The inelastic scattering then was fitted with a two-nucleon interaction of Gaussian form $V_0 g(r) = V_G \exp(-\gamma^2 r^2)$ having $V_G = 45$ MeV, but with the rather large range $\gamma^{-1} = 1.78$ fm.

In general, it may be hoped that realistic nucleon-nucleon interactions can be used in studies of nucleon inelastic scattering, provided enough detail is included in the wavefunctions ψ_A and ψ_B. This hope is in line with the experience with nuclear structure calculations that use realistic interactions [70].

* Subsequent work by Satchler and others makes a regular practice of carrying core-excitation effects in inelastic scattering analyses.

The strong 1^-, 2^+, 3^-, etc., collective excitations of nuclei are induced very easily by the inelastic scattering of nucleons. The macroscopic model of these excitations, based on the concept of a deformed optical potential (Section 5.3), generally gives form factors that yield quite good fits to the shapes and magnitudes of the observed cross section [42, 54, 63, 74]. Several authors have developed equivalent microscopic theories of these form factors [38–41, 46–48], along the lines discussed above, based upon elaborate studies of the wavefunctions ψ_A and ψ_B. It is gratifying that these microscopic theories have had a considerable record of success, and that the form factors that the microscopic theories yield for the enhanced transitions strongly resemble those yielded by the macroscopic theories.* One simplification that appears in the calculation of the enhanced transitions is that the contributions yielded by the poorly determined V_1 term of (5.77) do not tend to combine constructively [38, 46].

The microscopic theory of form factors for inelastic scattering, described above, has been applied for other projectiles than nucleons [40, 41, 43, 44, 46, 47, 48]. Indeed, most of the papers that discuss this microscopic theory treat several projectile types. However, as mentioned in Section 5.3, its use for strongly-absorbed projectiles is questionable. For these projectiles the sensitivity to the shapes of radial wavefunctions is enhanced. Also, because these projectiles are composite there is little a priori guidance as to the interaction V_β that should be used.

There is one further difficulty with the use of microscopic form factors in a distorted-waves theory that greatly affects their application for strongly-absorbed projectiles and also somewhat affects their application for nucleons: Because direct reaction theories are constructed in a truncated Hilbert space, in the continuum, all the potentials in these theories should have anti-Hermitian parts. However, V_β of the microscopic DW theory does not include any such anti-Hermitian parts. This omission of operators that would compensate for truncation errors is inevitable in a procedure that merely extrapolates the shell model into the continuum, and it is an essential limitation of the microscopic DW theory. The macroscopic theory is more complete because it is based on the full, complex optical potential. (For further discussion of the difficulties of microscopic form factors see [37] in Chapter 9.)

5.6 Distorted-waves Impulse Approximation

If their relative kinetic energy is very high ($\gtrsim$100 MeV), two nucleons inside a nucleus scatter as if they were in free space. This occurs because the

* For nonenhanced transitions the form factors yielded by the macroscopic and microscopic theories often differ greatly.

spectrum of intermediate states available for the scattering of high-energy nucleons is so extensive that the nuclear-structure effects modifying the low-energy end of this spectrum become of no importance. For small separations the wavefunction for the relative motion of the two nucleons then becomes identical with the free-space wavefunction. As a result, the effective two-body interactions v_{0i} in V_β of (5.74) reduce to the free-space t-matrix operator that describes nucleon-nucleon scattering. Thus

$$v_{0i} \approx t_{0i}, \tag{5.85}$$

where, by definition,

$$t_{0i}e^{i(\mathbf{k}\cdot\mathbf{r}_{0i})}\,\phi(0, i) = V_{0i}\psi\,(\mathbf{r}_{0i}, \boldsymbol{\sigma}_0, \boldsymbol{\sigma}_i, \boldsymbol{\tau}_0, \boldsymbol{\tau}_i)\,\phi(0, i). \tag{5.86}$$

Here $\phi(0, i)$ is the spin and i-spin wavefunction for the two colliding nucleons, V_{0i} is the exact nucleon-nucleon interaction, and $\psi(\mathbf{r}_{0i}, \boldsymbol{\sigma}_0, \boldsymbol{\sigma}_i, \boldsymbol{\tau}_0, \boldsymbol{\tau}_i)$ is the exact two-nucleon wavefunction. The definition of t_{0i} is chosen so that its Born approximation matrix elements are identical with the exact matrix elements of V_{0i}.

The definiteness of V_β at high energy suggests that at high energy quite definite and reliable microscopic calculations of nucleon inelastic scattering should be available. Such calculations promise to be even more reliable because distortion of the projectile wavefunctions $\chi_\alpha^{(+)}$ and $\chi_\beta^{(-)}$ becomes reduced as the energy increases, so that these wavefunctions become less dependent upon vagaries of optical potentials. It therefore appears that high-energy inelastic scattering should be a valuable means for the investigation of the wavefunctions ψ_A and ψ_B. Many authors have organized calculations based on use of the t-matrix operator. Most of these calculations handle $\chi_\alpha^{(+)}$ and $\chi_\beta^{(-)}$ in WKB approximation. These calculations are described in Chapter 7. The present section describes those few [75–77] calculations that handle the projectile wavefunctions straightforwardly, by numerical solution of the Schrödinger equation in partial wave expansion, as in earlier sections of this book. It is interesting that these DWIA calculations automatically include the important knockon exchange term (see Section 10.3).

Authors occasionally use t-matrix operators that describe only a limited portion of the known nucleon-nucleon scattering data. In such work it is customary to incorporate only very crude treatments of the known strong spin-dependence of the nucleon-nucleon interaction. Because such work discards the clarity and definiteness that provide the principal interest of a t-matrix approach it is not discussed further.

The operator t_{0i} is not a simple, local function of the displacement $\mathbf{r}_{0i}$, of the type used to build V_β of Sections 5.3–5.5. Instead, it possesses the full generality allowed by quantum mechanics; it is a matrix, labelled by the two variables $\mathbf{r}_{0i}$ and $\mathbf{r}'_{0i}$. For this reason, a DW calculation based on the t-matrix does not possess the zero-range form that is automatic in the earlier treatments of inelastic scattering. Instead of (5.40), the DW amplitude for inelastic scattering is

$$T_{\alpha\beta}{}^{DW} = \int d^3r_0\, d^3r_0'\chi_\beta^{(-)*}(\mathbf{k}_\beta, \mathbf{r}_0)\langle\psi_B\phi_{s0,\, m_b}|$$
$$\times \sum_{i=1}^{A} t(\mathbf{r}_0 - \mathbf{r}_i, \mathbf{r}_0' - \mathbf{r}_i')\, \delta(\mathbf{r}_0 + \mathbf{r}_i - \mathbf{r}_0' - \mathbf{r}_i')$$
$$\times\ |\psi_A\phi_{s0,\, m_a}\rangle\chi_\alpha^{(+)}(\mathbf{k}_\alpha, \mathbf{r}_0'), \qquad (5.87)$$

where the variables $\mathbf{r}_i$ appear in ψ_B, the variables $\mathbf{r}_i'$ appear in ψ_A, and the scalar-product notation indicates integration over these variables. However, it is possible to introduce a "pseudo" zero-range approximation that greatly simplifies the evaluation of (5.87).

The reduction to pseudo zero-range approximation is described at length by Kawai, Terasawa, and Izumo [75]. The t-matrix is discussed in momentum representation, the representation in which the results of scattering experiments are quoted most easily. It is noted that the momenta of the incident and emerging projectiles are so much higher than the momenta of bound nucleons that the latter momenta may be omitted from the matrix elements of t. It is also noted that the projectile momenta are much higher than the momenta that can be transferred by the optical potentials. Because of this the distributions of the arguments of the t-matrix cluster closely around the values of the asymptotic incoming and outgoing momenta $\mathbf{k}_\alpha$ and $\mathbf{k}_\beta$, and it is a good approximation to replace these distributions by the values of the asymptotic momenta. In this *asymptotic approximation* the t-matrix is then independent of location in the nuclear interior, and is equivalent, in coordinate space, to an interaction whose range is much smaller than the interval over which there is significant variation of ψ_A and ψ_B. However, this approximate interaction still cannot be said to have "zero range," because its range could be large compared with the wavelength k_α^{-1}. Such a departure from zero range would introduce a dependence of the scattering amplitude on the momentum transfer,

$$\mathbf{q} = \mathbf{k}_\alpha - \mathbf{k}_\beta. \qquad (5.88)$$

In fact, a strong $\mathbf{q}$-dependence of the amplitude is well known. This $\mathbf{q}$-dependence is most conveniently handled by treating the simplified interaction as a zero-range interaction (in coordinate space), multiplied by an explicitly $\mathbf{q}$-dependent factor (in momentum space).†

† Haybron [78] remarks that this strong $\mathbf{q}$-dependence also affects the use of asymptotic momenta in the arguments of the t-matrix. Avoidance of the asymptotic approximation introduces additional "finite-range effects" that are essential at 50 MeV and important over the entire range 50–100 MeV.

The above approximation restores the DW amplitude for inelastic scattering to the form of (5.40), with V_β a sum of interactions v_{0i}, and with

$$v_{0i} = t_{0i}(\mathbf{q}, E)\, \delta(\mathbf{r}_0 - \mathbf{r}_i). \tag{5.89}$$

Here t_{0i} is an operator in the spins of the two colliding nucleons. The form of this operator, compatible with various invariance principles, is given by

$$t_{0i} = \frac{2\hbar^2}{(2\pi)^2 m_p} [A + B(\boldsymbol{\sigma}_0 \cdot \hat{n})(\boldsymbol{\sigma}_i \cdot \hat{n}) + C(\boldsymbol{\sigma}_0 \cdot \hat{n} + \boldsymbol{\sigma}_i \cdot \hat{n}) + E(\boldsymbol{\sigma}_0 \cdot \hat{q})(\boldsymbol{\sigma}_i \cdot \hat{q}) + F(\boldsymbol{\sigma}_0 \cdot \hat{p})(\boldsymbol{\sigma}_i \cdot \hat{p})]. \tag{5.90}$$

The unit vectors $\hat{q}$, $\hat{n}$, and $\hat{p}$ form an orthogonal coordinate system, given by

$$\hat{q} = \frac{\mathbf{q}}{q}, \quad \hat{n} = \frac{[\mathbf{k}_\alpha \times \mathbf{k}_\beta]}{|\mathbf{k}_\alpha \times \mathbf{k}_\beta|}, \quad \hat{p} = \hat{q} \times \hat{n}. \tag{5.91}$$

The coefficients A, B, C, E, and F parametrize the dependence of t_{0i} on E and q, the incident energy and the momentum transfer. These coefficients are discussed and tabulated by Kerman, McManus, and Thaler [79], on the basis of analyses of nucleon-nucleon scattering data. The quantity m_p in (5.90) is the proton rest mass.

Naturally, nucleon-nucleon scattering only determines A, B, C, E, and F *on the energy shell*, that is, for $k_\beta = k_\alpha$. However, in the application to inelastic scattering we have $k_\beta \neq k_\alpha$. In principle we therefore should have to use a t-matrix that depends on three variables: (a) the initial energy of the projectile, (b) the momentum transfer, and (c) the final energy of the projectile. Because $t_{0i}(E, \mathbf{q})$ of (5.89) is indicated as depending on only two of these three variables, it tacitly incorporates the assumption that departures from the energy shell are so slight they cause only negligible alterations of the operator. This assumption probably is correct. Calculations with potentials that serve as models of the off-energy-shell behavior of the t-matrix suggest that its dependence upon $(k_\alpha - k_\beta)$ is comparable with its rather slow dependence upon k_α alone.

With the insertion of (5.89) into (5.87) the DW amplitude for inelastic scattering becomes

$$T_{\alpha\beta}{}^{DW} = \int d^3r_0 \chi_\beta^{(-)*}(\mathbf{k}_\beta, \mathbf{r}_0) \times \langle \psi_B \phi_{s_0, m_b} | \sum_{i=1}^{A} t_{0i}(\mathbf{q}, E)\, \delta(\mathbf{r}_0 - \mathbf{r}_i) \,| \psi_A \phi_{s_0, m_a} \rangle \chi_\alpha^{(+)}(\mathbf{k}_\alpha, \mathbf{r}_0). \tag{5.92}$$

Despite its apparently simple structure this expression differs greatly from (5.40), the equivalent DW expression that utilizes local potentials v_{0i}. Because the operator $t_{0i}(\mathbf{q}, E)$ depends on the unit vectors $\hat{q}$, $\hat{n}$, and $\hat{p}$, based on the asymptotic incoming and outgoing momenta, it is in general not scalar in the spin and position coordinates of the two interacting nucleons. As a result the mechanism by which the projectile and the target nucleus share their angular momentum is more complicated than in (5.40). One interesting complication is that the appearance of unequal values for the coefficients B, E, and F in (5.90) corresponds to the presence of the tensor force in the nucleon-nucleon interaction. It also is important that a multipole expansion of (5.92), an expansion that parallels the reduction from (5.40) to (5.41)–(5.43), must depend explicitly on the directions of $\hat{q}$, $\hat{n}$, and $\hat{p}$. Articles [75–77] that discuss the distorted-waves impulse approximation present the modified multipole expansion.

Applications of the distorted-waves impulse approximation for studies of (p, p') experiments in the energy range 100–200 MeV have been encouraging. For example, Kawai, Terasawa, and Izumo [75] studied the spin-flip excitation of strongly-excited M1 levels of nuclei in the $(2s, 1d)$ shell. Not only does the DWIA give the correct forward-peaked angular distributions for the excitation of these levels, but the introduction of standard high-quality models for the wavefunctions ψ_A and ψ_B immediately gives agreement with the magnitudes of the (p, p') cross sections. An even more elaborate study of the excitation of 2^+ and 3^- levels of C^{12}, O^{16}, and Ca^{40} was carried through by Haybron and McManus [77] using wavefunctions ψ_A and ψ_B that had previously been calculated by Gillet and collaborators [80]. These wavefunctions diagonalize among particle-hole excitations of the closed-shell core and incorporate extensive configuration mixing. Figure 5.13 shows an example of the collective enhancement that the configuration mixing introduces into the calculation of $C^{12}(p, p')$. The lowest curve is obtained if the simplest *j-j* coupled shell model wavefunctions are used. The middle curve is obtained if configuration mixing is carried in the excited state. The uppermost curve, which agrees with experiment, is obtained if configuration mixing is carried both in the excited state and in the "closed shell" ground state.

Except at diffraction minima the above calculations do not show much sensitivity to the distortion effects in the wavefunctions $\chi_\alpha^{(+)}$ and $\chi_\beta^{(-)}$. In particular, omission of the real part of the optical potential causes almost no alterations of the results; this indicates that refraction by the optical potential has little effect on the projectile wavefunctions. Absorption is more important; however, it causes only a fifty percent reduction of the peak cross section.

Agreement with the polarization of the inelastically scattered protons is not as good as with the differential cross section. It is embarrassing that the simple

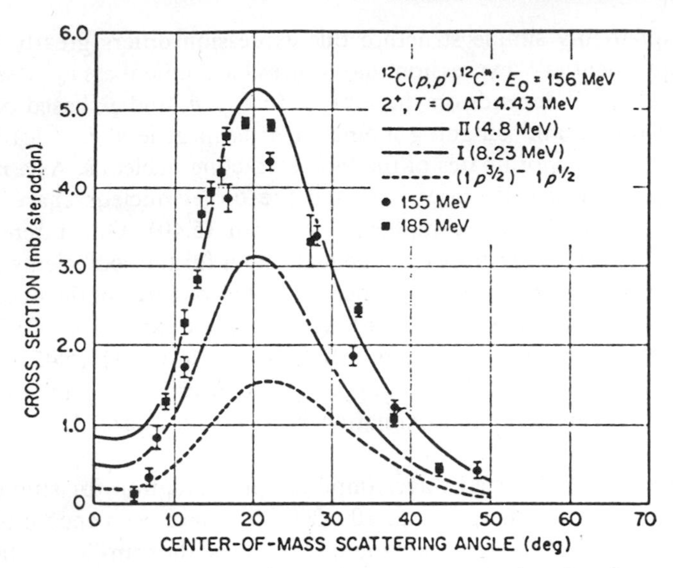

Fig. 5.13 Collective enhancement of the (p, p') cross section for the first excited state of C^{12}, at 156 MeV bombarding energy; described in the text. (From [78], Fig. 1.)

collective model (Sections 5.3 and 5.4) gives as good agreement with the differential cross sections as do the elaborate microscopic calculations just described, and that it gives better agreement with the polarizations. Of course, in this comparison of these two types of calculations, perfect agreement would serve to demonstrate the reasonableness of the more fundamental, microscopic interpretation of the collective excitations of nuclei. The disagreement that is found seems to suggest, as noted several times previously, that present microscopic models of nuclear reactions are affected to a serious extent by their omission of absorptive effects in the reaction mechanism.

5.7 Charge Exchange Scattering

Charge exchange is like inelastic scattering in that the reaction causes no change of the mass number of the projectile. However, the projectile does suffer a change in the projection of its *i*-spin. As a result the charge of the projectile may change by one or more units. Several common reactions, such as (p, n) or (He^3, H^3), change the charge by one unit. Several other reactions that would change the charge by two units are under active consideration. These would be of the type (π^+, π^-) or (C^{14}, O^{14}). Several other common projectiles of nuclear physics, such as deuterons or alpha particles, have *i*-spin zero and do not participate in charge exchange scattering.

The general distorted-waves discussion of Section 4.5 simplifies for charge exchange scattering in much the same fashion that is found for ordinary inelastic scattering at the beginning of Section 5.3. The simplifications are most straightforward if the *i*-spin notation is used. In this case the projectiles in channels α and β may be regarded as identical except for the projection of *i*-spin. Then the channel displacement variables of (4.53) become identical, so that

$$\mathbf{r}_\alpha = \mathbf{r}_\beta = \mathbf{r}_0,$$

and the Jacobian reduces to $\mathscr{J} = 1$. The projectile wavefunctions become identical except for spin and *i*-spin orientations, so that

$$\psi_a = \phi_{s_0,\, m_a}\xi_{t_0,\, \tau_a}, \qquad \psi_b = \phi_{s_0,\, m_b}\xi_{t_0,\, \tau_b}, \tag{5.93}$$

where τ_a and τ_b are the initial and final projections of the *i*-spin t_0. The distorting potential U_β may be chosen isoscalar; with this choice it has no matrix elements that link ψ_A and ψ_B, and it drops out of the calculation. Upon inserting these simplifications (4.53) takes the zero-range form

$$T_{\alpha\beta}{}^{DW} = \int d^3r_0 \chi_\beta^{(-)*}(\mathbf{k}_\beta, \mathbf{r}_0)\langle \psi_B \phi_{s_0,\, m_b}\xi_{t_0,\, \tau_b} | V_\beta | \psi_A \phi_{s_0,\, m_a}\xi_{t_0,\, \tau_a}\rangle \chi_\alpha^{(+)}(\mathbf{k}_\alpha, \mathbf{r}_0). \tag{5.94}$$

Here V_β is the net effective interaction between the (labelled) nucleons of ϕ and those of ψ. It consists of all those terms of $(H_{PP} + \overline{\mathscr{O}})$ that vanish as $r_0 \rightarrow \infty$. Nonvanishing contributions to the amplitude are obtained from those terms of V_β that are able to change the *i*-spin orientation of the projectile.

It must be emphasized that (5.94) is the *direct* charge-exchange amplitude. Although the incoming projectile may be a proton and the outgoing projectile a neutron, we have given these two projectiles the same set of spatial coordinates. Identical sets of spatial coordinates also appear in the wavefunctions of the target and residual nuclei ψ_A and ψ_B. Therefore the wavefunctions of the two projectiles, and the wavefunctions of the target and residual nuclei, have good spatial overlap. Antisymmetrization would require combining (5.94) with the *knockon amplitude*† (sections 4.8 and 5.12, and Chapter 10), the amplitude in which $\mathbf{r}_\alpha \neq \mathbf{r}_\beta$, and in which $\mathbf{r}_\alpha$ appears in ψ_B and $\mathbf{r}_\beta$ appears in ψ_A.

The interaction V_β reorients the *i*-spin of the projectile but does not change its magnitude. On the other hand, not only does V_β change the orientation of

† A number of published articles reverse the terminology, and refer to the knockon terms as "direct" and to (5.94) as an "exchange" term. This usage is based on the observation that (5.94) is a matrix element of an exchange term of V_β. In any case, (5.94) is the large term of the amplitude.

the target nucleus *i*-spin, but it also can change the total *i*-spin of this nucleus. Thus V_β can couple ψ_A to product nuclei ψ_B whose total *i*-spin either is or is not identical with that of ψ_A; for example, with projectiles that have $t_0 = \frac{1}{2}$ the total *i*-spin of the target nucleus can change by one unit. Particular interest nevertheless attaches to reactions in which the total *i*-spin of the target nucleus does not change, because in such cases there is often so great a similarity between ψ_A and ψ_B that large cross sections are obtained. For such reactions, except for the *i*-spin orientation, it may be either that ψ_B is identical with ψ_A, in which case ψ_B is known as the *analogue* of ψ_A, or that ψ_B is identical with an excited state of ψ_A. In the former case the charge-exchange scattering is the analogue of elastic scattering. In the latter case ψ_B is the *analogue of the excited state* of ψ_A, and the reaction is the analogue of inelastic scattering. Large coherent cross sections are just as likely in these cases as in cases of ordinary elastic scattering or ordinary inelastic excitation of low-lying collective levels.†

The nuclear matrix element in (5.94) may be expressed in terms of form factors of definite multipolarity, as is done for ordinary inelastic scattering. Nuclear structure theories may be devised that lead either to microscopic or macroscopic calculations of these form factors. The microscopic theories are based on the $(\boldsymbol{\tau}_0 \cdot \boldsymbol{\tau}_i)$ terms in the two-nucleon interaction v_{0i} of (5.75). The macroscopic theories are based on an *i*-spin dependent term of the optical potential [79, 81–83] in much the same way that inelastic scattering by deformed nuclei is based on deformation-dependent terms of the optical potential.

As a first example of charge-exchange reactions we consider the microscopic theory of the (p, n) reaction. The insertion of (5.74) and (5.75) in (5.94) yields

$$T_{\alpha\beta}{}^{DW} = -\int d^3r_0 \chi_\beta^{(-)*}(\mathbf{k}_\beta, \mathbf{r}_0) \langle \psi_B \phi_{s_0, m_b} \xi_{\frac{1}{2}, -\frac{1}{2}} | \times \sum_{i=1}^{A} (\boldsymbol{\tau}_0 \cdot \boldsymbol{\tau}_i)[V_{01} + V_{11}(\boldsymbol{\sigma}_0 \cdot \boldsymbol{\sigma}_i)] g(r_{0i}) | \psi_A \phi_{s_0, m_a} \xi_{\frac{1}{2}, \frac{1}{2}} \rangle \chi_\alpha^{(+)}(\mathbf{k}_\alpha, \mathbf{r}_0) \tag{5.95}$$

† Of course, the analogues of discrete states of nucleus A may not be discrete states of nucleus B. Nucleus A has Z protons and N neutrons, and its low-lying states have *i*-spin $\frac{1}{2}(N - Z)$. Nucleus B has $(Z + 1)$ protons and $(N - 1)$ neutrons, and its low-lying states have *i*-spin $\frac{1}{2}(N - Z - 2)$. The analogues of the states of nucleus A lie high in the spectrum of nucleus B, where the proton single-particle spectrum of B begins to overlap the neutron single-particle spectrum of A. At this high excitation the analogue states lie amid a dense set of states of the lower *i*-spin. Hence an analogue "state" really is an *excitation* distributed over many states of nucleus B. For heavy nuclei the energy of the analogue excitation is high enough to allow emission of protons from nucleus B.

Because the charge states of the incident and emerging projectiles are specified we immediately obtain a partial evaluation of the matrix element for the i-spin operator. We have

$$\boldsymbol{\tau}_0 \cdot \boldsymbol{\tau}_i = \tfrac{1}{2}(\tau_0^+\tau_i^- + \tau_0^-\tau_i^+) + \tau_0^z\tau_i^z,$$

where $\tau^\pm \equiv (\tau^x \pm i\tau^y)$ are the raising and lowering operators. In terms of these operators (5.95) reduces to

$$T_{\alpha\beta}^{DW} = -\int d^3r_0 \chi_\beta^{(-)*}(\mathbf{k}_\beta, \mathbf{r}_0)\langle \psi_B \phi_{s_0, m_b}|$$
$$\times \sum_{i=1}^{A} \tau_i^+[V_{01} + V_{11}(\boldsymbol{\sigma}_0 \cdot \boldsymbol{\sigma}_i)]g(r_{0i})\,|\psi_A \phi_{s_0, m_a}\rangle \chi_\alpha^{(+)}(\mathbf{k}_\alpha, \mathbf{r}_0). \quad (5.96)$$

The structure of (5.96) is seen to be identical with that of (5.40), therefore the multipole expansion of (5.43) can be applied without modification. Radial form factors are obtained by the same series of steps by which (5.77) led to (5.83), (5.84). The radial form factors for spin transfers $s = 0$ and $s = 1$ are found to be

$$A_{l0j}F_{l0j}(r_0) = -4\pi\sqrt{2}V_{01}(2J_B + 1)^{-1/2}\langle B\| \sum_i \tau_i^+ g_l(r_0, r_i) i^l Y_l(i) \,\|A\rangle\, \delta_{l,j}, \quad (5.97)$$

$$A_{l1j}F_{l1j}(r_0) = -4\pi\sqrt{2}V_{11}(2J_B + 1)^{-1/2}\langle B\| \sum_i \tau_i^+ g_l(r_0, r_i) T_{lsj}(i) \,\|A\rangle. \quad (5.98)$$

The nuclear matrix elements in these form factors are much the same as the ones discussed in connection with (5.83), (5.84). Once again these are matrix elements of single-particle operators, and they are related closely to matrix elements that appear in electromagnetic transitions and β decay. Once again the properties of the form factors are dominated by the properties of those few single-particle orbitals by which ψ_A and ψ_B differ. Closed shells common to the configurations that make up ψ_A and ψ_B do not contribute to these form factors.

Because i-spin and permutation symmetry are related closely, we must expect the role of the τ_i^+ operators in the nuclear matrix elements to be simple only if the space and spin properties of these matrix elements are simple under permutations. Monopole excitations provide a case with clearly marked simplicity [84]. For this case $l = j = 0$ and $s = 0$, and (5.97) reduces to

$$A_{000}F_{000}(r_0) = -4\pi\sqrt{2}V_{01}(2J_B + 1)^{-1/2}\langle B\| \sum_i \tau_i^+ g_0(r_0, r_i) \,\|A\rangle. \quad (5.99)$$

Further reduction occurs if nuclei A and B possess only one open shell, so that the active particles that contribute to (5.99) all have radial wavefunctions of the same form. When $g_0(r_0, r_i)$ is folded in with these (identical) radial wavefunctions the result is independent of i and factors out from the summation. Equation 5.99 reduces to

$$A_{000}F_{000}(r_0) = -4\pi\sqrt{2}V_{01}G_0(r_0)(2J_B + 1)^{-1/2}\langle B\| \, 2T^+ \, \|A\rangle, \quad (5.100)$$

where

$$\mathbf{T} \equiv \tfrac{1}{2}\sum_{i=1}^{A} \boldsymbol{\tau}_i. \quad (5.101)$$

The only operator that now remains in the matrix element is the raising operator for the total i-spin. The matrix elements of this operator depend on the i-spin projection of nucleus A. However, the projection of T_A is determined by the neutron excess of nucleus A, and normally is equal to $-T_A$. For this case we find

$$A_{000}F_{000}(r_0) = -16\pi V_{01}G_0(r_0)(2J_B + 1)^{-1/2}\sqrt{T_A}. \quad (5.102)$$

Equation 5.102 is the form factor for a (p, n) transition to the state of nucleus B that is the analogue of ψ_A. We see that this form factor depends on the interaction V_{01} and on the radial overlap integral $G_0(r_0)$. The properties of the overlap integral depend, in turn, on the properties of the radial wavefunctions of the orbitals active in the (p, n) transition. These radial wavefunctions often have properties that cause $G_0(r_0)$ to be surface peaked. Evidence exists that the charge-exchange form factor does tend to be surface peaked [85–87].

Monopole excitations of analogue states of even-even nuclei are very strong and are well known experimentally [88, 89]. Strong analogue excitations also appear with odd-A nuclei, however, the presence of nonzero spin allows other multipoles to contribute (incoherently) to the cross section, along with the monopole term [87, 90, 91].

If ψ_A contains two or more open shells we can obtain monopole (p, n) transitions to other states of nucleus B besides the analogue state. Each open shell α contributes its own characteristic radial overlap integral $G_0^{(\alpha)}(r_0)$ in the matrix element of (5.99). Although all these overlap integrals are positive they may have different magnitudes. As a result they may not factor out from the summation over i, and we may not necessarily obtain, as in (5.100), an operator proportional to the total i-spin operator of the nucleus. What happens in this more general case is that although the monopole (p, n) transition separately conserves the total i-spin of each open shell it may recouple these open shells to yield an altered value for the total i-spin of the nucleus. French and Macfarlane [84] give a thorough discussion of "unfavored states" produced in this fashion by the monopole (p, n) reaction. In Lane's terminology these are called "configuration states" [83, 89].

Excitations of higher multipolarity do not reduce in quite such general terms as the monopole excitations; for example, let us take $l = j \neq 0$ and examine the $l0j$ form factor for the case in which nuclei A and B possess only one open shell. Once again the radial overlap integral factors out from the summation, but now we are left with

$$A_{l0j}F_{l0j}(r_0) = -4\pi\sqrt{2}V_{01}G_l(r_0)(2J_B+1)^{-1/2}\langle B\| \sum_i \tau_i^+ Y_l(i) \|A\rangle. \quad (5.103)$$

Because the operators $Y_l(i)$ can cause angular momentum recoupling, they can link ψ_A with states ψ_B whose permutation symmetry is different from that of ψ_A, states whose total i-spin therefore is different. This is appropriate, because a given shell can contain many states that have the same multipolarity but different i-spins. Equation 5.103 connects ψ_A with these different states ψ_B. Obviously the analysis of individual cases is straightforward. Theoretical discussions are given, for example, in [47, 92].

The value used for V_{01} in (5.97) was discussed by Satchler [72] as part of a general survey of the use of effective interactions in nucleon inelastic scattering. In Section 5.5 we see that for (p, p') reactions this survey favors a Yukawa interaction with a range of about 1 fm and a depth of about $V_0 = 200$ MeV. (Here $V_0 \equiv V_{00} + V_{01}$.) For monopole excitations of analogue states Satchler finds $V_{01} \sim 20$ MeV, an order of magnitude weaker than V_0.

Similar values for V_{01} were found by Une, Yamaji, and Yoshida [92] in their study of the analogue transitions $C^{13}(p, n)N^{13}$ and $B^{11}(p, n)C^{11}$. Their antisymmetrized calculation employed for $g(r_0, r_i)$ a Gaussian form with a range of 1.8 fm.

Une, et al. [92] also applied their antisymmetrized theory to the study of the quadrupole transition $Al^{27}(n, p)Mg^{27}$, in which a bound $\frac{5}{2}^+$ proton is considered to be converted into a $\frac{1}{2}^+$ neutron. Nilsson orbitals were used. A force of Gaussian shape and of depth $V_{01} = 44.7$ MeV fits the observed cross section. This work resembles earlier work by Agodi and Schiffrer [67] who studied the reaction $Si^{28}(n, p)Al^{28}$. For the Si^{28} reaction a bound $d_{5/2}$ proton was considered to become converted into an $s_{1/2}$ neutron; the observed cross section was fitted by use of a force of Yukawa shape with range 1.15 fm and depth $V_{01} \approx 25$ MeV.

Less successful DW calculations were reported by Hodgson and Rook [91] and by Hansen, et al. [90]. The former authors [91] studied $V^{51}(p, n)Cr^{51}$ at 12.1 MeV bombarding energy. They considered that the reaction converts a bound $f_{7/2}$ single-particle neutron state into a bound $f_{7/2}$ proton state. This reaction admits angular momentum transfers $l = 0, 2, 4, 6$. Unfortunately the calculated angular distributions did not resemble experiment. This may be because use of a zero-range form for $g(r_{0i})$ gave excessive emphasis to the higher multipoles. However, Satchler (private communication) suggests that the difficulties with the $V^{51}(p, n)$ reaction may be caused by CN admixtures.—In the work of Hansen, et al. [90] a poor fit was obtained for the angular distributions for $Al^{27}(p, n)Si^{27}$, over the energy range 6.03–6.81 MeV. This reaction admits a mixture of multipoles having $l = 0, 2, 4$. In this case use of single-particle states derived from the spherical shell model may have given a poor picture of the l-mixing.

The question of collective enhancement in the charge-exchange form factors may now be raised. Collective enhancement is said to occur if the wavefunctions ψ_A and ψ_B are more complicated than those allowed by the simple shell model, and if the complications are such as to allow many nucleons to cooperate in the reaction. For the present application two cases must be distinguished. These are (a) the monopole excitation of the analogue of the target nucleus ground state; and (b) excitations of higher multipolarity, for example, excitations of states that are analogues of the excited states of nucleus A. For the ground state monopole transition we have already seen that the strength of the transition, hence the number of cooperating nucleons, is completely determined by the total i-spin. Adding complications to the wavefunctions only adds equal numbers of positive and negative expectations of the τ^+ operators to the form factor. (It is true that the addition of complications also causes orbitals of more than one radial shape to participate in the form factor, however, these all lead to radial integrals $G_0^{(\alpha)}(r_0)$ that are positive, with the result that no major change of the argument can ensue.) It appears that the ground state analogue transition necessarily has maximum "collective enhancement," and that complications of ψ_A and ψ_B do not affect the strength of this transition. (This fact may explain the small value that Satchler found [72] for V_{01}, in his analysis of monopole analogue transitions. This value may be "small" only by comparison with the artificially large value that V_{00} takes on to simulate collective effects in ordinary inelastic scattering.)

Excitations of higher multipolarity have strengths that are limited both by the total i-spin, as above, and by the angular momentum coupling of ψ_A and ψ_B. The angular momentum coupling determines whether these excitation strengths are large or small compared with the strength of the monopole analogue transition; for example, it is clear that transitions to the analogues of the collective excited states of the target nucleus are obtained by rotation of the total i-spin. This rotation preserves the special angular momentum coupling that makes these states collective. Therefore, the (p, n) excitation of the analogues of collective excited states may be expected to show the same pattern of strengths seen, say, in the (p, p') excitation of the collective states. (There is some inhibition of excited isobar formation in heavy odd-A nuclei [87, 93].)

Macroscopic theories of charge-transfer reactions have been very popular, largely since the initiative given by Lane [82, 83]. In these theories the nuclear matrix element of V_β is replaced by a one-body operator assumed to have a slow, smooth dependence upon energy and mass number; this operator then is used as a basis for phenomenological studies. This one-body operator affects the motions of the incoming and outgoing projectiles, and changes one into the other. There is an appealing generality about the use of such a phenomenological operator to represent the channel coupling. Such an operator may be used in distorted-waves approximation as in this chapter, or in a coupled-channels calculation as in Chapter 6. Furthermore, it may be used equally well either with projectiles that are nucleons, as an alternative to microscopic studies, or with projectiles that are composite, like H^3 or He^3, for which microscopic analyses are more questionable.

In our microscopic treatment of monopole (p, n) reactions we see that the nuclear matrix element of V_β has only a weak dependence upon detailed properties of ψ_A and ψ_B. It is this type of observation that makes plausible the use of a phenomenological one-body operator to represent the channel coupling. More detailed remarks along these lines are given in [94]. It is not clear that higher multipoles of V_β should be so nearly independent of details of ψ_A and ψ_B. For this reason, the phenomenological approach may be of interest primarily for monopole excitations. There have been almost no attempts to apply this approach for higher multipoles. (However, see [87].)

In the macroscopic approach, the monopole nuclear matrix element is replaced with an operator of the form

$$\langle\psi_B| V_\beta |\psi_A\rangle \to A^{-1} V_1(r_0)\langle\psi_B| (\mathbf{t}_0 \cdot \mathbf{T}) |\psi_A\rangle. \tag{5.104}$$

Here A is the mass number of the target nucleus, $\mathbf{t}_0$ is the i-spin operator for the projectile, $\mathbf{T}$ is the i-spin operator for the target nucleus, and $V_1(r_0)$ is a smoothly varying potential function. We have already seen that $V_1(r_0)$ probably tends to be surface-concentrated. Our general DI theory also suggests that $V_1(r_0)$ is likely to be complex.

Nonvanishing matrix elements of (5.104) arise in elastic scattering as well as in charge-exchange reactions. Thus (5.104) contributes a charge-dependent term of magnitude $\frac{1}{4}V_1(r_0)$ to the elastic optical potentials. The existence of such a term is already indicated by studies of the systematics of optical potential parameters (Section 5.2, [85]). Lane suggested that for neutron and proton scattering an estimate of $V_1(r_0)$ might be obtained [82] by identifying $\frac{1}{4}V_1(r_0)$ with the term V_1 of (5.14) and (5.15) that represented symmetry effects in elastic scattering. This identification probably is correct. In any case, such an identification is made as a matter of course in coupled-channels calculations of charge-exchange reactions, because these calculations require both diagonal and off-diagonal matrix elements of (5.104). Therefore it is interesting that coupled-channels calculations of charge-exchange reactions have been notably successful (Chapter 6).

While Lane's early estimates of V_1 for (p, n) reactions [82] could only suggest that its strength should lie in the range of 10–120 MeV, most later fits of (p, n) cross sections have yielded $V_1 \sim 100$ MeV. The macroscopic theory of analogue transitions has on the whole yielded a satisfactory description of experiment, much as for ordinary inelastic scattering.

Figures 5.14–5.16 illustrate the level of success reached in applications of the DW theory to charge-exchange experiments. Figure 5.14 was computed by Satchler [72], using the microscopic theory with $V_{01} = 21$ MeV, and describes (p, n) reactions induced by 18.5 MeV protons. Figure 5.15 also describes (p, n) reactions induced by 18.5 MeV protons, in work performed by Anderson, et al. [86]. The DW calculations were performed by Satchler,

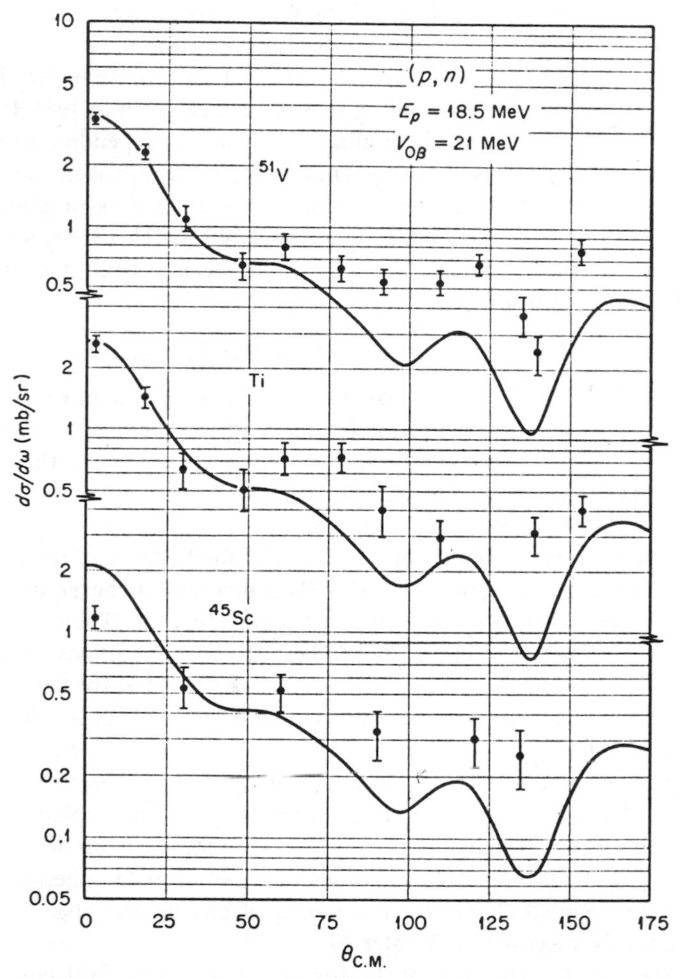

Fig. 5.14 Comparison of the microscopic DW theory with experimental data for analogue-state (p, n) reactions at 18.5 MeV bombarding energy. (From [72], Fig. 5.)

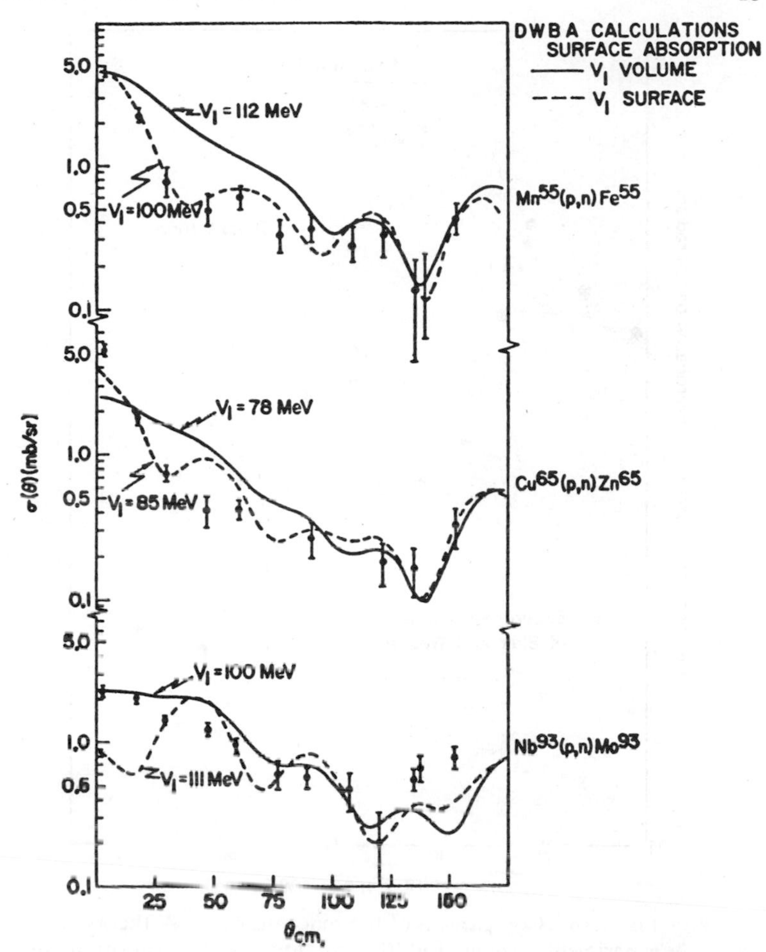

Fig. 5.15 Comparison of the macroscopic DW theory with experimental data for analogue-state (p, n) reactions at 18.5 MeV bombarding energy. A surface interaction (dashed line) is superior to a volume interaction (solid line). (From [86], Fig. 13.)

et al. [87] using the macroscopic theory. Figure 5.16 shows a further application of the macroscopic theory, given by Rook and Hodgson [95] for the 25 MeV (He^3, H^3) reaction on Fe^{56}.

5.8 Deuteron Stripping

Deuteron stripping is the best known direct reaction. In this reaction one nucleon of a projectile deuteron transfers to a target nucleus, while the other nucleon is a spectator and plays no role in the transfer process. Significant

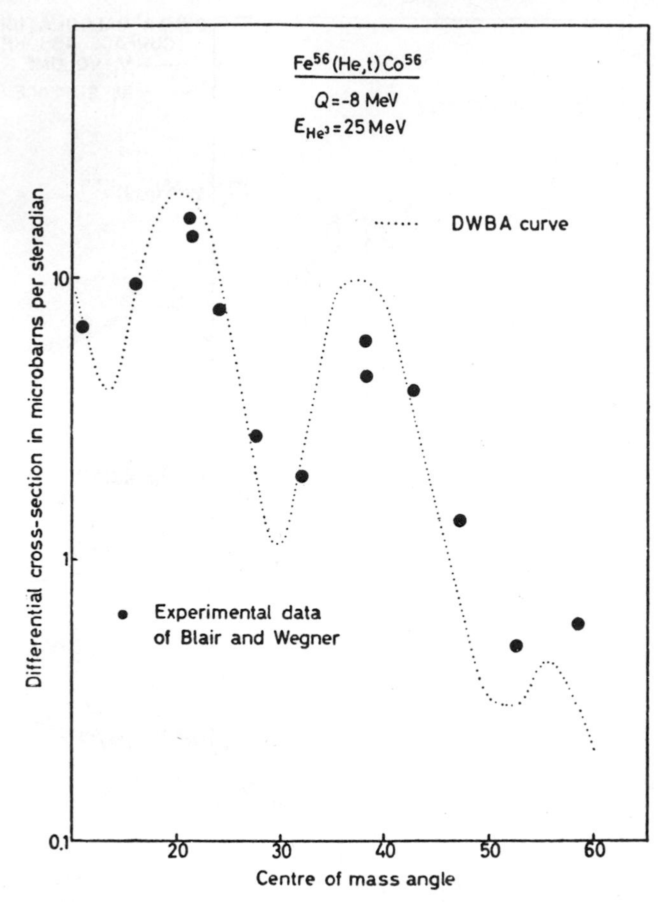

Fig. 5.16 Comparison of the macroscopic DW theory with experimental data for the analogue-state transition in $Fe^{56}(He^3, H^3)$ at 25 MeV. (From [95], Fig. 2.)

early treatments of stripping were given by Oppenheimer and Phillips [96] and by Serber [97], using semiclassical methods. A fully quantum-mechanical theory was not developed until the work of S. T. Butler [98] in 1951. Butler's paper emphasized the role played by angular momentum transfer, and thereby initiated the use of stripping as a decisive tool of nuclear spectroscopy. The distorted-waves theory, developed later by Daitch and French [1], by Horowitz and Messiah [1], and by Tobocman [2], supplanted the boundary-matching methods [98, 99] used by Butler, but retained the emphasis on angular momentum.

Because there is rearrangement, that is, different nuclear species appear in the entrance and exit channels, the dynamics of stripping is more complicated than that of inelastic scattering. Deuteron stripping is the simplest nuclear rearrangement reaction. Now a straightforward DI model of deuteron stripping obviously should be a three-body model of the type mentioned in Sections 4.1, 4.7, in which the incident neutron and proton are regarded as two interacting particles moving under the influence of complex potentials presented by the target nucleus. Very little understanding of such models has been achieved. We therefore employ the distorted-waves method and use product wavefunctions in both the entrance and exit channels. To the order of accuracy for which this method is used it is not relevant that the two product wavefunctions belong to nonorthogonal function spaces.

For definiteness we treat the (d, p) stripping reaction. Then the distorted-waves discussion of Section 4.5 yields

$$T_{\alpha\beta}{}^{DW} = (N_A + 1)\,\mathscr{J}\int d^3r_\alpha \int d^3r_\beta \chi_\beta^{(-)*}(\mathbf{k}_\beta, \mathbf{r}_\beta)\,\langle B, p_0|\,V\,|A, d\rangle\,\chi_\alpha^{(+)}(\mathbf{k}_\alpha, \mathbf{r}_\alpha), \tag{5.105}$$

where

$$\langle B, p_0|\,V\,|A, d\rangle = \langle \psi_B \chi_{\frac{1}{2}, m_p}(p_0)|\,V_\beta - U_\beta(\mathbf{r}_\beta)\,|\psi_A \phi_{d,\,m_d}\rangle. \tag{5.106}$$

Here $\chi_{\frac{1}{2}, m_p}(p_0)$ is the internal wavefunction of the proton, namely the spin wavefunction, with spin projection m_p. Also $\phi_{d,\,m_d}$ is the internal wavefunction of the deuteron, with projection m_d. The factor $(N_A + 1)$ is a result of particle identity as treated in Section 4.8, and correctly counts up all the equivalent contributions to the direct term of stripping. In this factor N_A is the neutron number of nucleus A. The cross section is obtained by substituting (5.105) in (4.61), using $(N_\beta/N_\alpha) = (N_A + 1)^{-1}$ for the channel weight factor. (An antisymmetrized DW amplitude would be obtained by adding the exchange terms of (4.97–99) to (5.105). It is seen in Chapter 10 that in (d, p) stripping these terms are small.) The Jacobian in (5.105) is discussed later in this section.

The interaction in (5.106) is composed of V_β, the sum of all the interactions between the two separating nuclei of channel β, and of U_β, the optical interaction between these nuclei. For the present application we have

$$V_\beta = \sum_{i=1}^{A} V_{ip_0} + V_{n_0 p_0}, \tag{5.107}$$

namely, V_β is the interaction between p_0 and the individual nucleons of the residual nucleus. In (5.107) V_β is written as the sum of the interactions between p_0 and the target nucleons, plus the interaction between p_0 and the stripped neutron. As usual (see Section 4.5) the potential U_β is not very clearly defined, and is chosen on grounds of convenience and definiteness. It has been customary in deuteron stripping calculations to choose U_β to be the optical potential that fits elastic scattering of a proton by nucleus B, and to replace the entire operator $(V_\beta - U_\beta)$ by the one term $V_{n_0 p_0}$. In this approximation

$$\langle B, p| V |A, d\rangle \approx \langle \psi_B \chi_{\frac{1}{2}, m_p}| V_{n_0 p_0} |\psi_A \phi_{d, m_d}\rangle. \tag{5.108}$$

The omission of the proton, target-nucleus interactions is understandable. Equation 5.106 sums the contributions of all the individual target nucleons. Such a sum is of the nature of an average over the V_{ip_0}, and the largest effect of the averaged interaction must be to cause elastic scattering of p_0. This effect is easily cancelled by proper choice of an optical potential U_β. The V_{ip_0} also induce small core excitations in ψ_A and ψ_B. However, although such inelastic effects make additional contributions to stripping, their contributions are likely to be small because (a) the excitations of ψ_A and ψ_B have small amplitudes, (b) the inelastic matrix elements would require n_0 and p_0 to have simultaneous good overlap with ψ_A and ψ_B, and (c) the overlaps are complicated. Excitations caused by the V_{ip_0} are treated in Section 6.5.

It is less understandable that the U_β that accords with the reduction to (5.108) should be the optical potential that fits elastic scattering of protons by nucleus B. In going from (5.106) to (5.108) we omit the terms $(\sum V_{ip_0} - U_\beta)$. From this fact it might seem that U_β should be the optical potential that fits the scattering of protons by nucleus A [100]. The difference between the two choices for U_β comes about because the interaction $V_{n_0 p_0}$ contributes to the scattering of p_0 by B but not to its scattering by A. However, it is not possible to choose U_β simply by inspection of (5.106). In the discussion in Section 4.5 it is pointed out that U_β must be chosen to minimize the contributions from those excited-state terms of Ψ_{model} ignored in the DW approximation. Because the largest of these terms probably correspond to rescattering of the proton in stripping exit channels, it is appropriate that the U_β that minimizes such terms should include "diagonal" parts of $V_{n_0 p_0}$. In any case, as noted before, questions of this sort are of the same order as the uncertainties of the DW method.

It is also necessary to go back to the origins of the DW method to understand why a *complex potential* U_β accords with the reduction to (5.108). The DW method omits higher terms of Ψ_{model}, the wavefunction that diagonalizes the direct reaction part of Hilbert space. However, because the direct reaction part of Hilbert space is incomplete, all the operators governing Ψ_{model} are complex. Among these operators are the V_{ip_0}. Therefore it is correct that a complex U_β must be used to cancel effects of the V_{ip_0}.

In view of the above remarks we may expect that a satisfactory zero-order DW theory of deuteron stripping is obtained by substituting (5.108) into (5.105), using the wavefunction that fits elastic scattering of deuterons by A for $\chi_\alpha^{(+)}$, and the time reverse of the wavefunction that fits elastic scattering of protons by B for $\chi_\beta^{(-)}$. This expectation implies that any major disagreements with experiment are not to be attributed to failures of $\chi_\alpha^{(+)}$ and $\chi_\beta^{(-)}$, but rather to failures of the DW method itself.

In the absence of spin-orbit terms in the optical potentials the internal functions ϕ_{p_0} and ϕ_d are decoupled from the distorted waves, as indicated in (5.105), and it is straightforward to obtain the form factors required for the DW calculation. The nuclear matrix element of (5.108) implies integration over only the internal coordinates of ψ_A and the spins of n_0 and p_0. The integration over the coordinates of the target nucleus is most meaningful if we introduce a generalized fractional parentage expansion of ψ_B that expresses this function in terms of the complete set of (antisymmetrized) eigenstates of the target nucleus $\psi_{\alpha_p J_p M_p}$. Here α_p are any additional quantum numbers required for a complete specification of the state. Then ψ_B becomes [101]

$$\psi_{\alpha_B J_B M_B} = \sum_{\alpha_p J_p j M_p} \langle J_B M_B \mid J_p j;\, M_p,\, M_B - M_p\rangle \psi_{\alpha_p J_p M_p} \phi^{\alpha_p J_p}_{j,\, M_B - M_p}(n_0), \tag{5.109}$$

where for each parent state $\psi_{\alpha_p J_p M_p}$ the set of coefficients $\phi^{\alpha_p J_p}_{j, M_B - M_p}(n_0)$ are functions of the coordinates of the transferred neutron, and have definite angular momentum j and projection $M_B - M_p$. Because the $\phi^{\alpha_p J_p}_{j, M_B - M_p}$ are introduced only as coefficients in (5.109) they are not necessarily normalized to unity. Substitution of (5.109) into (5.108) yields

$$\begin{aligned}\langle B, p|\, V\, |A, d\rangle = \sum_j \langle J_B M_B \mid J_A j;\, M_A,\, M_B - M_A\rangle \\ \times \langle \phi^{\alpha_A J_A}_{j,\, M_B - M_A} \chi_{\frac{1}{2} m_p}|\, V_{n_0 p_0}\, |\phi_{d,\, m_d}\rangle\end{aligned} \tag{5.110}$$

for the nuclear matrix element. This matrix element has selected from (5.109) those few terms that have ψ_A as a parent state. It is clear that the $\phi^{\alpha_A J_A}_{j, M_B - M_A}(n_0)$ are closely related to the stripping form factors.

To perform the spin sums that remain in (5.110) we now introduce more specific notation: The neutron functions $\phi^{\alpha_p J_p}_{j\mu}$ have definite angular momentum and parity, therefore they take the factored form

$$\phi^{\alpha_p J_p}_{j\mu}(n_0) = \mathscr{R}^{\alpha_p J_p}_{lj}(n_0) \sum_m \langle l\tfrac{1}{2};\, m,\, \mu - m \mid j\mu\rangle Y_l^m(\hat{n}_0) i^l \chi_{\frac{1}{2},\, \mu - m}(n_0). \tag{5.111}$$

The operator V_{np} is a linear combination of central and tensor interactions. However, because V_{np} multiplies the deuteron wavefunction, we may use the Schrödinger equation and replace the spin-dependent V_{np} operator in terms of the spin-independent kinetic energy operator

$$V_{np}\phi_{d,\, m_d} = \frac{\hbar^2}{M}(\nabla^2 - \gamma^2)\, \phi_{d,\, m_d}, \tag{5.112}$$

where $(\hbar^2\gamma^2/M)$ = deuteron binding energy. The deuteron wavefunction is itself a linear combination of S and D terms,

$$\phi_{d,m}(\mathbf{r}) = Nr^{-1}\{u(r)\mathscr{Y}^m_{011} + w(r)\mathscr{Y}^m_{211}\}, \tag{5.113}$$

where the normalized spin-angle wavefunctions are

$$\mathscr{Y}^m_{lsj} \equiv \sum_\nu \langle ls; m-\nu, \nu \mid jm\rangle Y_l^{m-\nu}(\hat{r})\, \chi_s{}^\nu.$$

In (5.113) we use effective range normalization

$$u(r) \to e^{-\gamma r}, \qquad \text{as} \qquad r \to \infty.$$

It is well known that with this normalization effective range theory gives

$$N = \left[\frac{2\gamma}{1 - \gamma\rho_t}\right]^{1/2} = 2.83 \times 10^6 \text{ cm}^{-1/2},$$

where ρ_t is the triplet effective range. Insertion of (5.111–5.113) in (5.110) now easily allows the sums on neutron and proton spins to be completed.

At this stage we only treat the contribution of the deuteron S-state. The D-state contribution in (5.110) is much smaller and tends to be incoherent with the S-state contribution when the cross section is computed. We therefore defer treatment of the D-state contribution until near the close of Section 5.8. The S-state part of the amplitude is

$$\begin{aligned}\langle B, p|\, V\, |A, d\rangle_S = \sqrt{\tfrac{3}{2}}\, D(r_{n_0p_0})(-)^{1/2 - m_p} \\ \times \sum_j \langle J_B M_B \mid J_A j; M_A, M_B - M_A\rangle\langle 1\tfrac{1}{2}; m_d, -m_p \mid \tfrac{1}{2}, m_d - m_p\rangle \\ \times \langle l\tfrac{1}{2}; m, m_d - m_p \mid j, M_B - M_A\rangle i^{-l} Y_l^{m*}(\hat{r}_{n_0A})\, \mathscr{R}_{lj}^{\alpha_A J_A}(r_{n_0A}),\end{aligned} \tag{5.114}$$

with

$$D(r_{np}) = \frac{N\hbar^2}{M\sqrt{4\pi}}(\nabla^2 - \gamma^2)\frac{u(r)}{r}. \tag{5.115}$$

Comparison with (4.59) and (4.60), which give the standard form for the DW amplitude, discloses that the S-state form factors for deuteron stripping are

$$A_{lsj} f_{lsj,m} = \sqrt{N_A + 1}\, \mathscr{J} \sqrt{\tfrac{3}{2}}\, D(r_{n_0p_0}) Y_l^{m*}(\hat{r}_{n_0A}) \mathscr{R}_{lj}^{\alpha_A J_A}(r_{n_0A})\, \delta_{s,1/2}. \tag{5.116}$$

Here $m = M_B - M_A - m_d + m_p$. The Jacobian of (5.105) is included in (5.116). For reasons of later convenience the square root of the weight factor $(N_A + 1)$ is also included in (5.116).

We observe above that the S-state part of the amplitude has the property that l, s, j are the quantum numbers of the transferred nucleon in nucleus B. In particular $s = \frac{1}{2}$, even though the general DW formalism also allows

$s = \frac{3}{2}$. We see later that the D-state contribution is not this simple. However, because $s = \frac{1}{2}$ for the dominant S-state part and because each $\phi^{\alpha_p J_p}_{j, M_B - M_p}$ in (5.109) has a definite parity, there is a unique value of l associated with each j. Therefore a typical stripping calculation requires consideration of only a very small number of form factors. For a spin-zero target nucleus, for example, $j = J_B$ and only one form factor appears.

Spectroscopic properties of nuclei A and B enter the form factor through the function $\mathscr{R}^{\alpha_A J_A}_{lj}$. This function is the radial part of $\phi^{\alpha_A J_A}_{jm}$, the "wavefunction of the transferred neutron." In practical applications of stripping theory this $\phi^{\alpha_A J_A}_{jm}$ has usually been taken to be a product of a *spectroscopic coefficient* times a normalized *shell-model single-particle state* [102, 103]:

$$\sqrt{N_A + 1}\, \phi^{\alpha_A J_A}_{jm}(n_0) = \mathscr{S}^{1/2}(\alpha_A J_A, lj \mid \alpha_B J_B)\, \psi_{nljm}(n_0). \tag{5.117}$$

The standard normalization of $\mathscr{S}$ includes the weight factor $\sqrt{N_A + 1}$, as shown. With this normalization, $\mathscr{S} = 1$ if ψ_{nljm} is orthogonal to the single-particle states in ψ_A, the wavefunction of the target nucleus; in particular $\mathscr{S} = 1$ if a nucleon is added to a closed-shell target nucleus.

Unfortunately (5.117) is not an exact consequence of the definition of $\phi^{\alpha_A J_A}_{jm}$. The definition in (5.109) does not suggest that the $\phi^{\alpha_A J_A}_{jm}$ are proportional to shell model eigenstates, nor does it suggest any other simple single-particle problem of which they are eigenstates. The $\phi^{\alpha_A J_A}_{jm}$ are joint properties of two nuclei, and they are influenced by all the effects that cause one nucleus to differ from another. Further discussion of the functions $\phi^{\alpha_A J_A}_{jm}$ and of the coefficients $\mathscr{S}$ is given in Chapter 8. For definiteness, the applications in the present chapter are based on (5.117).

To evaluate the Jacobian of (5.105) we introduce a set of variables that give unit Jacobian, and we transform to the variables $\mathbf{r}_\alpha$ and $\mathbf{r}_\beta$. One set known to have unit Jacobian is $\mathbf{r}_{n_0 A}$ and $\mathbf{r}_{p_0 B}$, which are shown in Fig. 5.17:

$$\mathbf{r}_{n_0 A} = \mathbf{r}_{n_0} - \mathbf{r}_A,$$
$$\mathbf{r}_{p_0 B} = \mathbf{r}_{p_0} - \frac{A\mathbf{r}_A + \mathbf{r}_{n_0}}{A + 1}.$$

Here A is the mass number of the target nucleus. This set of variables appears in a natural manner when the overall center-of-mass coordinate is eliminated. Transformation to the standard variables of the DW calculation then is described by the equations

$$\begin{aligned} \mathbf{r}_\alpha &= \frac{1}{2}\left[\mathbf{r}_{p_0 B} + \left(\frac{A+2}{A+1}\right)\mathbf{r}_{n_0 A}\right], \\ \mathbf{r}_\beta &= \mathbf{r}_{p_0 B}, \\ \mathscr{J} &= 8\left(\frac{A+1}{A+2}\right)^3. \end{aligned} \tag{5.118}$$

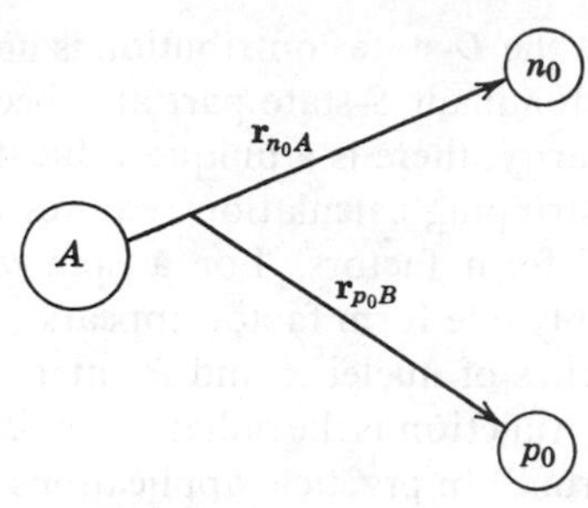

Fig. 5.17 Integration variables suitable for deuteron stripping analysis.

The variables $\mathbf{r}_\alpha$ and $\mathbf{r}_\beta$ are natural variables of a DW calculation because they are the arguments of the distorted waves. However, the variables in the form factor are $\mathbf{r}_{n_0p_0}$ and $\mathbf{r}_{n_0A}$, which are linear combinations of $\mathbf{r}_\alpha$ and $\mathbf{r}_\beta$. The presence of four distinct vector variables causes integration difficulties. General "finite-range" techniques for the integration of (5.105) are discussed in Section 5.13.

In our present discussion (5.105) is simplified by the introduction of *zero-range approximation* (see Section 5.1). In Section 5.13 we see that this approximation is very accurate for medium-energy deuteron stripping. Zero-range approximation is based on the observation that $D(r_{np})$ of (5.115) has a range of the order of the range of V_{np}. This range tends to be much smaller than the ranges in which there are significant variations of the other functions in the DW integrand.

In zero-range limit we have $u(r) = e^{-\gamma r}$ for all values of r, and (5.115) immediately gives

$$D_0(r) = -\frac{\sqrt{4\pi}\, N\hbar^2}{M}\,\delta(\mathbf{r}). \tag{5.119}$$

It must be understood that N in (5.119) still is the asymptotic normalization constant of the S-state term of the physical deuteron, as discussed previously, and is not to be generated by normalizing the zero-range wavefunction. Under this physical choice of N our wavefunctions have correct magnitudes in the regions of overlap that are important for stripping; also, under this choice $D(r)$ and $D_0(r)$ have the same volume integral.

The zero-range approximation requires use of the variables $\mathbf{r}_\alpha$ and $\mathbf{r}_{n_0p_0}$. The transformation to these variables

$$\begin{aligned}\mathbf{r}_\alpha &= \frac{1}{2}\left[\left(\frac{A+2}{A+1}\right)\mathbf{r}_{n_0A} + \mathbf{r}_{p_0B}\right],\\ \mathbf{r}_{n_0p_0} &= \left(\frac{A}{A+1}\right)\mathbf{r}_{n_0A} - \mathbf{r}_{p_0B},\end{aligned} \tag{5.120}$$

has unit Jacobian. With these variables the reduced DW amplitude for deuteron stripping is finally obtained by substituting (5.119) into (5.116), and by substituting (5.116) into (4.60). The result is

$$(2l+1)^{\frac{1}{2}} i^l A_{lsj} \beta_{sj}{}^{lm} = -\sqrt{6\pi}\left(\frac{N\hbar^2}{M}\right)\delta_{s,\frac{1}{2}}$$
$$\times \int d^3r \chi_\beta^{(-)*}\left(\mathbf{k}_\beta, \frac{A\mathbf{r}}{A+1}\right) Y_l^{m*}(\hat{r})[\sqrt{N_A+1}\, \mathscr{R}_{lj}^{\alpha_A J_A}(r)]\chi_\alpha^{(+)}(\mathbf{k}_\alpha, \mathbf{r}). \quad (5.121)$$

The cross section is obtained by substituting (5.121) in (4.62). The channel weight factor, N_β/N_α is conveniently cancelled by the factor $\sqrt{N_A+1}$ that is omitted from (5.116). We obtain

$$\frac{d\sigma}{d\Omega} = \frac{\mu_\alpha \mu_\beta}{(2\pi\hbar^2)^2}\left(\frac{k_\beta}{k_\alpha}\right)\left(\frac{2J_B+1}{2J_A+1}\right)\left(\frac{2\pi N\hbar^2}{M}\right)$$
$$\times \sum_{ljm}\left|\int \chi_\beta^{(-)*}\left(\mathbf{k}_\beta, \frac{A\mathbf{r}}{A+1}\right) Y_l^{m*}(\hat{r})[\sqrt{N_A+1}\, \mathscr{R}_{lj}^{\alpha_A J_A}(r)]\chi_\alpha^{(+)}(\mathbf{k}_\alpha, \mathbf{r})\, d^3r\right|^2. \quad (5.122)$$

Equation 5.122 is the expression calculated in standard zero-range DW treatments of deuteron stripping. [It may be noted that finite-range effects (Section 5.13) tend to reduce the magnitude of (5.122) by ten or fifteen percent without introducing any other major modifications.]

If it is assumed that stripping inserts a neutron into nucleus A in a shell-model single-particle state of known properties but of unknown normalization, then the approximation of (5.117) may be employed. Let the radial wavefunction of the shell model state be $\mathscr{R}_{nlj}(r)$, so that

$$\psi_{nlj,m} = \mathscr{R}_{nlj}\{i^l Y_l, \chi_{\frac{1}{2}}\}_{jm},$$
$$\int_0^\infty r^2 \mathscr{R}_{nlj}^2\, dr = 1. \quad (5.123)$$

The introduction of (5.117) in (5.122) then allows the display of $\mathscr{S}$, the spectroscopic coefficient:

$$\frac{d\sigma}{d\Omega} = \frac{\mu_\alpha \mu_\beta}{(2\pi\hbar^2)^2}\left(\frac{k_\beta}{k_\alpha}\right)\left(\frac{2J_B+1}{2J_A+1}\right)\left(\frac{2\pi N\hbar^2}{M}\right)\sum_{l,j}\mathscr{S}(\alpha_A J_A, lj \mid \alpha_B J_B)$$
$$\times \sum_m \left|\int \chi_\beta^{(-)*}\left(\mathbf{k}_\beta, \frac{A\mathbf{r}}{A+1}\right) Y_l^{m*}(\hat{r})\, \mathscr{R}_{nlj}(r)\, \chi_\alpha^{(+)}(\mathbf{k}_\alpha, \mathbf{r})\, d^3r\right|^2. \quad (5.124)$$

It should be recalled that the parity of the transition associates a definite l with each j, and that j can have more than one value only if $J_A \neq 0$.

The angular distributions yielded by (5.124) usually show a characteristic dependence upon l, such that measurement of an angular distribution gives an immediate indication of the l of the transition. This l-dependence is clearest for exothermic reactions performed at medium energies, that is, for bombarding energies in the range 8–20 MeV. Under these conditions $k_\beta \approx k_\alpha$; therefore it is only for fairly large scattering angles that the reaction can transfer angular momentum to the nucleus. The reaction tends to have a "main peak" that lies near 0° if $l = 0$ but is displaced to progressively larger angles if $l = 1, 2$, etc. The reaction often is said to "measure" l. Knowledge of l determines the parity of ψ_B and provides important *partial information* about the value of J_B. (Polarization or angular correlation measurements, discussed in Chapter 9, provide supplementary information about j and J_B.)

By fitting (5.124) to experiment it is also possible, in principle, to measure values of the coefficient $\mathscr{S}$ and thereby determine the probability with which a given single-particle state appears in a given state of nucleus B. Such detailed comparison of theory with experiment receives encouragement from the fact that (5.124) often shows very good fits with experiment.

Figures 5.18–23 illustrate the kinds of fits that can be achieved: Fig. 5.18 depicts the ground-state transition [104] $B^{10}(d, p)B^{11}$ taken at a bombarding energy of 13.5 MeV with $l = 1$. It is interesting that the DW angular distribution has a prominent peak at 180°. DW calculations frequently show such back-angle peaks [105, 116]. Figures 5.19 and 5.20 are taken from the very detailed study of $Ca^{40}(d, p)$, given by Lee, et al. [106], and depict the strong

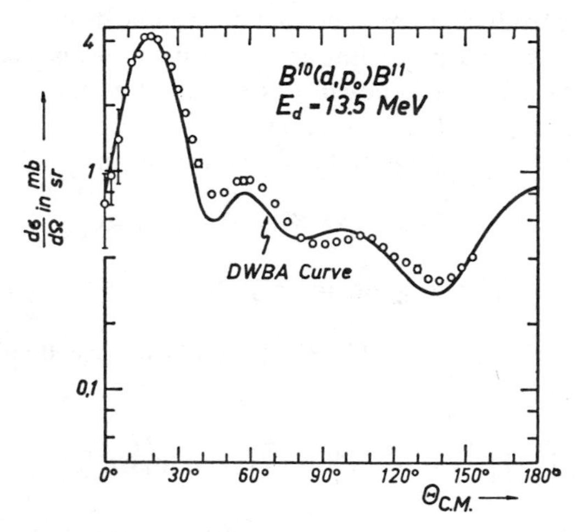

Fig. 5.18 Comparison of the DW theory with experimental data for the ground-state transition $B^{10}(d, p)B^{11}$, at 13.5 MeV bombarding energy. Note the peak in the calculated angular distribution at 180°. (From [104], Fig. 2.)

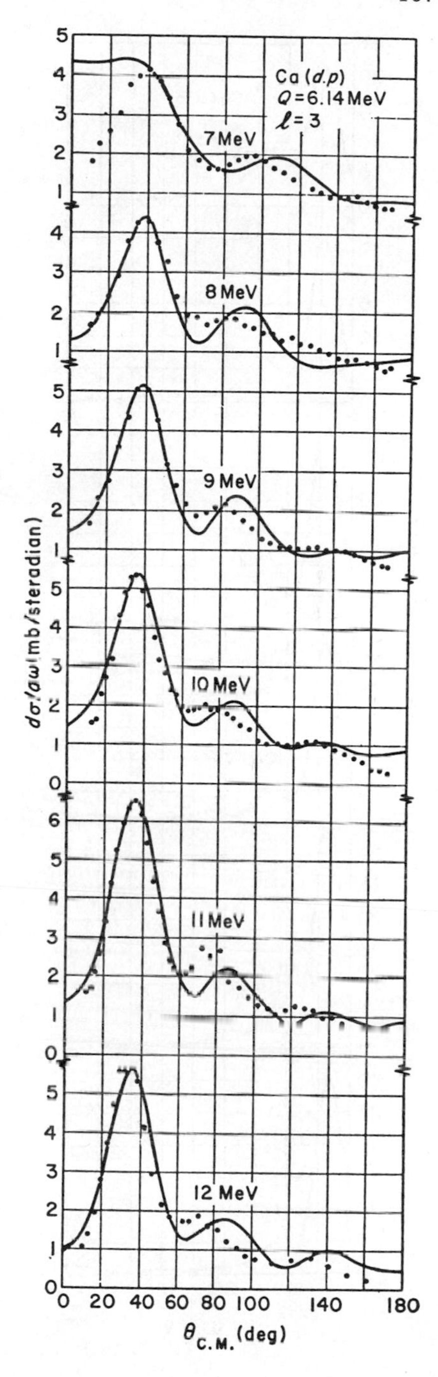

Fig. 5.19 Comparison of the DW theory with experimental data for the $Q = 6.14$ MeV, $l = 3$ transition in $Ca^{40}(d, p)$ for bombarding energies in the range 6.14–12 MeV. (Taken from [106], Fig. 4.)

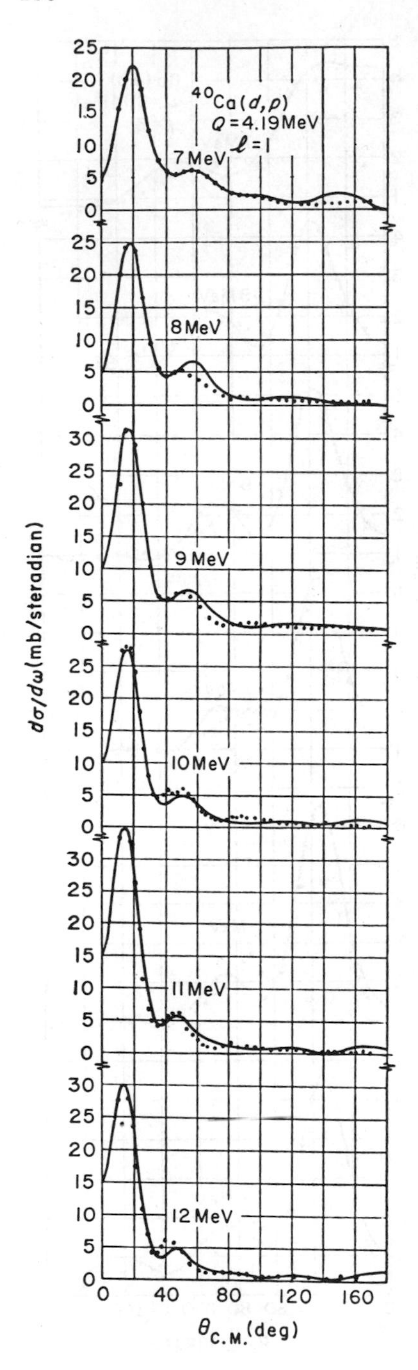

Fig. 5.20 This figure is the same as Fig. 5.19 for the $Q = 4.19$ MeV, $l = 1$ transition. (From [106], Fig. 5.)

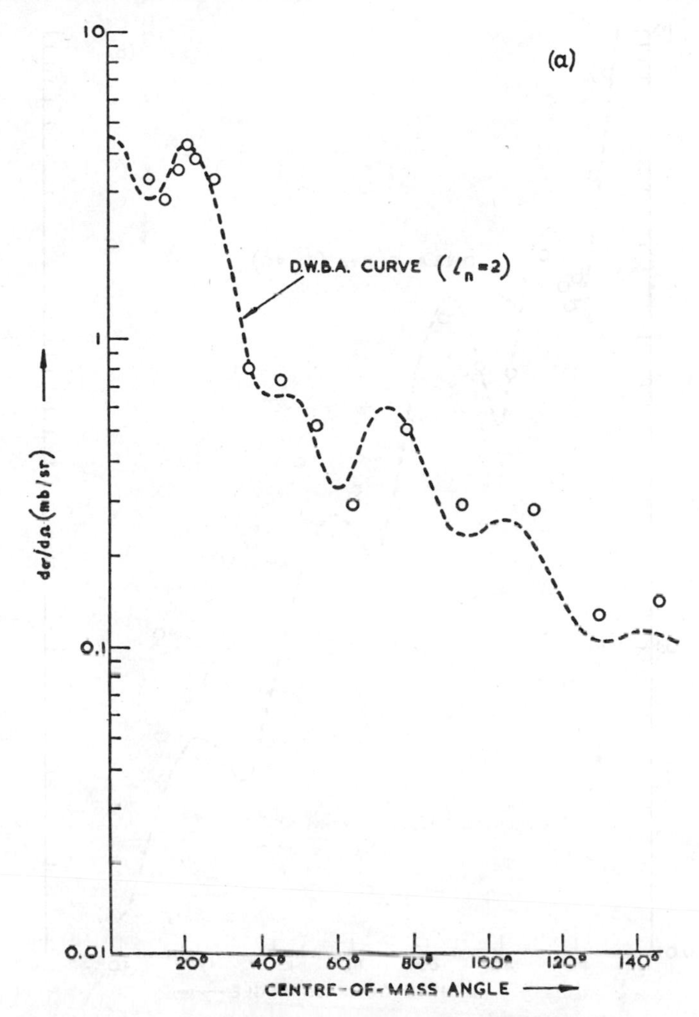

Fig. 5.21 Comparison of the DW theory with experimental data for the $Q = -13.67$ MeV, $l = 2$ transition and for the $Q = -16.04$ MeV, $l = 0$ transition in $Ca^{40}(p, d)$ at 30.3 MeV bombarding energy. (From [108], Fig. 6.)

$l = 3$ and $l = 1$ transitions that have $Q = 6.14$ MeV and $Q = 4.19$ MeV, respectively. As expected, because of the double-closed-shell nature of Ca^{40} the $\mathscr{S}$ factors for these transitions are large. To obtain the excellent fits shown in these figures, extending over 7–12 MeV bombarding energy, the authors needed to make a companion study of deuteron and proton elastic scattering. We note that the $l = 3$ cross sections are about fivefold smaller

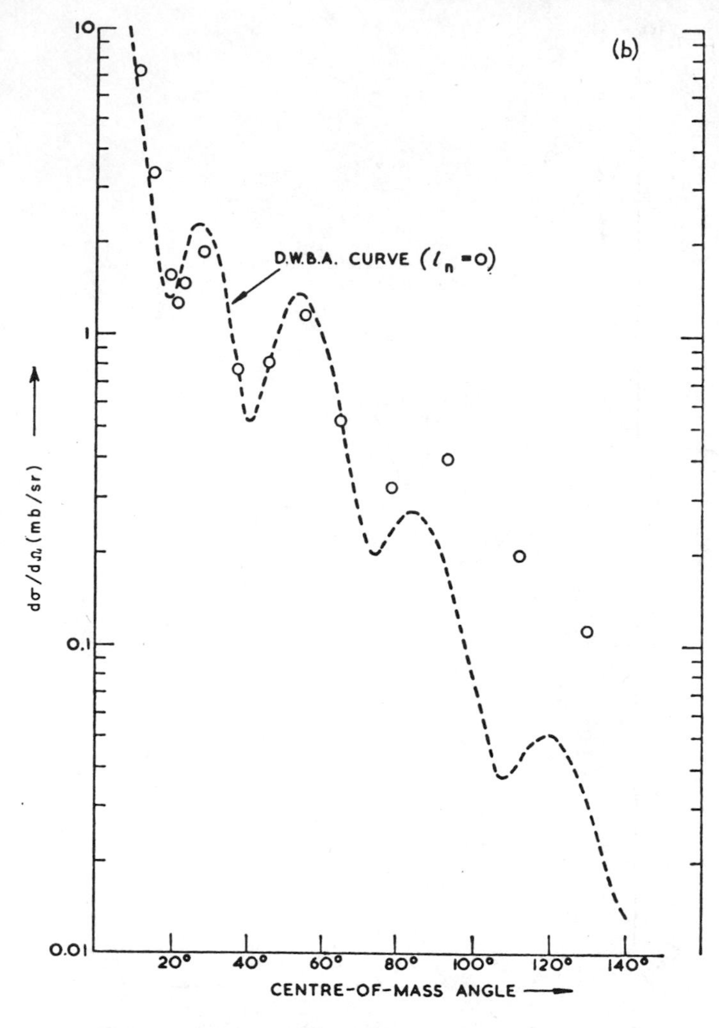

Fig. 5.21 (*contd.*)

than the $l = 1$ cross section. The authors of this work assigned ± 20 percent confidence limits to the $\mathscr{S}$ factors they extracted from their analysis. A similar careful study of the application of the DW theory to deuteron stripping was published by Forster, et al. [107] for the reaction $Zr^{90}(d, p)$ at 10 MeV and 12 MeV bombarding energy.

Figure 5.21 depicts $l = 0$, 2 groups in the pickup transition $Ca^{40}(p, d)$, taken [108] at a bombarding energy of 30 MeV. Here we note that the $l = 2$ cross section is several times weaker than that for $l = 0$.

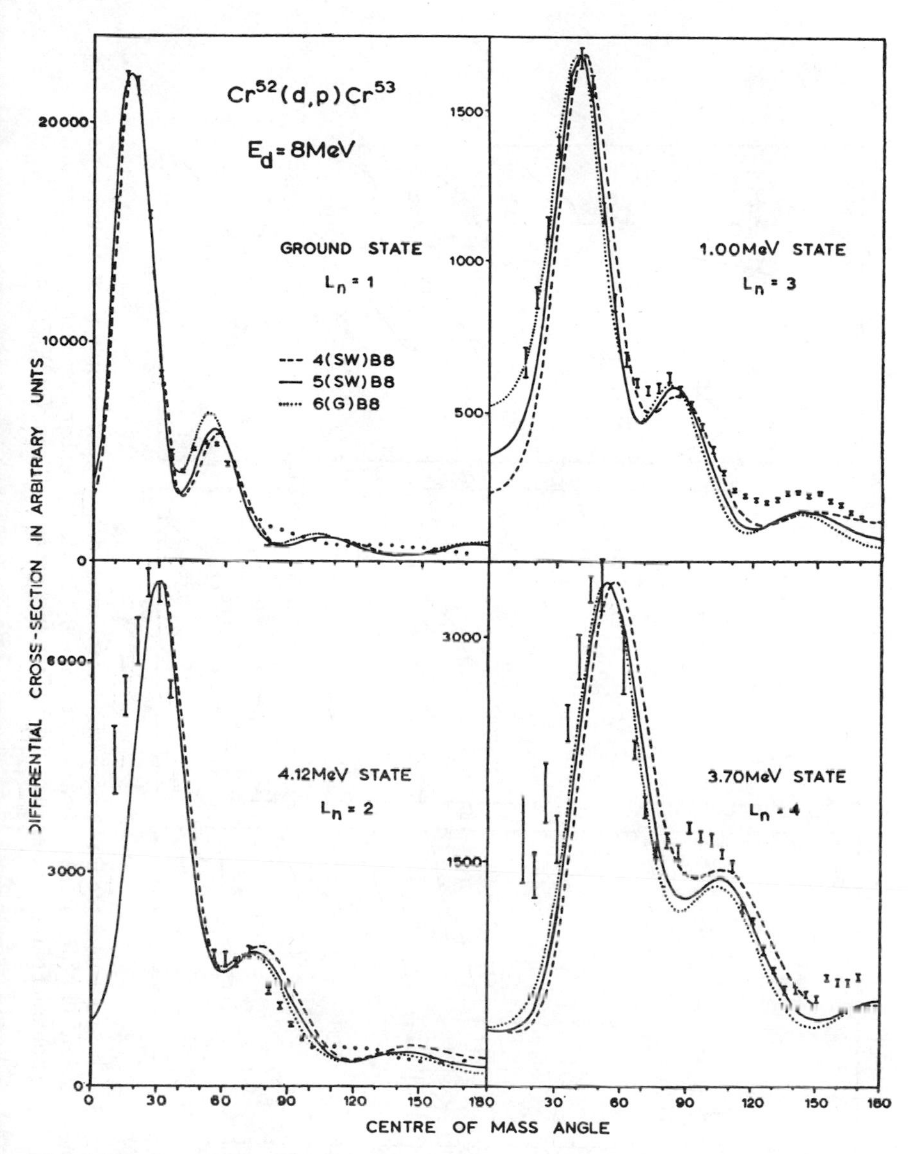

Fig. 5.22 Comparison of the DW theory with experimental data for $l = 1, 2, 3, 4$ transitions in $Cr^{52}(d, p)$ at 8 MeV bombarding energy. Here the three sets of curves arise from tests of sensitivity to variations of the deuteron optical potential. (From [109], Fig. 2.)

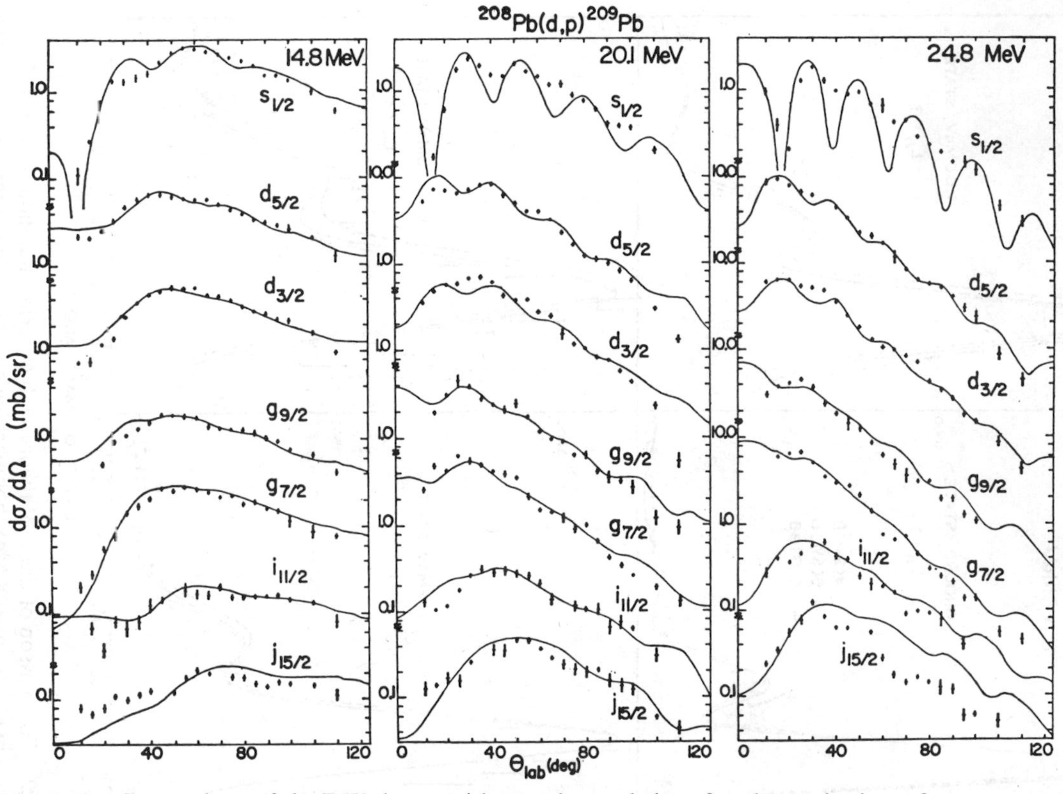

Fig. 5.23 Comparison of the DW theory with experimental data for the excitation of seven states in the reaction $Pb^{208}(d, p)$, as indicated, at 15, 20, 25 MeV bombarding energy. The optical potential parameters are held fixed as the energy varies. (From [113], Fig. 4.)

Figure 5.22 is taken from the study of $Cr^{52}(d, p)$, by P. T. Andrews, et al. [109], and depicts $l = 1, 2, 3, 4$ transitions obtained at 8 MeV bombarding energy. The stripping work was accompanied by studies of deuteron and proton elastic scattering. Several curves are shown on each graph to indicate the sensitivity of the (d, p) cross sections to variations of the deuteron optical potential. The various deuteron potentials used all yielded equally good fits to deuteron elastic scattering. Cross sections having high l values are seen to be more sensitive to the variations of the potentials than those having low l values, in agreement with usual experience. Acceptable variations of the proton potential do not affect the stripping cross section, again as usual.*

Legg, et al. [112] studied the reaction $Cr^{52}(d, p)$ over a 4.3–6.3 MeV range of bombarding energy, and obtained cross sections for the formation of four states of Cr^{53} having $l = 1, 3$. Despite the somewhat featureless cross sections at this low energy the zero-range DW theory was not able to provide as satisfactory fits to experiment as had been obtained at 8 MeV by Andrews, et al.

Figure 5.23 shows a study [113] of $Pb^{208}(d, p)$ at 15, 20, 25 MeV, for seven states of Pb^{209}. The transitions have $l = 0, 2, 4, 6, 7$. Use of optical potentials that fit elastic scattering is seen to give acceptable DW fits, in contradiction with earlier work [114] at somewhat lower energies. The angular distributions peak at backward angles and are only weakly l-dependent. Angular distributions of this type are expected when Coulomb effects are strong (Section 5.9).

Figures 5.24–5.26 illustrate the manner in which changes of bombarding energy affect the DW predictions for (d, p) reactions on heavy nuclei. To compute these cross sections Macefield [115] arbitrarily chose a typical set of optical potential parameters, and selected U^{238} as a target, with $Q = 3$ MeV for all cases. The most prominent feature of these calculations is the transition from Coulomb-dominated cross sections at low energy, to forward-peaked, diffraction-like cross sections at high energy. At both extremes of the energy range it is difficult to distinguish l values. Macefield recommends 20 MeV as presenting cross sections with maximum l-dependence.

* Most authors who apply the DW theory make some study of the sensitivity to parameter variation. Two papers that emphasize this topic are those of Dantzig and Tobocman [110] and of Smith [111]. The former article treats the $l = 1$ transition $C^{12}(d, p)C^{13}$ at a bombarding energy of 25.9 MeV. The sensitivity to the radial form of the imaginary optical potential was studied.

The article of Smith [111] studies the cases $Zr^{90}(d, p)$, $Cr^{52}(d, p)$, $Pb^{206}(d, p)$, and $Zn(d, p)$, and finds the $\mathcal{S}$ factors to be particularly sensitive to parameter variation. Unfortunately, the parameter variations that Smith introduced for $\chi_\alpha^{(+)}$ and $\chi_\beta^{(-)}$ were in each case also inserted into the calculation of the bound-state radial function $\mathcal{R}_{nlj}$. Therefore the meaning of Smith's results is not clear.

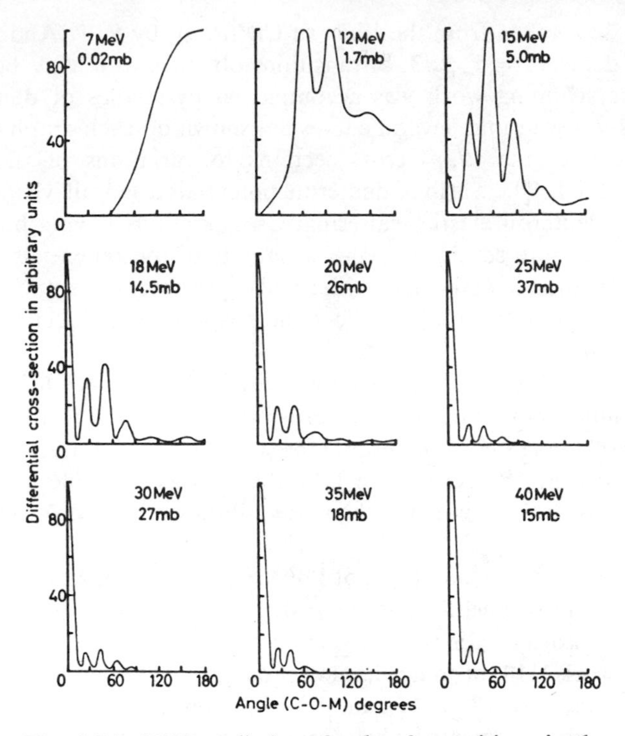

Fig. 5.24 DW predictions for $l = 0$ transitions in the reaction $U^{238}(d, p)$, with $Q = 3$ MeV, at the bombarding energies indicated. The appropriate peak value of the differential cross section is given beside each curve. (From [115], Fig. 1.)

Another interesting study of energy dependence was given by B. A. Robson [116], for several states formed in the reaction $N^{14}(d, p)$, for the rather low bombarding energies in the range 0.8–3.20 MeV. The DW stripping theory gives a good description of the data. In some cases it yields striking back-angle peaks. It is interesting that the weak imaginary potentials Robson used allowed a prominent single-particle resonance to appear at 1.2 MeV.

The above comparisons with experiment illustrate the level of success that can be obtained with (5.124). Obviously the fits with experiment are excellent but not perfect. Uncertainties in the choice of optical potential parameters (Section 5.2) stand in the way of accurate knowledge of $\chi_\alpha^{(+)}$ and $\chi_\beta^{(-)}$, and have noticeable effects in the (d, p) cross section. These uncertainties preclude too great confidence in $\mathscr{S}$ factors extracted with (5.124).

Other uncertainties of the DW stripping theory appear because the shape of $\mathscr{R}_{nlj}$ is uncertain. The barrier penetration effects that displace stripping

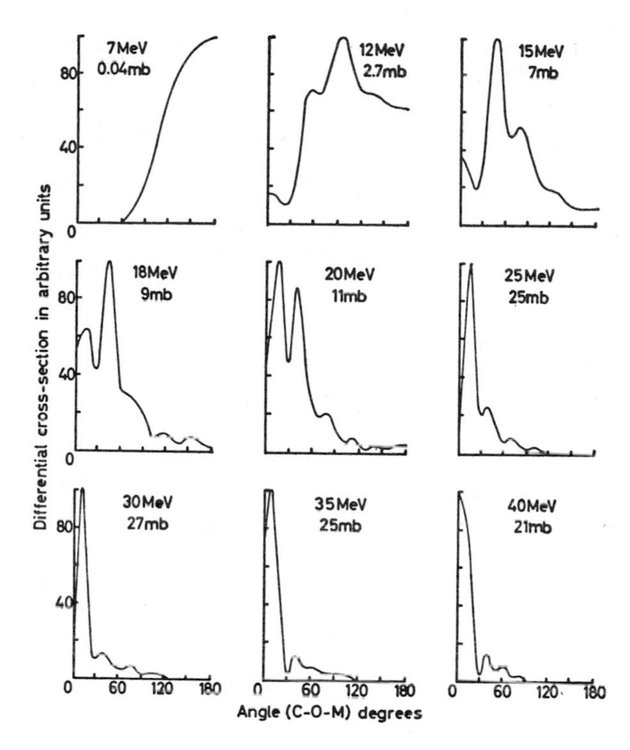

Fig. 5.25 This figure is the same as Fig. 5.24 for $l = 2$ transitions. (From [115], Fig. 2.)

peaks to large scattering angles also emphasize large values of r in the stripping integral. To achieve accuracy in the calculation of this integral it is therefore necessary to have accurate knowledge of $\mathscr{R}_{nlj}$ in the nuclear surface region. However, it is just in this region that weak residual interactions introduce considerable departures in shape between $\mathscr{R}_{nlj}$ of (5.124) and the more-accurate $\mathscr{R}_{lj}^{\alpha_A J_A}$ of (5.122). The so-called "well-depth procedure" (WD) chooses $\mathscr{R}_{nlj}$ so that it should at least have a correct shape at far-asymptotic values of r. In this procedure $\psi_{nlj,m}$ is calculated as an eigenfunction of a finite potential well (generally of Woods-Saxon shape), and the depth of this well is adjusted so that the eigenenergy of $\psi_{nlj,m}$ equals the experimental value of the energy $(E_B - E_A)$. Most stripping calculations use the WD procedure to determine $\mathscr{R}_{nlj}$. However, the WD procedure is fairly arbitrary. Other procedures are discussed in Chapter 8.

Another uncertainty of stripping calculations is in a sense based on an anachronism carried over into DW calculations from the original Butler theory. Authors occasionally "exclude the nuclear interior" by introducing

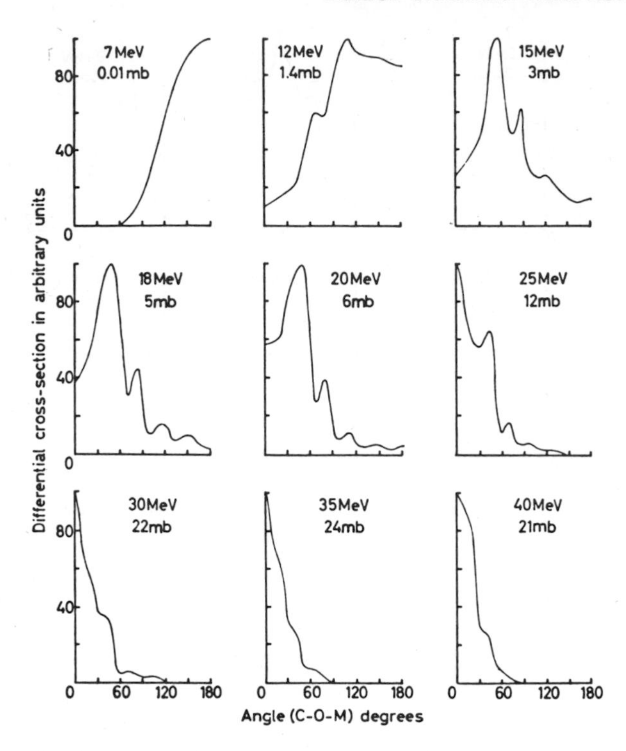

Fig. 5.26 This figure is the same as Fig. 5.24 for $l = 4$ transitions. (From [115], Fig. 3.)

rather arbitrary lower radial cutoffs in the overlap integral of (5.124). Although such cutoffs are meaningless in the DW theory, they tend to improve the agreement with experiment. It seems likely that the use of cutoffs simulates physical effects that are omitted from usual DW calculations; for examples, both the finite-range effect (Section 5.13) and the rather stronger Perey effect (Section 5.2) are known to reduce contributions from the nuclear interior. However, these effects can easily be inserted in (5.124) by means of (5.36) and (5.172). Systematic inclusion of these two effects may reduce the need for cutoffs.

In view of the above-mentioned theoretical uncertainties, cautious experimentalists often "calibrate" (5.124) by studying (d, p) reactions with nuclei of known spectroscopic properties; they then apply their calibrated formulas for a limited range of neighboring nuclei. In this way a systematics of $\mathscr{S}$ factors is developed [103, 117].*

* This point of view was adopted extensively in applications of the early Butler theory [98], and probably explains its many successes.

Similar techniques can be used to unravel the *mixtures of l values* that appear in (5.124) if $J_A \neq 0$. The uncertainties of the theoretical cross sections prevent using them in a reliable fashion to break down an experimental cross section into partial cross sections for individual l values. The difficulties in using the theory for this purpose are illustrated by the fact that the main peak of an $l = 3$ angular distribution is likely to lie at the same angle as the parameter-sensitive second peak of an $l = 1$ angular distribution. Sorenson, Lin, and Cohen [118] overcome this problem by using nearby nuclei with $J_A = 0$ to calibrate the angular distributions for individual l values. Other experiments that show l-mixing are found in [119]. It is interesting that these experiments tend to be sensitive indicators of admixtures of *lower l values* [120], as seen in Figs. 5.19–5.21. The magnitude of the stripping integral in (5.124) falls rapidly with l if the reaction conforms to the usual condition $k_\alpha \approx k_\beta$.

Of course, the integral in (5.124) is evaluated in partial waves expansion, following the general procedure discussed in Section 5.1. It is interesting to examine the structure of the partial waves series, following the procedure used in Section 5.3 for inelastic scattering of strongly-absorbed projectiles. In the earlier discussion a strongly marked L-space localization is observed to occur. The (d, p) reaction does not entirely meet the conditions of that earlier discussion, not only because the proton is not a strongly-absorbed projectile, but also because the small masses of the two projectiles limit the magnitudes of k_α and k_β, and thereby limit the range of angular momenta in which localization can take place. Nevertheless, the medium-energy (d, p) reaction generally shows $k_\alpha \approx k_\beta$, and possesses fairly strong absorption in the deuteron channel. As a consequence contributions from low partial waves tend to be suppressed, and the reaction tends to be dominated by partial waves that make grazing impacts with the nucleus [121]. By virtue of this effect, the properties of the main stripping peak tend to be stable with respect to small variations of parameters. Hooper [122] has discussed additional consequences of L-space localization in stripping reactions.

Because the important partial waves in medium-energy stripping do not have strong overlap with the nuclear interior their properties are somewhat independent of the less-certain aspects of the DW theory. Several authors have suggested that this fact might be exploited [123–125] by using the DW theory to calculate the coefficients of only the higher partial waves, and by treating the coefficients of the low partial waves as parameters to be fitted to experiment. However, such an approach has not been fruitful; not only does it introduce many new adjustable parameters, but also the division between "high" and "low" partial waves is vague.

At high energy ($\gtrsim$100 MeV) the bombarding energy is large compared with $|E_B - E_A|$ and we have $k_\alpha \approx \sqrt{2}k_\beta$. This considerable momentum mismatch reduces the magnitude of the cross section, and washes out the effects that emphasize the nuclear-surface region. Therefore, high-energy stripping theory may be expected to be quite different from medium-energy stripping theory. It is clear that finite-range effects (Section 5.13) must be of enhanced importance at high energy [126]. WKB methods (see Chapter 7) may be used at high energy.

A given orbital angular momentum l is associated with two j values, $j = l \pm \frac{1}{2}$. If $J_A = 0$, these two j values excite different states of nucleus B, having $J_B = l + \frac{1}{2}$, $J_{B'} = l - \frac{1}{2}$. Under these circumstances it is clear that experiment would yield unique determinations of the total angular momenta of (d, p)-excited states of nucleus B, if only there were differences between the angular distributions for these two j values. However, although (5.124) shows a strong dependence on l, the dependence on j is mild; it only comes in through details of the shape of $\mathscr{R}_{nlj}$ and through the energy dependence of $\chi_\beta^{(-)}$. It is interesting that this theoretical prediction is contradicted by experiment. Small systematic differences between stripping angular distributions for $l \pm \frac{1}{2}$ have been found empirically, and these are reliable indicators of j values [127–129].

Two obvious effects related to j-dependence are omitted from the derivation of (5.124). These are (a) the spin-dependent terms of the optical potentials, and (b) the deuteron D-state. Inclusion of the spin-dependence of the optical potentials is straightforward. Although abbreviated DW calculations [130] suggested this spin dependence might explain the empirical j-dependence, more complete DW calculations [128, 131] have not been able to confirm this explanation. (However, see [29] in Chapter 9.)

Inclusion of the D-state should produce a number of interesting spin-dependent effects in the (d, p) angular correlation and the (d, p) polarization (Chapter 9). However, it has long been known [132] that the D-state has little influence on the stripping angular distribution. So long as the optical potentials are spin independent, the deuteron S-state and D-state *contribute incoherently* to the angular distribution [133]. Therefore the contribution to the D-state is proportional to the square of the D-state amplitude and is very much reduced. To understand the origin of this incoherence we go back to (5.110) and evaluate the D-state contribution to the nuclear matrix element. The insertion of (5.111–113) in (5.110) gives

$$\langle B, p| V |A, d\rangle_D = \sqrt{4\pi}\, D_2(r) \sum_{j\mu} \langle J_B M_B \,|\, J_A j; M_A \mu\rangle \times \langle \phi_{j\mu}^{\alpha_A J_A}(n)\, \chi_{\frac{1}{2} m_p}(p) \,|\, \mathscr{Y}_{211}^{m_d}(n, p)\rangle,$$

with

$$D_2(r) = \frac{N\hbar^2}{M\sqrt{4\pi}} (\nabla^2 - \gamma^2) \frac{w(r)}{r}.$$

To put the above expression in standard form it is necessary to display the spin transfer s and the orbital angular momentum transfer l.

The spin transfer is determined by summing over the proton spin coordinate. It is necessary to recouple the D wavefunction to isolate the proton spin function. In the notation of Brink and Satchler [134] this step is written

$$\langle \phi_{j\mu}\chi_{\frac{1}{2}m_p} \mid (\tfrac{1}{2}\tfrac{1}{2})1, 2; 1m_d\rangle = \sum_{s=\frac{3}{2},\frac{5}{2}} \sqrt{3(2s+1)} \times \langle \phi_{j\mu}\chi_{\frac{1}{2}m_p} \mid \tfrac{1}{2}, (\tfrac{1}{2}\,2)s; 1m_d\rangle W(\tfrac{1}{2}\,\tfrac{1}{2}\,12; 1s),$$

in which W is the usual Racah coefficient. Clearly the vector sum of s and $\frac{1}{2}$ is required to equal unity, therefore only $s = \frac{3}{2}$ is allowed. The one remaining Racah coefficient has the value $W(\frac{1}{2}\,\frac{1}{2}\,12; 1\,\frac{3}{2}) = (12)^{-\frac{1}{2}}$, and the spin matrix element reduces to

$$\langle \phi_{j\mu}\chi_{\frac{1}{2}m_p} \mid \mathscr{Y}^{m_d}_{211}\rangle = (\tfrac{1}{2}\sqrt{3})(-)^{\frac{1}{2}-m_p} \times \langle 1, \tfrac{1}{2}; m_d, -m_p \mid \tfrac{3}{2}, m_d - m_p\rangle\langle \phi_{j\mu} \mid (\tfrac{1}{2}\,2)\tfrac{3}{2}, m_d - m_p\rangle.$$

We now see that the D-state form factor has $s = \frac{3}{2}$ uniquely, whereas the S-state form factor has $s = \frac{1}{2}$ uniquely. Therefore (Section 4.5) these two form factors contribute incoherently to the cross section. (Spin-orbit terms in the optical potential to some extent destroy this incoherence.)

To display the l transfer it is necessary to substitute (5.111) for $\phi_{j\mu}$, and to sum over the neutron spin coordinate. However, the notation of (5.111) must be generalized a little to allow the l transfer to differ from the orbital angular momentum of $\phi_{j\mu}$. We use

$$\phi_{j\mu}(n) = \mathscr{R}_{\lambda j}(r_n)\{i^\lambda Y_\lambda(\hat{r}_n), \chi_{\frac{1}{2}}(n)\}_{j\mu}.$$

The D part of the stripping form factor is finally found to be proportional to

$$W(l\lambda\,\tfrac{3}{2}\,\tfrac{1}{2}; 2j)\mathscr{Y}^{m*}_{2\lambda l}(\hat{r}_{np}, \hat{r}_n),$$

where

$$\mathscr{Y}^m_{2\lambda l} \equiv \{Y_2(\hat{r}_{np}), Y_\lambda(r_n)\}_{lm}.$$

It is interesting that j controls the allowed values of l, so that $l = \lambda, \lambda + 2$ if $j = \lambda + \frac{1}{2}$, and $l = \lambda, \lambda - 2$ if $j = \lambda - \frac{1}{2}$. Johnson and Santos [135] find that this effect enables the D state contribution to explain a part of the empirical j-dependence of cross sections.

Some additional j-dependence is obtained by inclusion of target excitation effects (see discussion following (5.108), and Section 6.5) and by the introduction of the correct radial function $\mathscr{R}^{\alpha_A J_A}_{lj}$ for the transferred neutron, in place of $\mathscr{R}_{nlj}$ of (5.124). However, these effects do not seem to lead to the systematic and orderly j-dependence found empirically. A full analysis of the observed j-dependence remains to be given.

One further topic of interest concerns the question of stripping to unbound states of nucleus B. If these states are sharp, the (d, p) reaction is effectively independent of the subsequent neutron emission, and the usual stripping analysis may be employed. Reactions that proceed in such a sequential fashion are not affected by the specifically three-body aspects of the breakup channels. Fortunately it is this well-understood stripping to sharp states of nucleus B that is of greatest practical interest. Stripping to other unbound states might not be subject to standard DW approximations.†

A sharp unbound state of nucleus B is a scattering resonance of the system (A + neutron). The radial function $\mathscr{R}_{lj}^{\alpha_A J_A}$ in (5.122), or $\mathscr{R}_{nlj}$ in (5.124), is the radial wavefunction for the resonant partial wave.‡ Because of the resonance this radial function has a large amplitude at small r, near and inside nucleus A. However, although this function becomes small outside the nucleus it does not go to zero exponentially as does a radial function for a bound state. Instead, as r grows large the radial function for an unbound state oscillates indefinitely with constant amplitude. Such a function does not limit the volume that contributes to the DW integral of (5.122) or (5.124), hence the procedure for calculating this integral becomes complicated.

There is no risk that the DW integral might diverge. Because of energy conservation and recoil it is impossible for the asymptotic momenta of $\mathscr{R}_{lj}$, $\chi_\alpha^{(+)}$, and $\chi_\beta^{(-)}$ to add up to zero. Therefore the DW integrand is badly behaved only in the sense that toward large r it oscillates and does not vanish.

A well-defined value is obtained for the integral if a convergence factor is introduced to average over the oscillations. One suitable version of this procedure replaces the DW integral by the expression

$$\lim_{\varepsilon \to 0} \int \chi_\beta^{(-)*} Y_l^{m*} \sqrt{N_A + 1}\, e^{-\varepsilon r} \mathscr{R}_{lj}^{\alpha_A J_A} \chi_\alpha^{(+)}\, d^3 r. \tag{5.125}$$

The physical origin of the convergence factor may be traced back to the wave-packet analysis of scattering (Chapter 1) in which we see that physical results depend on energy averages of stationary-state scattering amplitudes, these averages being taken over the width of the incident packet. In cases of three-body breakup such energy averages destroy the asymptotic oscillations of the integrand in the same manner as the convergence factor of (5.125). A different argument for use of a convergence factor was given by Huby and Mines [136], who first introduced this method in stripping calculations.

† An interesting early analysis of stripping to unbound states was given by Friedman and Tobocman [99] using Butler boundary-matching methods.

‡ This emphasis upon a single partial wave in the neutron-nucleus relative motion is the principal difference between the present discussion and a general discussion of deuteron breakup.

5.9 Coulomb Stripping

For stripping reactions that have bombarding energies below the Coulomb barrier of the target nucleus and also have low Q values, the distorted waves analysis simplifies and loses some of its sensitivity to uncertainties of the theory. Considerable attention has been given to this special case [137–145].

Because both the entrance and exit channels are required to have energies below the Coulomb barrier, the distorted waves $\chi_\alpha^{(+)}$ and $\chi_\beta^{(-)}$ have little overlap with the region of small r values in which the theory is least reliable. This weak overlap inhibits most nuclear reactions. However, the (d, p) reaction is not inhibited as strongly as most others. For low Q values, the radial function $\mathscr{R}_{nlj}$ falls off slowly with respect to r. Therefore this function can overlap with $\chi_\alpha^{(+)}$ and $\chi_\beta^{(-)}$ at large r, well outside the target nucleus, where these functions are large. Therefore the stripping integral of (5.124) is dominated by large r. It is this fact that reduces the sensitivity to uncertainties of the theory. Below the Coulomb barrier, and with low Q, the distorted-waves stripping integral is dominated by the best-known portions of the wavefunctions.

The kinematics of Coulomb stripping is clarified by recognizing that Q appears in the wavefunctions in two distinct roles. These are seen in the equations

$$Q = E_\beta - E_\alpha,$$
$$Q = -(E_B - E_A) - (\text{deuteron binding energy}).$$

In the first equation Q is the difference between the kinetic energies in the exit and entrance channels. To have simultaneous strong Coulomb repulsion in both channels requires $E_\beta \approx E_\alpha$. In the second equation Q is related to $-(E_B - E_A)$, the energy required to separate a neutron from nucleus B. Slow decrease of the function $\mathscr{R}_{nlj}$ requires $(E_B - E_A) \approx 0$. Obviously the two conditions required to have Coulomb stripping are slightly inconsistent, because (deuteron binding energy) $\neq 0$. Because this binding energy is small, the inconsistency is not important. We note that minimum fall off of $\mathscr{R}_{nlj}$ is achieved if $E_B - E_A = 0$, hence if $Q = -2.226$ MeV.*

Strong repulsion in both the entrance and exit channels appears to imply very small cross sections. However, Goldfarb has emphasized [140, 141] that

* The condition $Q \approx 0$ compels the study of highly excited states of nucleus B, and requires the detection of protons whose energy is nearly equal to that of elastically scattered incident deuterons. These facts limit the practicality of experimental studies of Coulomb stripping.

Goldfarb [140, 141] has mentioned that for (t, d) reactions, the condition $Q \approx 0$ is associated with lowlying states of nucleus B. However, in this case there is a rather greater inconsistency between the conditions $E_\beta \approx E_\alpha$ and $E_A \approx E_B$, and reduced cross sections may occur.

with low Q the stripping cross sections obtained even several MeV below the Coulomb barrier are comparable in magnitude with those obtained at medium energies, many MeV above the barrier. Apparently the Coulomb-dominated cross sections can be large because poor overlap with the target nucleus saves the functions $\chi_\alpha^{(+)}$ and $\chi_\beta^{(-)}$ from being quenched out by absorption. At medium energies the partial waves that penetrate deeply are absorbed, and do not contribute to the (d, p) reaction.

Under limiting conditions of Coulomb stripping the distorted waves $\chi_\alpha^{(+)}$ and $\chi_\beta^{(-)}$ reduce to Coulomb wavefunctions, and the relevant part of $\mathscr{R}_{nlj}$ is its Hankel-function tail. With wavefunctions of this simplicity, the stripping integral of (5.124) can be evaluated to some extent by analytic means [124, 137–139]. However, the resulting approximate analytic expressions for the cross section turn out to be either too complicated or too crude for practical analyses of experiment. We therefore recall that Coulomb stripping is only a special case of the general DW stripping theory, for particular choices of energies. Accordingly, practical analyses are conducted most simply and reliably by use of standard computer codes designed for the zero-range DW theory. Analytic evaluations of (5.124) serve chiefly to clarify the properties of the numerical calculation. They are not discussed in this book.

The analysis of Coulomb stripping requires no new theoretical development. Standard DW computer codes are used, and standard choices of optical potentials are inserted into these codes just as at higher energies. One merely hopes that under Coulomb-dominated circumstances the DW results will not be sensitive to either the choice of the optical potentials or such complications in the nuclear interior as nonlocality, shape of $\mathscr{R}_{lj}^{\alpha_A J_A}$, lower cutoffs, etc. It is clear that the sensitivity to the optical potentials and the nuclear interior is best tested by performing DW calculations with and without the optical potentials, and with and without lower radial cutoffs. Goldfarb [140] performed several calculations of this sort. Figure 5.27 shows a typical set of stripping angular distributions, computed for the reaction $\mathrm{Sn}^{116}(d, p)\,\mathrm{Sn}^{117}$ with $Q = l = 0$, and for eleven different bombarding energies in the range 3.5–12.0 MeV. The solid curves are computed with the optical potentials included. The dashed curves are computed with the optical potentials omitted. For energies as high as 5.5 MeV we see that the presence of the optical potentials makes little difference, therefore the precise values of optical parameters must make even less difference. We also see that at 5.5 MeV the cross section at 180° lies within an order of magnitude of peak cross sections obtained at medium energies. By calculations of the kind just described both Goldfarb [140] and Smith [142] defined ranges of Q and bombarding energy that give reliable and intense Coulomb stripping. These ranges are larger for heavy target nuclei than for light ones.

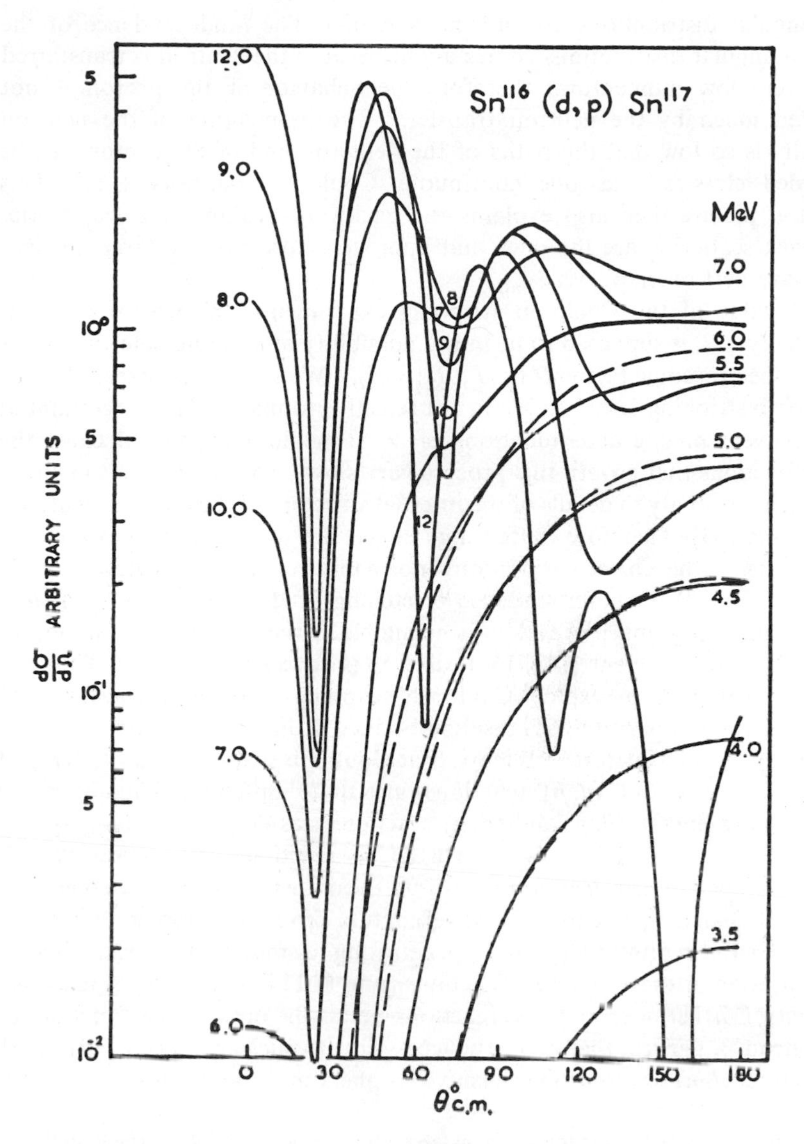

Fig. 5.27 DW predictions for the $Sn^{116}(d, p)Sn^{117*}$ reaction for various deuteron energies from 3.5 − 12.0 MeV, with $Q = 0$, $l = 0$. The dashed curves are computed with the optical potentials omitted. (From [141], Fig. 7.)

Two important general properties of Coulomb stripping are (a) the angular distributions peak at backward angles as seen in Fig. 5.27, and (b) the angular distributions are independent of l. The l-independence of the proton angular distributions comes about because the neutron is transferred with very low momentum; therefore the behavior of the proton is not affected much by the neutron transfer.* The momentum of the neutron actually is so low that the paths of the deuteron and of the proton can be regarded classically as one continuous Coulomb trajectory [137]. This classical picture then also explains the backward peaking: the trajectories for back scattering are the ones that approach closest to the target nucleus and have best overlap with $\mathscr{R}_{nlj}$.

The value of the Coulomb stripping experiments, in view of their insensitivity to l, is supposed to lie in the possibility of accurate determinations of the spectroscopic factor $\mathscr{S}(\alpha_A J_A, lj \mid \alpha_B J_B)$. We have already seen that the Coulomb stripping cross section is not sensitive to many of the uncertainties that prevent precise determinations of $\mathscr{S}$ at medium energies. Because the wavefunctions that govern this process vary slowly with $\mathbf{r}$, there also cannot be much sensitivity to details of the internal structure of the incident deuteron [124, 141, 143]. Therefore finite-range effects are almost entirely accounted for by using the correct asymptotic normalization of the deuteron wavefunction (5.113). Coulomb-induced stretching of the deuteron may at first sight seem important, however, it is negligible in the "prior" formulation of the DW matrix element [143, 144], used in the present discussion. Thus it is indeed of interest to exploit Coulomb stripping to determine accurate $\mathscr{S}$ factors,† for comparison with results obtained at higher energies.

Finally, it is of interest to remark that Coulomb stripping is a property of (d, p) reactions and (d, n) reactions are quite different. The difference between low-energy (d, p) and (d, n) reactions was emphasized long ago by Oppenheimer and Phillips [96]. In a (d, n) reaction the function $\mathscr{R}_{nlj}$ is the wavefunction of a *transferred proton*. Even for low binding energies, the Coulomb barrier prevents this wavefunction from penetrating to large r. The barrier also prevents $\chi_\alpha^{(+)}$ from penetrating to small r. As a result there is little overlap of these two wavefunctions, and (5.124) yields very small cross sections. Furthermore, in (d, n) reactions $\chi_\beta^{(-)}$ is the optical wavefunction for a neutron, a particle that has easy access to the nuclear interior [141, 145]. Therefore (d, n) reactions are sensitive to phenomena at small r.

* An alternative way to state this is that at large r the function $Y_l^{m*}\mathscr{R}_{nlj}$, which carries the l-dependence of (5.124), has very widely spaced nodes. Because the product $\chi_\beta^{(-)*}\chi_\alpha^{(+)}$ is large only in a limited region of space the overlap is not sensitive to the location of the nodes of the l-dependent function.

† It is more precise to state that the quantity measured in a Coulomb-stripping experiment is the asymptotic amplitude of the function $\mathscr{R}_{lj}^{\alpha_A J_A}$. This asymptotic amplitude leads to an $\mathscr{S}$ factor only if $\mathscr{R}_{lj}^{\alpha_A J_A}$ is known well enough at small r so that a meaningful overall normalization can be made. Considerable uncertainties can be caused by this effect [146].

5.10 Other Single-nucleon Transfer

It has been found from experiment that most other single-nucleon transfer reactions, in addition to (d, p) and (d, n), proceed predominantly by the stripping mechanism. The DW theory for these reactions is developed in the same fashion as for deuteron stripping, and it concerns the same spectroscopic properties of nuclei A and B. Of course, the approximations of the DW theory are not so justifiable for these reactions as they are for deuteron stripping. Nevertheless the theory is found to work well for reactions that involve light projectiles, such as (H^3, H^2), (He^3, H^2), (He^4, He^3), and (He^4, H^3). Many experiments that utilize these reactions have been performed. The details of this application of the DW theory are described in this section.

The use of projectiles more complicated than H^2 is suggested by practical experimental reasons. For example, although (d, n) and (He^3, d) test the same single-proton transfer, it is very helpful that both projectiles of the (He^3, d) reaction are charged. Also, the large binding energies of these projectiles are sometimes convenient. Thus the energy released in pickup reactions such as (H^2, H^3) and (H^3, He^4) facilitates the formation of excited states of the residual nucleus. The association of high kinetic energy with high projectile mass has another interesting experimental consequence. It causes the momentum transfer $|k_\beta - k_\alpha|$ to be large, with the result that the stripping mechanism favors large l transfers (see Section 5.1, and [5]). It is also of some value that both projectiles are strongly absorbed. Some reduction of sensitivity to optical potential parameters may be expected from this fact. It is also interesting that He^4 has spin zero. This facilitates understanding of the spin dependence of the stripping reaction.

Just as definiteness of discussion is obtained for deuteron stripping by treating the (d, p) reaction, it is obtained for the present cases by treating the neutron-transfer reactions (H^3, H^2) and (He^4, He^3).

The nonexchange distorted-waves amplitude for these reactions has the same structure as (5.105) of the (d, p) discussion:

$$T_{\alpha\beta}{}^{DW} = (N_A + 1)\mathcal{J}\int d^3r_\alpha \int d^3r_\beta \chi_\beta^{(-)*}(\mathbf{k}_\beta, \mathbf{r}_\beta)\langle B, b|\, V\, |A, a\rangle \chi_\alpha^{(+)}(\mathbf{k}_\alpha, \mathbf{r}_\alpha). \quad (5.126)$$

Antisymmetrization of the neutrons in nucleus B yields the same factor $(N_A + 1)$ that appeared for deuteron stripping. In this factor, N_A is the neutron number of nucleus A. However, when (5.126) is inserted in (4.61) to compute the cross section, it is necessary to use for the channel weight factor $N_\beta/N_\alpha = 2(N_A + 1)^{-1}$. The factor 2 appears both for H^3 and for He^4 because each of these nuclei has two equivalent neutrons available for the transfer process.

The nuclear matrix element is

$$\langle B, b\,|V|\,A, a\rangle = \langle \psi_B\phi_b(y)|\; V_\beta - U_\beta(\mathbf{r}_\beta)\;|\psi_A\phi_a(n_0, y)\rangle, \qquad (5.127)$$

in which y is the set of internal coordinates of nucleus b. Then ϕ_a depends on y and on n_0, the coordinates of the transferred neutron. By arguments similar to those used to reduce from (5.106) to (5.108), we are led to approximate (5.127) by

$$\langle B, b|\; V\;|A, a\rangle \approx \langle \psi_B\phi_b|\sum_k^{(b)} V(n_0k)\;|\psi_A\phi_a\rangle. \qquad (5.128)$$

Here we sum over all the interactions of n_0 with the nucleons of b. Equation 5.128 yields the form factors for the (H^3, H^2) and (He^4, He^3) reactions.

The spectroscopic aspects of (5.128) are identical with those of (5.108), and are treated by inserting for ψ_B the same generalized fractional parentage expansion, (5.109), used for the (d, p) case. The nuclear matrix element becomes

$$\langle B, b|\; V\;|A, a\rangle = \sum_j \langle J_BM_B\,|\,J_A\,j; M_A, M_B - M_A\rangle$$

$$\times\; \langle \phi^{\alpha_A J_A}_{j,\,M_B-M_A}\phi_{(b)s_b,\,m_b}|\sum_k^{(b)} V(n_0k)\;|\phi_{(a)s_a,\,m_a}\rangle. \qquad (5.129)$$

We insert the factored expression of (5.111) for the $\phi^{\alpha_A J_A}_{j,\mu}$, just as in deuteron stripping.

The projectile wavefunctions could be treated in a similar fashion by inserting a fractional parentage expansion of $\phi_{(a)s_a,m_a}$. However, such a procedure would be of little interest. Because we wish to use nuclei a and b to study spectroscopic properties of A and B it is more valuable to introduce suitable, known wavefunctions for ϕ_a and ϕ_b, and to explicitly carry out the integrations implied by (5.129). As an example, we now treat in detail the reaction (H^3, H^2).

In agreement with the procedure of the (d, p) analysis, we carry only the dominant S-state terms for both ϕ_a and ϕ_b. Suitable radial wavefunctions for these S-state terms are discussed by Bassel [147] as part of a careful study of spectroscopic applications of the reactions (H^3, d) and (He^3, d). For H^3 and He^3 he recommends use of the Irving-Gunn wavefunction [148]; for H^3 this function specializes to

$$\phi_{(a)s_a,\,m_a} = (A/R)e^{-\delta R/2}\chi_{0,0}(n_0, n_1)\chi_{\frac{1}{2},\,m_a}(p_1). \qquad (5.130)$$

Here p_1, n_0, and n_1 are the proton and the two neutrons, respectively, and $\chi_{0,0}$ is the singlet spin wavefunction. The radial wavefunction in the above expression depends on the variable

$$R = (r^2_{n_0n_1} + r^2_{n_0p_1} + r^2_{n_1p_1})^{\frac{1}{2}},$$

and has the normalization

$$A = 3^{1/4}\delta^2(2\pi^3)^{-1/2},$$

with $\delta = 0.768\ \text{fm}^{-1}$. For the deuteron radial wavefunction, to be inserted in (5.112), Bassel uses the Hulthén approximation

$$u(r) = e^{-\gamma r} - e^{-\zeta r}. \tag{5.131}$$

It is necessary to use (5.131) rather than the zero-range approximation for $u(r)$, because of the possible importance of finite-range effects.

The insertion of (5.111), (5.112), and (5.130) into (5.128) now allows the integration over spins to be done, with the result

$$\begin{aligned}\langle B, d|\ &V(n_0n_1) + V(n_0p_1)\,|A, t\rangle \\ &= \tfrac{1}{2}\sqrt{3}\ D^{(3,2)}(r_{n_0d})(-)^{1-m_d} \\ &\times \sum_j \langle J_BM_B \mid J_Aj;\ M_A, M_B - M_A\rangle\langle\tfrac{1}{2}\ 1;\ m_t, -m_d \mid \tfrac{1}{2}, m_d - m_t\rangle \\ &\times \langle l\ \tfrac{1}{2};\ m, m_t - m_d \mid j, M_B - M_A\rangle i^{-l}Y_l^{m*}(\hat{r}_{n_0A})\ \mathscr{R}_{lj}^{\alpha_AJ_A}(r_{n_0A}),\end{aligned} \tag{5.132}$$

with

$$D^{(3,2)}(r_{n_0d}) = \int d^3r_{n_1p_1}\frac{Nu}{r\sqrt{4\pi}}\,[V(n_0n_1) + V(n_0p_1)]\,\frac{Ae^{-\delta R/2}}{R}. \tag{5.133}$$

The superscripts indicate that $D^{(3,2)}$ refers to the (H^3, H^2) reaction. Comparison with (4.59) and (4.60) discloses that the form factors for the (H^3, H^2) reaction are

$$A_{lsj}f_{lsj,m} = (N_A + 1)^{1/2}\mathscr{J}\tfrac{1}{2}\sqrt{3}\ D^{(3,2)}(r_{n_0d})\,Y_l^{m*}(\hat{r}_{n_0A})\ \mathscr{R}_{lj}^{\alpha_AJ_A}(r_{n_0A})\,\delta_{s,1/2}. \tag{5.134}$$

Here $m = M_B - M_A + m_d - m_t$. The product $(N_A + 1)^{1/2}\mathscr{J}$ has been included in (5.134) for the same reasons that it is included in (5.116). Once again we observe that s has the unique value $s = \tfrac{1}{2}$.

We see that the form factors depend on two displacement variables, $\mathbf{r}_{n_0d}$ and $\mathbf{r}_{n_0A}$. In general, when a function of multipolarity l depends on more than one displacement there is no definite relation between l and parity. This question is treated at greater length in Section 5.13 in which finite-range integrations are discussed. However, the parity transfer in the present analysis is not complicated. Our use of S-state wavefunctions for ϕ_a and ϕ_b causes $D^{(3,2)}(r_{n_0d})$ to be a scalar function of its argument. Therefore both the multipolarity and the parity of the form factors are determined by $Y_l^{m*}(\hat{r}_{n_0A})$, which is a function of only a single displacement vector. Therefore the form factors of (5.134) possess normal parity.

The Jacobian for a finite-range (H^3, H^2) reaction is obtained by methods similar to those used in connection with (5.118). It is

$$\mathcal{J} = 27\left(\frac{A+1}{A+3}\right)^3.$$

The internal structure of H^2 plays no role in the coordinate transformation that leads to this $\mathcal{J}$.

Zero-range approximation is not so obvious for the (H^3, H^2) reaction as for (d, p). The range of the function $D^{(3,2)}(r_{n_0d})$ is determined by the triton and deuteron radii rather than the range of the two-nucleon interaction. To study the significance of the range of $D^{(3,2)}$, Bassel [147] evaluates (5.133) in two stages. In the first stage he computes the volume integral

$$D_0^{(3,2)} = \int D^{(3,2)}(r_{n_0d})\, d^3r_{n_0d}. \tag{5.135}$$

He then introduces the notation

$$D^{(3,2)}(r) \equiv D_0^{(3,2)} f(r), \tag{5.136}$$

and in the second stage of calculation he evaluates the second moment of the (normalized) function $f(r)$. Then the extent to which the subsequent DW calculation is sensitive to the range of $f(r)$ is tested by replacing $f(r)$ by a normalized gaussian that has the same second moment (see Section 5.13). It is found that this inclusion of finite range causes a sixteen percent reduction of the cross section at the main stripping peak and modifies the angular distribution at large angles. Although the departures from zero range are not severe, Bassel does conclude that finite-range form factors should be used in DW treatments of (H^3, d) and (He^3, d).

For the (H^3, H^2) reaction Bassel finds that $D_0^{(3,2)}$ has the numerical value

$$D_0^{(3,2)}(t, d) = -183.6(\text{MeV fm}^{3/2}). \tag{5.137a}$$

For the (He^3, H^2) reaction he computes two values for the normalization coefficient, because it is not clear whether the repulsive Coulomb interaction between H^2 and the transferred proton should or should not be included in the calculation of the form factor. These two values are

$$\begin{aligned} D_0^{(3,2)}(\text{He}^3, d)_{\text{coulomb}} &= -161\ (\text{MeV fm}^{3/2}), \\ D_0^{(3,2)}(\text{He}^3, d)_{\text{noncoulomb}} &= -172.8\ (\text{MeV fm}^{3/2}). \end{aligned} \tag{5.137b}$$

The uncertainty of $D_0^{(3,2)}(\text{He}^3, d)$ is of the same order as the more familiar uncertainties of the DW approach.

Bassel applied his normalized, finite-range form factor to an analysis of the reaction $Ca^{48}(He^3, d)Sc^{49}$, at 22 MeV bombarding energy. It is highly probable that this reaction proceeds as an $f_{7/2}$ proton transfer, with $\mathscr{S} = 1$. The analysis indeed yielded excellent predictions of the observed angular distribution, and yielded measured $\mathscr{S}$ values that lay close to $\mathscr{S} = 1$. Figure 5.28 shows the comparison of theory with experiment.

Bassel also studied the modifications caused by inclusion of the Perey effect (Section 5.2). Although strong absorption of the distorted waves apparently prevents the Perey effect from having much influence upon the cross section, Bassel does prefer the results obtained if this effect is ignored, that is, he prefers the results obtained by treating the optical potentials as local. Similar conclusions were reached by Hiebert, Newman, and Bassel [149] in a careful study of (d, He^3) reactions on the closed-shell nuclei O^{16} and Ca^{40}, at 34.4 MeV bombarding energy.

Bassel's a priori absolute normalization of form factors for (H^3, d) and (He^3, d) reactions also agrees well [147, 149, 150] with normalizations obtained empirically in sets of experiments designed to explore systematics of $\mathscr{S}$ factors. It must be concluded that these reactions are reliable tools for

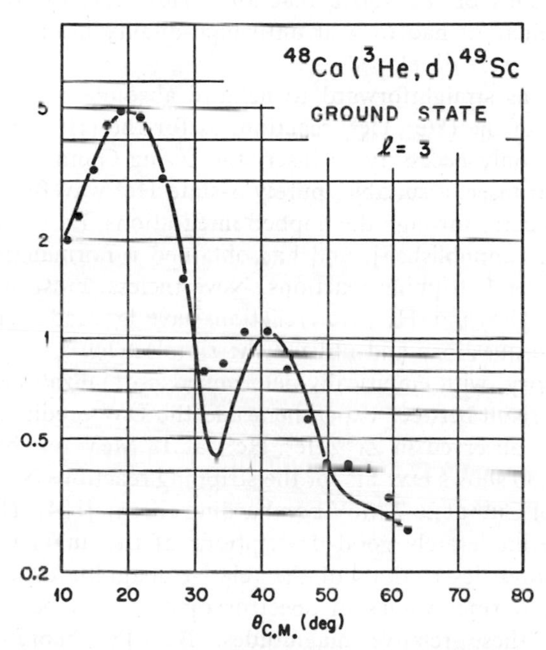

Fig. 5.28 Comparison of the DW theory with experimental data for the ground state $l = 3$ transition in $Ca^{48}(He^3, d)Sc^{49}$, at 22 MeV bombarding energy. The calculation incorporates finite range and the Perey effect. (From [147], Fig. 1*d*.)

the study of nuclear spectroscopy, and that their level of reliability is comparable with that of deuteron stripping. These reactions differ from deuteron stripping only in the marginal need to incorporate a finite-range treatment of the integrations.

Other theoretical studies of the absolute normalization of (H^3, H^2) and (He^3, H^2) reactions have been published by Glover, Jones, and Rook [151] and by Rook [152]. While [151] finds a normalization that agrees with Bassel, Rook [152] finds that quite different normalizations can be obtained by using wavefunctions ϕ_a and ϕ_b governed by nucleon–nucleon interactions that possess a repulsive core.

Naturally, the most reliable way to obtain $\mathscr{S}$ factors from H^3 and He^3 stripping experiments, as with deuteron stripping, is to establish comparisons among reactions that involve a series of neighboring nuclei.

The reactions (He^4, He^3), (He^4, H^3) have not received as much attention as have the reactions (H^3, H^2), (He^3, H^2). This is understandable. The high Q of the pickup reactions (H^3, He^4), (He^3, He^4) causes these to be the interesting versions of the above reactions. However, the exploitation of these pickup reactions had to wait until high-quality beams of 12–20 MeV H^3 and He^3 ions were developed.

It should be as straightforward to achieve absolute a priori theoretical normalizations of the (He^4, He^3) reaction, as for the (H^3, H^2) reaction just described. It is only necessary to insert the Irving-Gunn wavefunction for ϕ_b in (5.129), to insert a suitable, purely S-state He^4 wavefunction for ϕ_a in (5.129), and to carry through the implied integrations. Bassel has performed this calculation [unpublished], and has obtained a normalized finite-range form factor for He^4 stripping reactions. Nevertheless, most distorted-waves analyses of (H^3, He^4) and (He^3, He^4) reactions have ignored the normalization and finite-range questions and utilized the simplest local, zero-range forms of the DW theory, with empirically determined normalizations. Figure 5.29 shows a comparison between experiment and the DW predictions for $l = 1$, 3, 4 transitions observed in Zr^{90}(He^3, He^4) at 18 MeV bombarding energy [153]. Figure 5.30 shows DW fits for the stripping reactions Ni^{64}(He^4, H^3), to various states of Cu^{65} at 26.7 MeV bombarding energy [154]. The DW theory is seen to give moderately good descriptions of the angular distributions. It also gives good descriptions of the relative magnitudes of the different cross sections, normal values of spectroscopic factors being used in the calculation of these relative magnitudes. Bassel's theoretically derived normalization agrees with the normalizations found from these experiments.

It is a little surprising that the elementary stripping theory should be applicable for, say, the (He^4, He^3) reaction. The removal of a neutron from He^4 disrupts a tightly bound structure and leaves a loosely bound, easily

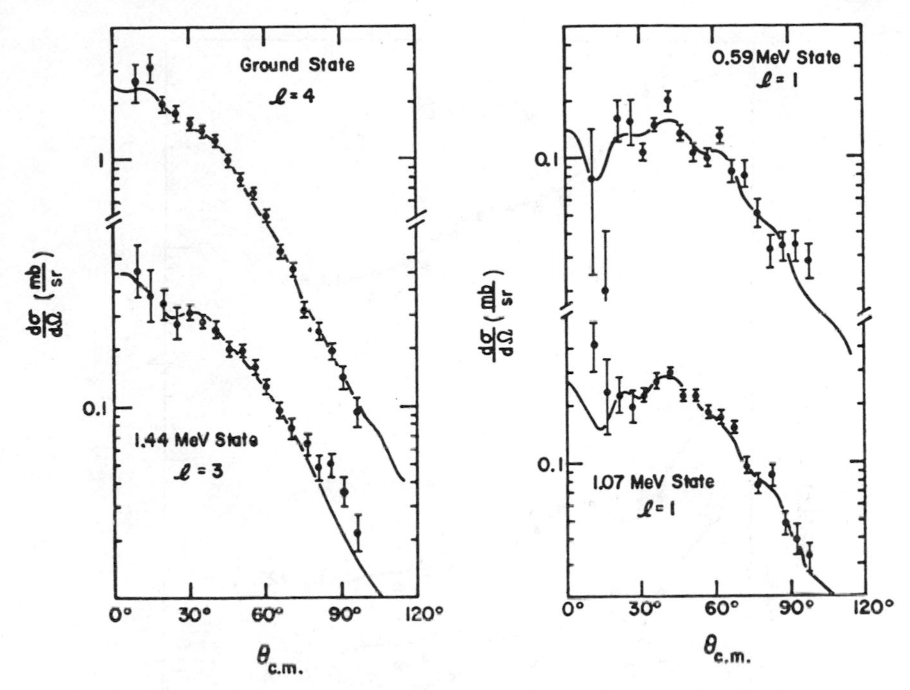

Fig. 5.29 Comparison of the DW theory with experimental data for $l = 1, 3, 4$ transitions in $Zr^{90}(He^3, He^4)$ at 18 MeV bombarding energy. (From [153], Fig. 3.)

deformable structure. It is not clear that usual stripping approximations are correct under these conditions [155].

J-dependent angular distributions are found experimentally, in both the reactions (H^3, d), (He^3, d) and (He^4, He^3), (He^4, H^3). Unfortunately, despite the convenient spin values in the He^4-induced reactions, the inclusion of spin-orbit optical potentials in the DW theory does not explain the observed *j*-dependence [154, 156]. This failure may be related to the deformability of the trinucleon systems, mentioned above.

Finally, it is interesting to remark again that the energy released in a (He^3, He^4) reaction often causes *preferential population* of higher l transfers. A striking example of this effect was found by Blair and Wegner [5] in a study of (He^3, He^4) reactions on isotopes of iron at 14.3 and 24.9 MeV bombarding energy. Because these reactions entail both a mass increase and a positive Q value, the momenta in the initial and final channels are very different. The authors point out that in their experiment this causes $(k_\beta R - k_\alpha R) \approx 5$, where R is the nuclear radius. As a result, as is explained in Section 5.1, the DW theory predicts that l transfers having $l \approx 5$ should be favored over lower

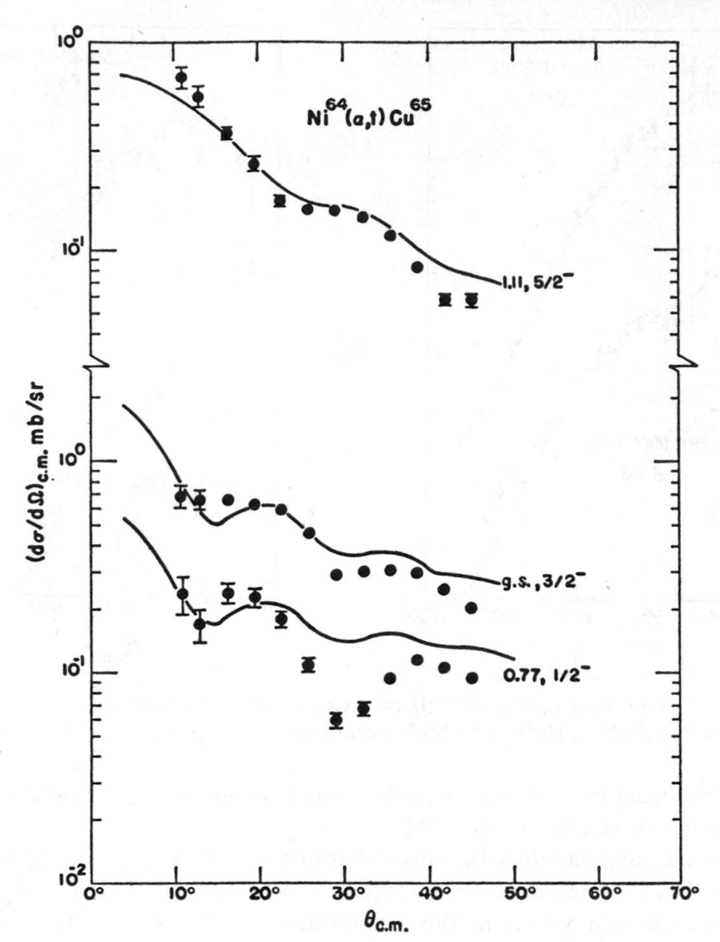

Fig. 5.30 Comparison of the DW theory with experimental data for $l = 1$, 3 transitions in $Ni^{64}(He^4, H^3)$ at 26.7 MeV bombarding energy. (From [154], Fig. 4.)

l transfers. Blair and Wegner found $l = 3$ to be favored over $l = 1$, in agreement with DW calculations.

5.11 Multiple-nucleon Transfer

Direct reaction mechanisms also are appropriate for a large variety of reactions in which two or more nucleons are transferred. Examples of two-nucleon transfer are (α, d), (He^3, p), (He^3, n), (H^3, p), and (H^3, n). Of a similar nature are the reactions $(\alpha, 2p)$ and $(\alpha, 2n)$, which lead to three-body breakup. Examples of three-nucleon transfer are (α, p), (α, n), (He^3, Li^6), (α, Li^7), etc. Examples of four- and five-nucleon transfer are (d, Li^6) and (d, Li^7). These reactions all seem to yield DI excitation of low-lying states of product nuclei if suitable bombarding energies are used.

A uniform theoretical treatment of the above reactions must not be expected. In several cases both the incoming and outgoing projectiles are strongly-absorbed particles; however, in other cases we note that one of the projectiles is a nucleon, a weakly-absorbed particle. In several cases the momentum transfer $|k_\alpha - k_\beta|$ is small, and in other cases it is very large. Under these conditions we may expect that several of the above reactions take place primarily at the nuclear surface, with consequent simplification of the reaction mechanism, and that others involve the entire nuclear interior very strongly. Finite-range effects are important in some cases, but tend to be quenched in other cases.

Despite the variety of conditions that the above reactions present, authors generally consider that all these reactions are dominated by stripping mechanisms. In other words, in the analysis of the reaction (a, b), with a more massive than b, the nuclide b is treated as a subunit of a. As an alternative, a knockout or heavy-particle stripping analysis (Sections 4.8, 5.12, Chapter 10) treats a as a subunit of B, and treats b as a subunit of A. The stripping mechanism apparently yields the larger amplitude, because it requires least rearrangement of coordinates. Unfortunately it is not easy to establish from experiment that the reaction mechanism is (light-particle) stripping, rather than knockout or heavy-particle stripping. We recall (Section 5.1) that in zero-range approximation the angular distribution predicted by a DW theory tends to be independent of the reaction mechanism. Therefore, valid studies of reaction mechanisms tend to require investigations of the systematics of cross section magnitudes [157–160]. In the present section we assume the (light-particle) stripping mechanism for multi-nucleon transfer, and only consider in detail how the present application of this stripping mechanism differs from the applications treated in Sections 5.8–5.10.

Only two nucleon transfer is treated explicitly. This case is an adequate illustration of the other cases of multi-nucleon transfer. It is also the case that has received the most widespread application [161–165]. For definiteness, we concentrate attention on the reactions (α, d), (H^3, n) and (H^3, p). The (α, d) and (H^3, n) reactions transfer a pair of unlike nucleons; the (H^3, p) reaction transfers a pair of like nucleons.

The spectroscopy of two-nucleon transfer is governed by a few simple selection rules and dynamical properties. We note first that the internal wavefunctions of H^2, H^3, He^3, and He^4 are dominated by symmetric S-state terms, in which the relative motion of each pair of nucleons is itself an S-state. In addition, the Pauli principle requires like nucleons to move in *singlet* relative S-states. Unlike nucleons are allowed to move in either *singlet* or *triplet* relative S-states. It is clear that a two-nucleon stripping reaction inserts into nucleus A a pair of nucleons whose relative motion is compelled to be either ${}^1S$ or ${}^3S$. It is also clear that the reaction often selects only one of these two spin states. For example, the (H^3, p) reaction uniquely selects ${}^1S$. The (α, d) reaction uniquely selects ${}^3S$; therefore it excites only states of nucleus B whose i-spin is the same as that of nucleus A. The (H^3, n) and (He^3, p) reactions allow both ${}^1S$ and ${}^3S$ transfer. We see that two-nucleon transfer reactions can be used to study S-state pair correlations in ψ_B. The two-nucleon transfer reactions also give some indication of the *spatial correlation* of these S-state pairs. Best overlap is obtained if the separation of the two transferred nucleons is about the same in B as in a.

Correct normalization of the amplitude for two-nucleon transfer requires the determination of the number of equivalent ways in which the two transferred nucleons appear in the antisymmetrized wavefunction of nucleus B. To determine this number we follow the discussion of the antisymmetrized amplitude for single-nucleon transfer given in Section 4.8. In that case the discussion yields the factor $(N_A + 1)$ that multiplies the amplitude for single-nucleon transfer [see (4.96), (5.105), (5.126)]. We must now conduct a similar discussion of the antisymmetrized amplitude for two-nucleon transfer. It is sufficient to consider the H^3-induced reactions, (H^3, p) and (H^3, n).

We therefore return to the general antisymmetrized amplitude of (4.86) and substitute explicit interchanges in place of the sum over permutations. For H^3-induced reactions this gives

$$T_{\alpha\beta}{}^{\mathrm{DW}} = \Bigg(\psi_b \psi_B \chi_\beta^{(-)}(\mathbf{r}_\beta), [V_\beta - U_\beta] \times \frac{1}{2}\left[1 - \sum_{i=1_p}^{n_p} P_{i0_p}\right]\left[1 - \sum_{j=1_n}^{n_n+1} P_{j0_n}\right]\left[1 - \sum_{k=2_n}^{n_n+1} P_{k1_n}\right] \times \phi_t(0_p, 0_n, 1_n)\, \psi_A(1_p, \ldots, n_p; 2_n, \ldots, n_n + 1)\, \chi_\alpha^{(+)}(\mathbf{r}_\alpha)\Bigg). \qquad (5.138)$$

In cases of two-nucleon transfer two of the three single-particle antisymmetrizers in (5.138) can be commuted through the interaction and allowed to operate on the already antisymmetrized function ψ_B. For the (H^3, p) reaction it is the neutron antisymmetrizers that commute through, and we obtain

$$T_{\alpha\beta}{}^{\mathrm{DW}} = \tfrac{1}{2}(N_A + 2)(N_A + 1)\big(\psi_b(0_p)\psi_B(1_p, \ldots, n_p; 0_n, \ldots, n_n + 1)\, \chi_\beta^{(-)}, \times [V_\beta - U_\beta][1 - Z_A P_{0_p 1_p}]\phi_t(0_p, 0_n, 1_n) \times \psi_A(1_p, \ldots, n_p; 2_n, \ldots, n_n + 1)\, \chi_\alpha^{(+)}\big). \qquad (5.139)$$

For the (H^3, n) reaction we obtain in a similar fashion

$$T_{\alpha\beta}{}^{\mathrm{DW}} = (N_A + 1)(Z_A + 1)\Big(\psi_b(0_n)\psi_B(0_p, \ldots, n_p; 1_n, \ldots, n_n + 1)\,\chi_\beta^{(-)},$$

$$\times\,[V_\beta - U_\beta]\left[1 - \frac{N_A}{2}\,P_{0_n 1_n}\right]\phi_t(0_p, 0_n, 1_n)$$

$$\times\,\psi_A(1_p, \ldots, n_p; 2_n, \ldots, n_n + 1)\,\chi_\alpha^{(+)}\Big) \tag{5.140}$$

Here the notation N_A, Z_A has been introduced in the normalization coefficients in place of n_n, n_p, in conformity with the usage of the present chapter.

Of course, this chapter only treats the direct terms of the above DW amplitudes. From the discussion given it is clear that $\nu = \frac{1}{2}(N_A + 2)(N_A + 1)$ is the normalizer of the direct DW amplitudes for all reactions that transfer two neutrons and that $\nu = (N_A + 1)(Z_A + 1)$ is the normalizer for all transfers of a neutron and a proton. We see above that these normalizers are consequences of the antisymmetrization of ψ_B, and that they are independent of the type of the projectile a that is incident. Evidently a reaction that transfers two protons would have a normalizer $\nu = \frac{1}{2}(Z_A + 2)(Z_A + 1)$.—Use of the i-spin formalism would give a different set of normalizers; however, this difference would be compensated by changes in the normalization of ψ_A and ψ_B.

To calculate cross sections the DW amplitudes must be inserted in (4.61). For reactions (H^3, p) and (He^3, n), in which two like particles are transferred, the channel weight factors in (4.61) are $(N_\beta/N_\alpha) = \nu^{-1}$. However, the channel weight factors for the reactions (α, d), (H^3, n) and (He^3, p) are $2\nu^{-1}$. The additional factor 2 appears because there are two independent ways to select the n, p pair transferred in each in these reactions. As in single-nucleon transfer, we see that the channel weight factor partially cancels the normalizer of the amplitude.

The nonexchange distorted-waves amplitude for two-nucleon transfer now has the same structure as (5.105) and (5.126) in the discussion of single nucleon transfer. It is

$$T_{\alpha\beta}{}^{\mathrm{DW}} = \nu\mathscr{J}\int d^3r_\alpha \int d^3r_\beta \chi_\beta^{(-)*}(\mathbf{k}_\beta, \mathbf{r}_\beta)\langle B, b|\, V\, |A, a\rangle\, \chi_\alpha^{(+)}(\mathbf{k}_\alpha, \mathbf{r}_\alpha), \tag{5.141}$$

with ν the normalization coefficient just introduced. The nuclear matrix element is of the usual form, with $V = V_\beta - U_\beta$ the difference between the net B, b interaction and the optical interaction. By arguments similar to those used to reduce from (5.106) to (5.108), we are led to approximate this V by just the sum of two-body interactions between b and the two transferred nucleons. For the (H^3, n) reaction the nuclear matrix element becomes

$$\langle B, b| V |A, a\rangle = \langle \psi_B \phi_b(n_0)| V(n_0 n_1) + V(n_0 p_0) |\psi_A \phi_t(n_0, n_1, p_0)\rangle. \quad (5.142)$$

This matrix element may now be reduced by the same methods followed for single-nucleon transfer. We introduce a parentage expansion of ψ_B, and we integrate explicitly over the product $\phi_b^* V \phi_t$.

The parentage expansion for two-nucleon transfer is generated by methods similar to those used to construct (5.109) for single-nucleon transfer. We expand the antisymmetrized function ψ_B in terms of the complete set of antisymmetrized eigenstates of the target nucleus $\psi_{\alpha_P J_P M_P}$. These "parent states" are functions of the coordinates $1_p, \ldots, n_p; 2_n, \ldots, n_n + 1$. Then ψ_B becomes

$$\psi_{\alpha_B J_B M_B} = \sum_{\alpha_P J_P j M_P} \langle J_B M_B \,|\, J_P j; M_P, M_B - M_P\rangle \times \psi_{\alpha_P J_P M_P} \phi_{j,\, M_B - M_P}^{\alpha_P J_P}(n_1, p_0), \quad (5.143)$$

where for each parent state the set of coefficients $\phi_{j,\, M_B - M_P}^{\alpha_P J_P}(n_1, p_0)$ are functions of the coordinates of the transferred nucleons, and have definite angular momentum j and projection $M_B - M_P$. Because the $\phi_{j,\mu}^{\alpha_P J_P}(n_1, p_0)$ are introduced only as coefficients in (5.143) they are not necessarily normalized to unity.

The coefficients $\phi_{j,\mu}^{\alpha_P J_P}$ depend on the coordinates of n_1 and p_0 relative to the center of mass of the target nucleus. Because ψ_B and ψ_P have definite parities the coefficient functions have definite parities. However, because the $\phi_{j,\mu}^{\alpha_P J_P}$ are functions of two displacement vectors their structure is not determined by j and parity to anything like the extent found for the corresponding expansion coefficients in (5.109). Articles about two-nucleon transfer generally give extensive discussions of properties of the $\phi_{j,\mu}^{\alpha_P J_P}$ and of methods for calculating them. Normal techniques of spectroscopy require that the $\phi_{j,\mu}^{\alpha_P J_P}$ be expanded in products of single-particle shell-model eigenfunctions for n_1 and p_0 separately. Then the $\phi_{j,\mu}^{\alpha_P J_P}$ are *coherent linear combinations* of such products. The magnitudes and phases of the coefficients in these linear combinations provide a rich subject for theoretical investigation [161–167]. However, the limitations of the present book compel us to sketch this spectroscopic investigation only in somewhat general terms.

To proceed with the calculation of the nuclear matrix element we substitute (5.143) in (5.142), to obtain

$$\langle B, b| V |A, a\rangle = \sum_j \langle J_B M_B \,|\, J_A j; M_A, M_B - M_A\rangle \times \langle \phi_{j, M_B - M_A}^{\alpha_A J_A}(n_1, p_0)\, \phi_b(n_0)| V(n_0, n_1) + V(n_0 p_0) |\phi_t(n_0, n_1, p_0)\rangle. \quad (5.144)$$

Orthogonality of the ψ_P has reduced the extent of contributions from the parentage expansion. Only those few terms of (5.143) based on ψ_A as a parent state contribute to the nuclear matrix element. Further reduction now originates from our assumption that each pair of nucleons in ϕ_t moves in a relative S-state. To be consistent with this assumption we must treat the interactions $V(ij)$ as central effective two-body interactions of the sort presented in (5.75). Under these conditions only the parts of $\phi_{j\mu}^{\alpha_A J_A}$ that are S-states in the vector $\boldsymbol{\rho}$ can contribute to (5.144), where

$$\begin{aligned} \boldsymbol{\rho} &\equiv \mathbf{r}_{n_1 A} - \mathbf{r}_{p_0 A}, \\ \mathbf{r} &\equiv \tfrac{1}{2}(\mathbf{r}_{n_1 A} + \mathbf{r}_{p_0 A}), \end{aligned} \tag{5.145}$$

are an alternative set of coordinate vectors for the two transferred nucleons.

The relative S-state part of $\phi_{j\mu}^{\alpha_A J_A}$ can now be expanded in terms of the orbital angular momentum in the vector $\mathbf{r}$, and the spin angular momenta of n_1 and p_0. The relative S-state part of $\phi_{j\mu}^{\alpha_A J_A}$ becomes

$$\sum_{ls} \theta_{lsj}^{A}(r, \rho)\{i^l Y_l(\hat{r}), \chi_s(n_1, p_0)\}_{j\mu}. \tag{5.146}$$

Here $Y_l^m(\hat{r})$ is an ordinary spherical harmonic function of one variable. The function $\chi_{s,\mu-m}$ is a singlet or triplet spin function. Thus, $s = 0, 1$. If $s = 0$ then $l = j$, but if $s = 1$ then $l = j \pm 1$ or $l = j$, depending on the parity. Evidently the sum in (5.146) is very limited.

We now substitute the above explicit functions into (5.144) and perform the spin sums. The functions to be inserted are (5.146) for the relative S-part of $\psi_{j\mu}^{\alpha_A J_A}$; a single-particle spin function for $\phi_b(n_0)$; (5.75) for the interactions; and the Irving-Gunn wavefunction of (5.130) for ϕ_t. We obtain

$$\begin{aligned} \langle B, b|\, V\, |A, a\rangle = 2^{-1/2}(-)^{1/2 - m_b} & \\ &\times \sum_{lsj} \langle J_B M_B \,|\, J_A j; M_A, M_B - M_A\rangle \\ &\times \langle \tfrac{1}{2}\tfrac{1}{2}; m_a, -m_b \,|\, s, m_a - m_b\rangle \\ &\times \langle ls; m, m_a - m_b \,|\, j, M_B - M_A\rangle (-)^{l}\, i^{l}\, Y_l^{m*}(\hat{r}) \\ &\times [(4V_{10} + 2V_{11} - 2V_{00}) - 4(V_{10} - V_{11})\,\delta_{s,0}] I_{lsj}^{A}(r, z), \end{aligned} \tag{5.147}$$

where I_{lsj}^{A} is an integral over the radial wavefunctions and the two-body interaction,

$$I_{lsj}^{A} = \int d^3\boldsymbol{\rho}\, \theta_{lsj}^{A}(r, \rho)\, g(r_{n_0 p_0})(A/R)e^{-\delta R/2} \tag{5.148}$$

Because the triton wavefunction is treated as a symmetric S-state, the integral I_{lsj}^{A} is not changed if $g(r_{n_0 p_0})$ is replaced by $g(r_{n_0 n_1})$. It is for this reason that only one radial integral I^A_{lsj} appears in (5.147). Clearly, because we only integrate over $\boldsymbol{\rho}$, the integral I^A_{lsj} remains a function of both r and an internal vector of the triton. Figure 5.31 defines in a symmetrical fashion a displacement $\mathbf{z}$ that can be used as the additional argument of I^A_{lsj}. Naturally, I^A_{lsj} is a scalar function of $\mathbf{z}$.

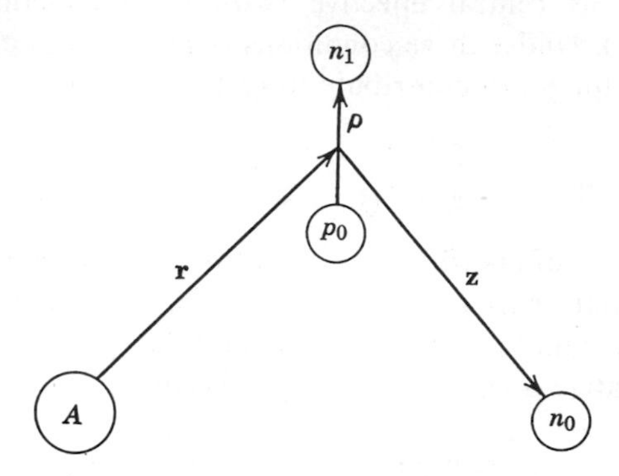

Fig. 5.31 Integration variables for two-nucleon transfer.

Comparison with (4.59) and (4.60), which give the standard form for the DW amplitude, discloses that the form factors for the (H^3, n) reaction are

$$A_{lsj}f_{lsj,m} = \sqrt{\nu}\mathscr{J}2^{-1/2}(-)^{1-s}[(4V_{10} + 2V_{11} - 2V_{00}) - 4(V_{10} - V_{11})\,\delta_{s,0}]$$
$$\times\ Y_l^{m*}(\hat{r})\, I^A_{lsj}(r, z). \tag{5.149}$$

Here $m = M_B - M_A - m_t + m_n$. As usual, the Jacobian and the square root of the normalizer ν are included in (5.149).

Because the triton wavefunction is treated as a symmetric S-state the function I^A_{lsj} is scalar, and the angular momentum and parity transfer are carried by the spherical harmonic $Y_l^{m*}(\hat{r})$. Therefore the parity transfer is *normal*. This spherical harmonic is a function of the displacement between the target nucleus and the center of mass of the two transferred nucleons. Thus, the kinematics of the reaction is identical with that for the transfer of a two-nucleon cluster.†

Because the triton is loosely bound we expect the function I^A_{lsj} to be of fairly long range in the variable z. Therefore finite-range problems present

† The various exchange stripping processes (Section 4.8) are sufficiently complicated as to cause violations of this parity rule. Therefore, experimental observations [168] that excitations with non-normal parity occur with small probability provide important confirmation that the mechanism of two-nucleon transfer is dominated by light-particle stripping.

themselves when the form factors of (5.149) are inserted into the DW integration.

The form factors for the (H^3, p) reaction differ in two respects from those of (5.149). First, the $s = 1$ terms vanish. Second, the interactions in the nuclear matrix element are $V(n_0 p_0)$ and $V(n_1 p_0)$. Therefore, the $(\boldsymbol{\tau}_i \cdot \boldsymbol{\tau}_j)$ operators of (5.75) all take the value -1. We obtain for the (H^3, p) reaction the set of form factors

$$A_{lsj} f_{lsj} = -\sqrt{\bar{\nu}}\, \mathscr{J}\, 2^{-1/2}\, 2(V_{01} - V_{00}) Y_l^{m*}(\hat{r})\, I^A_{l0l}(r, z)\, \delta_{s,0}. \qquad (5.150)$$

The radial integrals of (5.150) have the same structure as those of (5.149), as given in (5.148). However, different spectroscopic considerations are required to obtain the functions $\theta^A_{lsj}(r, \rho)$ used in the (H^3, n) and (H^3, p) reactions.

A similar set of form factors is obtained for the (α, d) reaction, with the one interesting difference that $s = 1$, uniquely. These form factors are not exhibited here. Naturally, the form factors for the (α, d) reaction contain radial overlap integrals that are a little more complicated than those of (5.148).

The potentials in the above form factors originated as interactions among constituent nucleons of H^3. Therefore the magnitudes of these potentials are not subject to the uncertainties, due to collective effects, that beset the discussion of inelastic scattering. Therefore these magnitudes should not be treated as adjustable normalization constants, but should be fitted to properties of H^3. Probably this is best accomplished if the potentials are replaced by kinetic energy operators in the fashion followed in the discussions of single-nucleon transfer. In any case, absolute normalization of the cross sections for two-nucleon transfer should be both attempted and expected, as with other stripping reactions.

On the whole the theory of two-nucleon stripping is identical with that of single-nucleon stripping. The angular distributions are characterized by the l of the center-of-mass motion of the transferred pair, and may be expected to resemble the angular distributions obtained for single-nucleon transfers with the same l value. Mixtures of l transfers are incoherent (in the absence of spin-orbit terms in the optical potentials) and should be as infrequent as with single-nucleon transfer. A further reason why one- and two-nucleon transfer should give similar angular distributions is that strong absorption partially suppresses the nuclear interior, where the form factors for these two reaction types are most likely to differ. Similar conclusions can be reached with regard to the DW stripping theory for the transfer of three or more nucleons.

The spectroscopic properties of the integrals $I^A_{lsj}(r, z)$ present the most noteable new aspects of two-nucleon transfer. These properties of the I^A_{lsj}

enter through the coefficient functions $\theta_{lsj}^{A}(r, \rho)$ that describe relative S-states in nucleus B. We first note that a given θ_{lsj}^{A} does not select contributions from only one configuration of nucleus B, as with single-nucleon transfer. Instead, the θ_{lsj}^{A} are coherent linear combinations of contributions from all configurations that have pairs in relative S-states, subject only to the requirement that the center of mass of each pair have angular momentum l. Because the low-lying states of nuclei possess a strong pairing structure, classified by the *seniority quantum number* [169], we may expect strong enhancements of the θ_{lsj}^{A} that connect to states of low seniority. Yoshida [170] first pointed out that collective enhancement of the cross sections for two-nucleon transfer is to be expected. Lane and Soper [171] pointed out that the (d, α) reaction should excite isobaric analogues of the preferred states produced in the (n, He^3) reaction, and that the (α, d) reaction should excite analogues of preferred states produced in the (H^3, p) reaction.*

We recall that for single-nucleon transfer the overlap of ψ_B with ψ_A leads to a radial function $\mathscr{R}_{lj}{}^{A}$ that controls the properties of the stripping form factor. It is often a good approximation to replace $\mathscr{R}_{lj}{}^{A}$ by the product of a spectroscopic coefficient times a normalized radial function determined according to some simple prescription.

Unfortunately, as Glendenning has emphasized [163], the radial functions θ_{lsj}^{A} are rarely of the form of a ρ-dependent factor times an r-dependent factor. Therefore the internal properties of nucleus B and of nucleus a are intermingled in the I_{lsj}^{A}, with the result that it is rarely possible to characterize the magnitude of a double-stripping cross section by means of a single spectroscopic coefficient $\mathscr{S}_{lsj}$. Partial factorization of the θ_{lsj}^{A} can be obtained if nucleus B can be described by products of single-particle orbitals that are harmonic-oscillator eigenfunctions. However, these eigenfunctions are defective in the region of the nuclear surface [173, 174], just where transfer reactions require the greatest accuracy. Figure 5.32 illustrates the extent to which a double-stripping cross section is changed when the bound single-particle orbitals are taken to be Woods-Saxon eigenfunctions rather than harmonic-oscillator eigenfunctions.

Broglia and Riedel [175] presented an extensive analysis of form factors for the reaction $Pb^{206}(H^3, p)Pb^{208}$. They considered four alternative approximate methods by which to estimate the r-dependence of $I_{lsj}^{A}(r, z)$. *Method* (*a*) regards the two transferred neutrons as a point dineutron that has no internal structure and is bound in a Woods-Saxon potential well whose depth is adjusted to fit the energy $E_B - E_A$ [176]. This method seems to give unreasonable results. *Method* (*b*) uses harmonic-oscillator single-particle

* An interesting collective enhancement of (d, α) transitions to states of low excitation was apparently found in a series of experiments by Mead and Cohen [172].

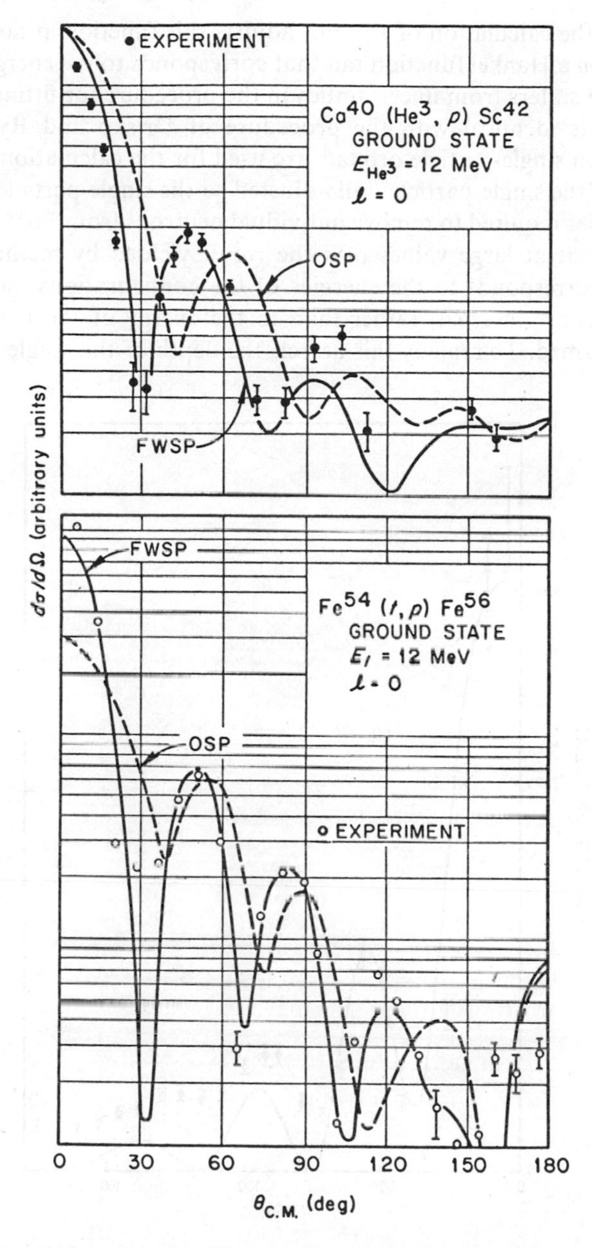

Fig. 5.32 Modification of DW prediction for two-nucleon transfer, caused by use of finite-well bound orbitals (FWSP) instead of oscillator-well bound orbitals (OSP). (From [173], Fig. 1.)

orbitals for the calculation of I^A_{lsj}, but adjusts this function at large values of r by fitting on a Hankel function tail that corresponds to the energy $E_B - E_A$. This method suffers from uncertainties in the procedure for fitting on the tail. *Method* (*c*) is identical with the procedure of Drisko and Rybicki [173]. Woods-Saxon single-particle orbitals are used for the calculation of I^A_{lsj}, with the depth of the single-particle well adjusted so the single-particle eigenvalues fit the energies required to remove individual neutrons from Pb^{208}. *Method* (*d*) recognizes that at large values of r the result yielded by method (*c*) has a shape that corresponds to the energies of the configurations into which the two neutrons are inserted, rather than to the energy of the particular state ψ_B that is formed. To remedy this defect, the depth of the single-particle well

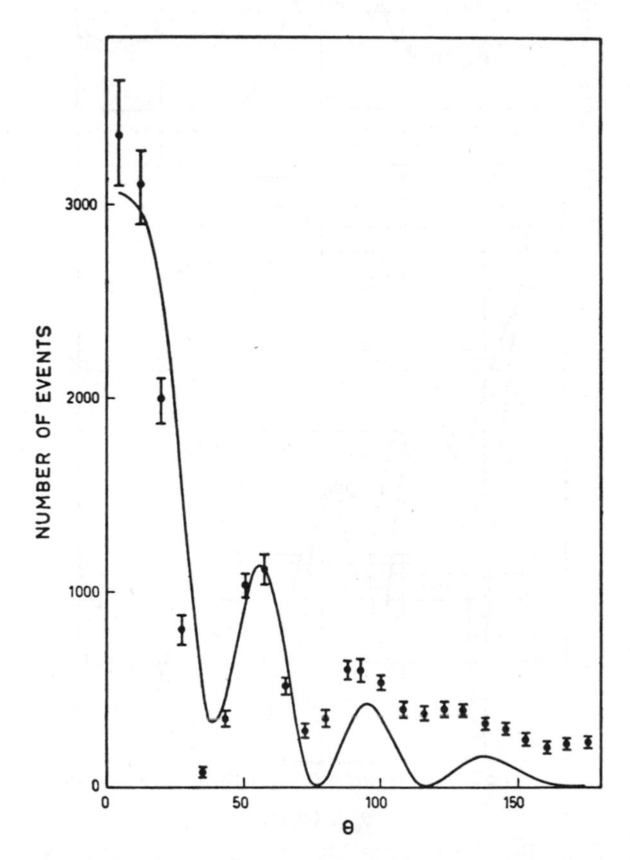

Fig. 5.33 Comparison of the DW theory with experimental data for two $l = 0$ transitions in $Pb^{206}(H^3, p)$ at 12 MeV bombarding energy. (Ground state on p. 202, excited state on p. 203.) Note the modifications caused by the different structures of the two 0^+ bound states of Pb^{208}. (From [175], Figs. 8, 9.)

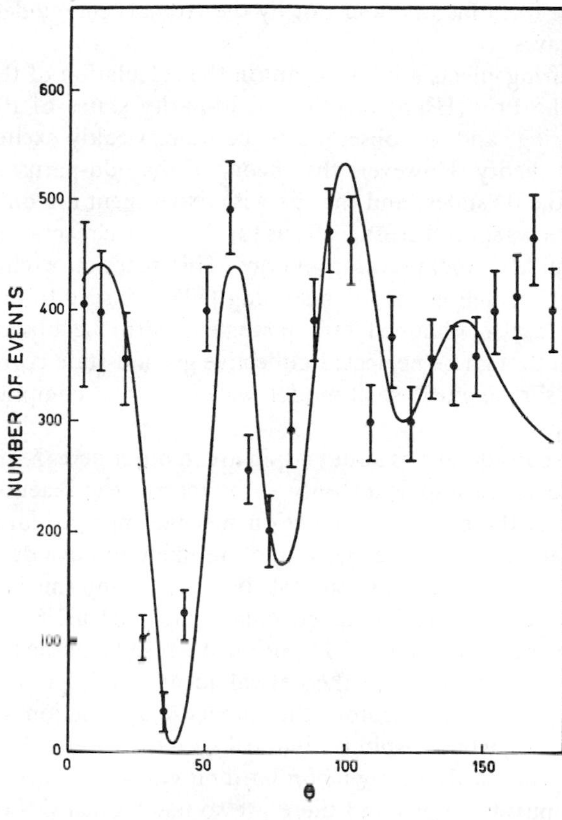

Fig. 5.33 *(contd.)*

is adjusted separately for each state ψ_B formed, so that the binding energy of each orbital is forced to take the value $\frac{1}{2}(E_B - E_A)$. It appears that method (*d*) gives maximum feasible accuracy. Method (*d*) conforms with the spirit of the well-depth (WD) method used to calculate form factors for single-nucleon transfer.

The most remarkable result found by Broglia and Riedel [175] is that details of the shape of the form factor influence to a major extent the angular distributions that are obtained. Therefore contributions from the interior of the nucleus are important. Figure 5.33 shows the cross sections for the $l = 0$ excitations of the ground state of Pb^{208} and the first excited 0^+ state at 4.87 MeV. The data were taken at 12 MeV bombarding energy. The two angular distributions are very different. The theory fits both angular distributions and also fits the relative normalization. The authors are careful to demonstrate that the difference between the calculated angular distributions is caused by the form factors, and not by the (trivial) energy dependence of the distorted waves.

Collective pairing effects are important in the calculation of the two $l = 0$ transitions in the $Pb^{206}(H^3, p)$ reaction. Odd-parity states of Pb^{208} are less affected by pairing, and are observed to be more weakly excited, again in agreement with theory. However, the theory of the odd-parity states is not as clear as for the 0^+ states, and the fits with experiment are only fair.

Figure 5.34 shows several cross sections for the inverse reaction $Pb^{208}(p, H^3)$ performed at 40 MeV bombarding energy. This reaction excites low-lying states of Pb^{206}. Calculations by Glendenning [177] are seen to fit the angular distributions. He also obtained fair agreement with the observed relative intensities, even though he neglected collective ground-state correlations and used only the simple Pb^{206} shell-model wavefunctions computed by True and Ford [178].

Some further considerations about *i*-spin are in order here. *I*-spin conservation is expected to be important only in direct nuclear reactions. Only in these reactions is there a simple reaction mechanism that connects initial and final nuclear states that are sharp and are likely to have definite *i*-spins. The CN reaction mechanism, in contrast, brings in as intermediate states the dense set of states of a highly-excited compound nucleus. Such CN states should show strong *i*-spin mixing.* Experimental results for the (d, α) reaction are in good accord with these theoretical ideas. Both projectiles in this reaction have *i*-spin zero, therefore the direct (d, α) reaction should excite only those states ψ_B whose *i*-spin is identical with that of ψ_A. This prediction should be valid primarily at high bombarding energies, where the DI exit channels are completely open and there are so many open CN exit channels that not much flux goes to any one of them. A typical high-energy experiment is that of Zafiratos, et al. [179], who studied the $C^{12}(\alpha, d)$ reaction at 42 MeV bombarding energy. There was no detectable excitation of the $T = 1$ levels of N^{14}. A typical low-energy experiment is that of Jänecke [180], who studied $Ca^{40}(d, \alpha)$ at 7.7 MeV bombarding energy. In this case both $T = 1$ and $T = 0$ states of K^{38} were excited. The statistical CN theory gave an adequate prediction of the $T = 1$ cross sections, but underestimated the $T = 0$ cross sections by a factor of about 3. Subtraction of the CN predictions left $T = 0$ cross sections that resembled the predictions of double-stripping theory.

* It has been remarked [see J. V. Noble, *Phys. Rev.* **173,** 1034 (1968)] that poor energy resolution tends to suppress the consequences of *i*-spin mixing in the compound states. The CN intermediate states tend to enter the *energy-averaged amplitude* as a locally-complete set, with the result that *i*-spin mixing among them is irrelevant; the average transition amplitude is controlled primarily by the *i*-spin conserving matrix elements that couple the intermediate states to the open channels. However, the significance of this observation is not clear, because the CN part of the cross section (Chapter 2) is determined by the fluctuating part of the amplitude, not by its average.

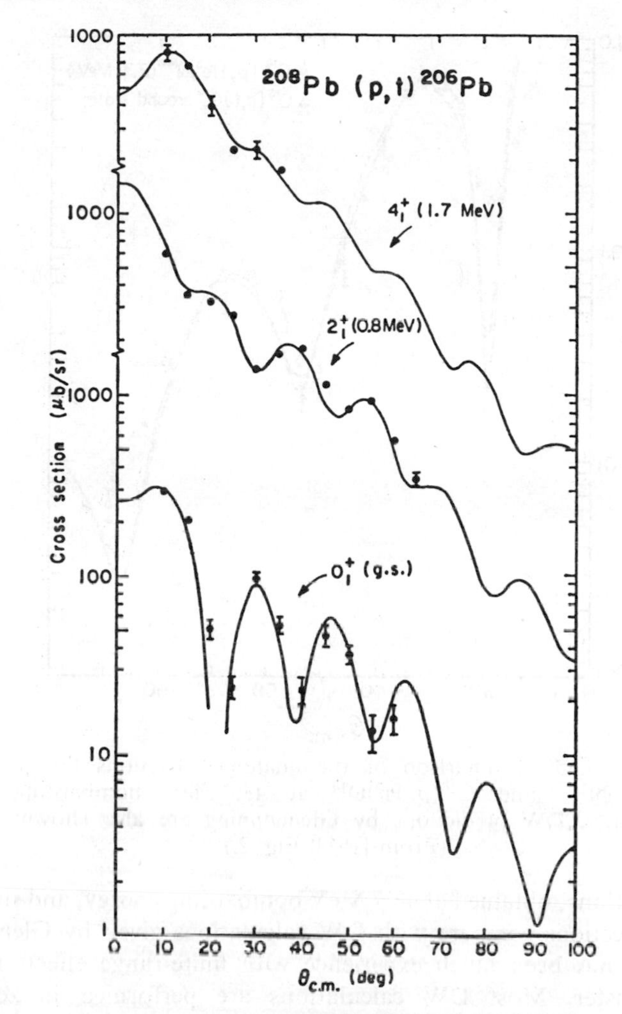

Fig. 5.34 Comparison of the DW theory with experimental data for $l = 0$, 2, 4 transitions in $Pb^{208}(p, H^3)$, at 40 MeV bombarding energy. (From [177], Fig. 1.)

Another interesting example of the role of *i*-spin in double-stripping reactions was studied by Cerny and Pehl [181]. These authors noted that the direct reaction $O^{16}(p, He^3)N^{14*}$, 2.31 MeV excitation, is the analog of the reaction $O^{16}(p, H^3)O^{14}$, g.s. Therefore the angular distributions for these reactions should be identical except for energy effects, and the magnitudes should bear a simple relation. Figure 5.35 shows a comparison between the

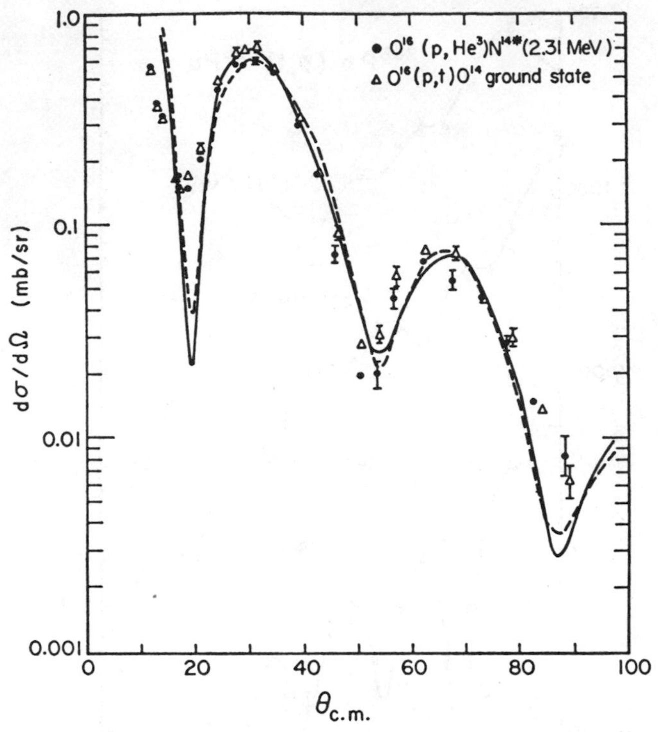

Fig. 5.35 Comparison of the analogue reactions $O^{16}(p, He^3)N^{14*}$ and $O^{16}(p, H^3)O^{14}$ at 43.7 MeV bombarding energy. DW predictions by Glendenning are also shown. (From [181], Fig. 2.)

two cross sections, obtained at 43.7 MeV bombarding energy, and shows how these cross sections compare with DW calculations given by Glendenning.

There has not been much experience with finite-range effects in multinucleon transfer. Most DW calculations are performed in zero-range approximation. However, important finite-range effects might be expected in the (H^3, n) reaction (see Fig. 5.31 and the accompanying discussion) because the dependence of the form factor on the variable z is not localized strongly about $z = 0$. We see from Fig. 5.31 and from (5.148) that even if the nuclear force should have zero range, the value of z can still become as large as about half the internucleon spacing in H^3. If the channel coordinates of the DW integration are expressed in terms of the variables in Fig. 5.31, it is found that

$$\mathbf{r}_\alpha = \mathbf{r} + \tfrac{1}{3}\mathbf{z},$$

$$\mathbf{r}_\beta = \left(\frac{A-2}{A}\right)\mathbf{r} + \mathbf{z},$$

where A is the mass number of nucleus A. From these expressions it follows that

$$\mathbf{z} = \frac{3}{2}\left(\frac{A}{A+1}\right)\left[\mathbf{r}_\beta - \left(\frac{A-2}{A}\right)\mathbf{r}_\alpha\right].$$

Evidently a 1 fm range for the variable z implies that $\mathbf{r}_\alpha$ and $\mathbf{r}_\beta$ can differ by about $\frac{2}{3}$ fm. These estimates do not suggest major departures from the zero-range limit. Furthermore, absorption partially suppresses the nuclear interior, where the distorted waves are most rapidly varying and finite-range modifications should be most important. Therefore the introduction of zero-range approximation seems to be reasonable. Less optimistic conclusions are reached in [194].

Preliminary investigations (R. M. Drisko, private communication) disclose that finite-range effects may be more important in (α, p) reactions. In these reactions $\mathbf{r}_\alpha$ and $\mathbf{r}_\beta$ may differ by a range that approximates the diameter of the α particle. Furthermore, the distorted wave that describes the motion of the α particle is rapidly oscillatory because the α particle has a high momentum.

Multi-nucleon transfer reactions not only tend to have large channel momenta k_α and k_β, but also tend to have a large momentum difference $|k_\alpha - k_\beta|$. As a result there are strong *preferential population* effects, as explained previously (Sections 5.1, 5.10 and [5]). With large values for $|k_\alpha - k_\beta|$ the distorted waves have good overlap only for l transfers of the order $l \approx |k_\alpha - k_\beta|R$, where R is the nuclear radius. Because the spectroscopy of nucleus B also easily allows the center of mass of the transferred cluster to have high angular momentum, we see that cross sections with large l values should be more frequent for multi-nucleon transfer than for single-nucleon transfer.

Other effects caused by large values for the difference $|k_\alpha - k_\beta|$ (see also Section 5.1) are an overall reduction of cross section magnitudes and an increase in the importance of "volume" contributions to the reaction amplitude as compared with "surface" contributions. These effects appear to be of greatest importance in the (α, p) reaction, in which the value of $|k_\alpha - k_\beta|$ tends to be particularly large. It is not surprising that this reaction and its inverse show extremely varied properties. Angular distributions often show their most prominent peaks at 180° scattering angle. The CN reaction mechanism tends to be strong, even for light nuclei and bombarding energies as high as 15–20 MeV. On the other hand, the zero-range DW theory was able to give a good description of $Y^{89}(p, \alpha)$ and $Zr^{90}(p, \alpha)$ reactions studied by Fulmer and Ball [182] at 20.2 and 22.5 MeV bombarding energies.

Another interesting property of the DI reaction mechanism for the (α, p) reaction is that so many nucleons are transferred that good overlap need not be limited to low-lying final states; good overlap may extend quite high up in the spectrum of nucleus B. Wall suggested [183] that for (α, p) reactions the average DI excitation of nucleus B may show a yield that is more strongly influenced by the level density of nucleus B than by special properties of the reaction mechanism. Just this effect may have been seen in (α, p) yield curves obtained by Swenson and Cindro [184] at 30.5 MeV bombarding energy, for a range of target nuclei that extended from Al^{27} to Ta^{181}. Although the yield curves in these experiments showed typical statistical shapes, the angular distributions showed strong forward peaking.

5.12 Knockon and Heavy-particle Stripping

An extended discussion of exchange terms of the DW theory is given in Chapter 10. That discussion presents details of the physical formulation of these terms. The present section is intended only as an illustration of the mathematical difficulties that prevent accurate assessments of the properties of these terms, however they may have been formulated. The present section provides an emphatic illustration of the importance of finite-range integrations of DW matrix elements, and as such it may be regarded as an introduction to the finite-range analysis in Section 5.13.

The amplitudes for knockon and heavy-particle stripping are encountered in Section 4.8, in which the exchange terms of the DW theory are classified on the basis of the type of interaction in which the outgoing particle engages. However, our nomenclature for these amplitudes reflects the fact that they not only are exchange terms of amplitudes otherwise of interest, but they also are descriptions of simple reaction mechanisms that are of independent interest. In all the exchange amplitudes for the reaction $A(a, b)B$ the nucleons of nuclide a enter into nucleus B, and the nucleons of nuclide b emerge from nucleus A. In the "knockon" term the interaction that governs the amplitude is the interaction between the set of labelled nucleons of a and the (distinct) set of labelled nucleons of b. In this collision between a and b the nuclide b is ejected from A and the nuclide a becomes bound in its stead. However, in the "heavy-particle stripping" term, first discussed by Madansky and Owen [185], nuclide a interacts with the *core* C of nucleus A, namely, a interacts with all nucleons of A other than the ones that compose b. In this collision between a and C the nuclide b is ejected, and a and C enter into a bound state. It may be considered that a has *stripped the* (*heavy*) *core* C from nucleus A. (In the HPS process there is some tendency for b to emerge preferentially at backward scattering angles.)

The mathematical difficulties of the exchange terms are seen already in the above verbal descriptions. Because a enters into B and b emerges from A, the evaluation of these terms requires fractional parentage expansions both of A and of B. Furthermore, because of recoil, these expansions use wavefunctions referred to different coordinate origins.

As an explicit illustration of the structure of the exchange terms we now consider the knockon term of the deuteron stripping amplitude, (4.97). This term is

$$T_{\alpha\beta}{}^{\mathrm{DW}}(\text{knockon}) = -Z_A(N_A+1)(\chi_\beta^{(-)}(0_p)\psi_B(1_p,\ldots,n_p;0_n,\ldots,n_n),$$
$$\times\ [V(0_p0_n)+V(0_p1_p)]\psi_A(0_p,2_p,\ldots,n_p;1_n,\ldots,n_n)$$
$$\times\ \phi_d(1_p,0_n)\,\chi_\alpha^{(+)}(1_p,0_n)), \tag{4.97}$$

where the notation of the present chapter is inserted in the normalization coefficient. Fractional parentage expansions suitable for use in (4.97) are those of (5.109) and (5.143). For the present application these become

$$\psi_A = \sum_{C,j}\langle J_AM_A\,|\,J_Cj;M_C,M_A-M_C\rangle\psi_C\zeta^C_{j,M_A-M_C}(0_p), \tag{5.151}$$

$$\psi_B = \sum_{C',j'}\langle J_BM_B\,|\,J_{C'}j';M_{C'},M_B-M_{C'}\rangle\psi_{C'}\phi^{C'}_{j',M_B-M_{C'}}(0_n,1_p), \tag{5.152}$$

where the symbols A, B, C, C' indicate quantum numbers from the sets $\{\alpha_A, J_A, M_A\}$, etc., as required. Substitution of these expansions in (4.97) yields

$$T_{\alpha\beta}{}^{\mathrm{DW}}(\text{knockon}) = -Z_A(N_A+1)\sum_{C,j,j'}\langle J_AM_A\,|\,J_Cj;M_C,M_A-M_C\rangle$$
$$\times\ \langle J_BM_B\,|\,J_Cj';M_C,M_B-M_C\rangle$$
$$\times\ (\chi_\beta^{(-)}(0_p)\phi^C_{j',M_B-M_C}(0_n,1_p),\,[V(0_p0_n)+V(0_p1_p)]$$
$$\times\ \zeta^C_{j,M_A-M_C}(0_p)\phi_d(1_p,0_n)\chi_\alpha^{(+)}(1_p,0_n)). \tag{5.153}$$

To avoid introduction of unnecessary complications due to recoil, we treat the core C as having infinite mass. Then the coordinates 0_p, 1_p, 0_n may be understood to be the displacements from C to the particles in question, along with the spin coordinates of the particles. These displacements are suitable integration variables for the matrix elements and have a Jacobian of value unity. However, the DW analysis requires introduction of the channel coordinates $\mathbf{r}_\alpha$, $\mathbf{r}_\beta$, which are the arguments of the distorted waves. A suitable coordinate transformation is

$$\begin{aligned}\mathbf{r}_\alpha &= \tfrac{1}{2}(\mathbf{r}_{1_p}+\mathbf{r}_{0_n}),\\ \mathbf{r}_\beta &= \mathbf{r}_{0_p},\\ \boldsymbol{\rho} &= \mathbf{r}_{1_p}-\mathbf{r}_{0_n},\end{aligned} \tag{5.154}$$

as illustrated in Fig. 5.36. This set of coordinates also has a unit Jacobian.

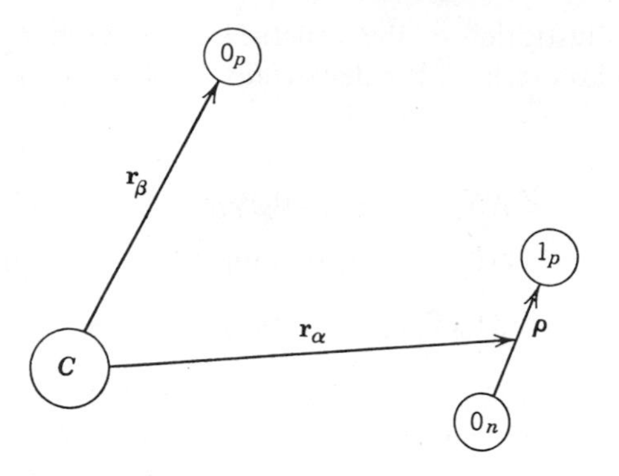

Fig. 5.36 Integration variables for the knockon term in deuteron stripping.

In terms of the coordinates of (5.154) the knockon matrix element becomes

$$T_{\alpha\beta}{}^{\mathrm{DW}}(\text{knockon}) = -Z_A(N_A+1)\sum_{C,j,j'}\langle J_A M_A \mid J_C j; M_C, M_A - M_C\rangle$$
$$\times \langle J_B M_B \mid J_C j'; M_C, M_B - M_C\rangle$$
$$\times \left(\chi_\beta^{(-)}(\mathbf{r}_\beta)\phi^C_{j',\,M_B-M_C}(\mathbf{r}_\alpha - \tfrac{1}{2}\boldsymbol{\rho}, \mathbf{r}_\alpha + \tfrac{1}{2}\boldsymbol{\rho}),\right.$$
$$\times [V(\mathbf{r}_\beta - \mathbf{r}_\alpha + \tfrac{1}{2}\boldsymbol{\rho}) + V(\mathbf{r}_\beta - \mathbf{r}_\alpha - \tfrac{1}{2}\boldsymbol{\rho})]$$
$$\left.\times \zeta^C_{j,\,M_A-M_C}(\mathbf{r}_\beta)\phi_d(\boldsymbol{\rho})\,\chi_\alpha^{(+)}(\mathbf{r}_\alpha)\right). \tag{5.155}$$

Clearly, the numerical evaluation of this matrix element presents great difficulties. Not only is it necessary to sum over C, j, j', but it is necessary to perform integrations over three vector variables. Furthermore, these integrations are not subject to zero-range approximation (Section 5.1). The interactions are the only short-ranged factors in the integrands. However, even if the interactions could be treated as δ-functions, the arguments of the two distorted waves would still differ by $\boldsymbol{\rho}/2$. Because the magnitude of ρ is controlled by the factor ϕ_D, the deuteron internal wavefunction, the average value of ρ may be expected to be 4.553 fm, the *radius of the deuteron*. Therefore the range of the $\boldsymbol{\rho}$-integration is so large that this integration must tend to average out the oscillatory factors of the integrand. Major reductions in the value of the matrix element, and major changes of the angular distribution must ensue. We note too that it is difficult to achieve adequate numerical evaluations of integrals that are subject to so much cancellation; careless approximations can give badly misleading results. For this reason there exists little practical knowledge of the exchange terms.

Further discussion of the exchange terms is given in Chapter 10, along with appropriate references to the literature.

5.13 Finite-range

We now return to the question first raised in Section 5.1: The reduced DW amplitude $\beta_{sj}{}^{lm}$ is a six-dimensional integral over the two channel-displacement variables $\mathbf{r}_\alpha$ and $\mathbf{r}_\beta$,

$$(2l+1)^{1/2} i^l \beta_{sj}{}^{lm} = \int d^3 r_\alpha \int d^3 r_\beta \chi_\beta^{(-)*}(\mathbf{k}_\beta, \mathbf{r}_\beta)\, f_{lsj,m}(\mathbf{r}_\beta, \mathbf{r}_\alpha)\, \chi_\alpha^{(+)}(\mathbf{k}_\alpha, \mathbf{r}_\alpha). \quad (4.60)$$

How should this integral best be done? Examination of the reactions treated in Chapter 5 shows that three cases must be distinguished: (a) The non-exchange amplitude for inelastic scattering has a reduced form factor $f_{lsj,m}$ that is proportional to $\delta(\mathbf{r}_\beta - \mathbf{r}_\alpha)$. Therefore the treatment of inelastic scattering does not even raise the finite-range question. (b) The nonexchange deuteron stripping amplitude has a form factor that is proportional to the neutron-proton interaction $V(r_{np})$, with $r_{np} = 2\,|(A+1)\mathbf{r}_\beta - A\mathbf{r}_\alpha|/(A+2)$. Because V is short ranged the zero-range approximation $(A+1)\mathbf{r}_\beta \approx A\mathbf{r}_\alpha$ should provide a good starting point for the development of an integration procedure. (c) Other rearrangement amplitudes have form factors that are longer-ranged functions of $\{(A+1)\mathbf{r}_\beta - A\mathbf{r}_\alpha\}$. The ranges of these form factors are controlled by the dimensions of composite projectiles a and b. A striking example appears in Section 5.12, in which the form factor for the exchange amplitude in deuteron stripping is seen to have a range equal to the radius of the deuteron. Zero-range approximation cannot be adequate for amplitudes of type (c). Therefore a general integration procedure must be developed.

What is really at issue in this discussion of integration procedures is the *sequence* in which wavefunctions of different types are brought into the integration. The complete matrix elements for inelastic scattering and deuteron stripping, for example, involve the same number of integrations. However, with inelastic scattering one stage of integration involves wavefunctions internal to the target nucleus, and does not concern the distorted waves at all. This first stage of integration produces a form factor $F_{lsj}(r)$ that is independent of bombarding energy and scattering angle and is easily folded in with whatever distorted waves are appropriate for the conditions of interest. In contrast to this, rearrangement collisions mix up bound and continuum wavefunctions in the same stages of integration. As a result, the calculation of the matrix element for a rearrangement collision does not have a first stage that is free from consideration of the complicated distorted waves. To develop a practical finite-range integration procedure it is necessary to achieve as much separation as possible between the treatment of the bound and continuum wavefunctions.

Finite-range integrations would be trivial if the distorted waves could be replaced by plane waves. With plane waves the transformation from the variables $\mathbf{r}_\alpha$, $\mathbf{r}_\beta$ to the variables in the form factor causes the six-dimensional integral to factor, giving a product of two three-dimensional integrals. Deuteron stripping provides a good example of the plane-wave approach. Equation 5.116 gives the form factor for deuteron stripping, $A_{lsj}f_{lsj,m}$. Let us take for the unnormalized form factor

$$f_{lsj,\,m} = D(r_{np})\, Y_l^{m*}\,(\hat{r}_{nA})\, \mathscr{R}_{lj}{}^A(r_{nA}). \tag{5.156}$$

We insert (5.156) in (4.60), along with the transformation

$$\begin{aligned} \mathbf{r}_\alpha &= \mathbf{r}_{nA} - \tfrac{1}{2}\mathbf{r}_{np}, \\ \mathbf{r}_\beta &= \left(\frac{A}{A+1}\right)\mathbf{r}_{nA} - \mathbf{r}_{np}. \end{aligned} \tag{5.157}$$

The plane-wave version of (4.60) then becomes

$$\begin{aligned} (2l+1)^{1/2} i^l \beta_{sj}{}^{lm} \approx & \int \exp i\mathbf{r}_{np} \cdot [\mathbf{k}_\beta - \tfrac{1}{2}\mathbf{k}_\alpha] D(r_{np})\, d^3 r_{np} \\ & \times \int \exp i\mathbf{r}_{nA} \cdot \left[\mathbf{k}_\alpha - \left(\frac{A}{A+1}\right)\mathbf{k}_\beta\right] Y_l^{m*}(\hat{r}_{nA}) \\ & \times \mathscr{R}_{lj}^A(r_{nA})\, d^3 r_{nA}. \end{aligned} \tag{5.158}$$

In this version of the reduced stripping amplitude the dependence on the range of $D(r_{np})$ is entirely contained in the first integral. Let us use the Hulthén wavefunction for the deuteron (5.131). Integration over $\mathbf{r}_{np}$ then gives

$$-\,(4\pi)^{1/2} N\hbar^2 M^{-1}\{(\zeta^2 - \gamma^2)/[\zeta^2 + (\tfrac{1}{2}\mathbf{k}_\alpha - \mathbf{k}_\beta)^2]\}, \tag{5.159}$$

instead of the zero-range result of (5.119). The factor in braces in (5.159) is the correction for finite range. This factor reduces the magnitude of the stripping amplitude. However, the momentum ζ is fairly large ($\zeta \approx 1.6\ \text{fm}^{-1}$); therefore the reduction caused by finite range is not large and should not be very dependent on scattering angle. The finite-range correction in (5.159) was carried automatically in the early, plane-wave theories of deuteron stripping. Because the subsequent DW theory was more complicated, it omitted the finite-range correction, and took advantage of such simplifications as are obtained from the zero-range approximation. However, a full treatment of the DW theory requires the inclusion of finite-range effects.

The finite-range correction factor in the plane-wave theory is proportional to the Fourier transform of $D(r_{np})$ evaluated for the argument $|\mathbf{k}_\beta - \frac{1}{2}\mathbf{k}_\alpha|$; this argument is the difference of asymptotic momenta of the distorted waves. It is interesting to consider adopting this plane-wave correction factor as a possible basis for an approximate finite-range correction of the DW theory. Clearly, this cannot be done just by multiplying the zero-range DW amplitude by the plane-wave correction factor, because the momenta important in the overlap of the distorted waves may be quite different from the asymptotic momenta. But perhaps the plane-wave correction could be inserted in a *local* fashion, in different radial regions of the zero-range DW integration. In any such approach we would recognize that the influence of the attractive optical potentials causes high momenta to be especially likely to appear in the nuclear interior; we therefore would be led to suspect that finite-range effects should give some suppression of interior contributions to the DW amplitude. Unfortunately, even such a local plane-wave analysis is misleading. Distortion of the wavefunctions introduces a broad distribution of both high and low momenta. Then the finite-range correction always weights the low momenta especially heavily, because the Fourier transform of $D(r_{np})$ is largest for low momenta. Under these conditions the plane-wave theory gives only a rough indication of the finite-range correction. It is necessary to develop methods that lie entirely within the DW context.

The exact numerical procedure for the finite-range integration is straightforward [186]. It follows the rule described earlier, that as many integrations as possible be performed before the distorted waves are brought into the calculation. Because the distorted waves are to be handled in spherical-harmonic expansion [(3.24) and (3.36)], it is clear that the angle integrations in (4.60) can all be made to precede the radial integrations over the numerically computed radial wavefunctions of the distorted waves. We exploit this and arrange a calculation that has two stages: (a) All integrals that involve internal variables of the form factor or that involve angles of $\hat{r}_\alpha$ and $\hat{r}_\beta$ are performed first. (b) The double radial integral with respect to r_α, r_β is performed second.

The first stage of calculation is carried through by expanding the form factor in a double series in spherical harmonics of $\hat{r}_\alpha$ and $\hat{r}_\beta$. Because $f_{lsj,m}$ transforms under rotations like Y_l^{m*}, this series takes the form

$$f_{lsj,m} = \sum_{L_\alpha L_\beta M} F^{lsj}_{L_\beta L_\alpha}(r_\beta, r_\alpha)\, Y^{M*}_{L_\beta}(\hat{r}_\beta)\, Y^{(m-M)*}_{L_\alpha}(\hat{r}_\alpha)\langle L_\beta L_\alpha; M, m - M \mid lm\rangle. \tag{5.160}$$

In terms of this expansion of the form factor, and in terms of the partial-wave expansions of the distorted waves [(3.24), (3.36)], (4.60) becomes

$$\beta_{sj}^{lm} = \frac{(4\pi)^{3/2}}{k_\beta k_\alpha} \sum_{L_\alpha L_\beta} i^{L_\alpha+L_\beta-l} \exp(i\sigma_{\alpha L_\alpha} + i\sigma_{\beta L_\beta}) I^{lsj}_{L_\beta L_\alpha}$$

$$\times \langle L_\beta l; m, -m \mid L_\alpha 0\rangle Y_{L_\beta}^{-m}(\Theta, 0), \tag{5.161}$$

where the z-axis is taken along $\mathbf{k}_\alpha$, the y-axis is taken along $\mathbf{k}_\alpha \times \mathbf{k}_\beta$, and Θ is the angle between $\mathbf{k}_\alpha$ and $\mathbf{k}_\beta$. The radial integrals in (5.161) are

$$I^{lsj}_{L_\beta L_\alpha} = \int_0^\infty r_\alpha \, dr_\alpha \int_0^\infty r_\beta \, dr_\beta f_{\beta L_\beta}(k_\beta, r_\beta) F^{lsj}_{L_\beta L_\alpha}(r_\beta, r_\alpha) f_{\alpha L_\alpha}(k_\alpha, r_\alpha). \tag{5.162}$$

Obviously the calculation of the coefficients $F^{lsj}_{L_\beta L_\alpha}$ is the most difficult step of the first stage, because this step must be done in part by numerical means. However, it is shown in [186] that each spherical-harmonic factor in $f_{lsj,m}$, whatever its argument, can be expanded analytically into a *finite* series of spherical harmonics whose arguments are $\hat{r}_\beta$ and $\hat{r}_\alpha$, as required in (5.160). This convenient expansion reduces the calculation of the $F^{lsj}_{L_\beta L_\alpha}$ to the extent that no more than one step of numerical integration is required. Gaussian approximations can sometimes be inserted in the $f_{lsj,m}$ to simplify even this one step of numerical integration.

In the second stage of calculation the double radial integrals of (5.162) are computed numerically. Thus the overall calculation requires the numerical evaluation of one set of single integrals and of one set of double integrals. The distorted radial wavefunctions only enter in the double integrals.

Comparison with (5.6) discloses that (5.161) does not contain the important Clebsch-Gordan coefficient $\langle L_\beta l; 00 \mid L_\alpha 0\rangle$, which compels the zero-range DW amplitude to have normal parity. The finite-range DW amplitude may have non-normal parity, as explained previously, because the form factor $f_{lsj,m}$ is a function of two vector variables. However, it is also seen (Sections 5.8–11) that in most cases of practical interest the finite-range form factor is a product of a scalar function times a function of a single vector variable. In all these cases the DW amplitude has normal parity. This property appears in (5.161) in the form of a selection rule,

$$L_\alpha + L_\beta + l = \text{even number},$$

that governs the coefficients $F^{lsj}_{L_\beta L_\alpha}$.

The initial applications of the exact finite-range integration procedure were to the study of deuteron stripping [186, 187]. An overall reduction of the cross section magnitude was found, much as predicted by the plane-wave analysis. However, contrary to plane-wave indications, finite-range effects were not found to cause any strong suppression of contributions from the nuclear interior. Any tendency for high-momentum Fourier components of $D(r_{np})$ to appear preferentially in the nuclear interior, as is mentioned earlier, is offset by the greater weighting given to the low Fourier components that also appear there. On the whole, all finite-range form factors whose Fourier transforms agree at low momenta give equivalent results in the distorted-waves calculation. This fact is of practical importance because it allows the use of mathematically-convenient Gaussian approximations.

A highly-successful approximate procedure for the finite-range integration has been developed [188, 189], based on an expansion about the zero-range limit. This expansion utilizes what is sometimes called the "local WKB approximation" or the "local-energy approximation (LEA)" (5.27). Because it is based on the zero-range limit, the approximate procedure is most applicable for reactions in which finite-range effects are not too severe. Deuteron stripping is a typical reaction of this kind, and is used to illustrate the approximate method. The insertion of (5.156) into (4.60) yields the DW amplitude for deuteron stripping. Let us use for integration variables $\mathbf{r} \equiv \mathbf{r}_{nA}$ and $\boldsymbol{\rho} \equiv \mathbf{r}_{np}$. Then the stripping amplitude is

$$(2l+1)^{1/2} i^l \beta_{sj}^{lm} = \int d^3r \int d^3\rho\, \chi_p^{(-)*}(\mathbf{k}_p, s^{-1}\mathbf{r} - \boldsymbol{\rho})$$

$$\times\, D(\rho)\, Y_l^{m*}(\hat{r})\, \mathscr{R}_{lj}^{A}(r)\, \chi_d^{(+)}(k_d, \mathbf{r} - \tfrac{1}{2}\boldsymbol{\rho}), \qquad (5.163)$$

where the symbol s is introduced for the ratio of masses, $s \equiv A^{-1}(A+1)$. The displacements $\boldsymbol{\rho}$ in the two distorted waves are conveniently represented with the aid of "shift operators," so that

$$\chi_p^{(-)*}(\mathbf{k}_p, s^{-1}\mathbf{r} - \boldsymbol{\rho}) = \exp(-s\boldsymbol{\rho}\cdot\nabla_p)\chi_p^{(-)*}(\mathbf{k}_p, s^{-1}\mathbf{r}),$$

$$\chi_d^{(+)}(\mathbf{k}_d, \mathbf{r} - \tfrac{1}{2}\boldsymbol{\rho}) = \exp(-\tfrac{1}{2}\boldsymbol{\rho}\cdot\nabla_d)\chi_d^{(+)}(\mathbf{k}_d, \mathbf{r}), \qquad (5.164)$$

where ∇_p and ∇_d differentiate with respect to $\mathbf{r}$ in functions $\chi_p^{(-)*}$ and $\chi_d^{(+)}$, respectively. We note that $-i\hbar\nabla_p$ and $-i\hbar\nabla_d$ are the *momentum operators* for the two wavefunctions, therefore these operators can be replaced by the *local momenta* of the wavefunctions at the point $\mathbf{r}$. Although the directions of these local momenta are not simply determined, their magnitudes are given by the local energies.

The insertion of (5.164) in (5.163) immediately allows the $\boldsymbol{\rho}$-integration to be done, with the operators ∇_p and ∇_d serving as constants in this stage of integration. Let us adopt $D(\rho)$ that is based on the Hulthén interaction. Then the $\boldsymbol{\rho}$-integration in the DW matrix element is formally identical with the $\mathbf{r}_{np}$-integration in the plane-wave matrix element of (5.158). The exact expression becomes

$$(2l+1)^{1/2} i^l \beta_{sj}{}^{lm} = -(4\pi)^{1/2} N \hbar^2 M^{-1}\left(1-\frac{\gamma^2}{\zeta^2}\right)$$
$$\times \int [1-\zeta^{-2}(s\nabla_p + \tfrac{1}{2}\nabla_d)^2]^{-1} \chi_p^{(-)*}(\mathbf{k}_p, s^{-1}\mathbf{r})\, \chi_d^{(+)}(\mathbf{k}_d, \mathbf{r})$$
$$\times Y_l{}^{m*}(\hat{r})\, \mathscr{R}_{lj}{}^A(r)\, d^3r. \tag{5.165}$$

In the limit $\zeta \to \infty$, (5.165) goes over to the zero-range result. Evidently the first factor of the integrand, a scalar function of the operators ∇_p and ∇_d, carries most of the finite-range correction. This factor is given meaning by expanding in powers of $\zeta^{-2}(s\nabla_p + \frac{1}{2}\nabla_d)^2$. Because ζ tends to be much larger than the local momenta, the higher terms of this expansion tend to be unimportant, and can be ignored *in part*.

Wherever the operator $(s\nabla_p + \frac{1}{2}\nabla_d)^2$ appears it is possible to substitute

$$(s\nabla_p + \tfrac{1}{2}\nabla_d)^2 = s^2\nabla_p{}^2 + s(\nabla_p \cdot \nabla_d) + \tfrac{1}{4}\nabla_d{}^2,$$
$$= \left(s^2 - \frac{s}{2}\right)\nabla_p{}^2 + \left(\frac{s}{2}\right)\nabla^2 + \left(\frac{1}{4} - \frac{s}{2}\right)\nabla_d{}^2, \tag{5.166}$$

where ∇^2 differentiates with respect to $\mathbf{r}$ both in $\chi_p^{(-)*}$ and in $\chi_d^{(+)}$. Repeated application of Green's theorem may then be used to transfer the operator ∇^2 so that it operates instead on the neutron wavefunction $Y_l{}^{m*}\mathscr{R}_{lj}{}^A$. This step may be indicated by giving a subscript n to the operator ∇^2. The exact DW amplitude becomes

$$(2l+1)^{1/2} i^l \beta_{sj}{}^{lm} = -(4\pi)^{1/2} N \hbar^2 M^{-1}\left(1-\frac{\gamma^2}{\zeta^2}\right)$$
$$\times \int \{1 - \tfrac{1}{2}\zeta^{-2}[(2s^2 - s)\nabla_p{}^2 + s\,\nabla_n{}^2 - (s-\tfrac{1}{2})\nabla_d{}^2]\}^{-1}$$
$$\times\, \chi_p^{(-)*}(\mathbf{k}_p, s^{-1}\mathbf{r})\chi_d^{(+)}(\mathbf{k}_d, \mathbf{r}) Y_l{}^{m*}(\hat{r})\mathscr{R}_{lj}{}^A(r)\, d^3r. \tag{5.167}$$

Each of the three Laplacian operators in (5.167) now operates on only one of the three wavefunctions. It is to be noted that (5.167) does not depend on the directions of the local momenta.

The three wavefunctions in (5.167) obey the Schrödinger equations

$$\left\{-(2s^2 - s)\nabla_p^{\,2} + \frac{2M}{\hbar^2}\,[U_p(s^{-1}r) - E_p]\right\}\chi_p^{(-)*}(\mathbf{k}_p, s^{-1}\mathbf{r}) = 0, \qquad (5.168)$$

$$\left\{-s\nabla_n^{\,2} + \frac{2M}{\hbar^2}\,[U_n(r) - E_n]\right\}Y_l^{m*}(\hat{r})\mathscr{R}_{lj}^{\,A}(r) = 0, \qquad (5.169)$$

$$\left\{-(s - \tfrac{1}{2})\nabla_d^{\,2} + \frac{2M}{\hbar^2}\,[U_d(r) - E_d]\right\}\chi_d^{(+)}(\mathbf{k}_d, \mathbf{r}) = 0, \qquad (5.170)$$

where correct reduced masses have been inserted, and where $(-E_n)$ is the binding energy of the transferred neutron. Equations 5.168–5.170 may be used to substitute out the Laplacian operators of (5.167), *wherever these Laplacians operate directly on the wavefunctions.* Unfortunately, (5.168)–(5.170) do not yield a general transformation of the inverse operator in (5.167). Consideration of the power series expansion of this operator discloses that (5.168)–(5.170) may be inserted only once in each term of the power series, to lower the degree of that term by one step. The remaining powers of the Laplacian operators do not commute with the potentials introduced in the first step, and cannot be substituted away.

However, the nonzero commutators of $\nabla_n^{\,2}$ with U_p, etc., only affect the higher terms in the power series expansion of the finite-range correction factor. Because ζ^2 is large, these higher terms are not important, and (5.168–5.170) may be used to derive the *approximate general result*

$$(2l + 1)^{1/2} i^l \beta_{sj}^{\;lm} \approx -(4\pi)^{1/2} N\hbar^2 M^{-1}\left(1 - \frac{\gamma'^2}{\zeta^2}\right)$$

$$+ \int (1 - \zeta^{-2}\Lambda^2)^{-1}\chi_p^{(-)*}(\mathbf{k}_p, s^{-1}\mathbf{r})$$

$$\times\, \chi_d^{(+)}(\mathbf{k}_d, \mathbf{r})\,Y_l^{\,m*}(\hat{r})\,\mathscr{R}_{lj}^{\,A}(r)\,d^3r, \qquad (5.172)$$

with

$$\Lambda^2 = \frac{M}{\hbar^2}\left[U_p(s^{-1}r) + U_n(r) - U_d(r) + B_d\right], \qquad (5.173)$$

and with $B_d = 2.226$ MeV the deuteron binding energy. Equation 5.172 differs from the zero-range DW expression only by the one additional factor $(1 - \zeta^{-2}\Lambda^2)^{-1}$. Therefore the finite range correction may be treated in practice as a modification of the form factor in a zero-range calculation.

Obviously, the properties of (5.173) are very sensitive to cancellations among the optical potentials. Therefore (5.173) is sensitive to the ambiguities discussed in Section 5.2. On the other hand, because d is a composite of n and

p, it is likely that the real parts of $(U_p + U_n - U_d)$ add up to zero (Section 5.2) The imaginary part of $(U_p + U_n - U_d)$ is less likely to vanish. Not only is there no imaginary part in U_n, the potential for the bound particle, but the imaginary potential for a deuteron always tends to be stronger than the sum of the imaginary potentials for the constituent neutron and proton. This net imaginary part of $(U_p + U_n - U_d)$ causes the magnitude of the finite-range correction in (5.172) to be smaller than unity. Outside the nucleus the optical potentials vanish, and the term B_a in (5.173) causes the finite-range correction factor to be slightly *larger* than unity.

Dickens, et al. [190] prepared a survey in which results obtained by the approximate and exact finite-range procedures were compared. The approximate procedure was found to give a good account of finite-range corrections in stripping reactions. The major effect of these corrections (at medium energy) was a change in the normalization of the cross section.

An interesting application of the approximate finite-range procedure was made by Towner [126] in a study of the $C^{12}(p, d)$ reaction at 155 MeV bombarding energy. Inclusion of the finite-range correction caused a major change of the shape of the angular distribution and brought the DW calculation into agreement with experiment.

5.14 Heavy-ion Transfer Reactions

There has been considerable interest in reactions $A(a, b)B$ in which the projectile a is a nucleus of appreciable size, perhaps as massive as the target nucleus A itself. The attention given to these reactions has been based on the hope that the strong Coulomb repulsion of a and A would so simplify the analysis as to allow these reactions to be reliable sources of spectroscopic information.

Although several special methods for the analysis of heavy-ion reactions have been developed (for a summary see [191]), the usual DW method seems to be as good as any of these [191, 192]. At first sight this use of DW may seem strange because the optical potentials in the DW method envisage free interpenetration of the colliding nuclei in channels α and β, without any internal distortion. Surely, this is an unreasonable description of the collision of two heavy ions. However, the apparently unphysical optical potentials lead to distorted waves $\chi_\alpha^{(+)}$ and $\chi_\beta^{(-)}$ that give no contributions from radii at which the heavy projectiles interpenetrate. This suppression of close collisions occurs either because Coulomb repulsion causes the distorted waves to be small at small radii, or because strong absorption prevents low partial waves from contributing to the DI matrix element. Therefore the transfer process tends to take place when the heavy ions are well separated, when only the asymptotic tails of their least-bound nucleons can overlap. The optical potentials serve only as devices to generate the wavefunctions $\chi_\alpha^{(+)}$ and $\chi_\beta^{(-)}$. These wavefunctions then are used well outside the region of interpenetration.

For definiteness we only treat heavy-ion reactions that transfer a single neutron from nucleus a to nucleus A.

The nonexchange distorted-waves amplitude for these reactions is identical with (5.126) of the general discussion of single-nucleon transfer:

$$T_{\alpha\beta}{}^{\mathrm{DW}} = (N_A + 1)\mathscr{J}\int d^3r_\alpha \int d^3r_\beta \chi_\beta^{(-)*}(\mathbf{k}_\beta, \mathbf{r}_\beta) \langle B, b| V |A, a\rangle \chi_\alpha^{(+)}(\mathbf{k}_\alpha, \mathbf{r}_\alpha). \tag{5.174}$$

Here N_A is as usual the neutron number of nucleus A. However, when (5.175) is inserted in (4.61) to compute the cross section, it is necessary to use for the channel weight factor

$$\frac{N_\beta}{N_\alpha} = N_a(N_A + 1)^{-1}.$$

Here N_a is the neutron number of nucleus a. The factor N_a accounts for the number of equivalent ways in which the transferred neutron can be selected from a.

The interaction V may be taken in the usual "post-interaction" form* $V = V_\beta - U_\beta(\mathbf{r}_\beta)$, where V_β is the sum of all interactions between the nucleons of B and the nucleons of b, and U_β is the optical interaction between the centers of mass of B and b. Our standard simplifications of transfer reactions (see (5.106), et seq.) lead to the approximation that U_β be considered to cancel all parts of V_β except the interaction between the transferred neutron n and the core of nucleus a. In this approximation V reduces to V_{nb}, because b is the core of a. Therefore the nuclear matrix element of (5.174) becomes

$$\langle B, b| V |A, a\rangle \approx \langle \psi_B\psi_b| V_{nb} |\psi_A\psi_a\rangle. \tag{5.175}$$

Equation 5.175 yields the form factors for the heavy-ion transfer reaction.

The present analysis differs from analyses of earlier transfer reactions by the introduction of a more symmetrical treatment of the spectroscopy. Fractional parentage expansions are used both for B and a. Equation 5.109 is used for ψ_B. The corresponding expansion for ψ_a is

$$\psi_{\alpha_a J_a M_a} = \sum_{\alpha_Q J_Q M_Q j'} \langle J_a M_a \mid J_Q j'; M_Q, M_a - M_Q\rangle \psi_{\alpha_Q J_Q M_Q} \phi_{j', M_a - M_Q}^{\alpha_Q J_Q}(n). \tag{5.176}$$

* We recall (Section 4.5) that identical DW results are obtained if either the post or prior form of V is inserted in (5.174). However, our subsequent approximations destroy this identity. To test these approximations, Buttle and Goldfarb [192] suggest that results obtained from the post and prior versions of (5.174) be compared.

Insertion of (5.109) and (5.176) in (5.175) then yields

$$\begin{aligned}\langle B, b| V |A, a\rangle = \sum_{jj'} &\langle J_B M_B \mid J_A j\,; M_A, M_B - M_A\rangle \\ &\times \langle J_a M_a \mid J_b j'\,; M_b, M_a - M_b\rangle \\ &\times \langle \phi^{\alpha_A J_A}_{j, M_B - M_A}(\mathbf{r}_2)| V_{nb}(\mathbf{r}_1) |\phi^{\alpha_b J_b}_{j', M_a - M_b}(\mathbf{r}_1)\rangle, \end{aligned} \qquad (5.177)$$

where $\mathbf{r}_2$ and $\mathbf{r}_1$ are, respectively, the displacements from A to the transferred neutron and from b to the transferred neutron. These displacements are shown in Fig. 5.37. One further scalar product remains on the RHS of (5.177); it involves integration over the spin coordinate of the transferred neutron. To perform this integration it is necessary that the single-particle wavefunctions be replaced by expressions of the form of (5.111). This step is deferred.

We now note that the DW amplitude for a heavy-ion transfer reaction presents a difficult finite-range integration problem. Let us consider the amplitude obtained by inserting only one term of (5.177) in (5.174):

$$\begin{aligned} t_{jj'} \equiv (N_A + 1)\mathscr{J} \int d^3 r_\alpha \int d^3 r_\beta \chi_\beta^{(-)*}(\mathbf{k}_\beta, \mathbf{r}_\beta) \\ \times \langle \phi_{j\mu}{}^A(\mathbf{r}_2)| V_{nb}(\mathbf{r}_1) |\phi^b_{j'\nu}(\mathbf{r}_1)\rangle\, \chi_\alpha^{(+)}(\mathbf{k}_\alpha, \mathbf{r}_\alpha). \end{aligned} \qquad (5.178)$$

It is necessary to choose a single set of coordinates to use in all the wavefunctions of (5.178). It is clear that the set $\mathbf{r}_1$, $\mathbf{r}_2$ would give a badly-convergent description of the distorted waves, inasmuch as the relation between $\mathbf{r}_\alpha$, $\mathbf{r}_\beta$ and $\mathbf{r}_1$, $\mathbf{r}_2$ is rather distant. Likewise, the set $\mathbf{r}_\alpha$, $\mathbf{r}_\beta$ would give a poor description of the bound-state wavefunctions. Let us therefore adopt as a compromise the coordinates $\mathbf{r}$, $\mathbf{r}_1$ used by Buttle and Goldfarb [192], as shown in Fig. 5.37. Here $\mathbf{r}$ is the displacement between A and b, the two "core

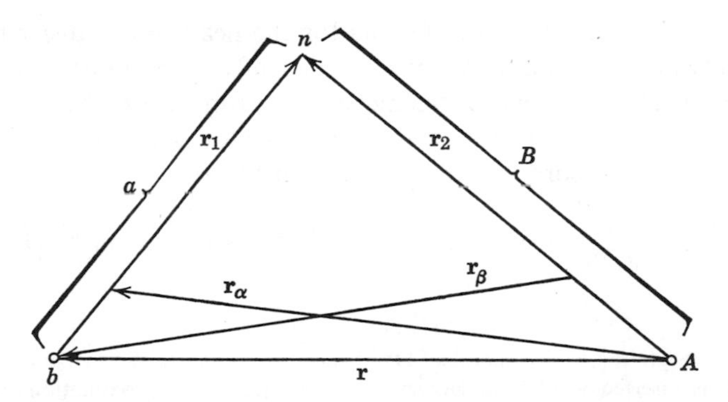

Fig. 5.37 Integration variables for a heavy-ion transfer reaction.

nuclei" of the transfer process. In terms of the coordinates $\mathbf{r}$, $\mathbf{r}_1$ we find

$$\begin{aligned}\mathbf{r}_2 &= \mathbf{r} + \mathbf{r}_1,\\ \mathbf{r}_\alpha &= \mathbf{r} + \mathbf{r}_1(b+1)^{-1},\\ \mathbf{r}_\beta &= s^{-1}\mathbf{r} - \mathbf{r}_1(A+1)^{-1},\\ \mathscr{I} &= 1,\end{aligned} \tag{5.179}$$

where

$$s = A^{-1}(A+1),$$

and where b and A are the mass numbers of nuclei b and A, respectively. We observe that the $\mathbf{r}_1$ recoil terms in $\mathbf{r}_\alpha$ and $\mathbf{r}_\beta$ differ in magnitude and have opposite signs.

The recoil terms in $\mathbf{r}_\alpha$ and $\mathbf{r}_\beta$ easily cause large variations of the phases of the distorted waves [193]. These phase variations are of the order (radius of nucleus a) × (momentum per nucleon of the colliding nuclei). Because a has an appreciable radius, these recoil effects are larger than the ones treated in Section 5.13, and can cause important modifications of the amplitude.

The recoil effects in the distorted waves can be made more explicit by the use of "shift operators," as in (5.164) of the general finite-range discussion. As in that discussion, these shift operators can be replaced by the "local momenta" of the two distorted waves. It is convenient to make this replacement immediately, using $\mathbf{K}_\alpha$ and $\mathbf{K}_\beta$ for the local momenta of $\chi_\alpha^{(+)}$ and $\chi_\beta^{(-)}$, respectively. Then the amplitude of (5.178) becomes

$$t_{\mathscr{I}\mathscr{I}'} = (N_A+1)\int \chi_\beta^{(-)*}(\mathbf{k}_\beta, s^{-1}\mathbf{r})F(\mathbf{r})\chi_\alpha^{(+)}(\mathbf{k}_\alpha, \mathbf{r})\, d^3r, \tag{5.180}$$

where

$$F(\mathbf{r}) = \int \exp\,[i(\mathbf{Q}\cdot\mathbf{r}_1)]\,\langle \phi_{\mathscr{I}\mu}{}^A(\mathbf{r}+\mathbf{r}_1)|\, V_{nb}(r_1)\,|\phi^b_{\mathscr{I}'\nu}(\mathbf{r}_1)\rangle\, d^3r_1, \tag{5.181}$$

and where

$$\mathbf{Q} = (A+1)^{-1}\mathbf{K}_\beta + (b+1)^{-1}\mathbf{K}_\alpha. \tag{5.182}$$

The function $F(\mathbf{r})$ yields the form factors for the transfer reaction. The exponential factor in $F(\mathbf{r})$ accounts for recoil effects. (We recall that the bra-ket notation in (5.181) refers only to integration over the spin of the transferred neutron.)

The recoil factor in (5.181) allows the vector $\mathbf{r}_1$ to take on a broad range of angular momenta in addition to the angular momentum of ϕ_b. Thereby, $F(\mathbf{r})$ becomes a rather smooth function, without much trace of the abrupt "surface fall off" that would be obtained just by folding ϕ^A with $V_{nb}\phi^b$. At energies above the Coulomb barrier this recoil modification of $F(\mathbf{r})$ tends to dominate the angular distributions obtained in heavy-ion transfer reactions [193].

Buttle and Goldfarb [192] give a thorough discussion of properties of $F(\mathbf{r})$ at low energy, where the recoil factor may be omitted. However, it is also of interest to carry this factor in a few steps of the low-energy discussion to see what modifications it causes.

At low bombarding energies ϕ^A and ϕ^b overlap only asymptotically, and therefore nonzero contributions to (5.181) arise only within nucleus b. Within this volume ϕ^A is the wavefunction of a free particle, with (negative) energy ε_A. These properties allow the elimination of V_{nb} by the substitution

$$\phi^{A*}V_{nb}\phi^b = (\varepsilon_b - \varepsilon_A)\,\phi^{A*}\phi^b + \frac{\hbar^2}{2M}[(\nabla^2\phi^{A*})(\phi^b) - (\phi^{A*})(\nabla^2\phi^b)], \quad (5.183)$$

where ε_b is the negative binding energy that the transferred neutron has in nucleus a. The insertion of (5.183) in (5.181), followed by integration by parts, then gives

$$\begin{aligned} F(\mathbf{r}) = (\varepsilon_b - \varepsilon_A)\int \exp\,[i(\mathbf{Q}\cdot\mathbf{r}_1)]\,\langle\phi^A\mid\phi^b\rangle\,d^3r_1 \\ + \frac{\hbar^2}{2M}\int_\Sigma \exp\,[i(\mathbf{Q}\cdot\mathbf{r}_1)]\{\langle\nabla\phi^A\mid\phi^b\rangle - \langle\phi^A\mid\nabla\phi^b\rangle\}\cdot d\mathbf{\Sigma} \\ - \frac{i\hbar^2}{2M}\int \exp\,[i(\mathbf{Q}\cdot\mathbf{r}_1)]\{\langle\mathbf{Q}\cdot\nabla\phi^A\mid\phi^b\rangle - \langle\phi^A\mid\mathbf{Q}\cdot\nabla\phi^b\rangle\}\,d^3r_1, \quad (5.184) \end{aligned}$$

where Σ is a closed surface that surrounds the region in which $V_{nb} \neq 0$ and $d\mathbf{\Sigma}$ a directed surface element of Σ.

To rough approximation only the second line of (5.184) need be retained. The first line is omitted on the grounds that $\varepsilon_b \approx \varepsilon_A$. The third line is omitted on the grounds that at low energy only glancing collisions need be considered. Therefore the first term of the third line vanishes for reasons of orthogonality: (a) $\mathbf{Q}$ tends to lie along the incident momentum, and (b) the important direction of $\nabla\phi^A$ is perpendicular to the incident momentum. The second term of the third line tends to vanish because it is averaged over all directions of $\nabla\phi^b$.

Under the above approximations, $F(\mathbf{r})$ becomes

$$F(\mathbf{r}) \approx \frac{\hbar^2}{2M}\int_\Sigma \exp\,[i(\mathbf{Q}\cdot\mathbf{r}_1)]\{\langle\nabla\phi^A\mid\phi^b\rangle - \langle\phi^A\mid\nabla\phi^b\rangle\cdot d\mathbf{\Sigma}. \quad (5.185)$$

The introduction of further approximations is now facilitated by noting that Σ may be *any* surface that encloses the region $V_{nb} \neq 0$. Therefore we take Σ to be the plane that bisects the interval between A and b. This plane is perpendicular to $\mathbf{r}$, and passes through the point $\mathbf{r}/2$. A hemisphere that completes this plane into a closed surface may also be imagined, however, it is not relevant. What is important is that $\mathbf{r}$ is sufficiently large so that everywhere on the plane Σ the functions ϕ^A and ϕ^b may be treated as asymptotic. Therefore the radial dependence of these two functions is as $\exp(-\lambda_A r_2)$ and $\exp(-\lambda_b r_1)$, respectively, where $\hbar^2\lambda_A{}^2/2M = -\varepsilon_A$ and $\hbar^2\lambda_b{}^2/2M = -\varepsilon_b$.

The manner in which these functions depend on angles is less important, because along the plane Σ the distances between angular nodes of ϕ^A and ϕ^b are much larger than λ_A^{-1} and λ_b^{-1}. We therefore approximate

$$\nabla\phi^A \approx -\lambda_A\hat{r}_2\phi^A, \qquad \nabla\phi^b \approx -\lambda_b\hat{r}_1\phi^b.$$

It is further reasonable to approximate

$$(\hat{r}_2 \cdot \hat{r}) \approx 1, \qquad (\hat{r}_1 \cdot \hat{r}) \approx 1,$$

because the important regions of Σ lie near $\mathbf{r}/2$. Now $F(\mathbf{r})$ has reduced to

$$F(\mathbf{r}) \approx -\frac{\hbar^2}{2M}(\lambda_A + \lambda_b)\int_\Sigma \exp\,[i(\mathbf{Q}\cdot\mathbf{r}_1)]\langle\phi^A \mid \phi^b\rangle\, d\Sigma, \tag{5.186}$$

where $d\Sigma$ is an ordinary surface element of the plane Σ.

The region of integration in the plane Σ is limited because $\exp\,(-\lambda_A r_2)$ and $\exp\,(-\lambda_b r_1)$ fall off as $\mathbf{r}_1$ and $\mathbf{r}_2$ depart from the center point $\mathbf{r}/2$. Let us introduce ϕ and θ, the polar angles of $\mathbf{r}_1$ with respect to $\mathbf{r}$. Then the fall off of $\langle\phi^A \mid \phi^b\rangle$ is as

$$\exp\,[-\tfrac{1}{2}(\lambda_A + \lambda_b)r \sec\theta].$$

Under the introduction of small-angle approximation for θ, $F(\mathbf{r})$ reduces to

$$F(\mathbf{r}) \approx -\frac{\hbar^2}{2M}(\lambda_A + \lambda_b)\langle\phi^A(\mathbf{r}/2) \mid \phi^b(-\mathbf{r}/2)\rangle$$

$$\times \int_0^{2\pi} d\phi \int_0^{\pi}\left(\frac{r}{2}\right)^2 \sin\theta\, d\theta \exp\,[i(\mathbf{Q}\cdot\mathbf{r}_1)]\exp\,[-\tfrac{1}{4}(\lambda_A + \lambda_b)r\theta^2]. \tag{5.187}$$

Integration with respect to ϕ only concerns the recoil factor, $e^{i(\mathbf{Q}\cdot\mathbf{r}_1)}$. However, it was argued earlier that $\mathbf{Q}$ tends to lie perpendicular to $\mathbf{r}$. Therefore the ϕ integration becomes

$$\int_0^{2\pi} \exp\,(iQ\,\frac{r}{2}\sin\theta\cos\phi)\, d\phi = 2\pi J_0(\tfrac{1}{2}Qr\sin\theta),$$

and (5.187) assumes the form

$$F(\mathbf{r}) \approx -\frac{\hbar^2}{2M}(\lambda_A + \lambda_b)\langle\phi^A(\mathbf{r}/2) \mid \phi^b(-\mathbf{r}/2)\rangle$$

$$\times\, \tfrac{1}{2}\pi r^2 \int_0^{\pi} \theta J_0(\tfrac{1}{2}Qr\theta)\exp\,[-\tfrac{1}{4}(\lambda_A + \lambda_b)r\theta^2]\,d\theta. \tag{5.188}$$

Because $\mathbf{Q}$ is small at low energy and θr is limited to values of the order of $\frac{1}{2}(\lambda_A + \lambda_b)$, the argument of the Bessel function does not become large. Therefore the Bessel function is approximated by its value at the origin and is not carried further. Integration with respect to θ now yields

$$F(\mathbf{r}) \approx -\pi r\frac{\hbar^2}{2M}\langle\phi^A(\mathbf{r}/2) \mid \phi^b(-\mathbf{r}/2)\rangle. \tag{5.189}$$

This result is equivalent to the expression obtained by Buttle and Goldfarb [192].

Computation of the transfer amplitude is completed by inserting (5.189) in (5.180) and using standard DW computer codes. In fact, because (5.189) is a rapidly decreasing function of r, the numerical integration tends to be dominated by the point of closest approach of the two heavy ions, as they move along a classical Coulomb trajectory. Therefore, the use of WKB methods for the evaluation of (5.180) is indicated. Such questions as these are fully discussed in the literature [191].

References

[1] P. B. Daitch, Ph.D. thesis, University of Rochester 1952, (unpublished). J. Horowitz and A. M. L. Messiah, *J. Phys. et Radium* **14,** 695, 731 (1953), J. Horowitz and A. M. L. Messiah, *J. Phys. et Radium* **15,** 142 (1954).

[2] W. Tobocman, *Phys. Rev.* **94,** 1655 (1954). W. Tobocman and M. Kalos, *Phys. Rev.* **97,** 132 (1955).

[3] W. Tobocman, *Phys. Rev.* **115,** 98 (1959).

[4] C. A. Levinson and M. K. Banerjee, *Ann. Phys.* **2,** 471 (1957); **2,** 499 (1957); **3,** 67 (1958). See also N. K. Glendenning, *Phys. Rev.* **114,** 1297 (1959).

[5] A. G. Blair and H. E. Wegner, *Phys. Rev.* **127,** 1233 (1962). R. H. Pehl, J. Cerny, E. Rivet, B. G. Harvey, *Phys. Rev.* **140,** B605 (1965). E. Rivet, R. H. Pehl, J. Cerny, B. G. Harvey, *Phys. Rev.* **141,** 1021 (1966). G. M. Reynolds, J. R. Maxwell, and N. M. Hintz, *Phys. Rev.* **153,** 1283 (1967).

[6] See Chapter 4, Ref. 11.

[7] F. G. Perey and B. Buck, *loc. cit.*, Chapter 4, Ref. 12.

[8] G. W. Greenlees, G. J. Pyle, and Y. C. Tang, *Phys. Rev. Letters* **17,** 33 (1966), and further work. Also see D. Slalina and H. McManus, *Nucl. Phys.* **A116,** 271 (1968).

[9] R. D. Woods and D. S. Saxon, *Phys. Rev.* **95,** 577 (1954); *Phys. Rev.* **101,** 506 (1956).

[10] P. E. Hodgson, *The Optical Model of Elastic Scattering*, Oxford University Press, Oxford, 1963.

[11] E. Fermi, *Nuovo Cim.* **11,** 407 (1954); see Ref. 10, pp. 18–19; W. B. Riesenfeld and K. M. Watson, *Phys. Rev.* **102,** 1157 (1956); S. Fernbach, W. Heckrotte and J. V. Lepore, *Phys. Rev.* **97,** 1059 (1955); G. E. Brown, *Proc. Phys. Soc.* (London) **A70,** 351 (1957).

[12] R. M. Drisko, G. R. Satchler, R. Bassel, *Phys. Letters* **5,** 347 (1963).

[13] C. M. Perey and F. G. Perey, *Phys. Rev.* **132,** 755 (1963).

[14] G. Igo, *Phys. Rev. Letters* **1,** 72 (1958); **3,** 308 (1959); *Phys. Rev.* **115,** 1665 (1959).

[15] F. G. Perey, *Phys. Rev.* **131,** 745 (1963).

[16] D. J. Baugh, J. A. R. Griffith, and S. Roman, *Nucl. Phys.* **83,** 481 (1966); P. Kossanyi-Demay, et al., *Nucl. Phys.* **A94,** 513 (1967), **A108,** 577 (1968); L. Rosen, J. G. Beery, A. S. Goldhaber and E. H. Auerbach, *Ann. Phys.* **34,** 96 (1965).

[17] G. W. Greenlees and G. J. Pyle, *Phys. Rev.* **149,** 836 (1966); G. R. Satchler *Nucl. Phys.* **A92,** 273 (1967).

[18] M. P. Fricke, E. E. Gross, B. J. Morton, and A. Zucker, *Phys. Rev.* **156,** 1207 (1967); see also M. P. Fricke, et al., *Phys. Rev.* **163,** 1153 (1967).
[19] F. Bjorklund and S. Fernbach, *Phys. Rev.* **109,** 1293 (1958).
[20] F. G. Perey and G. R. Satchler, *Nucl. Phys.* **A97,** 515 (1967), also see S. Mukherjee, *Nucl. Phys.* **A118,** 423 (1968).
[21] F. Hinterberger, et al., *Nucl. Phys.* **A111,** 265 (1968).
[22] J. C. Hafele, E. R. Flynn, and A. G. Blair, *Phys. Rev.* **155,** 1238 (1967); R. N. Glover and A. D. W. Jones, *Nucl. Phys.* **81,** 268 (1966).
[23] E. F. Gibson, et al., *Phys. Rev.* **155,** 1194 (1967); D. J. Baugh, et al., *Nucl. Phys.* **A95,** 115 (1967).
[24] L. McFadden and G. R. Satchler, *Nucl. Phys.* **84,** 177 (1966).
[25] A.M. Lane, *Rev. Mod. Phys.* **29,** 191 (1957); P.C. Sood, *Nucl. Phys.* **89,** 553 (1966).
[26] C. R. Bingham, M. L. Halbert, and R. H. Bassel, *Phys. Rev.* **148,** 1174 (1966).
[27] K. T. R. Davies, S. J. Krieger, and M. Baranger, *Nucl. Phys.* **84,** 545 (1966).
[28] F. G. Perey, in *Direct Interactions and Nuclear Reaction Mechanisms*, E. Clementel and C. Villi, Ed., Gordon and Breach, Science Publishers, New York, 1963, p. 125.
[29] N. Austern, *Phys. Rev.* **137,** B752 (1965).
[30] F. G. Perey and D. S. Saxon, *Phys. Letters* **10,** 107 (1964).
[31] E. S. Rost and N. Austern, *Phys. Rev.* **120,** 1375 (1960).
[32] N. Austern and J. S. Blair, *Ann. Phys.* **33,** 15 (1965).
[33] N. Austern, *Ann. Phys.* **15,** 299 (1961).
[34] E. Rost, *Phys. Rev.* **128,** 2708 (1962).
[35] N. Austern, in *Selected Topics in Nuclear Theory*, F. Janouch, Ed., IAEA, Vienna, 1963.
[36] J. S. Blair, *Phys. Rev.* **115,** 928 (1959).
[37] B. L. Cohen and A. G. Rubin, *Phys. Rev.* **111,** 1568 (1958).
[38] W. T. Pinkston and G. R. Satchler, *Nucl. Phys.* **27,** 270 (1961).
[39] E. Rost, Ph.D. Thesis, University of Pittsburgh 1961, (unpublished).
[40] E. A. Sanderson and N. S. Wall, *Phys. Letters* **2,** 173 (1962).
[41] L. S. Kisslinger, *Phys. Rev.* **129,** 1316 (1963).
[42] H. O. Funsten, N. R. Roberson and E. Rost, *Phys. Rev.* **134,** B117 (1964).
[43] D. F. Jackson, *Phys. Letters* **14,** 118 (1965).
[44] V. A. Madsen and W. Tobocman, *Phys. Rev.* **139,** B864 (1965).
[45] M. B. Johnson, L. W. Owen, and G. R. Satchler, *Phys. Rev.* **142,** 748 (1966).
[46] N. K. Glendenning and M. Veneroni, *Phys. Rev.* **144,** 839 (1966).
[47] V. A. Madsen, *Nucl. Phys.* **80,** 177 (1966).
[48] R. Ceuleneer, *Nucl. Phys.* **81,** 339, 349 (1966).
[49] A. Bohr and B. Mottelson, *Kgl. Danske Videnskab. Selskab, Mat.-Fys. Medd.* **27,** No. 16 (1953).
[50] J. S. Blair, in *Lectures on Nuclear Interactions, Herceg-Novi* 1962, Vol. II, p. 63. Federal Nuclear Energy Commission of Yugoslavia, Belgrade, 1964; and J. S. Blair, in *Proceedings of the Conference on Direct Interactions and Nuclear Reaction Mechanisms*, E. Clementel and C. Villi, Ed., Gordon and Breach, Science Publishers, New York, 1963, pp. 669–694.
[51] N. Austern, R.M. Drisko, E. Rost, and G.R. Satchler, *Phys. Rev.* **128,** 733 (1962).

[52] M. Sakai, et al., *Phys. Letters* **8,** 197 (1964).
[53] K. Yagi, et al., *Phys. Letters* **10,** 186 (1964).
[54] M. P. Fricke and G. R. Satchler, *Phys. Rev.* **139,** B567 (1965).
[55] H. W. Brock, J. L. Yntema, B. Buck and G. R. Satchler, *Nucl. Phys.* **64,** 259 (1965).
[56] J. K. Dickens, F. G. Perey and G. R. Satchler, *Nucl. Phys.* **73,** 529 (1965).
[57] H. Niewodniczański, J. Nurzyński, A. Strzatkowski, and G. R. Satchler, *Phys. Rev.* **146,** 799 (1966).
[58] C. R. Bingham, M. L. Halbert, and R. H. Bassel, *Phys. Rev.* **148,** 1174 (1966).
[59] H. L. Wilson and M. B. Sampson, *Phys. Rev.* **137,** B305 (1965).
[60] E. F. Gibson, et al., *Phys. Rev.* **155,** 1208 (1967).
[61] R. H. Siemssen and J. R. Erskine, *Phys. Rev.* **146,** 911 (1966).
[62] B. T. Lucas, S. W. Cosper, and O. E. Johnson, *Phys. Rev.* **144,** 972 (1966).
[63] T. Stovall and N. M. Hintz, *Phys. Rev.* **135,** B330 (1964).
[64] K. Alder and A. Winther, *Coulomb Excitation*, Academic, New York, 1966; L. C. Biedenharn and P. J. Brussard, *Coulomb Excitation*, Clarendon, Oxford, 1965.
[65] R. H. Bassel, G. R. Satchler, R. M. Drisko, and E. Rost, *Phys. Rev.* **128,** 2693 (1962).
[66] D. Robson, *Phys. Rev.* **137,** B535 (1965).
[67] A. Agodi and G. Schiffrer, *Nucl. Phys.* **50,** 337 (1964).
[68] K. A. Amos, V. A. Madsen, and I. E. McCarthy, *Nucl. Phys.* **A94,** 103 (1967).
[69] A. B. Clegg, *Nucl. Phys.* **57,** 509 (1964).
[70] T. T. S. Kuo and G. E. Brown, *Nucl. Phys.* **85,** 40 (1966), T. T. S. Kuo, E. Baranger, and M. Baranger, *Nucl. Phys.* **81,** 241 (1966).
[71] K. A. Amos, *Nucl. Phys.* **A103,** 657 (1967).
[72] G. R. Satchler, *Nucl. Phys.* **A95,** 1 (1967).
[73] J. T. Lamarsh and H. Feshbach, *Phys. Rev.* **104,** 1633 (1965).
[74] R. L. Robinson, J. L. C. Ford, P. H. Stelson, and G. R. Satchler, *Phys. Rev.* **146,** 816 (1966).
[75] M. Kawai, T. Terasawa, and K. Izumo, *Nucl. Phys.* **59,** 289 (1964).
[76] R. M. Haybron and H. McManus, *Phys. Rev.* **136,** B1730 (1964).
[77] R. M. Haybron and H. McManus, *Phys. Rev.* **140,** B638 (1965).
[78] R. M. Haybron, *Phys. Rev.* **160,** 756 (1967).
[79] A. K. Kerman, H. McManus, and R. M. Thaler, *Ann. Phys.* **8,** 551 (1959).
[80] V. Gillet and N. Vinh Mau, *Nucl. Phys.* **54,** 321 (1964), V. Gillet and E. A. Sanderson, *Nucl. Phys.* **54,** 472 (1964), V. Gillet and M. A. Melkanoff, *Phys. Rev.* **133,** B1190 (1964).
[81] J. B. French and M. H. Macfarlane, *Nucl. Phys.* **26,** 168 (1961).
[82] A. M. Lane, *Phys. Rev. Letters* **8,** 171 (1962).
[83] A. M. Lane, *Nucl. Phys.* **35,** 676 (1962).
[84] J. B. French and M. H. Macfarlane, *Phys. Letters* **2,** 255 (1962).
[85] T. Terasawa and G. R. Satchler, *Phys. Letters* **7,** 265 (1963). Also see A. Langsford, et al., *Nucl. Phys.* **A113,** 433 (1968).

[86] J. D. Anderson, C. Wong, J. W. McClure, and B. D. Walker, *Phys. Rev.* **136,** B118 (1964).
[87] G. R. Satchler, R. M. Drisko, and R. Bassel, *Phys. Rev.* **136,** B637 (1964); also see C. Wong, et al., *Phys. Rev.* **156,** 1266 (1967).
[88] J. D. Anderson and C. Wong, *Phys. Rev. Letters* **7,** 250 (1961).
[89] J. D. Anderson, C. Wong, and J. W. McClure, *Phys. Rev.* **129,** 2718 (1963).
[90] L. F. Hansen, M. L. Stelts, and J. J. Wesolowski, *Phys. Rev.* **143,** 800 (1966).
[91] P. E. Hodgson and J. R. Rook, *Nucl. Phys.* **37,** 632 (1962).
[92] T. Une, S. Yamaji, and H. Yoshida, *Progr. Theoret. Phys.* **35,** 1010 (1966).
[93] J. D. Anderson and C. Wong, *Phys. Rev. Letters* **8,** 442 (1962).
[94] C. Mahaux and H. A. Weidenmüller, *Nucl. Phys.* **89,** 33 (1966).
[95] J. R. Rook and P. E. Hodgson, *Nucl. Phys.* **52,** 603 (1964).
[96] J. R. Oppenheimer and M. Phillips, *Phys. Rev.* **48,** 500 (1935).
[97] R. Serber, *Phys. Rev.* **72,** 1008 (1947).
[98] S. T. Butler, *Proc. Roy. Soc.* (London) **A208,** 559 (1951).
[99] F. L. Friedman and W. Tobocman, *Phys. Rev.* **92,** 93 (1953).
[100] K. R. Greider and G. L. Strobel, *Bull. Am. Phys. Soc.* **11,** 304 (1966). W. R. Smith, *Nucl. Phys.* **A130,** 657 (1969).
[101] W. T. Pinkston and G. R. Satchler, *Nucl. Phys.* **72,** 641 (1965).
[102] J. B. French, in *Nuclear Spectroscopy*, Part B, Fay Ajzenberg-Selove, Ed., Academic, New York, 1960.
[103] M. H. Macfarlane and J. B. French, *Rev. Mod. Phys.* **32,** 567 (1960).
[104] H. W. Barz, et al., *Nucl. Phys.* **73,** 473 (1965).
[105] G. L. Strobel, *Nucl. Phys.* **86,** 535 (1966). P. R. Almond and J. R. Risser, *Nucl. Phys.* **72,** 436 (1965); also see B. T. Lucas, D. R. Ober, and O. E. Johnson, *Phys. Rev.* **167,** 990 (1968).
[106] L. L. Lee, et al., *Phys. Rev.* **136,** B971 (1964).
[107] J. S. Forster, et al., *Nucl. Phys.* **A101,** 113 (1968).
[108] P. E. Cavanagh, et al., *Nucl. Phys.* **50,** 49 (1964).
[109] P. T. Andrews, et al., *Nucl. Phys.* **56,** 465 (1964).
[110] V. Dantzig and W. Tobocman, *Phys. Rev.* **136,** B1682 (1964).
[111] W. R. Smith, *Phys. Rev.* **137,** B913 (1965).
[112] J. C. Legg, H. D. Scott, and M. K. Mehta, *Nucl. Phys.* **84,** 398 (1966).
[113] G. Muehllener, A. S Poltorak, W. C. Parkinson, and R. H. Bassel, *Phys. Rev.* **159,** 1039 (1967).
[114] D. W. Miller, H. E. Wegner, and W. S. Hall, *Phys. Rev.* **125,** 2054 (1962).
[115] B. E. F. Macefield, *Nucl. Phys.* **59,** 573 (1964).
[116] B. A. Robson, *loc. cit.*, Chap. 2, Ref. 21.
[117] J. Bardwick and R. Tickle, *Phys. Rev.* **161,** 1217 (1967); B. L. Cohen, *Phys. Rev.* **125,** 1358 (1962); R. H. Fulmer and A. L. McCarthy, *Phys. Rev.* **131,** 2133 (1963); B. L. Cohen and R. E. Price, *Phys. Rev.* **121,** 1441 (1961).
[118] R. A. Sorenson, E. K. Lin, and B. L. Cohen, *Phys. Rev.* **142,** 729 (1966).
[119] C. Daum, *Nucl. Phys.* **45,** 273 (1963); R. B. Weinberg, G. E. Mitchell, and L. J. Lidofsky, *Phys. Rev.* **133,** B884 (1964).
[120] H. A. Bethe and S. T. Butler, *Phys. Rev.* **85,** 1045 (1952).
[121] I. Filosofo, et al., *Nucl. Phys.* **58,** 522 (1964).

[122] M. B. Hooper, *Nucl. Phys.* **76,** 449 (1966).
[123] J. E. Bowcock, *Proc. Phys. Soc.* (London) **A68,** 512 (1955).
[124] C. F. Clement, *Nucl. Phys.* **66,** 241 (1965).
[125] F. B. Morinigo, *Nucl. Phys.* **77,** 289 (1966); **A90,** 113 (1967); **A95,** 571 (1967).
[126] I. S. Towner, *Nucl. Phys.* **A93,** 145 (1967).
[127] L. L. Lee, Jr., and J. P. Schiffer, *Phys. Rev.* **136,** B405 (1964).
[128] J. P. Schiffer, et al., *Phys. Rev.* **147,** 829 (1966).
[129] R. Sherr, E. Rost, and M. E. Rickey, *Phys. Rev. Letters* **12,** 420 (1964).
[130] K. R. Greider, *Phys. Rev.* **136,** B420 (1964).
[131] R. M. Drisko, private communication; C. Glashausser and M. E. Rickey, *Phys. Rev.* **154,** 1033 (1967).
[132] R. H. Dalitz, *Proc. Phys. Soc.* (London) **A66,** 28 (1953).
[133] R. C. Johnson, *Nucl. Phys.* **A90,** 289 (1967).
[134] D. M. Brink and G. R. Satchler, *Angular Momentum*, Oxford University Press, Oxford, 1962.
[135] R. C. Johnson and F. D. Santos, *Phys. Rev. Letters* **19,** 364 (1967).
[136] R. Huby and J. R. Mines, *Rev. Mod. Phys.* **37,** 406 (1965).
[137] R. H. Lemmer, *Nucl. Phys.* **39,** 680 (1962).
[138] A. Dar, A. de-Shalit, and A. S. Reiner, *Phys. Rev.* **131,** 1732 (1963).
[139] F. B. Morinigo, *Phys. Rev.* **134,** 1243 (1964); *Nucl. Phys.* **50,** 136 (1964); *Nucl. Phys.* **62,** 373 (1965); H. D. Zeh, *Nucl. Phys.* **75,** 423 (1966).
[140] L. J. B. Goldfarb, *Nucl. Phys.* **72,** 537 (1965); L. J. B. Goldfarb and E. Parry, *Nucl. Phys.* **A116,** 289 (1968).
[141] L. J. B. Goldfarb, in *Lectures in Theoretical Physics*, Vol. VIIIC, University of Colorado Press, Boulder, 1966.
[142] W. R. Smith, *Nucl. Phys.* **72,** 593 (1965).
[143] F. P. Gibson and A. K. Kerman, *Phys. Rev.* **145,** 758 (1966).
[144] L. J. B. Goldfarb and M. B. Hooper, *Phys. Letters* **19,** 299 (1965).
[145] L. J. B. Goldfarb and K. K. Wong, *Nucl. Phys.* **A90,** 361 (1967).
[146] G. M. Crawley, B. V. N. Rao, and D. L. Powell, *Nucl. Phys.* **A112,** 223 (1968).
[147] R. H. Bassel, *Phys. Rev.* **149,** 791 (1966).
[148] J. C. Gunn and J. Irving, *Phil. Mag.* **42,** 1353 (1951).
[149] J. C. Hiebert, E. Newman, and R. H. Bassel, *Phys. Rev.* **154,** 898 (1967).
[150] D. D. Armstrong and A. G. Blair, *Phys. Rev.* **140,** B1226 (1965).
[151] R. N. Glover, A. D. W. Jones, and J. R. Rook, *Nucl. Phys.* **81,** 289 (1966).
[152] J. R. Rook, *Nucl. Phys.* **A97,** 217 (1967).
[153] C. M. Fou, R. W. Zurmühle, and J. M. Joyce, *Phys. Rev.* **155,** 1248 (1967).
[154] D. D. Armstrong, A.G. Blair, and H. C. Thomas, *Phys. Rev.* **155,** 1254 (1967).
[155] L. F. Hansen, et al., *Phys. Rev.* **158,** 917 (1967).
[156] M. K. Brussel, D. E. Rundquist, and A. I. Yavin, *Phys. Rev.* **140,** B838 (1965).
[157] L. Colli, I. Iori, S. Micheletti, and M. Pignanelli, *Nuovo Cimento* **20,** 94 (1961).
[158] B. F. Bayman, F. P. Brady, and R. Sherr, in proceedings of *Rutherford Jubilee Int. Conf.*, J. B. Birks, Ed., Academic, New York, 1961; R. Sherr in *Direct Interactions and Nuclear Reaction Mechanisms*, E. Clementel and C. Villi, Eds., Gordon and Breach, Science Publishers, New York, 1963.
[159] C. B. Fulmer and J. B. Ball, *Phys. Rev.* **140,** B330 (1965); see also C. R. Bingham and M. L. Halbert, *Phys. Rev.* **158,** 1085 (1967).

[160] J. R. Priest and J. S. Vincent, *Phys. Rev.* **167,** 933 (1968).
[161] H. C. Newns, *Proc. Phys. Soc.* (London) **A76,** 489 (1960).
[162] E. M. Henley and D. U. L. Yu, *Phys. Rev.* **133,** B1445 (1964).
[163] C. L. Lin and S. Yoshida, *Prog. Theoret. Phys.* **32,** 885 (1964).
[164] N. K. Glendenning, *Phys. Rev.* **137,** B102 (1965).
[165] C. L. Lin, *Prog. Theoret. Phys.* **36,** 25 (1966).
[166] B. F. Bayman, in *Nuclear Spectroscopy with Direct Reactions*, F. E. Throw, Ed., Argonne National Laboratory report ANL-6878, 1964; also in *Proceedings of the Summer Study Group*, G. Holland and R. Steiglitz, Eds., Brookhaven National Laboratory report BNL 948(C-46), 1965.
[167] J. C. Hardy and I. S. Towner, *Phys. Letters* **25B,** 577 (1967) and subsequent publications.
[168] J. H. Bjerregaard, O. Hansen, O. Nathan, and S. Hinds, *Nucl. Phys.* **86,** 145 (1966). Also see L. F. Hansen, et al., *Nucl. Phys.* **A98,** 25 (1967).
[169] A. de-Shalit and I. Talmi, *Nuclear Shell Theory*, Academic, New York, 1963.
[170] S. Yoshida, *Nucl. Phys.* **33,** 685 (1962).
[171] A. M. Lane and J. M. Soper, *Nucl. Phys.* **37,** 506 (1962).
[172] J. B. Mead and B. L. Cohen, *Phys. Rev.* **125,** 947 (1962).
[173] R. M. Drisko and F. Rybicki, *Phys. Rev. Letters* **16,** 275 (1966).
[174] B. F. Bayman and A. Kallio, *Phys. Rev.* **156,** 1121 (1967).
[175] R. A. Broglia and C. Riedel, *Nucl. Phys.* **A92,** 145 (1967).
[176] J. R. Rook and D. Mitra, *Nucl. Phys.* **51,** 96 (1964).
[177] N. K. Glendenning, *Phys. Rev.* **156,** 1344 (1967).
[178] W. W. True and K. W. Ford, *Phys. Rev.* **109,** 1675 (1958).
[179] C. D. Zafiratos, et al., *Phys. Rev.* **154,** 887 (1967).
[180] J. Jänecke, *Nucl. Phys.* **48,** 129 (1963).
[181] J. Cerny and R. H. Pehl, *Phys. Rev. Letters* **12,** 619 (1964); also see D. G. Fleming, J. Cerny, and N. K. Glendenning, *Phys. Rev.* **165,** 1153 (1968).
[182] C. B. Fulmer and J. B. Ball, *Phys. Rev.* **140,** B330 (1965).
[183] N. S. Wall, private communication.
[184] W. Swenson and N. Cindro, *Phys. Rev.* **123,** 910 (1961).
[185] L. Madansky and G. E. Owen, *Phys. Rev.* **99,** 1608 (1955).
[186] N. Austern, R. M. Drisko, E. C. Halbert, and G. R. Satchler, *Phys. Rev.* **133** B3 (1964).
[187] R. M. Drisko and G. R. Satchler, *Phys. Letters* **9,** 342 (1964).
[188] P. J. A. Buttle and L. J. B. Goldfarb, *Proc. Phys. Soc.* (London) **83,** 701 (1964).
[189] Gy. Bencze and J. Zimányi, *Phys. Letters* **9,** 246 (1964).
[190] J. K. Dickens, R. M. Drisko, F. G. Perey, and G. R. Satchler, *Phys. Letters* **15,** 337 (1965).
[191] K. R. Greider, in *Advances in Theoretical Physics*, Vol. 1, K. A. Brueckner, Ed., Academic, New York, 1965.
[192] P. J. A. Buttle and L. J. B. Goldfarb, *Nucl. Phys.* **78,** 409 (1966). Also see P. J. A. Buttle and L. J. B. Goldfarb, *Nucl. Phys.* **A115,** 461 (1968); L. J. B. Goldfarb and J. W. Steed, *Nucl. Phys.* **A116,** 321 (1968).
[193] L. R. Dodd and K. R. Greider, *Phys. Rev. Letters* **14,** 959 (1965); J. Birnbaum, J. C. Overly, and D. A. Bromley, *Phys. Rev.* **157,** 787 (1967).
[194] Gy. Bencze and J. Zimányi, *Nucl. Phys.* **81,** 76 (1966).

CHAPTER 6

More About Coupled Channels

In the method of coupled channels the DI wavefunction Ψ_{model} is calculated by exact numerical integration of the reduced Schrödinger equation. This method is made possible by our basic physical assumption that Ψ_{model} spans a very small part of Hilbert space. A brief description of this method is given in Section 4.6. The many-body Schrödinger equation is seen to reduce to a finite system of coupled equations in only a single vector variable. This simplification took place because each term of the assumed Ψ_{model} contains only one undetermined factor, and this factor depends on only one displacement vector. Therefore, only a few coefficient functions in Ψ_{model} need to be determined, much as in a variational calculation.

Sometimes we not only carry strongly-coupled terms in Ψ_{model}, but we also are led to carry some terms that are coupled in rather weakly. When this is done only the strongly-coupled terms of Ψ_{model} need to be computed with coupled equations; the other terms may be brought in by perturbation theory. This procedure may be thought of as a *generalized distorted waves method* as described in Section 3.3.

The present chapter presents several applications of these coupled-channels methods.

6.1 Inelastic Scattering

The coupled equations take their simplest form for nonexchange inelastic scattering. In this case

$$\Psi_{\text{model}} = \phi \sum_i \psi_i \xi_i(\mathbf{r}), \tag{6.1}$$

where ϕ is the internal wavefunction of the projectile particle, the ψ_i are the internal wavefunctions of the relevant states of the target nucleus, and the ξ_i are coefficient functions that describe the motion of the projectile relative to the target nucleus. Equations 4.75 are the set of coupled equations for the ξ_i that are obtained by substituting (6.1) into the reduced Schrödinger equation.

However, (6.1) and its associated coupled equations is a poor starting point for practical calculation. Distinct terms appear in (6.1) for each spin projection of each included energy eigenstate of the target nucleus. Additional degenerate terms appear if the projectile has nonzero spin. Therefore (6.1) is a confused mixture of geometry and dynamics, with each spin projection of each energy state having a separate coefficient function $\xi_i(\mathbf{r})$. We must introduce a change of representation to separate the geometry from the dynamics.

Practical coupled-channels calculations of inelastic scattering were first described by Yoshida [1] and by Margolis, et al., and Chase, et al. [2], following suggestions by Bohr and Mottelson [3]. The present development closely follows the presentation in Tamura's reviews [4, 5].

In place of (6.1) we immediately introduce partial waves, to obtain the expansion

$$\Psi_{\text{model}} = r^{-1} \sum_{JMl_0j_0} A_{JMl_0j_0} \sum_{nlj} R^{l_0j_0}_{Jnlj}(r)\{\mathscr{Y}_{lj}, \psi_{I(n)}\}_{JM}, \tag{6.2}$$

in which the wavefunction is expressed as a sum over terms that have definite J and M. Here the functions

$$\mathscr{Y}_{ljm} = \{i^l Y_l(\hat{r}), \phi_s\}_{jm} \tag{6.3}$$

are formed by vector coupling the spin of the projectile with the orbital angular momentum of the projectile-nucleus relative motion. Spin-angle states that have definite total J and M then are constructed by vector coupling the $\mathscr{Y}_{ljm}$ with the $\psi_{I(n)}$, the internal wavefunctions of the target nucleus, for energy state n and angular momentum $I(n)$. These total spin-angle wavefunctions distinguish individual channels. Because the spin-angle functions incorporate the angular part of the projectile-nucleus relative motion, the coefficient functions $R^{l_0j_0}_{Jnlj}$ are functions of only the radial coordinate r. A distinct radial coefficient function is associated with each spin-angle wavefunction. The Schrödinger equation for Ψ_{model} reduces to a set of coupled differential equations for these radial functions. (See below.)

The interactions in the Schrödinger equation couple the various terms in the nlj sum in (6.2), but they of course do not couple terms with different J, M, or parity. Hence each term in the JM sum is an independent eigenfunction of the Schrödinger equation, and is governed by an independent set of coupled differential equations.

The solutions of the coupled equations must be subjected to scattering boundary conditions; this implies that only the radial functions $R^{l_0j_0}_{J0lj}$ associated with the target nucleus ground state may contain incoming parts. Unfortunately, imposition of these boundary conditions may not suffice to define unique sets of solutions of the coupled equations. We therefore impose uniqueness by defining solutions that have nonzero incoming parts only in individual ground-state channels, labelled by the superscripts l_0j_0.

It is now clear that the nlj sum in (6.2) is needed to construct eigensolutions of the Schrödinger equation, and that the JMl_0j_0 sum is needed to generate from these Schrödinger eigenfunctions that one linear combination whose incoming part correctly matches the incoming plane-wave part of Ψ_{model}. Because the unique aspects of the coupled-channels theory are bound up primarily with the calculation of the Schrödinger eigenfunctions, rather than with their use to compute cross sections, there need be no detailed discussion of the JMl_0j_0 sum. Details of the calculation of cross sections may be found, for example, in [4]. Nevertheless, it is interesting to carry the discussion of the $A_{JMl_0j_0}$ one step further, simply to have a clear understanding of the meaning of these coefficients and of the JMl_0j_0 sum. Let us arbitrarily assign unit normalization to the incoming part of each radial wave $R^{l_0j_0}_{J0l_0j_0}$, thereby obtaining for the net incoming part of Ψ_{model} the expression

$$r^{-1} \sum_{JMl_0j_0} A_{JMl_0j_0}(i/2)H^*_{l_0}(r)\{\mathscr{Y}_{l_0j_0}, \psi_{I(0)}\}_{JM}.$$

The coefficients $A_{JMl_0j_0}$ then are determined by the requirement that this expression for the incoming part of Ψ_{model} must be equivalent to the more familiar expression

$$r^{-1}\sum_{l_0} i^{l_0}\sqrt{4\pi(2l_0+1)}\,\frac{i}{2}\,H^*_{l_0}(r)Y_{l_0}{}^0(\hat{r})\phi_{sm_s}\psi_{I(0)m_I},$$

in which the z-axis is taken to lie along the incident beam, and the projectile and target-nucleus spins are taken to have the projections m_s and m_I. —At this stage we drop the discussion of the $A_{JMl_0j_0}$ and turn to the coupled equations. In the discussion of these equations the superscripts l_0j_0 are redundant, and they are henceforth omitted.

The coupled differential equations for the radial coefficient functions are derived, as in Section 4.6, by substituting (6.2) into the reduced Schrödinger equation and then left multiplying this equation by each spin-angle wavefunction in turn. Because radial functions associated with different J and M are not coupled the detailed form of the coupled equations is

$$\left\{E_n + \frac{\hbar^2}{2u}\left[\frac{d^2}{dr^2} - l(l+1)r^{-2}\right] - \langle ljI(n)|\, V + \overline{\mathscr{O}}\,|ljI(n)\rangle_J\right\}R_{Jnlj}(r)$$
$$= \sum_{n'l'j'}{}' \langle ljI(n)|\, V + \overline{\mathscr{O}}\,|l'j'I(n')\rangle_J R_{Jn'l'j'}(r) \quad (6.4)$$

where

$$\langle ljI(n)|\, V + \overline{\mathscr{O}}\,|l'j'I(n')\rangle_J = \langle\{\mathscr{Y}_{lj}, \psi_{I(n)}\}_{JM}|\, V + \overline{\mathscr{O}}\,|\{\mathscr{Y}_{l'j'}, \psi_{I(n')}\}_{JM}\rangle, \quad (6.5)$$

and where

$$E_n = E - \varepsilon - \varepsilon_n$$

as in (4.74). Here $V + \overline{\mathcal{O}}$ is the net projectile-nucleus interaction, with $\overline{\mathcal{O}}$ the operator that accounts for energy-averaged effects due to parts of the physical system that are not included in Ψ_{model}. Naturally, the sum on the RHS in (6.4) omits the diagonal term $n'l'j' = nlj$.

The decoupling of geometry and dynamics is completed by introducing a multipole expansion of the interaction,

$$V + \overline{\mathcal{O}} = \sum_{t,\lambda} v_\lambda^{(t)}(r)(Q_\lambda^{(t)} \cdot Y_\lambda(\hat{r})), \tag{6.6}$$

where the superscript t distinguishes terms of different character but the same tensorial rank λ, and $Q_\lambda^{(t)}$ is an operator that operates only on the internal coordinates of the target nucleus. Introduction of (6.6) in (6.5) separates angular coordinates of the projectile from internal coordinates of the target nucleus, and yields for the matrix elements

$$\langle ljI(n)|\, V + \overline{\mathcal{O}}\, |l'j'I(n')\rangle_J = \sum_{t,\lambda} v_\lambda^{(t)}(r)\langle I(n)\|\, Q_\lambda^{(t)}\, \|I(n')\rangle A(ljI(n), l'j'I(n'); \lambda Js), \tag{6.7}$$

with

$$\begin{aligned} A(ljI(n), l'j'I(n'); \lambda Js) = {} & (4\pi)^{-1/2}(-)^{J-s-I(n')+l+l'+1/2(l'-l)} \\ & \times \sqrt{2l+1}\sqrt{2l'+1}\sqrt{2j+1}\sqrt{2j'+1} \\ & \times \langle ll'; 00 \mid \lambda, 0\rangle W(jI(n)j'I(n'); J\lambda) \\ & \times W(ljl'j'; s\lambda). \end{aligned} \tag{6.8}$$

The factor $A(ljI(n), l'j'I(n'), \lambda Js)$ is entirely geometrical. Tamura [4] gives simplified expressions for this factor for the special projectile spin values $s = 0$ and $s = \frac{1}{2}$.

The reduced matrix elements $\langle I(n)\|\, Q_\lambda^{(t)}\, \|I(n')\rangle$ contain all the dynamics of the target nucleus. These are the same reduced matrix elements that appeared in the DW treatment of inelastic scattering, and that are found, for example, in (5.61), (5.83), (5.84). (They are based on the symmetrical form of the Wigner-Eckart theorem, as remarked in the footnote on p. 140.) The products formed by multiplying the reduced matrix elements by the $v_\lambda^{(t)}(r)$ are a full description of our DI model of the inelastic scattering, as based on the structures of the target and projectile nuclei and their interaction. The calculation of these products is affected by such questions as whether the nuclear excitations are vibrational, rotational, or single-particle, and whether the projectile-target interaction is pictured as microscopic or macroscopic. Tamura [4] gives a number of explicit expressions for the reduced matrix elements for cases of collective excitations of nuclei. Additional model-dependent expressions for these matrix elements may be found in the references cited in Chapter 5.

Thus the calculation of the $v_\lambda^{(t)}(r)\langle I(n)\| Q_\lambda^{(t)} \|I(n')\rangle$ is a standard problem of nuclear structure theory, of the type required for all nuclear reaction calculations. Once these channel-coupling functions have been calculated they may be used in first order in DW calculations, for example, or they may be inserted in the multipole expansion of (6.7) and used in (6.4) in the coupled differential equations. Because the coupled equations use the channel-coupling functions to infinite order, effects not known in DW approximation can be generated. Excitations forbidden in first order can take place, and excitations allowed in first order can be modified. As an example of a higher-order excitation we note that a 4^+ state of an even-even nucleus might be reached from the 0^+ ground state as a direct transition induced by a multipole of order $\lambda = 4^+$, or as a two-step transition through the $I = 2^+$ state induced by $\lambda = 2^+$, or as a two-step transition through $I = 3^-$ induced by $\lambda = 3^-$, and so forth. The coupled equations take all these mechanisms of excitation into account, and allow them to interfere. In the same fashion the coupled equations allow excited states to react on the ground-state elastic scattering.

Although the exact coupled equations carry interesting higher-order effects of the types just mentioned, it must not be forgotten that they are based on only a limited number of states n of the target nucleus. This limitation of the number of intermediate states means that great accuracy can be expected for only the few lower coupled states. In the same vein, great accuracy cannot be expected if too few of the multipoles $v_\lambda^{(t)}$ are carried. While the coupled equations allow operators of low multipolarity to act repeatedly, and thereby to excite states of high I, omission of competition with higher multipoles can cause gross errors [6].

In applications of the coupled equations it is customary to treat the diagonal matrix elements $\langle ljI(n)|\ V + \bar{\mathcal{O}}\ |ljI(n)\rangle_J$ as phenomenological potentials, to be adjusted so that the solution of the coupled equations gives a correct fit to observed elastic scattering. One then only asks whether the theory can correctly predict the inelastic scattering. This procedure resembles the DW approach, which is based on optical potentials fitted to the observed elastic scattering. However, the diagonal matrix elements in the coupled-channels procedure often differ very much from the optical potentials that fit the same elastic scattering data. Because the coupled equations explicitly carry all the interactions among the strongly-coupled states of Ψ_{model}, the optical operator $\bar{\mathcal{O}}$ need only account for the rather weak coupling to other parts of Hilbert space. An operator similar to this $\bar{\mathcal{O}}$ appears in the optical potential (see Sections 4.1, 4.2), however, it then accounts for the coupling to *all states* that lie outside the entrance channel; in particular, $\bar{\mathcal{O}}$ in the full optical potential accounts for the strong coupling to excited-state terms in Ψ_{model}. Under these conditions it is not surprising that the diagonal matrix elements in the coupled-channels procedure have much smaller imaginary parts than the single-channel optical potentials.

Techniques for solving the coupled equations exploit two basic properties: (a) the equations are differential, that is, local in r; (b) the equations are linear. Because of property (a) the equations can be solved by step-by-step numerical integration, working outward from $r = 0$ until r is large enough so that all nuclear potentials have become negligibly small. To perform such a step-by-step integration it is necessary to use as input an assumed set of relative values of the R_{Jnlj} at $r = 0$. Each linearly independent set of these starting values yields a linearly independent solution of the differential equations. Because of property (b), we are then able to construct, from the complete set of linearly independent solutions, those few solutions that fulfill scattering boundary conditions at large r: Hence the technique of solution requires solving N coupled equations N times, using N different sets of starting values. From these N solutions the method then selects those linear combinations that have incoming waves in ground-state channels. It is clear that such calculations are very laborious. Iterative methods of solution have also been used [7]. It is noteworthy that at each stage of an iterative process we only treat functions that obey the boundary conditions.*

Recognition of the great labor involved in coupled-channels calculations points to the need for a clear understanding of the circumstances under which such calculations should be performed. Such calculations extend into the continuum the shell-model procedure of diagonalizing H within a small, chosen part of Hilbert space. However, we want to do more than just pursue an analogy. It is noted in Chapter 4 that for reactions allowed in first order the coupled-channels method gives only minor improvements† over the DW method, provided correct optical potentials are used in the DW calculations (Chapter 4 [20, 21]). Hence we see that the principal role of coupled-channels

* Coupled-channels eigenfunctions may be computed by step-by-step integration, as described here, and subjected to energy-independent boundary conditions at some *matching radius*. Eigenfunctions generated in this fashion possess the properties of the interior eigenfunctions of an R-matrix procedure. A rather inclusive description of the energy dependence of a nuclear reaction may then be obtained by using these coupled-channel interior functions in the R-matrix formalism [8]. By this procedure the energy dependence is obtained in good approximation without solving the full coupled-channels problem at every energy.

† Glendenning and collaborators [9] have noted that weak single-particle excitations may require analysis by coupled equations, even if they are allowed in first order. Weak first-order excitations can be modified to a significant extent by competition with higher-order transitions through collective intermediate states. From this point of view the DW method only would be reliable for collective first-order excitations.

calculations must be to treat excitations forbidden in first order, excitations caused by repeated interaction with the projectile. In the DW method such excitations are treated in higher Born approximation, and require the use of awkward, distorted-waves Green's functions [10]. Coupled-channels treatments of such higher excitations are simpler (!) than the DW treatments, and allow careful assessments of the mechanism of excitation [4, 5, 11].

Applications of the coupled-channels method to the analysis of experimental data require adjusting the magnitudes of the channel-coupling functions,

$$v_\lambda^{(t)}(r)\langle I(n)\| \, Q_\lambda^{(t)} \, \|I(n')\rangle,$$

to achieve simultaneous good fits to the cross sections for all states that are coupled strongly. By this means *coupling constants that link excited states to each other can be determined*. Such coupling constants often can be determined with good sensitivity because their contributions to particular cross sections interfere with contributions derived from known excitations of the ground state. In a similar fashion, unique *spin assignments* for higher excited states can sometimes be made [4].

Because the formation of higher-excited states may involve interference among different orders of excitations the cross sections for such states can show interesting energy-dependent effects. One example is an energy-dependent phase reversal [12, 13] that appears in the cross section for excitation of the 4^+ state of Ni^{58}.

The DW theory of first-order excitations uses optical potentials that are fitted to observed elastic scattering. For strongly deformed target nuclei these optical potentials are affected by coupling to collective excited states, and as a consequence may exhibit unusual parameter values. Even under these conditions the DW theory usually gives good fits to the inelastic cross section, and gives reliable values for spins and coupling constants. Nevertheless it may be desirable to analyze such reactions with a coupled-channels theory, to eliminate channel-coupling effects from the ground-state optical potential and thereby to obtain optical-potential parameters that are more usual and more understandable.

It is noted previously that the explicit treatment of channel-coupling effects leads to reduced values for the imaginary parts of the diagonal potentials. It is likely that the explicit representation of coupling effects also reduces the need for imaginary parts in the off-diagonal operators. One may hope that exploitation of this fact will overcome the difficulty (Sections 5.5, 5.6) that microscopic theories lead to operators that have real form factors.

An elaborate analysis by Hendrie, et al., illustrates several of the above ideas [14]. These authors studied the inelastic scattering of 50 MeV alpha particles by permanently deformed even-even nuclei in the rare-earth region. Differential cross sections were observed for the excitation of as many as five members of the ground-state rotational band, hence for spins as high as $I = 8^+$. The analysis proceeded on the basis that $V + \bar{\mathcal{O}}$, the projectile-nucleus interaction, could be regarded as a rotating deformed optical potential of the type discussed in Section 5.3. Coulomb excitation was included, and coupling to states not belonging to the ground-state band was ignored. This physical picture allowed all the channel-coupling functions to be expressed in terms of the deformation parameters β_2, β_4, β_6, etc., that characterize successive multipoles in the shape of the deformed nuclear surface. Figures 6.1 and 6.2 illustrate how fits to the higher excited states of Yb^{176} require nonvanishing values for β_4 and β_6. This analysis provides the first measured values of these higher deformation parameters.

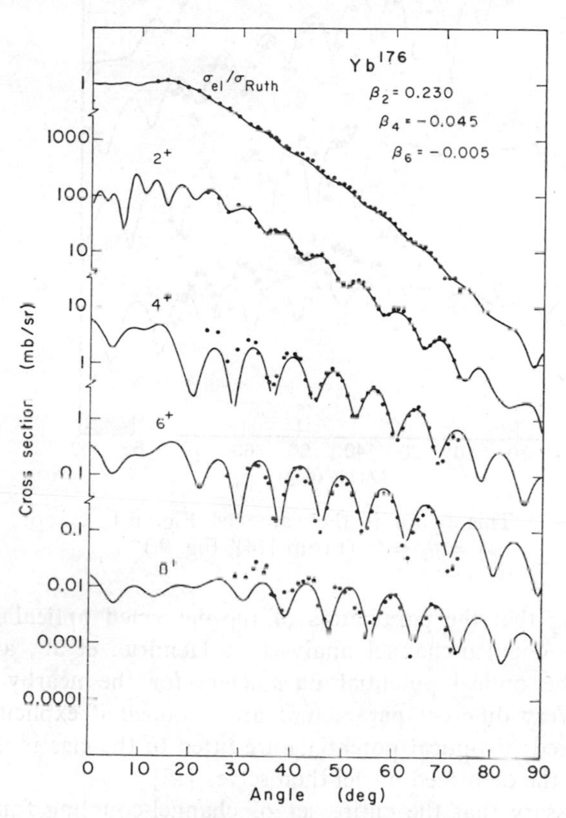

Fig. 6.1 Comparison of the coupled-channels theory with experimental data for $Yb^{176}(\alpha, \alpha')$ to five levels in the ground-state rotational band, at 50 MeV bombarding energy. Deformations of multipolarities 2^+, 4^+, 6^+ are used, as shown. (From [14], Fig. 8.)

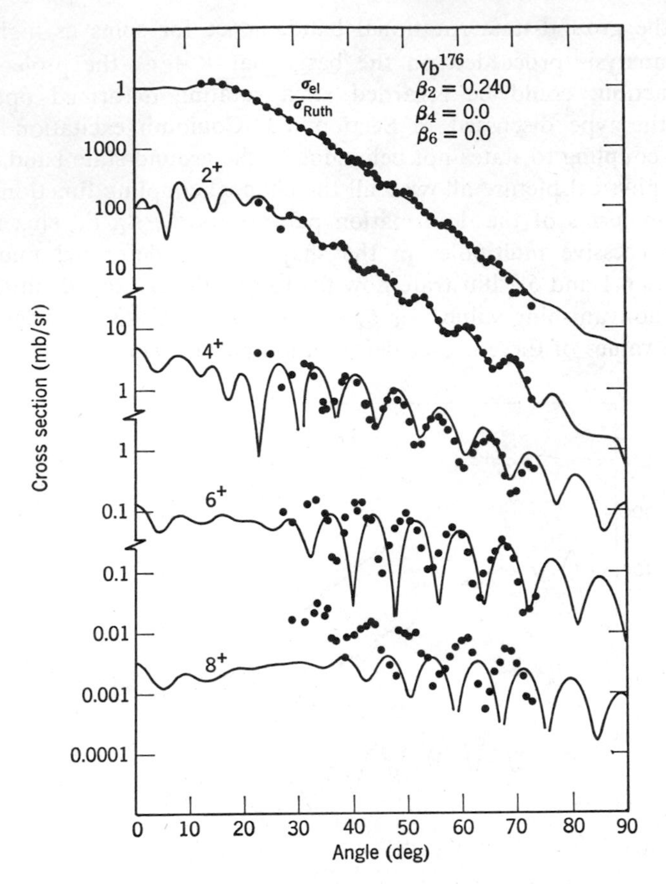

Fig. 6.2 This figure is the same as Fig. 6.1, except $\beta_4 = \beta_6 = 0$. (From [14], Fig. 9.)

It is interesting that the parameters of the deformed optical potentials required for the coupled-channel analysis of Hendrie, et al., are almost identical with the optical potential parameters for the nearby spherical nucleus Sm^{148}. Very different parameters are obtained if explicit channel-coupling is ignored; if optical potentials are fitted to the elastic differential cross sections of the deformed nuclei themselves [15].

It is not necessary that the entire set of channel-coupling functions be derived in terms of the β_λ parameters of an underlying phenomenological model, as in the above-described analysis. In a number of cases Tamura was able to obtain independent determinations of β_{24}, a quadrupole "deformation parameter" that measures the strength with which the first 2^+ state of a vibrational nucleus is coupled to the first 4^+ state. He noted [5] that the analysis of the reaction $Cd^{112}(p, p')$, for example, yields a much smaller value of β_{24} than is obtained from studies of higher-order Coulomb excitation. It is interesting to see whether microscopic theories of vibrational nuclei can predict both the value of β_{24} and the reason it depends on the nature of the projectile-nucleus interaction [8, 9].

Tamura cites [5] another important example of an excitation forbidden in first-order and fitted by a coupled-channels calculation. This is the excitation of the ("non-normal parity") 3^+ state of Mg^{24} by 42 MeV alpha particles. Success in fitting the cross section for this state verifies the physical model that the 3^+ state is the second member of the γ-vibrational band. Use of this model allowed the 3^+ excitation to be predicted in terms of parameters fitted to the (second) 2^+ state at the head of the γ-vibrational band.

Another intriguing application of coupled-channels calculations to first-order excitations is for the prediction of characteristic differences between cross sections for vibrational and rotational excitations. It should be recalled that the DW theory makes no distinction between these two types of excitations (Sections 5.1, 5.3). Both use the same distorted waves and similar, surface-peaked form factors. However, in the coupled-channels theory there are characteristic differences between the diagonal matrix elements found for excited states of vibrational and deformed nuclei ([16], also [4] p. 682), and in some cases these give rise to large differences between the calculated cross sections. Figure 6.3 shows how the inelastic scattering of 16 MeV protons yields angular distributions for the 2^+ excitation of the vibrational nucleus Sm^{148} that differ greatly from those for the rotational nucleus Sm^{154}. Corresponding differences between the two theoretical fits are seen [5].

Many authors have discussed resonances that appear in cross sections obtained from coupled-channels calculations. We recall two resonance types that may certainly be expected. (a) The coupling of open and closed channels gives familiar compound-nucleus resonances, as pointed out in Chapters 2, 4, and 11, and in Section 6.2. (b) Resonances also can appear if all the coupled channels are open, because individual partial waves may resonate in the single-particle potential that underlies the coupled-channels calculation. However, it is noted (Section 2.2–5) that nuclear single-particle potentials have diffuse surfaces and appreciable imaginary parts, therefore single-particle resonances only appear at low energies, close to thresholds.*

It is interesting that coupled-channels calculations in which open channels are coupled strongly apparently yield yet another type of resonance [17–19]. These resonances appear in states of low angular momentum, states which have good barrier penetration; they appear to be caused by coupling to

* It is the possible appearance of such single-particle resonances that causes the application of the DW theory near thresholds to be suspect.

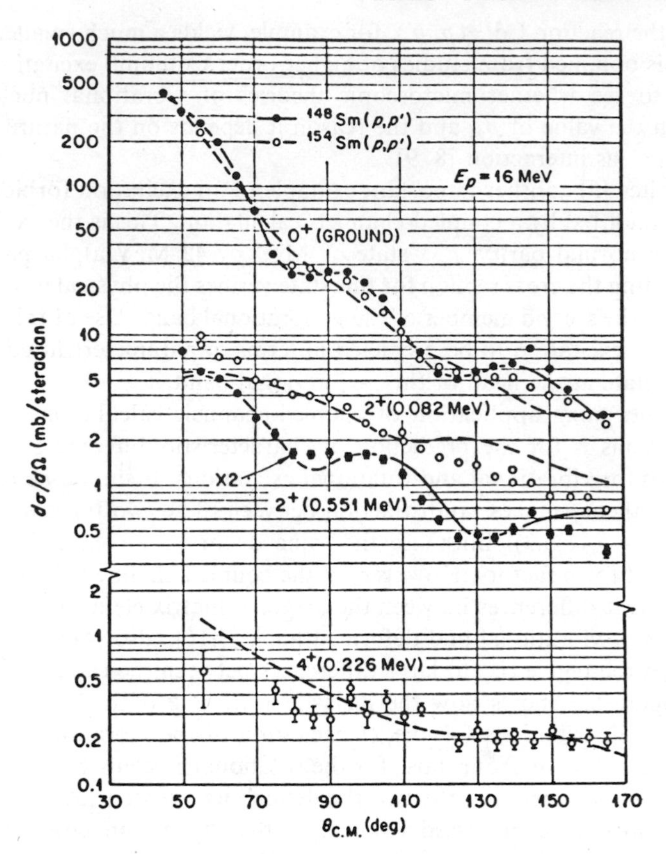

Fig. 6.3 Elastic and inelastic scattering of 16 MeV protons by $Sm^{148,154}$. Note the striking differences between the angular distributions for the "vibrational" nucleus Sm^{148} and the "rotational" nucleus Sm^{154}. Results of coupled-channels calculations by Tamura are also shown. (From [5], Fig. 3.)

single-particle states that have high angular momentum, states which the centrifugal barrier causes to be confined near $r = 0$. However, careful study of the calculations discloses that these resonances are not of a new type caused by the strong coupling, but rather they are modifications of single-particle resonances in the state that has high angular momentum. Coupling to a partial wave of low angular momentum displaces such a resonance to lower energy and renders it more visible. The state that has high angular momentum seems to serve as a "quasi" closed state. (It is unfortunate that investigations of this effect have used square-well single-particle potentials.) A resonance found in the "unified" calculation of Lemmer and Shakin [20] may have been of this type.

We also note in passing that the wavefunction Ψ_{model} of (6.2) probably is too limited in form to allow the definition of a good K quantum number for the compound system (projectile + target nucleus). Tamura discusses generalized wavefunctions for which a K quantum number for this compound system does have meaning [21].

6.2 Inclusion of Closed Channels

Most coupled-channels calculations are of the kind described in the preceding section, in which only low excited states of the target nucleus are treated, and in which as a consequence all the coupled channels are *open*; this means that all the ξ_j relative wavefunctions in (6.1) become oscillatory as r becomes asymptotically large. However it is also of interest to consider calculations in which some or all of the coupled channels are *closed*. For such closed channels the ξ_j decay exponentially as r becomes asymptotically large.

Closed terms appear in conjunction with open terms if the model wavefunction includes both low-lying excited states of the target nucleus and states whose excitation energy is greater than the energy of the incident projectile. For states ψ_j of the latter type the products $\psi_j\xi_j$ are "bound states in the continuum," such as are discussed in Chapter 2. Such states play the role of compound-nucleus states. A coupled-channels analysis that includes such states therefore describes a dynamical process that includes DI and CN reaction mechanisms in a unified fashion. Obviously, an analysis that carries many closed states would not be very practical, therefore such inclusive dynamical descriptions are employed either for a few, special compound states (e.g., analogue states), or for light systems at low incident energies. Nevertheless, unified calculations of this kind are of increasing popularity [22]. Often they take the form of "extended" shell-model calculations [20, 22, also see 23–26], in which the coupling among the closed, excited terms is of primary interest, and in which the additional coupling to open terms of the wavefunction only gives the shell-model states some decay width. While some further discussion of these unified calculations is given in Chapter 11, they are on the whole not within the range of this book.

Sometimes it is convenient to analyze nuclear bound states by means of coupled-channels calculations in which all the channels are closed. Such bound-state analyses by means of differential equations allow greater attention to details of radial wavefunctions than can be achieved by diagonalization within a finite shell-model function space. Theories of stripping form factors (see Sections 5.8–5.11), for example, require that just this kind of careful attention be given to the computation of radial wavefunctions. Therefore, further remarks about such calculations appear in Section 8.4, where an extended discussion of form-factor calculations is given.

6.3 Adiabatic Theory

The adiabatic theory of inelastic scattering is based on the same approximations used in unified theories of collective bound states [3, 27]. It is convenient to discuss these approximations by examining the *formal solution* (3.51) of the reduced Schrödinger equation. Thus we consider

$$\Psi_{\text{model}} = \{1 + (E^{+} - T - H_t - \varepsilon - V - \bar{\mathcal{O}})^{-1}(V + \bar{\mathcal{O}})\}\psi_A \exp[i(\mathbf{k}_\alpha \cdot \mathbf{r})], \tag{6.9}$$

where E^{+} indicates in a slightly modified notation that the usual infinitesmal positive imaginary term has been added to the total energy. The other quantities in (6.9) already appeared in the coupled-channels discussions of Sections 4.6 and 6.1. Thus T is the kinetic energy operator for the projectile-target nucleus relative motion, H_t is the internal Hamiltonian of the target nucleus, and ε is the binding energy of the incident projectile. It is helpful to emphasize the internal coordinates of the target nucleus, which we denote by the symbol y, and to indicate

$$H_t = H_t(y),$$
$$V + \bar{\mathcal{O}} \equiv \tilde{V}(\mathbf{r}, y).$$

It is clear that much of the complication of (6.9) lies in the fact that $H_t(y)$ does not commute with $\tilde{V}(\mathbf{r}, y)$.

We now introduce adiabatic approximation by omitting $H_t(y)$ from the Green's function of (6.9). Under this approximation the coordinates y no longer appear in noncommuting operators; therefore the approximate Green's function can be computed as if the coordinates y were c-numbers. This allows the approximate wavefunction to be computed in two steps: In the first step we compute a *one-body scattering wavefunction*

$$\chi^{(+)}(\mathbf{k}_\alpha, \mathbf{r}; y) \equiv \{1 + (E^{+} - T - \varepsilon - \tilde{V})^{-1}\tilde{V}\} \exp[i(\mathbf{k}_\alpha \cdot \mathbf{r})], \tag{6.10}$$

that depends parametrically on the coordinates y. In the second step we multiply this $\chi^{(+)}$ by ψ_α and obtain the *full adiabatic wavefunction*

$$\Psi(\text{adiab}) = \chi^{(+)}(\mathbf{k}_\alpha, \mathbf{r}; y)\, \psi_A(y). \tag{6.11}$$

The y-dependence of $\chi^{(+)}$ evidently causes Ψ(adiab) to contain a mixture of the states of the target nucleus; this y-dependence therefore implies inelastic scattering. We further observe that the $\chi^{(+)}$ one-body wavefunction of the adiabatic theory is closely analogous to the "intrinsic wavefunction" that appears in theories of collective bound states [3, 27].

From a time-dependent point of view [28], the above omission of $H_t(y)$ implies omission of propagation of target nucleus coordinates in intermediate states of the scattering process. The projectile is considered to be scattered so quickly that the properties of the target nucleus remain constant throughout the period of the scattering interaction. In this picture, although the coordinates y remain constant, the scattering process nevertheless delivers a y-dependent impulse to the target nucleus, and the state of the nucleus can therefore change. Naturally, it is only for limited portions of the spectrum of $H_t(y)$ that propagation in intermediate states can be disregarded in this fashion [29]. Fortunately, for most of the excited states of the target nucleus the inelastic excitation can be treated by perturbative methods. Only a few states are so strongly coupled that a nonperturbative analysis, with accurate intermediate-state propagators, must be employed.—Now, the most strongly coupled states are the members of the ground-state rotational band. Because of their low excitation energies, all the members of this band indeed tend to propagate at the same rate as the ground state. Therefore, an adiabatic analysis, in which $\tilde{V}$ only couples the members of the ground-state rotational band, is likely to give an accurate description of the excitation of this band.

In practical applications of the adiabatic theory the one-body wavefunction $\chi^{(+)}(\mathbf{k}_\alpha, \mathbf{r}; y)$ is calculated from the differential equation

$$[E - T - \varepsilon - \tilde{V}(\mathbf{r}, y)]\, \chi^{(+)} = 0, \tag{6.12}$$

which is equivalent to the integral equation of (6.10). We see that $\tilde{V}(\mathbf{r}, y)$ is to be regarded as a y-dependent optical potential. The particular coordinates y that we carry in this potential are the ones that describe the internal structure of the strongly coupled low-lying states of the target nucleus. For example, in an adiabatic theory of the excitation of the ground-state band of a deformed nucleus, the potential $\tilde{V}(\mathbf{r}, y)$ would be a deformed optical potential of the type discussed in Section 5.3. The coordinates y would be the deformation parameter β and the Euler angles that specify the nuclear orientation.

At this stage it should be clear that the adiabatic theory is not a general theory of inelastic scattering, but rather is a technique for working out the inelastic scattering associated with *macroscopic dynamical models* of nuclear collective states. It is only such macroscopic models that lead to convenient, parameter-dependent optical potentials whose parameters generate collective excitations of the nucleus. The importance of the adiabatic reaction theory thus derives from the importance of macroscopic models of nuclear collective states. On the other hand, although the adiabatic theory only applies for certain models, we should note that it provides a nonperturbative treatment of the *entire spectrum of states* predicted by those models. Because the adiabatic theory treats the collective spectrum in an overall fashion it sometimes yields better results than can be obtained from the coupled-channels theory (Section 6.1). The latter theory is inherently more accurate and flexible, but it only treats a few excited states at a time.

There are two points at which adiabatic calculations present mathematical difficulties. First, there is the difficulty of solving (6.12), to compute the one-body wavefunction $\chi^{(+)}(\mathbf{k}_\alpha, \mathbf{r}; y)$. In general the potential $\tilde{V}(\mathbf{r}, y)$ is nonspherical, and therefore the various partial waves and magnetic substates of the one-body wavefunction are coupled. The second point of difficulty lies in the expansion of the full adiabatic wavefunction (6.11) in eigenstates of the target nucleus. This expansion takes the form

$$\Psi(\text{adiab}) = \sum_B \psi_B(y) \int \psi_B^*(y')\chi^{(+)}(\mathbf{k}_\alpha, \mathbf{r}; y')\, \psi_A(y')\, dy'. \qquad (6.13)$$

Because the y-dependence tends to be buried implicitly in the technique of calculation of $\chi^{(+)}$, the procedure for calculating the integrals in (6.13) is not clear.

In analyses of the rotational excitation of a permanently deformed nucleus it is to a large extent possible to overcome the above two difficulties [4, 30]. The calculations are performed in the body-fixed coordinate system of the deformed nucleus. In this coordinate system the nuclear potential has cylindrical symmetry, and therefore there is no coupling of magnetic substates. The *coupled-partial-waves problem* therefore reduces to a (simplified) special case of the general coupled-channels problem. Furthermore, because β for a deformed nucleus is constant, the expansion in eigenstates merely involves the geometrical operations required to transform between the body-fixed and space-fixed coordinate systems. These geometrical operations reduce to analytic integrations over Euler angles.

In summary, the adiabatic theory is both appropriate and practical for analysis of inelastic scattering by deformed nuclei. The adiabatic theory not only yields a simplification of the coupled-channels calculation, but for higher members of the rotational band it yields more accurate results [4, 31].

6.4 Coupled Isobaric Channels

We saw in Section 5.7 that the nuclear interaction contains terms that can change the charge of the incident projectile. These terms can be treated in the usual, non-exchange theory of inelastic scattering, merely by recognizing that the projectile has a charge degree of freedom. In Chapter 5 the charge-dependent terms were introduced in the DW theory of inelastic scattering. We now introduce them in the coupled-channels theory of inelastic scattering.

For simplicity, the present discussion only considers projectiles that are single nucleons. Therefore the channels that are coupled are those composed of p plus the states of nucleus $A(Z, N)$, together with those composed of n plus the states of nucleus $B(Z + 1, N - 1)$. In case our interest should center on isobar analogue excitation we further limit this list of coupled channels, and we carry only the ground state of A and the one excited state of B that is the analogue of A. In this case only two isobaric channels are coupled, and we find a particularly simple application of the coupled-channels theory.

Let us consider (p, n) reactions that lead to isobar-analogue excitation. In this case both of the coupled channels are open, as in the treatment of inelastic scattering in Section 6.1. However, the present discussion is simpler than that in Section 6.1; because there are no excited-state spins to be considered it is not necessary to introduce an immediate partial-wave decomposition.

Let us use Lane's macroscopic model of the channel coupling [32]. This model suppresses the internal coordinates of nuclei A and B, as is noted in Section 5.7, and represents the charge-dependent interaction between the projectile and target nucleus simply by adding a monopole charge-dependent term in the optical potential. In this fashion the (p, n) reaction is described in terms of an equivalent one-body theory. Such a description is plausible because of the great similarity between ψ_A and the analogue state ψ_B. The coupled equations for this macroscopic theory were given by Lane [32]. We have

$$\Psi_{\text{model}} = \psi_A \xi_p(\mathbf{r}) + \psi_B \xi_n(\mathbf{r}), \tag{6.14}$$

$$(T + U_p + U_c - E)\xi_p = -\frac{\sqrt{2T_0}}{2A} V_1 \xi_n, \tag{6.15a}$$

$$(T + U_n + \Delta_c - E)\xi_n = -\frac{\sqrt{2T_0}}{2A} V_1 \xi_p, \tag{6.15b}$$

where V_1 is the charge-dependent term of the optical potential as defined in (5.104). In addition, U_p and U_n are the projections of the full (charge-dependent) optical potential on to the proton and neutron channels, respectively, U_c is the proton-nucleus Coulomb potential, A is the mass number of the target nucleus, $2T_0$ is the neutron excess of the target nucleus, and Δ_c is the difference between the projectile energies in the proton and neutron channels. Most of Δ_c is contributed by the Coulomb splitting of the two channels.

It is natural to express U_p and U_n in terms of V_1 and U_0, where U_0 is the charge-independent term of the optical potential. The single-channel projections of the optical potential then are found to be

$$\begin{aligned} U_p &= U_0 - \left(\frac{1}{2A}\right) T_0 V_1, \\ U_n &= U_0 + \left(\frac{1}{2A}\right)(T_0 - 1)V_1. \end{aligned} \tag{6.16}$$

The properties of U_0 and V_1 are expected to vary smoothly with energy and mass number.

However, (6.16) presents an oversimplified picture of the relation between U_p and U_n. As one indication of this oversimplification we note that the optical potential is energy dependent, and that the energy splitting of the proton and neutron channels tends to be large (of the order of 10 MeV). Therefore the difference between U_p and U_n should properly involve both charge-dependent effects and energy-dependent effects. Effects of the latter type have been ignored. The importance of the energy-dependent effects becomes more evident when we consider bombarding energies that are low enough so that the neutron channel is closed. On the whole, what we encounter here is simply another example of the difficulty of choosing optical potentials to use in DI reaction theories. The truncation of Hilbert space tends to be a physically-vague concept.

Exact coupled-channels solutions of the Lane equations were computed by Schwarz [33] for the reactions $Ti^{48}(p, n)V^{48}$, $V^{51}(p, n)Cr^{51}$, and $Cr^{52}(p, n)$-Mn^{52} at 18.5 MeV bombarding energy. The coupled-channels results agree fairly well with the results of DW approximate solutions of these equations, as mentioned in Section 5.7.

It is also interesting to solve the Lane equations at proton energies that lie below the threshold for the neutron emission that leaves nucleus B in the analogue state. Even though no neutrons are emitted the isobaric coupling between the open channel (proton + A) and the closed channel (neutron + analogue state) has observable consequences in the proton elastic scattering. The effects of the closed channel are especially strong whenever the energy in the closed neutron channel equals an eigenenergy of the neutron single-particle potential well. At each such energy the product (neutron eigenfunction) $\times$ (analogue state wavefunction) plays the role of a bound state in the continuum, and it causes a sharp "analogue state resonance" in the elastic scattering cross section.

An analogue resonance is a typical compound-nucleus resonance. However, the coupled equations provide an especially simple model for the wavefunction of the compound state; from this model we are able to appreciate why an analogue resonance has so large a reduced width for breakup into the channel $(p + A)$ that it stands out as an isolated resonance even in a region of high level density. Unfortunately, the Lane model of the isobar analogue resonance is a little too simple. It only includes the one excitation of the target nucleus that is represented by a rotation of the i-spin; this is the one excitation that does have good overlap with the channel $(p + A)$. Therefore this model overestimates the reduced width for the channel $(p + A)$. Better values for this reduced width can be obtained by making *ad hoc* changes in the strength of the channel coupling [34], or by introducing additional excitations of the target nucleus (and additional coupled equations [34]), or by using a formal boundary-matching theory to describe the reaction mechanism [35]. Since our present interest lies in the Lane equations themselves, this question of the value of the $(p + A)$ reduced width is not pursued.

Resonance solutions of the Lane equations have been computed by Tamura [34] and by Bondorf and collaborators [36]. A principal concern in both these calculations was the determination of the projected potentials U_p and U_n. To analyze the energy dependence of these potentials it was found best to make a change of representation and to discuss, not the physical channels $(p + A)$ and $(n + \text{analogue state})$, but rather the two linear combinations of these channels that have definite i-spin. For brevity, let us use $\tilde{A}$ to denote the isobar analogue of nucleus A. In this notation the physical coupled channels are $(p + A)$ and $(n + \tilde{A})$, and the wavefunction of the coupled-channels model is

$$\Psi_{\text{model}} = \psi_A \xi_p(\mathbf{r}) + \psi_{\tilde{A}} \xi_n(\mathbf{r}). \qquad (6.17)$$

The unitary transformation to the (projectile + target nucleus) internal wavefunctions that have definite i-spin is given by

$$|p + A\rangle = (2T_0 + 1)^{-1/2}\{\phi(T_0 + \tfrac{1}{2}, T_0 + \tfrac{1}{2}) - (2T_0)^{1/2}\phi(T_0 - \tfrac{1}{2}, T_0 + \tfrac{1}{2})\},$$
$$(6.18)$$
$$|n + \tilde{A}\rangle = (2T_0 + 1)^{-1/2}\{(2T_0)^{1/2}\phi(T_0 + \tfrac{1}{2}, T_0 + \tfrac{1}{2}) + \phi(T_0 - \tfrac{1}{2}, T_0 + \tfrac{1}{2})\},$$

where $2T_0$ is the neutron excess of the target nucleus, and where we make the normal assumption that the i-spin of ψ_A and the ζ-projection of the i-spin both equal T_0. The functions $\phi(T_0 + \frac{1}{2}, T_0 + \frac{1}{2})$ and $\phi(T_0 - \frac{1}{2}, T_0 + \frac{1}{2})$ are labelled by the total i-spin of the (projectile + target nucleus) system, and by its ζ-projection. For brevity, these functions are denoted $\phi_>$ and $\phi_<$. In terms of these vector-coupled internal functions the full wavefunction of the coupled-channels model is

$$\Psi_{\text{model}} = \phi_> \chi_>(\mathbf{r}) + \phi_< \chi_<(\mathbf{r}), \qquad (6.19)$$

and the coupled equations are

$$
\begin{aligned}
(T + U_{>} - E)\,\chi_{>}(\mathbf{r}) &= \left(\frac{\Delta_c - V_c}{2T_0 + 1}\right)[\chi_{>}(\mathbf{r}) - (2T_0)^{1/2}\chi_{<}(\mathbf{r})], \\
(T + U_{<} - E)\,\chi_{<}(\mathbf{r}) &= \left(\frac{\Delta_c - V_c}{2T_0 + 1}\right)[(2T_0)^{1/2}\chi_{>}(\mathbf{r}) + \chi_{<}(\mathbf{r})],
\end{aligned}
\tag{6.20}
$$

where $U_{>}$ and $U_{<}$ are the projections of the optical potential onto the $>$ and $<$ channels, respectively. We note that in this representation the coupling terms are proportional to $(\Delta_c - V_c)$, a quantity that has its largest values *external* to the target nucleus [35].

For heavy target nuclei $T_0 \gg 1$. For such cases it is clear from (6.18) that $\phi_{>}$ dominates the bound state $|n + \tilde{A}\rangle$. Now all the states $\phi_{>}$ have *i*-spin $(T_0 + \frac{1}{2})$, and are analogues of *low-lying states* of the system (neutron + target nucleus). As a consequence the states $\phi_{>}$ are widely spaced in energy; therefore the coupling among these states makes no significant contribution to their decay. All decay of a state $\phi_{>}$ takes place by virtue of coupling to $\phi_{<}$, and this is described in full by the Lane equations. It is therefore clear that the *imaginary part of the projected potential* $U_{>}$ *is zero.*

We now see, within the general limitations of this kind of model, how $U_{>}$ and $U_{<}$ in (6.20) may be chosen. The real parts of these potentials follow from V_1 and from the real part of the optical potential U_0, in the usual fashion required by Lane's charge-dependent optical model. The full imaginary part of U_0 is carried over into $U_{<}$ because the $\phi_{<}$ spectrum contains the usual dense distribution of open states; however, the imaginary part is omitted from $U_{>}$. By this procedure a strong *i*-spin dependence is introduced in the imaginary potential. This treatment of the imaginary potential strongly affects the shape of the resonance and is required to obtain agreement with experiment [34].

6.5 DW Applications of Coupled Channels

Weak coupling between two classes of terms in Ψ_{model} often must be treated in conjunction with strong coupling among the terms within each class. Two examples of reactions that require such a treatment are (a) deuteron stripping accompanied by inelastic excitations, (b) deuteron stripping accompanied by charge exchange between proton and neutron exit channels. These two processes are now discussed. The matrix elements that link the deuteron-target nucleus channels with the nucleon-residual nucleus channels are treated as perturbations.

For definiteness, only macroscopic models of the excitations of the target nucleus and the residual nucleus are considered. Therefore the projectile-nucleus interaction can be represented by a deformed, charge-dependent optical potential. It is by now clear that such an interaction correctly represents the strong excitations. For further definiteness, it is the (d, p) amplitude that is computed.

A DI model wavefunction that describes both the above-mentioned process is

$$\Psi_{\text{model}} = \phi_d \sum_i \psi_{Ai}(y)\xi_{Ai}(\mathbf{r}_\alpha) + \phi_p \sum_i \psi_{Bi}(y, n)\xi_{Bi}(\mathbf{r}_\beta) + \phi_n \sum_i \psi_{Ci}(y, p)\xi_{Ci}(\mathbf{r}_\beta), \tag{6.21}$$

where

$$r_\alpha = \mathbf{r}_d - \mathbf{r}_A,$$

and

$$\mathbf{r}_\beta = \mathbf{r}_p - \mathbf{r}_B \qquad \text{or} \qquad \mathbf{r}_\beta = \mathbf{r}_n - \mathbf{r}_C.$$

The functions ϕ_d, ϕ_p, and ϕ_n are internal wavefunctions of a deuteron, a proton, and a neutron, respectively, and the functions ψ_{Ai}, ψ_{Bi}, and ψ_{Ci} are eigenstates of the target nucleus and of the two alternative residual nuclei, respectively. The internal variable y describes the collective motion of the target nucleus. It is the coefficient functions ξ_{Ai}, ξ_{Bi}, and ξ_{Ci} that are to be calculated in DW coupled-channels approximation.

Weak coupling between the deuteron and nucleon channels implies that the $(d + A)$ terms of (6.21) do not depend on the other terms, and therefore that the ξ_{Ai} may be computed by solving the $(d + A)$ coupled-channels inelastic scattering problem. The part of Ψ_{model} that is thus obtained may be recognized as a *generalized distorted wave*,

$$X_\alpha^{(+)}(\mathbf{r}_\alpha, y; \mathbf{k}_\alpha, A) \equiv \sum_i \psi_{Ai}(y)\xi_{Ai}(\mathbf{r}_\alpha), \tag{6.22}$$

such as discussed in Section 3.3. The generalized distorting potential that governs $X_\alpha^{(+)}$ is the deformed optical potential of the $(d + A)$ system.

It is straightforward to set up coupled equations for the ξ_{Bi} and ξ_{Ci}, with the coupling to $X_\alpha^{(+)}$ appearing in these equations as known inhomogeneities. However, such an approach would imply solving these equations for ξ_{Bi} and ξ_{Ci} for all values of $\mathbf{r}_\beta$. On the other hand, to obtain the transition amplitudes for the various exit channels we need only the asymptotic amplitudes of the ξ_{Bi} and ξ_{Ci}. It is more direct to calculate these transition amplitudes from the generalized distorted waves expression of Section 3.3, rather than compute the ξ_{Bi} and ξ_{Ci} as such.

Our basic generalized distorted waves expression for the transition amplitude is (3.44) of Section 3.3. In the present application this equation becomes

$$T_{\alpha\beta} = (\hat{\Psi}^{(-)}(\mathbf{k}_\beta), \hat{V}^\dagger \phi_d \psi_{A0}(y) e^{i(\mathbf{k}_\alpha \cdot \mathbf{r}_\alpha)}) + (\hat{\Psi}^{(-)}(\mathbf{k}_\beta), (V - \hat{V}^\dagger)\Psi_{\text{model}}), \tag{6.23}$$

where $\hat{V}$ is the deformed, charge-dependent optical potential that couples nucleon channels with each other and $\hat{\Psi}^{(-)}$ is the time-reversed nucleon-nucleus scattering wavefunction that is governed by $\hat{V}$. We now introduce the DW approximations usual in treatments of deuteron stripping, in the spirit of the discussions in Sections 4.5 and 5.8. Equation 6.23 reduces to the approximate form

$$T_{\alpha\beta}{}^{\mathrm{DW}} = (\hat{\Psi}^{(-)}(\mathbf{k}_\beta), V_{np}\phi_d X_\alpha^{(+)}(\mathbf{r}_\alpha, y; \mathbf{k}_\alpha, A)). \tag{6.24}$$

Here the basic DW step is the replacement of Ψ_{model} by $\phi_d X_\alpha^{(+)}$, with the result that $T_{\alpha\beta}{}^{\mathrm{DW}}$ depends only linearly upon the coupling between deuteron and nucleon channels. Omission of the first term of (6.23) is based on the assumption that the excited terms that $\hat{V}$ couples into $\hat{\Psi}^{(-)}$ have negligible asymptotic overlap with $(d + A)$ channels. Use of V_{np} instead of $(V - \hat{V}^\dagger)$ parallels the discussion in Section 5.8, and is based on the assumption that the matrix elements of $\hat{V}^\dagger$ accurately represent all the more important excitations caused by the microscopic interactions $(V - V_{np})$. Equation 6.24 is the basic equation of the DW coupled-channels theory.

Let us first treat the special case in which charge exchange between neutron and proton exit channels is omitted but inelastic excitations of the target and residual nuclei are carried. Equation 6.24 becomes

$$T_{\alpha\beta}{}^{\mathrm{DW}} = (\phi_p X_\beta^{(-)}(\mathbf{r}_\beta, n, y; \mathbf{k}_\beta, B), V_{np}\phi_d X_\alpha^{(+)}(\mathbf{r}_\alpha, y; \mathbf{k}_\alpha, A)). \tag{6.25}$$

Here $X_\beta^{(-)}$ is the time-reverse of the scattering wavefunction,

$$X_\beta^{(+)}(\mathbf{r}_\beta, n, y; \mathbf{k}_\beta, B) = \sum_i \psi_{Bi}(y, n)\chi_{Bi}(\mathbf{r}_\beta), \tag{6.26}$$

that describes inelastic scattering of protons by the residual nucleus B. Equation 6.25 describes a DW transition between two generalized distorted waves, each of which is computed as a coupled-channels eigenfunction. Evidently, (6.25) allows a (d, p) transition between definite states of A and B to proceed either directly or through intermediate excited states of nuclei A and B. Because the indirect transitions through intermediate states are subject to less restrictive selection rules than the direct transitions (Section 5.8), it would seem that the indirect transitions could easily give important corrections to unfavored direct transitions.

A number of authors have performed calculations based on (6.25) or simple variants of this equation [37, 38, 39]. In all these (rather exploratory) calculations the contributions of indirect transitions were found to be "small," that is, to be of the same order as other well-known uncertainties (Section

5.8) of allowed direct stripping transitions. Although larger indirect contributions may still be found in future applications of the theory, it is interesting to keep in mind Schiffer's observation [40] that experimental studies of forbidden stripping transitions have shown no evidence for the presence of large indirect contributions to the cross section.

Iano and Austern [38] treated a variant of (6.25) in which $X_\alpha^{(+)}$ and $X_\beta^{(-)}$ were computed from the adiabatic theory of inelastic scattering. In their work the normal DW stripping theory is employed in the body-fixed coordinate system of a deformed unexcited target nucleus. The amplitude computed in this fashion depends on the Euler angles of the deformed nucleus. Projection onto definite states of the target and residual nuclei then yields the transition of physical interest. This adiabatic stripping theory extends the concept of "intrinsic wavefunction" to include particle-transfer processes; it is a natural companion to usual adiabatic theories of the bound states of deformed nuclei.

Let us now treat stripping that includes charge exchange between neutron and proton exit channels, but omits other inelastic excitations of the target and residual nuclei. For this case the DI wavefunction of (6.21) reduces to

$$\Psi'_{\text{model}} = \phi_d \psi_A(y) \chi_\alpha^{(+)}(\mathbf{r}_\alpha, \mathbf{k}_\alpha) + \phi_p \psi_B(y, n)\, \xi_B(\mathbf{r}_\beta) + \phi_n \psi_{\tilde{B}}(y, p)\, \xi_{\tilde{B}}(\mathbf{r}_\beta), \tag{6.27}$$

where ψ_A is the wavefunction of the target nucleus ground state, ψ_B is the wavefunction of the residual nucleus formed by stripping, and $\psi_{\tilde{B}}$ is the isobar analogue of ψ_B. Once again, DW approximation simplifies the treatment of the $(d + A)$ channel, such that $\chi_\alpha^{(+)}$ is now the usual optical-model eigenfunction for this channel. However, the $(p + B)$ and $(n + \tilde{B})$ channels are coupled by the charge-exchange interaction. Suitable specialization of (6.24) yields the generalized distorted waves formula for the transition amplitude,

$$T_{\alpha\beta}{}^{\text{DW}} = (X_\beta^{(-)}(\mathbf{r}_\beta, n, p, y; \mathbf{k}_\beta, B), V_{np} \phi_d \psi_A(y)\, \chi_\alpha^{(+)}), \tag{6.28}$$

where the generalized distorted wave is the time reverse of

$$X_\beta^{(+)} = \phi_p \psi_B(y, n)\, \chi_p(\mathbf{r}_\beta) + \phi_n \psi_{\tilde{B}}(y, p)\, \chi_n(\mathbf{r}_\beta). \tag{6.29}$$

The functions χ_p and χ_n are obtained by computing $X_\beta^{(+)}$ as an eigenfunction of the Lane equations. Equation 6.28 describes a (d, p) stripping amplitude that is composed of a direct (d, p) term plus an indirect term in which a (d, n) transition is followed by charge exchange. Such generalized DW expressions for the (d, p) amplitude have been discussed by Zaidi and von Brentano [41], by Tamura [5, 42], Tamura and Watson [43], and Rawitscher [44] (also J. Zimányi and B. Gyarmati, to be published).

A perturbative calculation of the indirect term in (6.29) provides a helpful guide to the importance of charge-exchange effects, and serves as an introduction to coupled-channels calculations of these effects. Thus, (6.15a) and

(6.15*b*) describe the coupling between χ_n and χ_p, under the assumption that ψ_B and $\psi_{\bar{B}}$ are isobar analogues. We assume χ_p is the optical wavefunction that describes $(p + B)$ elastic scattering in zero-order. Then χ_n is found in first order to be

$$\chi_n \approx (E^+ - T - U_n - \Delta_C)^{-1}\left(\frac{T_0}{2}\right)^{\frac{1}{2}} A^{-1}V_1\chi_p. \tag{6.30}$$

Because V_1 is a monopole operator, each proton partial wave of angular momentum l couples only to the neutron partial wave with the same angular momentum. Therefore a partial wave expansion of (6.30) has the convenient form

$$f_{nl}(r) = -\left(\frac{T_0}{2}\right)^{\frac{1}{2}} A^{-1}\frac{2\mu_n}{\hbar^2 k_n}\int_0^\infty f_{nl}^{(0)}(r_<)\, h_{nl}^{(0)}(r_>)\, V_1(r')\, f_{pl}(r')\, dr', \tag{6.31}$$

in which the DW radial Green's function of (3.30) is inserted. The functions $f_{nl}^{(0)}$ and $h_{nl}^{(0)}$ are regular and outgoing solutions, respectively, of the uncoupled Lane equation for the neutron. A rough estimate of the integral in (6.31) suggests that the Green's function picks up nonzero contributions from a range $\lambda\!\!\!^{-}_n$ around the jump point at $r = r'$. On this basis we obtain the very rough result

$$f_{nl}(r) \approx -\left(\frac{T_0}{2}\right)^{\frac{1}{2}} A^{-1}\frac{V_1(0)}{E_n} f_{pl}(r). \tag{6.32}$$

Because the coefficient in (6.32) is small and V_1 has a depth of only about 100 MeV, we see that f_{nl} tends to be only a few percent of f_{pl}. Therefore the indirect, charge-exchange contribution to stripping tends to be a small correction to the direct contribution.

Under special conditions the charge-exchange contribution can have enhanced importance, and can cause major modifications of the stripping cross section. One such special condition may be caused by the spectroscopic factors in the two terms of (6.28) for the DW stripping amplitude. According to the discussion of Section 5.8 these spectroscopic factors are proportional to the two overlaps

$$(\psi_B(y, n),\ V_{np}\phi_d\psi_A(y)),$$
$$(\psi_{\bar{B}}(y, p),\ V_{np}\phi_d\psi_A(y)).$$

Although ψ_B and $\psi_{\bar{B}}$ are analogues, their overlaps with the target nucleus wavefunction $\psi_A(y)$ can be quite different. For example, in the case $Be^9(d, n)$ that was treated by Tamura [42] the target nucleus is neutron rich. Therefore the residual nucleus in the charge-exchange channel is $(n + Be^9)$ and has a wavefunction of the form

$$\psi_{\bar{B}} = \psi_A \times \text{(single-particle wavefunction)};$$

this wavefunction has excellent overlap with ψ_A. However, the analogue residual nucleus in the direct channel is $(p + Be^9)$ and contains extensive admixtures of core excitations; these inhibit overlap with ψ_A and thereby enhance the relative importance of the indirect term. (Evidently, with neutron-rich target nuclei, the corrections to (d, n) reactions are bigger than those to (d, p) reactions.) Further enhancement of the indirect term in the $Be^9(d, n)$ reaction takes place because of the low energy that is treated, $E_d = 5.5$ MeV, as is suggested by (6.32). Taken together, these two effects allow the indirect term to become about twenty five percent of the direct term. The importance of this indirect term depends on the sign of its interference with the direct term [42].

An enhanced charge-exchange contribution to stripping also is found in the analysis [43] of the anomaly of the $Zr^{90}(d, p)Zr^{91}$ cross section at the threshold at which the analogue (d, n) channel first becomes open for neutron emission [45]. At threshold the neutron energy is low, and therefore [see (6.32)] the channel coupling is especially strong. In addition, the neutron $p_{3/2}$ wave in the charge-exchange channel is found to have a single-particle resonance near this threshold. Under these conditions the rapid threshold energy dependence of the neutron wavefunction is reflected into the (d, p) cross section.

6.6 Rearrangement and Exchange

Coupled equations may be used to calculate the relative wavefunctions ξ_γ of the DI model

$$\Psi_{\text{model}} = \sum_{\gamma} \xi_\gamma(\mathbf{r}_\gamma)\, \psi_{1\gamma}\psi_{2\gamma}, \tag{6.33}$$

even if the internal functions $\psi_{1\gamma}\psi_{2\gamma}$ are not all orthogonal.* The set of equations obtained by left multiplication by the various $\psi_{1\gamma}\psi_{2\gamma}$, for example, is given previously as (4.68):

$$\begin{aligned}[][E - \varepsilon_{1\gamma} - \varepsilon_{2\gamma} - T_\gamma - \langle\gamma|\, V_\gamma + \bar{\mathcal{O}}\, |\gamma\rangle]\xi_\gamma(\mathbf{r}_\gamma) \\ = -\sum_{\delta \neq \gamma} [\langle\gamma, \delta\rangle(E - \varepsilon_{1\delta} - \varepsilon_{2\delta} - T_\delta) - \langle\gamma|\, V_\delta + \bar{\mathcal{O}}\, |\delta\rangle]\xi_\delta(\mathbf{r}_\delta),\end{aligned} \tag{6.34}$$

where $|\gamma\rangle$, $|\delta\rangle$ denote the internal states

$$|\gamma\rangle = |\psi_{1\gamma}\psi_{2\gamma}\rangle, \qquad |\delta\rangle = |\psi_{1\delta}\psi_{2\delta}\rangle,$$

and where $\langle\gamma, \delta\rangle$ is the overlap of different internal states

$$\langle\gamma, \delta\rangle \equiv \langle\psi_{1\gamma}\psi_{2\gamma} \mid \psi_{1\delta}\psi_{2\delta}\rangle. \tag{6.35}$$

In the LHS of (6.34) a local, differential operator is applied to ξ_γ. This operator is characteristic of coupled equations in configuration-space representation; it appears prominently in the treatment of nonexchange inelastic scattering in Section 6.1. It is the RHS of (6.34) that exhibits all the consequences of nonorthogonality.

* It is interesting to remark here that the coupled equations not only provide a procedure for calculating the ξ_γ, but that, in general, it is only in these equations that we find a precise definition of H_{PP}, the projection of H on the space spanned by Ψ_{model}.

In all the terms of the RHS of (6.34) integration over $\mathbf{r}_\delta$ is understood. Hence these terms are averages over the function $\xi_\delta(\mathbf{r}_\delta)$, with respect to complicated weight factors that are functions of $\mathbf{r}_\gamma$. Thus for rearrangement reactions, for which some channels have $\mathbf{r}_\gamma \neq \mathbf{r}_\delta$, we see that (6.34) is an integro-differential equation. Two major mathematical difficulties then appear in this equation: One is that the functions $\langle\gamma, \delta\rangle$ and $\langle\gamma| V_\delta + \overline{\mathcal{O}} |\delta\rangle$ are difficult to evaluate. The other is that an equation that contains integral operators cannot be handled by the step-by-step numerical methods described in Section 6.1. Orthogonality would eliminate both these difficulties because it would eliminate the terms proportional to $\langle\gamma, \delta\rangle$ and would cause the terms $\langle\gamma| V_\delta + \overline{\mathcal{O}} |\delta\rangle$ to be proportional to $\delta(\mathbf{r}_\gamma - \mathbf{r}_\delta)$.

A further difficulty caused by nonorthogonality is that the terms proportional to $\langle\gamma, \delta\rangle$ are similar in structure to the terms on the LHS of the equation. Hence it is not clear which terms should be treated as zero-order quantities in iterative solutions of the equations.

Many authors have discussed the formulation of coupled equations for problems containing nonorthogonal channels (e.g., see [46, 47]). However, none of these discussions seem to yield equations that are more "correct" than (6.34), and all seem to yield equations that possess the mathematical difficulties of (6.34). Presumably, what might be obtained from new formulations would be clear guidance in the organization of iterative methods of solution. In any case, the exact solution of one system of equations for the ξ_γ is as good as the exact solution of another. Our discussion proceeds on the basis of (6.34).

Deuteron stripping is the simplest rearrangement reaction of nuclear physics. Therefore, it is interesting that several authors have applied (6.34) to the analysis of this reaction [44, 48, 49]. A discussion of this work illustrates the application of coupled equations for rearrangement reactions. For simplicity, we treat a (d, p) reaction in which spinless distinguishable nucleons impinge on an infinitely massive target nucleus that has no internal degrees of freedom. We assume there is only one bound state into which the neutron may be captured, and we neglect the charge-exchange (d, n) channel. The DI wavefunction for this model problem then is

$$\Psi_{\text{model}} = \phi_d(r)\xi_d(\mathbf{R}) + \phi_n(\mathbf{r}_n)\xi_p(\mathbf{r}_p), \tag{6.36}$$

where

$$\mathbf{r} = \mathbf{r}_p - \mathbf{r}_n, \qquad \mathbf{R} = \tfrac{1}{2}(\mathbf{r}_p + \mathbf{r}_n).$$

The Hamiltonian for the model problem is obtained by projecting the three-body Hamiltonian

$$H = T_n + T_p + U_n(r_n) + U_p(r_p) + V(r), \tag{6.37}$$

onto the space spanned by Ψ_{model}. For definiteness, one fairly arbitrary departure from the general notation of Chapter 4 is introduced in (6.37), namely, U_n, U_p, and V in (6.37) have been indicated as complex *two-body* interactions.

The coupled equations obtained from (6.36) and (6.37) are

$$[E - \varepsilon_n - T_p - U_p - \langle\phi_n| V |\phi_n\rangle]\xi_p$$
$$= -(E - \varepsilon_n - T_p - U_p)\langle\phi_n, \phi_d\rangle\xi_d + \langle\phi_n| V |\phi_d\rangle\xi_d. \quad (6.38a)$$

$$[E - \varepsilon_d - T_R - \langle\phi_d| U_n + U_p |\phi_d\rangle]\xi_d$$
$$= -\langle\phi_d| E - \varepsilon_n - T_p - U_p |\phi_n\rangle\xi_p + \langle\phi_d| V |\phi_n\rangle\xi_p. \quad (6.38b)$$

As an illustration of the complications of the RHS of (6.38a) and (6.38b), we note that

$$\langle\phi_n, \phi_d\rangle\xi_d = \int \phi_n^*(\mathbf{r}_n)\, \phi_d(|\mathbf{r}_p - \mathbf{r}_n|)\, \xi_d(\tfrac{1}{2}(\mathbf{r}_p + \mathbf{r}_n))\, d^3r_n. \quad (6.39)$$

Both Stamp [48] and Rawitscher [44] simplify the RHS of (6.38a) and (6.38b) by discarding the first terms and by treating V in the second terms as having zero range. Thereby these equations are reduced to the same kinds of coupled differential equations that appear for nonexchange inelastic scattering:

$$[E - \varepsilon_n - T_p - \tilde{U}_p(r)]\xi_p(\mathbf{r}) \approx V_0\phi_n^*(\mathbf{r})\, \xi_d(\mathbf{r}), \quad (6.40a)$$
$$[E - \varepsilon_d - T_R - \tilde{U}_d(r)]\, \xi_d(\mathbf{r}) \approx V_0\phi_n(\mathbf{r})\, \xi_p(\mathbf{r}). \quad (6.40b)$$

Here $\tilde{U}_p(r)$ and $\tilde{U}_d(r)$ are optical potentials chosen so that the solutions of the coupled equations fit proton and deuteron elastic scattering. Ohmura, et al. [49] do not introduce the above drastic simplifications in the RHS of (6.38a) and (6.38b), however, they do use optical potentials $\tilde{U}_p$ and $\tilde{U}_d$ on the LHS.

Rawitscher gives some partial justification for his omission of the first terms of the RHS of (6.38a) and (6.38b). His arguments seem to imply that in zero order of an iterative procedure these terms can be treated as being on the energy shell. It is interesting to see if the work of Ohmura, et al., can support these arguments.

Both Stamp and Rawitscher remark that even the extremely crude (6.40a) and (6.40b) give some insight regarding deuteron elastic scattering: It is known that deuterons that impinge on nuclei in partial waves of low angular momentum have a high probability of being reflected and elastically scattered (Sections 5.2, 5.3 and [43]), despite the small deuteron binding energy. This strong reflection of low partial waves has a qualitative influence upon the differential cross section for elastic scattering. What the coupled equations show is that the scattering of low partial waves can be attributed to "virtual stripping," that is, to a process of breakup and reassociation. Inside the nucleus the breakup amplitude $\xi_p(\mathbf{r})$ is found to be large, and the deuteron amplitude $\xi_d(\mathbf{r})$ is found to be small. Single-channel optical models that fit the same scattering data give much larger probabilities for finding intact deuterons inside the nucleus.

Antisymmetrization causes mathematical difficulties similar to those discussed above for rearrangement reactions. It is remarked in Section 4.9 that this similarity is not an accident. Antisymmetrization of Ψ_{model} introduces rearrangement terms. Although these rearrangement terms do not increase the number of coupled equations, they do introduce a coordinate dependence that makes the equations difficult.

References

[1] S. Yoshida, *Proc. Phys. Soc.* (London) **A69,** 668 (1956).

[2] B. Margolis and E. S. Troubetzkoy, *Phys. Rev.* **106,** 105 (1957); also D. M. Chase, L. Wilets and A. R. Edmonds, *loc. cit.*, Chap. 2, Ref. 11.

[3] A. Bohr and B. R. Mottelson, *loc. cit.*, Chap. 5, Ref. 49.

[4] T. Tamura, *Rev. Mod. Phys.* **37,** 679 (1965).

[5] T. Tamura, *Jour. Phys. Soc. Japan* **24** (*Supplement*), 288 (1967).

[6] J. S. Blair, in *Lectures in Theoretical Physics*, Vol. VIIIC, University of Colorado Press, Boulder, 1966; also in *Lectures on Nuclear Interactions*, Vol. II, Hercegnovi, 1962, The Federal Nuclear Energy Commission of Yugoslavia, Belgrade.

[7] For example, see W. R. Gibbs, *Helv. Phys. Acta* **35,** 437 (1962).

[8] P. J. A. Buttle, *Phys. Rev.* **160,** 719 (1967).

[9] N. K. Glendenning, *Phys. Letters* **21,** 549 (1966); N. K. Glendenning, et al., *Proceedings of the International Conference on Nuclear Structure*, Gatlinburg (1966). A. Faessler, N. K. Glendenning, and A. Plastino, *Phys. Rev.* **159,** 846 (1967).

[10] N. Austern, R. M. Drisko, E. Rost, and G. R. Satchler, *Phys. Rev.* **128,** 733 (1962).

[11] J. L. Yntema and G. R. Satchler, *Phys. Rev.* **161,** 1137 (1967).

[12] J. R. Meriwether, A. Bussiere de Nercy, B. G. Harvey, and D. J. Horen, *Phys. Letters* **11,** 299 (1964); D. J. Horen, et al., *Nucl. Phys.* **72,** 97 (1965).

[13] T. Tamura, *Nucl. Phys.* **73,** 81 (1965).

[14] D. L. Hendrie, et al., *Jour. Phys. Soc. Japan* **24** (*Supplement*), 306 (1967), and *Phys. Letters* **26B,** 127 (1968).

[15] N. K. Glendenning, D. L. Hendrie, and O. N. Jarvis, *Phys. Letters* **26B,** 131 (1968).

[16] B. Buck, *Phys. Rev.* **130,** 712 (1963); B. Buck, A. P. Stamp, and P. E. Hodgson, *Phil. Mag.* **8,** 1805 (1963).

[17] T. A. Tombrello and G. C. Phillips, *Nucl. Phys.* **20,** 648 (1960).

[18] S. Okai and T. Tamura, *Nucl. Phys.* **31,** 185 (1962); T. Tamura, *Phys. Letters* **6,** 111 (1963).

[19] A. C. L. Barnard, *Phys. Rev.* **155,** 1135 (1967); see also P. P. Delsanto, M. F. Roetter, and H. G. Wahsweiler, *Phys. Letters* **28B,** 246 (1968).

[20] R. H. Lemmer and C. M. Shakin, *Ann. Phys.* **27,** 13 (1964).

[21] T. Tamura, *Phys. Letters* **6,** 111 (1963).

[22] For example, see B. Buck and A. D. Hill, *Nucl. Phys.* **A95,** 271 (1967), and references therein.
[23] C. Bloch and V. Gillet, *Phys. Letters* **16,** 62 (1965); V. Gillet and C. Bloch, *Phys. Letters* **18,** 58 (1965); C. Bloch, *Proc. XXXVI, Int. School of Physics "Enrico Fermi"*, Academic, New York, 1966.
[24] J. Raynal, M. A. Melkanoff, and T. Sawada, *Nucl. Phys.* **A101,** 369 (1967).
[25] I. R. Afnan, *Phys. Rev.* **163,** 1016 (1967).
[26] A. Pfitzner, *Nucl. Phys.* **A100,** 673 (1967).
[27] D. L. Hill and J. A. Wheeler, *Phys. Rev.* **89,** 1102 (1953).
[28] A. Messiah, *loc. cit.*, Chap. 1, Ref. 2; R. J. Glauber, *loc. cit.*, Chap. 4, Ref. 8; also L. D. Landau and E. M. Lifschitz, *Quantum Mechanics*, Pergamon, Oxford, New York, 1965.
[29] D. M. Chase, *Phys. Rev.* **104,** 838 (1956).
[30] S. I. Drozdov, *Soviet Phys. JETP* **9,** 1335 (1959); R. C. Barrett, *Nucl. Phys.* **51,** 27 (1964).
[31] L. Zuffi, *Jour. Phys. Soc. Japan* **24** (*Supplement*), 703 (1967); W. J. Thompson, S. Edwards, and T. Tamura, *Nucl. Phys.* **A105,** 678 (1967).
[32] A. M. Lane, *loc. cit.*, Chap. 5, Ref. 83.
[33] E. H. Schwarz, *Phys. Rev.* **149,** 752 (1966).
[34] T. Tamura, in *Isobaric Spin in Nuclear Physics*, J. D. Fox and D. Robson Eds., Academic, New York, 1966; see also E. H. Auerbach, et al., *Phys. Rev. Letters* **17,** 1184 (1966).
[35] D. Robson, *loc. cit.*, Chap. 5, Ref. 66.
[36] J. P. Bondorf, et al., *Nucl. Phys.* **A101,** 338 (1967).
[37] S. K. Penney and G. R. Satchler, *Nucl. Phys.* **53,** 145 (1964); S. K. Penney, Ph.D. Thesis, University of Tennessee, 1966 (unpublished).
[38] P. J. Iano and N. Austern, *Phys. Rev.* **151,** 853 (1966), P. J. Iano, S. K. Penney, and R. M. Drisko, *Nucl. Phys.* **A127,** 47 (1969).
[39] B. Kozlowsky and A. de-Shalit, *Nucl. Phys.* **77,** 215 (1966); D. Dillenburg and R. A. Sorenson, *Bull. Am. Phys. Soc.* **10,** 40 (1965), and Sorenson, Dillenburg and Drisko, to be published, F. S. Levin, *Phys. Rev.* **147,** 715 (1966).
[40] J. P. Schiffer, in *Proceedings of the Summer Study Group*, pp 554–556, Brookhaven Nat. Lab. BNL 948 (C-46), 1965 (unpublished).
[41] S. A. A. Zaidi and P. von Brentano, *Phys. Letters* **13,** 151 (1967).
[42] T. Tamura, *Phys. Rev. Letters*, **19,** 321 (1967), see also T. Tamura, *Phys. Rev.* **165,** 1123 (1968). For corrections and further developments see R. Coker and T. Tamura, *Phys. Rev.* **182,** 1277 (1969) and subsequent articles.
[43] T. Tamura and C. E. Watson, *Phys. Letters* **25B,** 186 (1967).
[44] G. H. Rawitscher, *Phys. Rev.* **163,** 1223 (1967).
[45] C. F. Moore, C. E. Watson, S. A. A. Zaidi, J. J. Kent, and J. G. Kulleck, *Phys. Rev. Letters* **17,** 926 (1966).
[46] M. Coz, *Annals of Phys.* **35,** 53 (1965); *ibid*, **36,** 217 (1966).
[47] Y. Hahn, *loc. cit.*, Chap. 4, Ref. 5.
[48] A. P. Stamp, *Nucl. Phys.* **83,** 232 (1966).
[49] T. Ohmura, B. Imanishi, M. Ichimura, and M. Kawai, *Jour. Phys. Soc. Japan* **24** (*Supplement*), 683 (1967) and subsequent publications.

CHAPTER 7

Wavefunction Models, High Energy

Simplified (semi-analytic) discussions of optical wavefunctions are given in this chapter. These discussions are applied for several classes of reactions, especially for high bombarding energy.

7.1 Properties of Optical-model Wavefunctions

An understanding of optical-model wavefunctions is helpful for understanding the content of DI reaction theories. The importance of these wavefunctions is illustrated by the fact that the DW transition amplitude is proportional to optical wavefunctions in the entrance and exit channels. Wavefunctions of a similar nature play central roles in coupled-channels calculations. A preliminary discussion of properties of optical-model wavefunctions is given in Sections 5.2 and 5.3. This discussion is now extended by additional analyses in partial-wave expansion and by overall studies of the wavefunction in three-dimensional configuration space.

Accurate manipulations with optical wavefunctions are best performed in partial wave expansion, as remarked in Chapter 5. However, McCarthy noticed [1–3] that certain qualitative aspects of the wavefunction are revealed more clearly by summing up the partial waves and studying $\chi^{(+)}(\mathbf{k}, \mathbf{r})$ as a function of $\mathbf{r}$. The cylindrical symmetry of the wavefunction allows the preparation of three-dimensional graphs that display $\chi^{(+)}(\mathbf{k}, \mathbf{r})$ as a function of r, θ. Here θ is the polar angle of $\mathbf{r}$ with respect to $\mathbf{k}$, so that $(\hat{k} \cdot \hat{r}) = \cos\theta$. Figures 7.1, 7.2 are such graphs of the moduli $|\chi^{(+)}(\mathbf{k}, \mathbf{r})|$ of optical wavefunctions for the two cases $\alpha + Ca^{10}$ at 18 MeV, and $p + Ca^{40}$ at 40 MeV, computed using normal values of optical potential parameters. In each case the beam is incident from the left. The dark zone in each graph marks the 10%–90% region of the optical potential. We see that the most prominent feature of each wavefunction is a strong peak near the nuclear surface, at $\theta = 0°$. In the $p + Ca^{40}$ case the peak intensity is 3.8 times the intensity of the incident beam. In the $\alpha + Ca^{40}$ case the peak is weakened by absorption, and its intensity is only 1.5 times that of the incident beam. All optical wavefunctions show such peaks. Further examples appear in three-dimensional

Fig. 7.1 Three-dimensional model of $|\chi^{(+)}|$, the modulus of the optical model wavefunction, for 18 MeV alpha particles bombarding Ca^{40}. The beam is incident from the left. The dark zone is the 10%–90% region of the optical potential. (Computed by R. M. Drisko and N. Austern, unpublished.)

Fig. 7.2 This figure is the same as Fig. 7.1, for 40 MeV protons bombarding Ca^{40}.

graphs prepared by Amos [4]. Obviously an explicit awareness of the McCarthy peaks may help us to understand the origin of effects that emerge from numerical calculations in partial wave expansion.

McCarthy suggested that the optical potential produces the wavefunction peak by bringing the incident wave to a "focus" at this location. To support this interpretation, he showed that the trajectories of classical particles that encounter a spherical attractive potential are indeed refracted in such a fashion as to be brought to rough convergence at the location of the peak [1]. The principal failing of this classical interpretation is that the peak appears even if the projectile wavelength is not much smaller than the nuclear radius, when diffractive spreading of the wavefunction should be more important than refractive bending of classical trajectories. A more secure understanding of the McCarthy peak is obtained by examining the details of exact optical model calculations. It is found [3] that the "focus" is caused by constructive interference among the two or three partial waves that graze the nuclear surface. The interference is constructive because the optical potential interacts more strongly with the lower members of the important set of partial waves; the potential thereby causes a phase difference that offsets the normal i^L relative phase of successive terms in the partial-wave series.

There has not been much orderly investigation of the influence of the focus upon direct reaction calculations. However, McCarthy and collaborators [2, 5, 6] did discuss a number of ways in which the focus affects the overlap of entrance and exit channels in distorted wave calculations. Certain aspects of angular distributions could be associated with the focus. (a) The focus in the entrance-channel wavefunction $\chi_\alpha^{(+)}$ always lies at 0°. However the focus in the exit-channel wavefunction $\chi_\beta^{(-)}$ lies at $180° - \Theta$, where Θ is the scattering angle of the reaction. Therefore the foci overlap if $\Theta = 180°$. This overlap of foci probably explains [5] the very common occurrence of back-angle peaks in DW calculations. (b) The focus typically lies near the nuclear surface, but *in the interior*. Therefore focus effects in the angular distribution tend to be sensitive to details of the radial location of the direct interaction [2, 4–6].

The great labor required to perform DW calculations in partial wave expansion has led many authors to concoct analytic approximate optical wavefunctions that are expressed mathematically without the use of series expansions, or are obtained by briefer methods than the step-by-step solution of differential equations. However, the McCarthy focus and the other, smaller complications visible in Figs. 7.1 and 7.2 illustrate the intricacies that crude wavefunction models easily overlook. Considerable care is required to construct analytic wavefunction models that are not dangerously oversimplified.

It appears to be feasible to construct reliable wavefunction models for reactions dominated by strong absorption in entrance and exit channels (Section 5.3), provided the entrance and exit relative momenta are approximately equal, $k_\alpha \approx k_\beta$. Analyses conducted under these special physical conditions often only require knowledge of optical wavefunctions outside the nucleus. Because wavefunctions in the outside region are known in full in terms of the reflection coefficients η_L, any short-cut determination of the η_L determines the wavefunctions and allows the analyses to proceed. Hence, the task of constructing wavefunction models is thrown back on the much easier task of constructing analytic models of the η_L. It is also helpful that these easily constructed L-space wavefunction models retain the partial wave expansion; therefore they automatically incorporate correct treatments of diffractive spreading and of angular momentum quantization.

Wavefunction models based on the η_L are applied in Section 5.3 in discussions of the inelastic scattering of strongly-absorbed projectiles. It was pointed out that semiclassical considerations of the impact parameters associated with given partial waves lead to the picture that low partial waves encounter the target nucleus so strongly that $\eta_L = 0$, and high partial waves miss the nucleus to such an extent that $\eta_L = 1$. The simplest expression of this physical picture is the so-called *sharp-cutoff model*,

$$\begin{aligned} \eta_L &= 0, \qquad \text{if} \qquad L \leqslant L_0, \\ \eta_L &= 1, \qquad \text{if} \qquad L > L_0. \end{aligned} \tag{7.1}$$

Generally the *cutoff angular momentum* L_0 is associated with some *strong-absorption radius*, R_{abs}, so that

$$L_0 = kR_{abs}. \tag{7.2}$$

The sharp-cutoff model was first employed in nuclear physics by Bethe and Placzek and by Akhiezer and Pomeranchuk [7]. Later it was used by Blair in a series of analyses of alpha particle elastic scattering [8, 9]. These analyses gave excellent descriptions of cross sections at small scattering angles and correct qualitative indications of the way cross sections fall off at larger angles. It is interesting that R_{abs} of (7.2) was found to be consistently larger (by about 2 fm) than known matter radii of nuclei. Projectiles whose impact parameters are less than this (large) R_{abs} apparently penetrate the centrifugal barrier to a sufficient extent to be strongly absorbed.

The radial wavefunctions associated with the sharp-cutoff model are seen from (3.25) to be

$$\begin{aligned} f_L(k, r) &= \frac{i}{2} H_L^*(kr), \qquad \text{if} \qquad L \leqslant L_0, \\ f_L(k, r) &= F_L(kr), \qquad \text{if} \qquad L > L_0, \end{aligned} \tag{7.3}$$

where the $F_L(kr)$ are the regular Coulomb wavefunctions. Equations 7.3 are applicable if $r \geqslant R_{abs}$. Often the Coulomb potential is so weak at $r \geqslant R_{abs}$ that the Coulomb radial wavefunctions of (7.3) may be replaced by ordinary spherical Bessel and Hankel functions. The approximations just mentioned seem to be implied in theories that replace optical wavefunctions by plane waves that are cutoff (set equal to zero) for $r \leqslant R_{abs}$. However, it has already been seen that such cutoff plane wave theories are defective in their treatments of partial waves with $L < L_0$ (see Section 5.3).

It is reasonable to seek to improve the sharp-cutoff model by introducing a more gradual transition between $\eta_L = 0$ and $\eta_L = 1$; several authors have accordingly considered *smoothed-cutoff models* [10–14]. One such model [13, 14] is

$$\begin{aligned} \operatorname{Re} \eta_L &= \varepsilon + (1 - \varepsilon) g(L), \\ \operatorname{Im} \eta_L &= \mu \frac{dg}{dL}, \end{aligned} \tag{7.4}$$

with

$$g(L) = \left[1 + \exp\left(\frac{L_0 - L}{\Delta}\right)\right]^{-1}. \tag{7.5}$$

Here the parameter ε allows for a small amount of reflection (elastic scattering) in low partial waves, and the parameter Δ is the half width of the transition of η_L; generally $\Delta \sim 1$. A nonzero Im η_L appears in (7.4), with a magnitude measured by the parameter μ. This Im η_L allows nonzero phase shifts for $L \approx L_0$. It does not matter that the particular form of this Im η_L probably gives poor descriptions of the phases of partial waves with $L < L_0 - \Delta$, because $|\eta_L|$ for these low partial waves is small.

Applications of (7.4, 5) and of other, related smoothed-cutoff models are reviewed at length by Frahn [14]. It is seen that under suitable conditions the few additional parameters introduced in smoothed-cutoff models overcome all the shortcomings of the sharp-cutoff model, and give excellent overall fits to elastic cross sections. The principal conditions to be fulfilled are (a) strong absorption, and (b) high enough bombarding energy so that $L_0 \gg \Delta$. Condition (b) is fulfilled well enough if $L_0 = kR_{abs} \geqslant 10$. Unfortunately, the success of the smoothed-cutoff models probably should be regarded only as an indication of the very limited information content of the (rapidly-oscillatory) elastic angular distributions found for strongly-absorbed projectiles.

The smoothed-cutoff models of η_L are based on semiclassical considerations that associate angular momenta with impact parameters. Very few attempts have been made to develop models of η_L that have a more secure physical basis [15–16]. In particular, analyses of η_L as an analytic function in the complex L-plane [16, 17] have not yielded improved models [14]. It may be that further development of models of η_L will rest less on theory than on precise determinations of η_L from fits to elastic scattering experiments. The η_L determined from experiment can at least give accurate descriptions of effects peculiar to the system in question [18]. *The optical wavefunction based on these η_L can then be used with confidence in studies of associated nuclear reactions.*

Configuration-space models of optical wavefunctions are less reliable than the L-space models described above, for reasons already mentioned. Abandonment of the partial-wave expansion means abandonment of a secure kinematical guide to the quantum-mechanical aspects of scattering. Nevertheless, rough $\mathbf{r}$-space models that do not require series expansions have been used from time to time. The best-known of these models is a special case of the *high-energy approximation* (WKB) that is discussed in Section 7.3. Wavelengths that are small compared to the radius of the target nucleus are assumed. For small wavelengths and black target nuclei the function $\chi^{(+)}(\mathbf{k}, \mathbf{r})$ is described as identical with the undistorted incident plane wave $e^{i(\mathbf{k}\cdot\mathbf{r})}$ everywhere except in the region occupied by the geometric shadow of the target nucleus. In the region of this *shadow cylinder* the wavefunction is given the value zero. Then theories that use this model wavefunction only in the region outside the target nucleus sometimes achieve great accuracy, as is seen in Section 7.2. However, this model divides the nuclear surface into *bright* and *dark* hemispheres, separated by a sharp boundary. At this boundary the model wavefunction is extremely misleading. For this reason, theories that use the extreme $\mathbf{r}$-space model in overlaps that emphasize the nuclear surface may be expected to be very inaccurate, inasmuch as best overlap usually takes place at the boundary between the two hemispheres.*

McCarthy and Pursey [2] proposed a less extreme $\mathbf{r}$-space model of the optical wavefunction. Their model is a rough summary of wavefunctions found in exact optical-model calculations. It describes $\chi^{(+)}(\mathbf{k}, \mathbf{r})$ *as a function of angle at a definite radius* that lies just at or outside the nuclear surface. Therefore it is useful only for reactions that are localized to a thin radial shell at the nuclear surface.

The McCarthy-Pursey model proceeds from the observation that on the nuclear surface the phase of the optical wavefunction tends to be identical with the phase of the incident wave, and the amplitude tends to damp smoothly as one proceeds from the bright to the dark hemisphere. Superimposed on this regular behaviour is the McCarthy focus. A rough formula that expresses the entire picture then is

$$\chi^{(+)}(\mathbf{k}, \mathbf{r}) \approx A \exp\{(ik - \gamma)(\hat{k} \cdot \mathbf{r})\} + \text{focus term}. \qquad (7.6)$$

The structure of the first term of (7.6) allows easy evaluation of the integrals of the DW theory. It was with the aid of (7.6) that McCarthy and Pursey investigated the focus effects we discuss at the beginning of this section.

It would be possible to extrapolate (7.6) to radii outside the nuclear surface, by expanding (7.6) in Legendre functions of $(\hat{k} \cdot \hat{r})$, and then using

* We note here that the extreme $\mathbf{r}$-space model can of course be improved by substituting a Coulomb wave for the incident plane wave.

the Legendre expansion as a boundary condition on the partial-wave expansion. However, it would be unreasonable to devote this much labor to the extrapolation of a crude model.

It is interesting that the extrapolation of a carefully selected **r**-space wavefunction model can be developed into a technique for determining the η_L parameters of an L-space model. Thus, despite the difficulty of guessing $\chi^{(+)}(\mathbf{k}, \mathbf{r})$ throughout all **r**-space, there are regions of **r**-space in which reliable guesses can be made. For strongly absorbing nuclei one such region lies just outside the nuclear surface, in the middle of the bright hemisphere, at $(\hat{k} \cdot \hat{r}) = -1$. In this region diffractive spreading of the wavefunction is slight, reflection is zero by definition, and $\chi^{(+)}(\mathbf{k}, \mathbf{r})$ has a well-defined momentum, so that

$$\nabla \chi^{(+)} = i\mathbf{k}\chi^{(+)}. \tag{7.7}$$

Condition (7.7) can be exploited by choosing a trial wavefunction in partial-wave expansion, and by adjusting the η_L parameters of this wavefunction to minimize the integral

$$I = \int W(\mathbf{r})\, |\nabla \chi_T^{(+)} - i\mathbf{k}\chi_T^{(+)}|^2\, d^3r. \tag{7.8}$$

Here $W(\mathbf{r})$ is a weight function that selects the region of **r**-space in which (7.7) obtains. We see that the set of equations from which the η_L are to be determined are

$$\frac{\partial I}{\partial \eta_L} = 0. \tag{7.9}$$

Once the η_L are determined it is straightforward, as usual, to calculate $\chi^{(+)}(\mathbf{k}, \mathbf{r})$ throughout all space.

A calculation of the type just described [19] yields η_L parameters that are typical of optical-model results, and a wavefunction whose angular dependence (on the nuclear surface) conforms to the description of McCarthy and Pursey [2]. This wavefunction even displays a prominent "focus" at 0°.

The principal interest attached to models of the type expressed by (7.8) and (7.9) lies in the extent to which they generate $\chi^{(+)}(\mathbf{k}, \mathbf{r})$ without reference to a potential. For strongly-absorbed, composite projectiles this is appealing.

The principal difficulty with (7.8) and (7.9) is that the weight function $W(\mathbf{r})$ must be chosen to emphasize the limited region of space in which the wavefunction can be guessed reliably; in this region the radius r necessarily is large. Because r is large all the low partial waves are asymptotic and they affect the radial dependence of $\chi_T^{(+)}(\mathbf{k}, \mathbf{r})$ in the same fashion; therefore (7.8, 7.9) are not sensitive to the values of η_L for the low partial waves. For strongly-absorbing systems this lack of sensitivity is not damaging, because $\eta_L = 0$ for the low partial waves. However, we note it is easy enough to generalize (7.8) and (7.9) to cases of weak absorption by treating the expression

$$I' = \int W(\mathbf{r}) \,|\nabla(\chi_T^{(+)} - Ae^{-i(\mathbf{k}\cdot\mathbf{r})}) - i\mathbf{k}(\chi_T^{(+)} - Ae^{-i(\mathbf{k}\cdot\mathbf{r})})|^2 \, d^3r, \tag{7.10}$$

$$\frac{\partial I'}{\partial \eta_L} = 0, \tag{7.11}$$

where $Ae^{-i\mathbf{k}\cdot\mathbf{r}}$ is the reflected wave in the region selected by the weight function. Unfortunately, for weak absorption it matters that (7.10) and (7.11) do not determine η_L in the low partial waves, because now these η_L are nonzero. Therefore we do not have enough information to determine the wavefunction.

The author is hopeful that a useful theory that incorporates (7.10, 7.11) will be found.

7.2 Applications of r-space Models: Diffraction

Nuclear reaction theories that apply the analytic wavefunction models of Section 7.1 are called "diffraction theories." This term suggests an analogy with the familiar Fresnel analyses of the propagation of light waves past opaque screens. The Fresnel theory uses a simple model of the wavefunction in the vicinity of the screen, and it uses the free Green's function to treat the onward propagation of the wave. Much the same procedure is followed in the **r**-space diffraction theories discussed in this section. Further details of these theories may be found in various extensive review articles [14, 20–25].

Applications of L-space wavefunction models are treated elsewhere in this book; for example, their applications in the DW theory appear in Sections 5.3 and 7.4. Their application to elastic scattering is trivial and does not require any extended discussion at all, inasmuch as the models postulate the η_L parameters from which the elastic cross section is computed. It is already remarked in Section 7.1 that L-space models give excellent fits to the cross sections for elastic scattering of strongly-absorbed projectiles at moderate to high energies. It is even possible to obtain simple closed formulas for the elastic cross section by approximately summing the partial-wave series given by the L-space models [13, 14].

r-space models approximate the wavefunction in a fashion that is valid only in the immediate vicinity of the target nucleus. To obtain cross sections we use these models as source terms in Green's theorem, and use the free single-particle Green's function to describe the way outgoing waves propagate to the detector. This step may be performed in a number of ways [20, 21, 23]. The present discussion follows the derivation of [21], and therefore starts [see (3.20)] from the Green's-theorem formula for the elastic-scattered amplitude,

$$T = \int e^{-i(\mathbf{k}_\beta \cdot \mathbf{r})} V(\mathbf{r}) \, \chi_\alpha^{(+)}(\mathbf{k}_\alpha, \mathbf{r}) \, d^3r \tag{7.12}$$

Because $V(\mathbf{r})$ is short ranged this formula emphasizes the region near the target nucleus, in which $\chi_\alpha^{(+)}$ is assumed known. The use of the adiabatic approximation (Section 6.3) allows us to apply (7.12) both to elastic and inelastic scattering.

Let us assume that the region in which $V(\mathbf{r})$ is nonzero is bounded by an arbitrary surface Σ that is at least large enough to enclose the nucleus. We transform (7.12) by the following steps: The use of the Schrödinger equation for $\chi_\alpha^{(+)}$ gives

$$T = \frac{\hbar^2}{2\mu} \int_{\text{vol}} e^{-i(\mathbf{k}_\beta \cdot \mathbf{r})} [k_\alpha{}^2 + \nabla^2] \chi_\alpha^{(+)} \, d^3r. \tag{7.13}$$

Next, because elastic scattering conserves energy we have $k_\alpha{}^2 = k_\beta{}^2 = k^2$, and therefore $k_\alpha{}^2$ in (7.13) may be replaced with a Laplacian that operates to the left, giving

$$T = \frac{\hbar^2}{2\mu} \int_{\text{vol}} \{ e^{-i(\mathbf{k}_\beta \cdot \mathbf{r})} [\nabla^2 \chi_\alpha^{(+)}] - [\nabla^2 e^{-i(\mathbf{k}_\beta \cdot \mathbf{r})}] \chi_\alpha^{(+)} \} \, d^3r. \tag{7.14}$$

Next, before making the obvious transformation to a surface integral on Σ, it is helpful to insert

$$\chi_\alpha^{(+)} = e^{i(\mathbf{k}_\alpha \cdot \mathbf{r})} + \chi_{\text{scatt}},$$

and to notice that the plane-wave term of $\chi_\alpha^{(+)}$ does not contribute in (7.14). Therefore, the surface integral equivalent to (7.14) is

$$T = \frac{\hbar^2}{2\mu} \int_{\Sigma} \{ e^{-i(\mathbf{k}_\beta \cdot \mathbf{r})} [\nabla \chi_{\text{scatt}}] - [\nabla e^{-i(\mathbf{k}_\beta \cdot \mathbf{r})}] \chi_{\text{scatt}} \} \cdot d\mathbf{\Sigma} \tag{7.15}$$

Because the surface Σ is *outside the target nucleus*, it lies in the region of validity of the wavefunction models of Section 7.1.

In the extreme, short-wavelength $\mathbf{r}$-space model $\chi_\alpha^{(+)}$ differs from the incident plane wave only in the region of the *shadow cylinder*, downstream from the target nucleus. Therefore Σ is conveniently taken to be a cylinder with its axis in the direction $\mathbf{k}_\alpha$, as shown in Fig. 7.3. Then the shadow cylinder intercepts Σ only on the "shadow plane," on the downstream end of Σ. In the shadow region the wavefunction χ_{scatt} is

$$\chi_{\text{scatt}} = -e^{i(\mathbf{k}_\alpha \cdot \mathbf{r})}; \tag{7.16}$$

elsewhere on Σ both χ_{scatt} and its normal derivative are zero. The element of area on the shadow plane is

$$d\mathbf{\Sigma} = \hat{k}_\alpha \, dA.$$

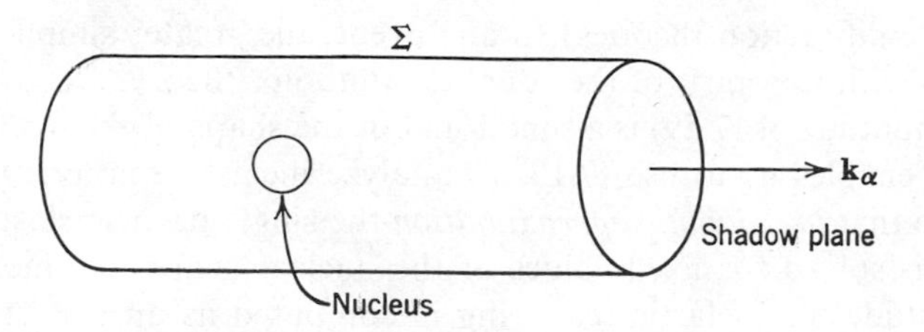

Fig. 7.3 The surface Σ on which we apply Green's theorem.

Upon substituting the above expressions, (7.15) becomes

$$T = -\frac{i\hbar^2 k}{2\mu}(1 + \cos \Theta) \int_{\text{shadow}} e^{i([\mathbf{k}_\alpha - \mathbf{k}_\beta]\cdot \mathbf{r})} \, dA, \tag{7.17}$$

where Θ is the scattering angle of the reaction, so that $\cos \Theta = (\hat{k}_\alpha \cdot \hat{k}_\beta)$. The coefficient $(1 + \cos \Theta)$ is the so-called *obliquity factor* of the diffraction model. By placing the origin for **r** on the shadow plane and introducing polar coordinates $\{b, \phi\}$ on this plane we find

$$(\mathbf{k}_\alpha \cdot \mathbf{r}) = 0, \qquad (\mathbf{k}_\beta \cdot \mathbf{r}) = kb \sin \Theta \cos \phi,$$

so that the amplitude finally reduces to

$$T = -\frac{i\hbar^2 k}{2\mu}(1 + \cos \Theta) \int_{\text{shadow}} e^{-ikb \sin \Theta \cos \phi} \, b \, db \, d\phi. \tag{7.18}$$

Equation 7.18 is the basic formula of the Fraunhofer theory of inelastic scattering, developed by Drozdov [26], Inopin [27], and Blair [28].

Toward large scattering angles the integrand of (7.18) becomes rapidly oscillatory, and T becomes small. Therefore it is at large angles that T is most sensitive to our use of a rough approximate wavefunction in (7.15). Therefore the Fraunhofer theory is essentially a small-angle theory, and the higher order Θ-dependence in (7.18) is irrelevant. We may make the approximations

$$(1 + \cos \Theta) \approx 2, \qquad \sin \Theta \approx \Theta,$$

to obtain

$$T \approx -\frac{i\hbar^2 k}{\mu} \int_{\text{shadow}} e^{-ikb\Theta \cos \phi} \, b \, db \, d\phi. \tag{7.19}$$

Blair remarks [23] that (7.19) often seems to be superior to (7.18) at large Θ because it yields results that are in better accord with those of DW theories or L-space diffraction theories! In any event, the greater simplicity of (7.19) conforms with the spirit of the rough Fraunhofer theory.*

The amplitude of (7.19) is a functional of the shape of the nuclear shadow. This fact enables us to use (7.19) to analyze inelastic scattering. In Section 6.3 we see that *in adiabatic approximation* the single-particle elastic scattering problem is solved for fixed values of the nuclear shape parameters y. Then the amplitude for inelastic scattering is computed as an expectation of the y-dependent single-particle amplitude, in the form

$$T_{\alpha\beta}(\mathbf{k}_\alpha, \mathbf{k}_\beta) \approx \langle \psi_B(y) | \, t(\mathbf{k}_\alpha, \mathbf{k}_\beta; y) \, | \psi_A(y) \rangle. \tag{7.20}$$

Equation 7.19 is precisely the single-particle amplitude required for (7.20), the y-dependence of the size and shape of the shadow causing the y-dependence of the amplitude.

The adiabatic boundary radius of the nuclear shadow differs from the strong-absorption radius of (7.2) by an angle-dependent increment $\delta(\phi)$, so that

$$b_{\text{shadow}}(\phi) = R_{abs} + \delta(\phi). \tag{7.21}$$

Inserting (7.21) into the Fraunhofer amplitude of (7.19) gives

$$t = -\frac{i\hbar^2 k}{\mu} \int_0^{2\pi} d\phi \int_0^{R_{abs}+\delta(\phi)} b \, db e^{-ikb\Theta \cos\phi}, \tag{7.22}$$

in which the notation for the amplitude has been changed to conform with (7.20). Now, although R_{abs} is rather greater than the half radius of either the optical potential or the matter distribution, it is plausible that $\delta(\phi)$ equals the deformation-dependent increment in the radius of the optical potential. To first order this increment may be taken from (5.57) and evaluated at the nuclear equator, giving

$$\delta(\phi) \approx \sum_{kq} \xi_{kq} Y_k^{q*}\left(\frac{\pi}{2}, \phi\right), \tag{7.23}$$

where k is the multipole order of the deformation and q is the z-projection.

To first order in $\delta(\phi)$, integration of (7.22) gives

$$t = -\frac{2\pi i\hbar^2}{\mu\Theta} J_1(kR_{abs}\Theta) - \frac{2\pi i\hbar^2 k}{\mu} \sum_{kq} \left(\frac{2k+1}{4\pi}\right)^{1/2} \left(\frac{q}{|q|}\right)^k \xi_{kq}[k:q] J_{|q|}(kR_{abs}\Theta), \tag{7.24}$$

* It is often remarked that (7.19) is not invariant under time reversal, because the entrance and exit channels have not been treated symmetrically. However, the time-reversal violation is just one aspect of the overall crudeness of (7.19), and hardly seems to require special attention.

where $[k:q]$ is the symbol defined in (5.49). We now insert (7.24) in (7.20) and evaluate the cross sections. The first term of (7.24) is the familiar semi-classical amplitude for elastic scattering by a black disc [7]. The corresponding cross section is

$$\left(\frac{d\sigma}{d\Omega}\right)_{\text{elastic}} = R_{abs}^2 \Theta^{-2} \, |J_1(kR_{abs}\Theta)|^2. \tag{7.25}$$

For a spin-zero target nucleus the cross sections for inelastic scattering are found to be

$$\left(\frac{d\sigma}{d\Omega}\right)_{\text{inelastic}} = (kR_{abs})^2 \left(\frac{2k+1}{4\pi}\right) |\langle \psi_B{}^q| \, \xi_{kq} \, |\psi_A{}^0\rangle|^2$$

$$\times \sum_{q=-k,-k+2,\ldots}^{k} [k:q]^2 \, |J_{|q|}(kR_{abs}\Theta)|^2, \tag{7.26}$$

where a final state ψ_B that has spin I is excited by the multipole of order $k = I$. Only alternate values of q contribute in (7.26), because*

$$[k:q] = 0 \qquad \text{if } k + q \text{ is odd.}$$

Because the matrix element $\langle \psi_B{}^q| \, \xi_{kq} \, |\psi_A{}^0\rangle$ is independent of q it may be regarded as the reduced matrix element $\langle \psi_B \| \, \xi_k \, \| \psi_A \rangle$. For a permanently deformed nucleus it is

$$\langle \psi_B \| \, \xi_k \, \| \psi_A \rangle = (\beta R_{abs})(2k+1)^{-1/2},$$

as in the discussion in Section 5.3.

The reduced matrix elements are the only adjustable parameters in the above theory. The radius R_{abs} is determined from experiment by fitting (7.25) to the observed elastic cross section. Then (7.26) predicts the shapes of the inelastic cross sections. These predicted shapes generally agree well with observed shapes, for example, for the collective inelastic scattering of medium-energy alpha particles. The reduced matrix elements determined by fitting (7.26) to experiment agree with those determined by electromagnetic means.

The most characteristic property of (7.26) is that for even k values the cross section is a linear combination of

$$J_0{}^2, J_2{}^2, \ldots, J_k{}^2,$$

whereas for elastic scattering and for odd k values the cross section is a linear combination of

$$J_1{}^2, J_3{}^2, \ldots, J_k{}^2.$$

* Blair remarks that the reason the amplitude vanishes if $(k + q)$ is odd is that nuclear deformations that have $(k + q)$ odd cast circular shadows [23].

Now Bessel functions very rapidly approach their asymptotic forms, such that

$$J_q^2 \to \sin^2\left(kR_{abs}\Theta + \frac{\pi}{4} - \frac{q\pi}{2}\right).$$

Therefore we see that beyond the first oscillation of the Bessel function of order $q = k$ the angular distributions of all even-parity transitions tend to be of the same shape, as do those of all odd-parity transitions. However, the two oscillate precisely 180° out of phase with each other. We also see that the elastic angular distribution oscillates in phase with the odd-parity transitions, and thereby it calibrates the parities of the inelastic transitions. These statements constitute Blair's *phase rule*. We already encountered the first part of the phase rule in Section 5.3, where it was a consequence of the DW calculation for strongly-absorbing nuclei. Observed angular distributions usually obey the phase rule very closely. Figure 7.4 shows some typical experimental results and their comparison with the Blair formulas.

It is seen in Fig. 7.4 that the Fraunhofer curves follow the period of the oscillations of the observed cross section, but that their magnitudes drop off too slowly with angle. This characteristic difficulty of the extreme **r**-space diffraction model can be remedied by smoothing the transition of $\chi_\alpha^{(+)}$ at the edge of the nuclear shadow. Blair [23] reviews several methods for "giving the shadow a fuzzy edge," and shows how they improve the fit to experiment. It is noteworthy that smoothing the edge of the shadow gives the diffraction curves a faster falloff but does not alter the phase rule.

The cross sections that (7.26) predicts at small angles have characteristics that may in principle be used to determine precise k values of transitions, as well as parities. Blair especially advocates careful study of the location of the first deep minimum of the angular distribution. Unfortunately it is difficult to measure low-Q inelastic cross sections at small angles. In addition, (7.26) becomes inaccurate at very small angles because it omits Coulomb excitation and it omits Q-dependent effects of order $(k_\alpha - k_\beta)$. These failings of the Fraunhofer theory may be overcome by performing DW calculations.

We note that it is straightforward to evaluate (7.19) to second order in ξ_{kq} to obtain Fraunhofer formulas for double excitation [23]. However, these formulas are tied to the adiabatic model (Section 6.3), which associates single- and double-excitation processes with the same collective mode of motion. Because the microscopic dynamics of nuclei seldom allow such a simple picture of double excitation it would seem better to treat this process by the more flexible coupled-channels method of Section 6.1.

It is difficult to generalize the diffraction theory in a fashion suitable for the treatment of rearrangement reactions. A major reason for this difficulty is the absence of any adiabatic approximation that relates rearrangement amplitudes to elastic amplitudes, similar to the approximation we use for the

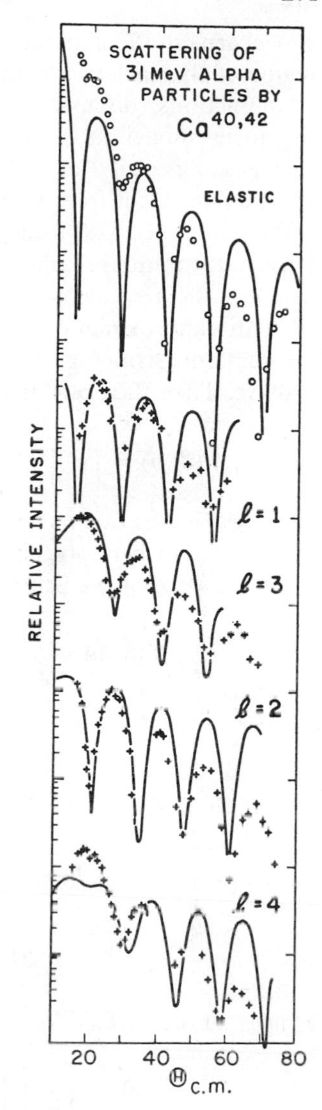

Fig. 7.4 Comparison of the Blair theory with experimental data for the elastic and inelastic scattering of 31 MeV alpha particles by $Ca^{40,42}$. The elastic and $l = 1$ data are for Ca^{40}, all others are for Ca^{42}. (Based on graphs provided through the courtesy of Professor A. M. Bernstein. For further information see E. P. Lippincott and A. M. Bernstein, *Phys. Rev.* **163**, 1170 (1967); see also A. M. Bernstein in *Advances in Nuclear Physics*, Vol. III, M. Baranger and E. Vogt, Eds., Plenum Press, New York, 1970.)

inelastic amplitude. Because rearrangement reactions excite the coordinates of individual nucleons, a discussion of these reactions goes beyond a mere discussion of the shape and diffuseness of the nuclear surface. Additional physical principles are required.

Several authors have constructed diffraction theories of rearrangement, by introducing **r**-space wavefunction models for the two distorted waves of the DW theory [22, 29, 30]. They regarded the limited region of overlap of two sharp-cutoff, extreme **r**-space wavefunctions as a "ring locus" in the region of the nuclear equator. In view of the known complications of optical wavefunctions, discussed in Section 7.1, there is no reason to believe the ring-locus model to be a correct description of the overlap. This model becomes especially doubtful for scattering angles that depart to any appreciable extent from $\Theta = 0$. It is not discussed further.

Dar attempted [29] to develop a diffraction approximation of the general Green's-theorem formula for rearrangement (3.20), using a method that parallels the treatment of inelastic scattering but that does not use the adiabatic approximation. Let us consider the application of Dar's method for deuteron stripping. The target nucleus is treated as inert and infinitely massive. Then the exact amplitude for the (d, p) reaction is

$$T(d, p) = \int e^{-i(\mathbf{k}_p \cdot \mathbf{r}_p)} \phi^*(\mathbf{r}_n)[V_{np} + V_{pA}]\psi^{(+)}(\mathbf{r}_p, \mathbf{r}_n)\, d^3r_n\, d^3r_p, \qquad (7.27)$$

where $\psi^{(+)}$ is the exact wavefunction for the three-body system, with a deuteron incident, and where $\phi(\mathbf{r}_n)$ is the wavefunction of the neutron bound state that is formed. Use of the Schrödinger equations that govern $\psi^{(+)}$ and ϕ gives

$$T(d, p) = \int e^{-i(\mathbf{k}_p \cdot \mathbf{r}_p)} \left[E - \varepsilon + \frac{\hbar^2}{2M} \nabla_p{}^2\right] \Gamma(\mathbf{r}_p)\, d^3r_p, \qquad (7.28)$$

where ε is the binding energy of state ϕ and Γ is the overlap

$$\Gamma(\mathbf{r}_p) \equiv \int \phi^*(\mathbf{r}_n)\, \psi^{(+)}(\mathbf{r}_n, \mathbf{r}_p)\, d^3r_n. \qquad (7.29)$$

Energy conservation then allows $(E - \varepsilon)$ to be replaced by $\nabla_p{}^2$ operating on the proton plane wave, with the result

$$T(d, p) = \frac{\hbar^2}{2M} \int \left\{ e^{-i(\mathbf{k}_p \cdot \mathbf{r}_p)}[\nabla^2 \Gamma] - [\nabla^2 e^{-i(\mathbf{k}_p \cdot \mathbf{r}_p)}]\Gamma \right\} d^3r_p. \qquad (7.30)$$

This expression transforms into the surface integral

$$T(d, p) = \frac{\hbar^2}{2M} \int_\Sigma \left\{ e^{-i(\mathbf{k}_p \cdot \mathbf{r}_p)}[\nabla \Gamma] - [\nabla\, e^{-i(\mathbf{k}_p \cdot \mathbf{r}_p)}]\, \Gamma \right\} \cdot d\,\mathbf{\Sigma}, \qquad (7.31)$$

where Σ is an arbitrary surface that encloses the target nucleus.

The key step of Dar's theory now lies in the selection of a diffraction model for $\Gamma(\mathbf{r}_p)$, valid on the surface Σ. This step is nontrivial. We note that straightforward cutoff models indicate that $\psi^{(+)}$ is proportional to the internal wavefunction of the incident deuteron $\phi_d(\mathbf{r}_p - \mathbf{r}_n)$. Then $\Gamma(\mathbf{r}_p)$ is proportional to the product $\phi^*(\mathbf{r}_n)\phi_d(\mathbf{r}_p - \mathbf{r}_n)$, and it vanishes strongly as r_p becomes large. However, such a dependence on r_p is unreasonable because

Σ in (7.31) is arbitrary. Evidently, valid models of Γ must take into account the breakup of the incident deuteron.

One way to construct a valid model of Γ is to introduce the "sudden approximation" (see Chapter 4, [21]) for $\psi^{(+)}$ in (7.29). In this approximation the incident deuteron is decomposed into momentum eigenstates for the neutron and proton, and these states are allowed to scatter separately from the target nucleus. Then the extreme **r**-space wavefunction model is introduced separately for the neutron and proton eigenstates. It might be interesting to develop a theory based on this idea.

7.3 WKB: High-energy Method

At high bombarding energies wavefunctions may be calculated in WKB approximation. WKB yields accurate results whenever the wavelength of a projectile is much smaller than the interval over which it undergoes significant change, so that reflection is negligible. This method has been used extensively in nuclear physics, and many excellent reviews are available [14, 20, 23–25].

In one group of applications the wavefunction is treated in partial-wave expansion, and WKB is used in one dimension to compute the required radial waves [9, 19, 32–34]. This procedure yields useful results even at medium and low energies, if it is employed to study only those radial waves that have low angular momentum and are strongly absorbed [9, 19, 32, 33]. It is obvious that WKB is applicable for these cases, because strong absorption implies absence of reflection. These studies of strong absorption emphasize composite projectiles, which are absorbed because they are composite, and which have short wavelengths (hence low reflection) because they have large masses.

At high energy two new considerations arise. The first is that all projectiles have short wavelengths and experience minimal reflection; this is true even of projectiles that are not composite and that have small masses. Therefore WKB may be used for nearly all nuclear reactions, whether or not they involve strong absorption. The second consideration is that the small wavelengths at high energy make the partial-wave expansion impractical. Therefore the WKB method must be used in its full three-dimensional form. In this form it becomes a convenient starting point for generalized diffraction theories.

In three dimensions the lowest WKB approximation to the wavefunction is

$$\chi^{(+)}(\mathbf{k}, \mathbf{r}) = e^{iS(\mathbf{r})}, \tag{7.32}$$

where the phase function $S(\mathbf{r})$ is computed as an integral along the classical trajectory that passes through the point $\mathbf{r}$,

$$S(\mathbf{r}) = \int^{\mathbf{r}} \left[k^2 - \frac{2\mu}{\hbar^2} U \right]^{1/2} d\tau. \tag{7.33}$$

The integrand, which is governed by the interaction $U(\mathbf{r})$, is the *local momentum* at each point along the trajectory. At high energy the interaction does not cause large changes of the local momentum, but may well cause important changes in S, because the effects in S are cumulative.

An important simplification of the phase integral $S(\mathbf{r})$ was introduced by Squires [35], McCauley and Brown [36], and Glauber [20]. These authors observed that at very high energy the classical trajectories of particles do not deviate much from straight lines in the direction $\mathbf{k}$. The deviation is particularly slight for the short segments of these trajectories that lie inside the target nucleus or in its immediate vicinity. When such straight-line trajectories are used, the phase integral reduces to

$$S(\mathbf{r}) = \int_{-\infty}^{0} \left[k^2 - \frac{2\mu}{\hbar^2} U(\mathbf{r} + \hat{k}z) \right]^{1/2} dz, \tag{7.34}$$

where $\hat{k}$ is the unit vector in the direction of $\mathbf{k}$. Equation 7.34 expresses $S(\mathbf{r})$ as an explicit quadrature, computed separately for each point $\mathbf{r}$ at which the wavefunction is desired. It is customary to simplify the integral further, and to expand the square root, on the basis that at high energy k^2 is large and U generally is small. We then obtain

$$S(\mathbf{r}) \approx (\mathbf{k} \cdot \mathbf{r}) - \frac{\mu}{\hbar^2 k} \int_{-\infty}^{0} U(\mathbf{r} + \hat{k}z)\, dz, \tag{7.35}$$

where a more customary zero of phase has been introduced. Because (7.35) is intended for use at high energy, it is worthwhile to remark that the same result is obtained if one starts from the Klein-Gordon equation [36]. However, in this case the reduced mass μ must be replaced by E/c^2.

Insertion of (7.35) in (7.32) yields a convenient approximate wavefunction that may be applied in a number of different ways. One obvious application of the high-energy approximation is to generate the distorted waves required for DW calculations. In this case $U(\mathbf{r})$ is an optical potential, and the distorted waves are

$$\chi_\alpha^{(+)}(\mathbf{k}_\alpha, \mathbf{r}) \approx \exp i\left\{(\mathbf{k}_\alpha \cdot \mathbf{r}) - \frac{\mu_\alpha}{\hbar^2 k_\alpha} \int_{-\infty}^{0} U_\alpha(\mathbf{r} + \hat{k}_\alpha z)\, dz\right\}, \tag{7.36a}$$

$$\chi_\beta^{(-)*}(\mathbf{k}_\beta, \mathbf{r}) \approx \exp i\left\{-(\mathbf{k}_\beta \cdot \mathbf{r}) - \frac{\mu_\beta}{\hbar^2 k_\beta} \int_{-\infty}^{0} U_\beta(\mathbf{r} - \hat{k}_\beta z)\, dz\right\}, \tag{7.36b}$$

where the time-reversal relation of (3.40) has been used to generate (7.36*b*). It is instructive to change the sign of the dummy variable in (7.36*b*), so that

$$\chi_\beta^{(-)*}(\mathbf{k}_\beta, \mathbf{r}) \approx \exp i\left\{-(\mathbf{k}_\beta \cdot \mathbf{r}) - \frac{\mu_\beta}{\hbar^2 k_\beta} \int_{0}^{\infty} U_\beta(\mathbf{r} + \hat{k}_\beta z)\, dz\right\}. \tag{7.36b'}$$

Now (7.36a, b') are inserted in the DW matrix element. We see that the optical potential U_α modulates the wavefunction of the incident particle, from $-\infty$ to the point $\mathbf{r}$ where the reaction takes place. Beyond the point $\mathbf{r}$ the wavefunction of the emerging particle is modulated by the potential U_β.

Further simplifications appear when the above distorted waves are used in the study of inelastic scattering. In this case $U_\alpha = U_\beta = U$, $\mu_\alpha = \mu_\beta = \mu$, and $k_\alpha \approx k_\beta = k$. Furthermore, because small scattering angles are of primary interest, the phase integrals may be computed under the approximation $\hat{k}_\alpha \approx \hat{k}_\beta$. Therefore the product of distorted waves becomes

$$[\chi_\beta^{(-)*}(\mathbf{k}_\beta, \mathbf{r})\, \chi_\alpha^{(+)}(\mathbf{k}_\alpha, \mathbf{r})] \approx \exp i\left\{([\mathbf{k}_\alpha - \mathbf{k}_\beta]\cdot\mathbf{r}) - \frac{\mu}{\hbar^2 k}\int_{-\infty}^{\infty} U(\mathbf{r} + \hat{k}z)\, dz\right\}, \quad (7.37)$$

where the modulations of the entrance and exit channel phases combine to give a single integral over the range $-\infty < z < \infty$. This integral depends only on the *impact parameter* of the classical trajectory, and may be incorporated in the form factor of the DW calculation. The DW calculation may then be completed by using plane wave Born approximation with the altered form factor. The inclusion of distortion effects by this method is quite easy and does not require the use of computing machines [35–37]. Sanderson [37], for example, used this method for inelastic scattering of 185 MeV protons by C^{12}, and found that distortion reduced the differential cross sections by factors of 2 or 3, and in some cases altered the angular distributions.

The generalizations required to include spin-orbit distorting potentials in the above method are straightforward, and are described in the papers of McCauley and Brown [36] and Glauber [20], already cited, and of Köhler [38]. The latter author computes the polarizations of inelastically scattered protons.

The accuracy of the distorted waves generated by the high-energy method has been tested by D. F. Jackson and collaborators [39] and by Lee and McManus [40]. These articles primarily concern (p, p') reactions at bombarding energies in the vicinity of 100 MeV. It is found that at 156 MeV (7.36a, b') give excellent agreement with the results of exact partial-wave analyses [40]. However, the calculations by Jackson suggest that the small-deflection approximation in (7.37) is of dubious value. Results obtained with this approximation are in only qualitative agreement with results of partial-wave analyses. Equation 7.37 overlooks both refraction and the fact that bent trajectories are longer than straight ones. Of course, these effects become less important as the energy becomes higher.

Another important application of the high-energy approximation is for the development of diffraction theories of nuclear reactions. Diffraction theories, as is noted in Section 7.2, exploit model wavefunctions that are good approximations in the near vicinity of the target nucleus. It is seen in (7.15) and (7.31) that the reaction amplitudes implied by particular model wavefunctions can conveniently be expressed as integrals over a surface Σ that encloses the nucleus. We now insert the wavefunction of the high-energy approximation in these surface integrals. In fact this is only done with (7.15). Insertion of the high-energy approximation in (7.31) yields much the same theory already discussed in connection with that equation.

Because the high-energy method assumes straight-line trajectories in the direction $\hat{k}_\alpha$ we once again assume Σ to be a cylinder with its axis parallel to $\hat{k}_\alpha$. Once again χ_{scatt} is nonvanishing only on the shadow plane on the downstream end of Σ. Once again we express χ_{scatt} on the shadow plane in terms of polar coordinates $\{b, \phi\}$, where b is the impact parameter of the classical trajectory through the point $\{b, \phi\}$. Then, because the shadow plane lies beyond the target nucleus, the expression for χ_{scatt} is found to be

$$\chi_{\text{scatt}} = e^{i(\mathbf{k}_\alpha \cdot \mathbf{r})}\{e^{i2\delta_\alpha(\mathbf{b})} - 1\}, \tag{7.38}$$

where the *phase-shift function* is

$$\delta_\alpha(\mathbf{b}) \equiv -\frac{\mu_\alpha}{2\hbar^2 k_\alpha}\int_{-\infty}^{\infty} U_\alpha(\mathbf{b} + \hat{k}_\alpha z)\, dz, \tag{7.39}$$

and where $\mathbf{b}$ is the vector whose components in polar coordinates are $\{b, \phi\}$. From (7.15), the scattered amplitude is found to be

$$T = \frac{-i\hbar^2 k}{\mu}\int e^{-i(\mathbf{k}_\beta \cdot \mathbf{b})}\{e^{i2\delta_\alpha(\mathbf{b})} - 1\}\, d^2b, \tag{7.40}$$

where the small-angle approximation for the obliquity factor is inserted. Equation 7.40 is the basic equation of the high-energy diffraction theory of elastic and inelastic scattering. (Inelastic scattering is of course obtained from (7.40) by the adiabatic approximation.)

If $U(\mathbf{r})$ is spherically symmetric there is a close correspondence between (7.40) and the partial-wave expansion of the amplitude [25]. In this correspondence $\delta(b)$ is associated with δ_L under the rule $L + \frac{1}{2} = kb$. The notation is chosen accordingly.

Equation 7.40 reduces to the extreme $\mathbf{r}$-space theory of elastic and inelastic scattering (Section 7.2) if $\delta_\alpha(b) \to i\infty$ for $b < b_{\text{shadow}}$, and if $\delta_\alpha(b) = 0$ for $b > b_{\text{shadow}}$. Under typical circumstances the transition at the edge of the shadow is not this sharp and therefore (7.40) yields the fuzzy-edge $\mathbf{r}$-space models (Section 7.2) reviewed by Blair [23]. Thus the high-energy method helps to clarify our understanding of $\mathbf{r}$-space diffraction models.

Most further applications of the high-energy method go beyond the simple discussions of optical wavefunctions given in the last several pages [20, 24, 25, 36]. A rather complete many-body theory of nuclear reactions may be derived if proper account is taken of the individual nucleons of the target nucleus. It is necessary to formulate the analysis in adiabatic approximation (see [20]), and to picture the target nucleons as remaining fixed in their positions while the high-energy projectile passes through the nucleus. Within this approximation the many-body theory differs from the optical-potential theory only in the rather elementary fact that the phase function on the classical trajectory is not computed by integrating the phase shift caused by a potential, as in (7.35) or (7.39), but rather it is computed by summing the phase shifts caused by encounters with individual nucleons. In this way $S(\mathbf{r})$ is computed in terms of the elementary projectile-nucleon interaction. It is particularly interesting that although in this approach the *phase function* $S(\mathbf{r})$ *is linear* in the contributions of the individual nucleons, the *scattered amplitude* [see (7.40)] *is an exponential* of the sum of these contributions. Therefore strong multiple-scattering effects appear in the amplitude. Characteristic effects of this multiple scattering are found in experiments performed in the GeV energy range [25, 41]. Because of the clarity and easy availability of the reviews already cited there is no point in presenting the many-body theory in this book

One rather crude application of the high-energy method, having to do with approximate DW calculations, may finally be mentioned. We recall that for inelastic scattering in the small-deflection approximation the product of entrance and exit channel distorted waves reduced to (7.37). We recognize in (7.37) the phase-shift function of (7.39), such that

$$[\chi_\beta^{(-)*}(\mathbf{k}_\beta, \mathbf{r})\chi_\alpha^{(+)}(\mathbf{k}_\alpha, \mathbf{r})] = e^{i\{([\mathbf{k}_\alpha - \mathbf{k}_\beta]\cdot\mathbf{r}) + 2\delta(b)\}}. \tag{7.41}$$

Therefore the DW matrix element for inelastic scattering is

$$\int \chi_\beta^{(-)*}(\mathbf{k}_\beta, \mathbf{r})\, F(\mathbf{r})\, \chi_\alpha^{(+)}(\mathbf{k}_\alpha, \mathbf{r})\, d^3r = \int M(\mathbf{k}_\alpha - \mathbf{k}_\beta, \mathbf{b}) e^{i2\delta(b)}\, d^2b, \tag{7.42}$$

where M is the Born approximation matrix element

$$M(\mathbf{k}_\alpha - \mathbf{k}_\beta, \mathbf{b}) = \int_{-\infty}^{\infty} e^{i([\mathbf{k}_\alpha - \mathbf{k}_\beta]\cdot\mathbf{r})} F(\mathbf{r})\, dz. \tag{7.43}$$

New results are now obtained if the integral over impact parameters is transcribed into the corresponding (see above) sum over angular momentum. Then (7.42) suggests that the L-space expansion of the DW matrix element be approximated in the following fashion: (a) Born approximation matrix elements $M_L(\mathbf{k}_\alpha - \mathbf{k}_\beta)$ are computed for each partial wave. (b) Distortion is introduced by multiplying the M_L by the phase shift factors $e^{i2\delta_L}$. This procedure has been used for several years to treat "peripheral scattering" of elementary particles [42, 43]. It has occasionally been advocated for use in nuclear physics [22, 44].

The crude procedure of (7.41–7.43) may be generalized to give an even cruder procedure for approximate DW analyses of rearrangement reactions. With rearrangement, the two phase factors of (7.36*a*, *b*′) do not combine to give the simple phase correction factor of (7.37). In principle, the **r**-dependence of the potentials U_α and U_β introduces an intricate **r**-dependence in the product of phase factors, and this complicates the DW integration. However, it may sometimes happen that only **r** values that lie in the equatorial plane of the target nucleus are of interest. In this case, because U_α and U_β are symmetric with respect to reflection in the equatorial plane, we again obtain a simple dependence on phase shifts, in the form

$$[\chi_\beta^{(-)*}(\mathbf{k}_\beta, \mathbf{r})\chi_\alpha^{(+)}(\mathbf{k}_\alpha, \mathbf{r})] \approx e^{i([\mathbf{k}_\alpha - \mathbf{k}_\beta]\cdot\mathbf{r})} e^{i[\delta_\alpha(b)+\delta_\beta(b)]}. \tag{7.44}$$

Equation 7.44 suggests that the L-space expansion of the DW matrix element be approximated by multiplying Born matrix elements by the distortion factors $\exp i[\delta_{\alpha L} + \delta_{\beta L}]$. Such approximations are discussed in [22, 43].

However, it seems pointless to derive approximate expressions in L-space by a succession of crude reductions of the **r**-space high-energy approximation. More direct L-space methods are discussed in Section 7.4.

7.4 Applications of *L*-space Models: Austern-Blair Theory

L-space wavefunction models are based on analytic models of the reflection coefficients η_L. In this section these models are used to develop approximate theories of inelastic scattering. Some remarks about rearrangement reactions are also given; however, the application of L-space models for rearrangement is more speculative than for inelastic scattering. Elastic scattering is not discussed. It is noted already (Section 7.1 and the beginning of 7.2) that the application of L-space models to elastic scattering is trivial and gives excellent agreement with experiment.

In developing the theory of inelastic scattering we follow the methods of Austern and Blair [45] as modified by Hahne [46]. The initial derivation of these methods used the adiabatic approximation and gave unified treatments of all orders in the projectile-nucleus interaction. The present discussion modifies this order of development somewhat. L-space models of first-order excitations are treated first, and are developed as approximations of the usual DW expressions. Models of higher-order excitations require additional approximations (such as the adiabatic approximation), therefore these are treated in a subsequent step.

L-space models are especially appropriate for macroscopic, collective theories of inelastic scattering like those discussed in Section 5.3. There a multipole expansion is used to express the general first-order DW amplitude of (5.40) in terms of reduced amplitudes $\beta_{sj}{}^{lm}$,

$$(2l+1)^{1/2} i^l \beta_{sj}{}^{lm} = \int d^3r \chi_\beta^{(-)*}(\mathbf{k}_\beta, \mathbf{r}) F_{lsj}(r)\, Y_l^{m*}(\hat{r}) \chi_\alpha^{(+)}(\mathbf{k}_\alpha, \mathbf{r}), \tag{5.42}$$

that depend on form factors $F_{lsj}(r)$. The calculation of these reduced amplitudes is discussed in partial-wave expansion,

$$\beta_{sj}{}^{lm} = \sum_{L_\alpha, L_\beta} i^{L_\alpha - L_\beta - l} e^{i\sigma_{\alpha L_\alpha} + i\sigma_{\beta L_\beta}} \sqrt{2L_\beta + 1}\, I^{lsj}_{L_\beta L_\alpha}$$
$$\times \langle L_\beta l; 00 \mid L_\alpha 0 \rangle \langle L_\beta l; -mm \mid L_\alpha 0 \rangle Y_{L_\beta}{}^{-m}(\Theta, 0), \tag{5.6}$$

in terms of radial integrals

$$I^{lsj}_{L_\beta L_\alpha}(k_\beta, k_\alpha) = \frac{4\pi}{k_\beta k_\alpha} \int_0^\infty f_{\beta L_\beta}(k_\beta, r) F_{lsj}(r) f_{\alpha L_\alpha}(k_\alpha, r)\, dr. \tag{5.5}$$

The form factors in the macroscopic theories always are derivatives of the optical potential with respect to some parameter h. (Usually this parameter is the nuclear radius.) Thus

$$F_{lsj}(r) = \frac{\partial U}{\partial h}, \tag{7.45}$$

where h is the optical potential parameter pertinent to the inelastic interaction of interest. We now consider approximate analytic methods for the evaluation of the radial integrals of (5.5). Because we are emphasizing spin-independent interactions, the indices sj are henceforth omitted.

The Austern-Blair theory is based on the observation that radial integrals of similar structure arise in perturbative analyses of *elastic scattering*. Let us again suppose h is a parameter in the optical potential, and let us suppose α is a scalar increment of h, so that the modification $h \to h + \alpha$ preserves spherical symmetry. Then the perturbed optical potential is

$$U(h + \alpha, r) = U_0 + \alpha \frac{\partial U}{\partial h} + \cdots. \tag{7.46}$$

The reflection coefficients η_L also are functions of h. It is straightforward to show that the expansion

$$\eta_L(h + \alpha) = \eta_L(h) + \alpha \left(\frac{\partial \eta_L}{\partial h}\right) + \cdots, \tag{7.47}$$

has the first-order coefficient

$$\left(\frac{iE}{2k}\right)\left(\frac{\partial \eta_L}{\partial h}\right) = \int_0^\infty f_L(k, r) \frac{\partial U}{\partial h} f_L(k, r)\, dr, \tag{7.48}$$

with the radial wavefunctions $f_L(k, r)$ in (7.48) computed with the unperturbed potential U_0. For convenience we express (7.48) in new notation

$$I_{LL}(k, k) \equiv \left(\frac{4\pi}{k^2}\right) \int_0^\infty f_L(k, r) \frac{\partial U}{\partial h} f_L(k, r)\, dr, \qquad (7.49)$$

to obtain

$$I_{LL}(k, k) = \left(\frac{i\pi\hbar^2}{k\mu}\right)\left(\frac{\partial \eta_L}{\partial h}\right). \qquad (7.50)$$

Equation 7.50 relates the η_L parameters of elastic scattering to radial integrals very much like those required for inelastic scattering.

It would be possible to obtain the complete set of radial integrals for inelastic scattering if some technique could be found to extrapolate the integrals $I_{LL}(k, k)$ to the conditions $L_\beta \neq L_\alpha$ and $k_\beta \neq k_\alpha$. Hahne discovered [46] that a remarkably accurate extrapolation could be based on the geometric mean

$$I^l_{L_\beta L_\alpha}(k_\beta, k_\alpha) \approx [I_{L_\beta L_\beta}(k_\beta, k_\beta) I_{L_\alpha L_\alpha}(k_\alpha, k_\alpha)]^{1/2}, \qquad (7.51)$$

with the result

$$I^l_{L_\beta L_\alpha}(k_\beta, k_\alpha) \approx \frac{i\pi\hbar^2}{\mu(k_\beta k_\alpha)^{1/2}} \left[\frac{\partial \eta_{L_\beta}(k_\beta)}{\partial h}\right]^{1/2} \left[\frac{\partial \eta_{L_\alpha}(k_\alpha)}{\partial h}\right]^{1/2}. \qquad (7.52)$$

Hahne subjected (7.51) to extensive numerical tests, and found it to be more than adequate for the development of models of inelastic scattering.

To understand the accuracy of Hahne's extrapolation we may imagine the inelastic scattering interaction to be localized in the vicinity of one radius, $r = R$. Under this condition the integrals are proportional to the integrands evaluated at $r = R$, and (7.51) becomes an identity! Therefore Hahne's approximation is exact under a condition that is fulfilled with great accuracy for strongly-absorbed projectiles, and that is often fulfilled fairly well for weakly-absorbed projectiles.

Equation 7.52 may be applied as it stands by performing optical model calculations of the coefficients η_L and by numerically differentiating these coefficients with respect to h. Such a procedure may occasionally be simpler than a full DW calculation of the integrals $I^l_{L_\beta L_\alpha}(k_\beta, k_\alpha)$.

More interesting applications follow if analytic models of the η_L are inserted in (7.52). Of course, our analytic models [see (7.4, 5)] give η_L as a function of L. However, for certain parameters h, differentiation with respect to h bears a close equivalence to differentiation with respect to L. Such an equivalence between $\partial/\partial h$ and $\partial/\partial L$ is found in cases of strong absorption if h is the nuclear radius. With strong absorption the shape of the η_L function tends to depend only on the displacement of L from the cutoff L_0. The R-dependence of η_L therefore tends to be determined by the R-dependence of

L_0 according to the rule of (7.2). Although (7.2) relates L_0 to R_{abs}, in the differential form of this equation we are free to replace increments of R_{abs} by increments of the optical-potential radius, as is noted in connection with (7.21) and (7.23). It therefore follows that for strong absorption the physics of the η_L function allows the relation

$$\frac{\partial \eta_L}{\partial R} \approx -k \frac{\partial \eta_L}{\partial L}. \tag{7.53}$$

The use of (7.53) in (7.52) allows the radial integrals of inelastic scattering to be derived from analytic models of the L-dependence of η_L. Thereupon we obtain *smoothed-cutoff L-space theories* of inelastic scattering.

The above approximate theory of inelastic scattering has much the same spirit as the DW theory, however, it does not make explicit use of potentials. To apply the theory we first carefully fit an η_L function to the observed elastic scattering. We then introduce (7.52) and (7.53) in the partial-wave expansion of (5.6), and compute the cross sections for inelastic scattering. An excellent example of this procedure was given by Springer and Harvey [47], who studied $Ca^{40}(\alpha, \alpha')$ at 50.9 MeV bombarding energy. They used the extrapolation procedure of [45], rather than Hahne's improved procedure. Nevertheless, their computed cross sections were in very good agreement with experiment. The agreement was as good as is normally yielded by a DW theory. Evidently a theory based directly on η_L correctly expresses the relation between elastic and inelastic scattering. There is no need to go through the complicated and ambiguous DW procedure of developing an optical potential fit to the elastic scattering.

If the sharp cutoff model of η_L [see (7.1)] is inserted in the Austern-Blair theory we of course recover [45] the Fraunhofer formulas of the extreme **r**-space theory.

L-space theories that directly use the partial-wave expansion for numerical calculations are said to be in "open form." However, it is often possible to sum the partial-wave series in good approximation to obtain compact expressions in "closed form", these allow easy understanding of certain aspects of inelastic cross sections. Such summations are developed and applied in [14, 44, 46, 48]. They are not discussed in this book.

Bassichis and Dar [44] derive their closed-form version of the Austern-Blair theory from the high-energy approximation (see Section 7.3) as well as by summation of an L-space theory. However, under most conditions of interest in nuclear physics the approximations of the high-energy method are more drastic than those of the above L-space discussion. Therefore the L-space derivation is the more fundamental of the two.

We now go on to higher-order inelastic scattering [45]: It is convenient to refer back to (3.38) for an exact expression for the transition amplitude for inelastic scattering,

$$T_{\alpha\beta} = \langle \chi_\beta^{(-)}(\mathbf{k}_\beta, \mathbf{r})\psi_B(y)|\, \Delta U\, |\Psi^{(+)}(\mathbf{r}, y)\rangle, \tag{7.54}$$

where

$$\Delta U \equiv U(h + \alpha, r) - U_0, \tag{7.55}$$

and α is a y-dependent increment of an optical-potential parameter h. To expand $T_{\alpha\beta}$ in orders of excitation, we use (3.61) to expand $\Psi^{(+)}$ in powers of ΔU, and we then expand ΔU in powers of α. The second-order term of $T_{\alpha\beta}$ is found to be

$$T_{\alpha\beta}(2) = \langle \chi_\beta^{(-)}(\mathbf{k}_\beta, \mathbf{r}_1)\psi_B(y)|\, \tau_2(\mathbf{r}_1, \mathbf{r}_2)\, |\chi_\alpha^{(+)}(\mathbf{k}_\alpha, \mathbf{r}_2)\, \psi_A(y)\rangle, \tag{7.56}$$

with

$$\tau_2(\mathbf{r}_1, \mathbf{r}_2) = \frac{1}{2}\left[\alpha^2(y, \hat{r}_1)\delta(\mathbf{r}_1 - \mathbf{r}_2)\frac{\partial^2 U}{\partial h^2}\right] + \left[\alpha(y, \hat{r}_1)\frac{\partial U}{\partial h}\right] G_0(\mathbf{r}_1, \mathbf{r}_2)\left[\alpha(y, \hat{r}_2)\frac{\partial U}{\partial h}\right], \tag{7.57}$$

and with the distorted-waves Green's function

$$G_0 = (E^+ - T - H_t(y) - U_0)^{-1}. \tag{7.58}$$

The y-dependence of G_0 may be eliminated immediately by use of the familiar adiabatic approximation. We replace $H_t(y)$ by the eigenvalues ε_A and ε_B for the states ψ_A and ψ_B. This is done in the symmetrical form

$$G_0 \approx \tfrac{1}{2}(G_A + G_B), \tag{7.59}$$

$$G_A = (E^+ - T - \varepsilon_A - U_0)^{-1}, \tag{7.60a}$$

$$G_B = (E^+ - T - \varepsilon_B - U_0)^{-1}. \tag{7.60b}$$

At this stage of reduction the second-order amplitude begins to resemble expressions found for elastic scattering.

The most complicated aspect of the second-order operator of (7.57) is that nuclear coordinates and particle coordinates are intermingled in the operator α. Under the adiabatic approximation the nuclear coordinates commute with the reduced G_0. However, the particle coordinates of α do not commute with G_0, and this prevents us from gathering together the nuclear operators and extracting them in a simple reduced matrix element. The noncommutativity of α and G_0 has the interpretation that the two steps by which the projectile excites nuclear surface modes take place at two different locations.

Great simplification is obtained if we disregard such propagation in intermediate states and allow α to commute with G_0. Equation 7.56 reduces to

$$T_{\alpha\beta}(2) \approx \tfrac{1}{2}\iint \chi_\beta^{(-)*}(\mathbf{k}_\beta, \mathbf{r}_1)\{\omega_{2B}\langle B|\, \alpha^2\, |A\rangle + \langle B|\, \alpha^2\, |A\rangle\omega_{2A}\}\, \chi_\alpha^{(+)}(\mathbf{k}_\alpha, \mathbf{r}_2)\, d^3r_1\, d^3r_2, \tag{7.61}$$

where

$$\omega_{2A,B} = \tfrac{1}{2}\delta(\mathbf{r}_1 - \mathbf{r}_2)\frac{\partial^2 U}{\partial h^2} + \frac{\partial U}{\partial h} G_{A,B} \frac{\partial U}{\partial h}. \tag{7.62}$$

It is now easy to see that (7.61) is the derivative with respect to h of the corresponding amplitude for single excitation

$$T_{\alpha\beta}(2) = \frac{1}{2}\frac{\partial}{\partial h}\int \chi_\beta^{(-)*}(\mathbf{k}_\beta, \mathbf{r})\langle B|\, \alpha^2\, |A\rangle \frac{\partial U}{\partial h}\, \chi_\alpha^{(+)}(\mathbf{k}_\alpha, \mathbf{r})\, d^3r. \tag{7.63}$$

If we assume nucleus A to have spin zero, the nuclear matrix element in (7.63) becomes

$$\langle B|\, \alpha^2\, |A\rangle = C_2(I)\, Y_I^{M_I *}(\hat{r}), \tag{7.64}$$

where I, M_I are the spin and projection of nucleus B, and $C_2(I)$ is a reduced matrix element. Therefore the amplitude for second-order excitation finally becomes

$$T_{\alpha\beta}(2) = \tfrac{1}{2}C_2(I)\frac{\partial}{\partial h}\int \chi_\beta^{(-)*}(\mathbf{k}_\beta, \mathbf{r})\, Y_I^{M_I *}\frac{\partial U}{\partial h}\, \chi_\alpha^{(+)}(\mathbf{k}_\alpha, \mathbf{r})\, d^3r. \tag{7.65}$$

Equation 7.65 may be evaluated by computing the DW amplitude for first-order excitation and differentiating with respect to h, or it may be expressed in terms of second derivatives of η_L by inserting (7.52) for the first-order DW amplitude. The latter procedure yields L-space models of double excitation.

The accuracy of (7.65) has been tested numerically [45] by comparing it with the results of coupled-channel calculations. The comparisons suggest that (7.65) is primarily of qualitative value. The origins of the errors of (7.65) can be seen very easily in the second-order amplitude of (7.57). The first term of (7.57) describes a "direct" second-order transition $A \rightarrow B$ that appears because the optical potential has a second-order dependence on α. The second term of (7.57) describes an "indirect" second-order transition caused by iteration of the first-order term of the optical potential. Now both the approximations used to reduce (7.56) to (7.65) concerned the indirect term of (7.57). Because there is considerable cancellation between the direct and indirect second-order terms (Chapter 6, [10]), the rather innocuous errors of the indirect term become of importance in the net second-order amplitude.

However, the qualitative merits of (7.65), such as they are, probably are all we should desire from any theory based on a macroscopic, collective dynamical model of the $A \to B$ excitation. It is noted in Section 6.3 that nuclear structure studies seldom substantiate such a restricted dynamical model of double excitation. Sometimes the macroscopic picture is very accurate (Chapter 6, [14]). However, in general this is not true, and to achieve accuracy it is necessary to conduct independent microscopic analyses of the direct and indirect modes of double excitation and to use a coupled-channels calculation to discuss their interference. Nevertheless, Springer and Harvey [49] were able to utilize the qualitative properties of (7.65) to disentangle competing single and double excitation in $Ne^{20}(\alpha, \alpha')$ transitions at 50.9 MeV bombarding energy.

Mutual excitation of two colliding heavy ions is a second-order process that is easily treated by the Austern-Blair theory, and that is described by a formula similar to (7.65). In this process the two colliding ions experience simultaneous single-excitations. Coupled-channels descriptions of this process are not easy to construct. Frahn [14] reviews the theory of mutual excitation and describes several applications. Bassel, Satchler, and Drisko [50] present several quantitative applications of this theory.

The *polarizability* modification of elastic scattering is another interesting second-order process that may be treated with (7.65). This process adds to the elastic amplitude a small term caused by virtual transitions to the collective excited states and back again. The modified elastic amplitude is most conveniently expressed in terms of modified reflection coefficients [45]

$$\eta_L' \approx \left\{1 + \langle\alpha^2\rangle\left[-\frac{1}{R}\frac{\partial}{\partial h} + \frac{1}{2}\frac{\partial^2}{\partial h^2}\right]\right\}\eta_L, \tag{7.66}$$

where

$$\langle\alpha^2\rangle = (4\pi)^{-1}\langle\psi_A|\sum_{k\neq 0,q} |\xi_{kq}|^2\,|\psi_A\rangle \tag{7.67}$$

is the ground-state expectation of the square of the nuclear deformation. To employ (7.66) we of course use (7.53) for $\partial/\partial h$, and we use an L-space model (7.4, 7.5) for η_L. Let us consider the simplest special case of (7.4), with $\varepsilon = \mu = 0$. It is found from (7.66) that for this case

$$\eta_L' \approx \left\{1 + k^2\langle\alpha^2\rangle\left[\left(\frac{1}{kR}\right)\frac{\partial}{\partial L} + \frac{1}{2}\frac{\partial^2}{\partial L^2}\right]\right\}\eta_L, \tag{7.68}$$

$$= \eta_L + \eta_L\Delta^{-2}k^2\langle\alpha^2\rangle e^x \operatorname{sech} x\{(2kR)^{-1} + \tfrac{1}{4}\tanh x\}, \tag{7.69}$$

where

$$x = \frac{L_0 - L}{2\Delta}.$$

It is clear that the correction term in (7.69) causes η'_L to be smoother than η_L, as if the zero-point deformations of the nucleus made the surface more diffuse.

It is necessary to estimate $\langle\alpha^2\rangle$ in order to estimate the numerical value of the correction term in (7.69). From other experimental data it is known [45] that $\langle\alpha^2\rangle$ is typically of the order 0.1 fm^2, or larger. Therefore, for the elastic scattering of 40 MeV alpha particles we have

$$k^2\langle\alpha^2\rangle \approx 1,$$

and it is clear that the polarizability correction to η_L is sizeable.

It is desirable to keep (7.68) in mind when making applications of L-space theories of nuclear reactions. It is the function η'_L that is obtained in fits to elastic scattering, but it is the function η_L that enters in the reaction theories. While (7.68) shares the quantitative difficulties of (7.65), it does allow an easy discussion of the discrepancy between η_L and η'_L.

The central property of the entire above discussion of L-space models of inelastic scattering is that in an L-space representation the components of the transition amplitude are confined to a *narrow band of orbital angular momenta* around a critical value L_0. This property holds accurately for the inelastic scattering of strongly-absorbed particles, and for this case its origin is clearly understood (Section 5.3). (The reaction generally is more sharply localized in L than in r.) We note that a similar localization in L is obtained by multiplying the Born-approximation components of T by the reflection coefficients η_L, as suggested in connection with (7.42) and (7.43). The product $(\eta_L M_L)$ vanishes for $L < L_0$ because η_L vanishes, and it vanishes for $L > L_0$ because centrifugal repulsion causes M_L to vanish. Hence, although the simple expression $(\eta_L M_L)$ lacks the quantitative accuracy of the approach discussed in this section, it does correctly express the L-space localization.

Rearrangement reactions that involve strongly-absorbed projectiles show the same L-space localization as is discussed for inelastic scattering. The localization is sharp when the energies and masses of the projectiles allow $k_\alpha \approx k_\beta$. It is true that rearrangement reactions excite coordinates of individual nucleons, and the discussion of these reactions therefore goes beyond a mere discussion of the shape and diffuseness of the nuclear surface. Thus the rearrangement amplitude cannot be derived as a consequence of the elastic scattering amplitude. However, the localization in L-space certainly is caused by the elastic scattering in the entrance and exit channels. For this reason it is plausible that the localization of the radial integrals for the rearrangement amplitude should be related to the η_L coefficients in much the same way that is found for inelastic scattering [14, 51]:

$$I^l_{L_\beta L_\alpha}(k_\beta, k_\alpha) \approx (\text{coeff})\left[\frac{\partial\eta_{\beta L_\beta}}{\partial L_\beta}\right]^{1/2}\left[\frac{\partial\eta_{\alpha L_\alpha}}{\partial L_\alpha}\right]^{1/2}. \tag{7.70}$$

Here the channel labels β, α refer to particle types as well as excitation energies. Frahn reviews [14] several applications of formulas like (7.70). A more explicit formula, based on high-energy approximations, is alluded to in the discussion following (7.42) and (7.43):

$$I^l_{L_\beta L_\alpha}(k_\beta, k_\alpha) \approx (\eta_{\beta L_\beta}\eta_{\alpha L_\alpha})^{1/2} M^l_{L_\beta L_\alpha}, \tag{7.71}$$

where $M^l_{L_\beta L_\alpha}$ is the Born approximation matrix element for the rearrangement. While (7.70) and (7.71) give equivalent descriptions of L-space localization of the radial integrals, (7.71) also gives information about their magnitudes. Both these formulas are presented on purely heuristic grounds.

References

[1] R. M. Eisberg, I. E. McCarthy, and R. A. Spurrier, *Nucl. Phys.* **10,** 571 (1959); I. E. McCarthy, *Nucl. Phys.* **10,** 583 (1959); I. E. McCarthy, *Nucl. Phys.* **11,** 574 (1959).

[2] I. E. McCarthy and D. L. Pursey, *Phys. Rev.* **122,** 578 (1961).

[3] I. E. McCarthy, *Phys. Rev.* **128,** 1237 (1962).

[4] K. A. Amos, *Nucl. Phys.* **77,** 225 (1966); see also Chap. 5, Ref. 14.

[5] A. J. Kromminga and I. E. McCarthy, *Nucl. Phys.* **24,** 36 (1961).

[6] K. A. Amos and I. E. McCarthy, *Phys. Rev.* **132,** 2261 (1963).

[7] G. Placzek and H. A. Bethe, *Phys. Rev.* **57,** 1075 (1940); A. Akhiezer and I. Pomeranchuk, *Jour. Phys.* (USSR) **9,** 471 (1945).

[8] J. S. Blair, *Phys. Rev.* **95,** 1218 (1954); D. D. Kerlee, J. S. Blair, and G. W. Farwell, *Phys. Rev.* **107,** 1343 (1957).

[9] J. S. Blair, *Phys. Rev.* **108,** 827 (1957).

[10] K. R. Greider and A. E. Glassgold, *Ann. Phys.* **10,** 100 (1960).

[11] J. A. McIntyre, K. H. Wang, and L. C. Becker, *Phys. Rev.* **117,** 1337 (1960).

[12] L. R. B. Elton, *Nucl. Phys.* **23,** 681 (1961).

[13] W. E. Frahn and R. H. Venter, *Ann. Phys.* **24,** 243 (1963).

[14] W. E. Frahn, in *Fundamentals in Nuclear Theory*, IAEA, Vienna (1967), and references therein.

[15] N. Austern, A. Prakash, and R. M. Drisko, *Ann. Phys.* **39,** 253 (1966).

[16] T. E. O. Ericson, in *Preludes in Theoretical Physics*, A. de-Shalit, H. Feshbach, L. Van Hove, Eds., North-Holland, Amsterdam, 1965.

[17] E. V. Inopin, *JETP* **48,** 1620 (1965), [Eng. transl: *Soviet Phys. JETP* **21,** 1090 (1965)].

[18] C. R. Gruhn and N. S. Wall, *Nucl. Phys.* **81,** 161 (1966).

[19] N. Austern, *loc. cit.*, Chap. 5, Ref. 33.

[20] R. J. Glauber, *loc. cit.*, Chap. 4, Ref. 8.

[21] N. Austern, *loc. cit.*, Chap. 5, Ref. 35.

[22] E. Henley, in *Proceedings of the Summer Study Group*, Brookhaven National Laboratory, BNL 948(C-46), Brookhaven (1965), (unpublished).
[23] J. S. Blair, in *Lectures in Theoretical Phys.*, Vol. VIII C, P. D. Kunz, D. A. Lind, W. E. Brittin, Eds., University of Colorado Press, Boulder, 1966.
[24] H. Feshbach, *International School of Physics "Enrico Fermi,"* XXXVIII Course, T. E. O. Ericson, Ed. Academic, New York, 1967.
[25] C. Wilkin, lectures at the C. A. P. Summer School, McGill University, Montreal (1967), to be published.
[26] S. I. Drozdov, *JETP* **28,** 734, 736 (1955) [Eng. transl.: *Soviet Phys. JETP* **1,** 791, 788 (1955)].
[27] E. V. Inopin, *JETP* **31,** 901 (1956) [Eng. transl.: *Soviet Phys. JETP* **4,** 764 (1957)].
[28] J. S. Blair, *Phys. Rev.* **115,** 928 (1959).
[29] A. Dar, *Nucl. Phys.* **55,** 305 (1964).
[30] E. M. Henley and D. U. L. Yu, *Phys. Rev.* **133,** B1445 (1964); *Phys. Rev.* **135,** B1152 (1964).
[31] H. A. Bethe, *Phys. Rev.* **57,** 1125 (1940).
[32] V. M. Strutinsky, *Nucl. Phys.* **68,** 221 (1965).
[33] W. Bierter, *Helv. Phys. Acta* **38,** 736 (1965).
[34] D. R. Yennie, F. L. Boos, Jr., and D. G. Ravenhall, *Phys. Rev.* **137,** B882 (1965).
[35] E. J. Squires, *Nucl. Phys.* **6,** 505 (1958). See also L. Egardt, *Nucl. Phys.* **11,** 349 (1959), *Nucl. Phys.* **12,** 84 (1959), and D. J. Hooton and G. R. Allcock, *Proc. Phys. Soc.* (London) **73,** 881 (1959).
[36] G. P. McCauley and G. E. Brown, *Proc. Phys. Soc.* (London) **71,** 893 (1958).
[37] E. A. Sanderson, *Nucl. Phys.* **26,** 420 (1961).
[38] H. S. Köhler, *Nucl. Phys.* **9,** 49 (1958).
[39] D. F. Jackson and L. R. B. Elton, *Nucl. Phys.* **43,** 136 (1963); D. F. Jackson, *Nucl. Phys.* **54,** 561 (1964); D. F. Jackson and T. Berggren, *Nucl. Phys.* **62,** 353 (1965).
[40] H. K. Lee and H. McManus, *Phys. Rev.* **161,** 1087 (1967).
[41] R. J. Glauber, in *High Energy Physics and Nuclear Structure*, North-Holland, Amsterdam, 1967.
[42] N. I. Sopkovich, *Nuovo Cim.* **26,** 186 (1962).
[43] K. Gottfried and J. D. Jackson, *Nuovo Cimento* **34,** 735 (1964).
[44] W. H. Bassichis and A. Dar, *Ann. Phys.* **36,** 130 (1966); S. Varma and A. Dar, *Ann. Phys.* **39,** 435 (1966).
[45] N. Austern and J. S. Blair, *Ann. Phys.* **33,** 15 (1965).
[46] F. J. W. Hahne, *Nucl. Phys.* **A104,** 545 (1967); *Nucl. Phys.* **A106,** 660 (1968).
[47] A. Springer and B. G. Harvey, *Phys. Letters* **14,** 116 (1965).
[48] J. M. Potgieter and W. E. Frahn, *Nucl. Phys.* **80,** 434 (1966); J. M. Potgieter and W. E. Frahn, *Phys. Letters* **21,** 211 (1966); F. J. W. Hahne, *Nucl. Phys.* **80,** 113 (1966); J. M. Potgieter and W. E. Frahn, *Nucl. Phys.* **A92,** 84 (1967).
[49] A. Springer and B. G. Harvey, *Phys. Rev. Letters* **14,** 316 (1965).
[50] R. H. Bassel, G. R. Satchler, and R. M. Drisko, *Nucl. Phys.* **89,** 419 (1966).
[51] W. E. Frahn and R. H. Venter, *Nucl. Phys.* **59,** 651 (1964).

CHAPTER 8

Spectroscopic Applications of Stripping

While nuclear reactions are interesting natural phenomena in their own right, they also are a source of much valuable data about nuclear bound states. Because direct reactions are simple, they are a particularly valuable source of bound-state data. As an example of this simplicity, we recall that in the DW theory the amplitude for a direct reaction is proportional to a single matrix element,

$$\langle \psi_B \psi_b | V_\beta - U_\beta | \psi_A \psi_a \rangle,$$

that links internal states of the colliding nuclei; it is practical to hope to measure this matrix element by fitting the DW theory to experiment. Coupled-channels theories involve several matrix elements of the above form; once again it is practical to hope to measure these matrix elements by reaction experiments.

The above matrix element contains bound-state wavefunctions of many-body systems. Therefore, interesting questions of *coherence* present themselves. For example, in the discussions of inelastic scattering in Chapter 5 we see that the bound-state matrix element contains a sum over all the nucleons of the target nucleus. Therefore inelastic scattering strongly favors collective excitations in which nucleons participate cooperatively. However, coherence effects in inelastic scattering are discussed extensively in Chapter 5, and there is no reason to continue this discussion in this chapter.

In nonexchange stripping reactions the nuclear matrix element only receives contributions from the few nucleons that are transferred. Coherent contributions from the other nucleons of the system are suppressed. For this reason stripping reactions have a close correspondence with shell-model wavefunctions. This correspondence is explored and developed in this chapter. Only single-nucleon transfer is treated. The special effects caused by the correlations in two-nucleon transfer are discussed at sufficient length in Section 5.11.

Sections 8.1–8.3 develop a purely shell-model theory of the bound-state matrix element. In this theory the wavefunction of the transferred nucleon is approximated, according to (5.117), as a normalized shell-model single-particle state multiplied by a spectroscopic coefficient. A number of simple formulas and sum rules for spectroscopic coefficients are developed.* Improved theories for the wavefunction of the transferred nucleon are given in Section 8.4.

8.1 $\mathscr{S}$ Factors for Simple Cases

Stripping and pickup reactions measure the amplitudes with which individual single-particle states appear in nuclear wavefunctions. These amplitudes, in turn, tend to measure the coefficients in the expansion of nuclear wavefunctions in products of shell-model single-particle states. It is understandable that stripping has been closely associated with the development of shell-model theories.

If it is assumed that we know the radial shapes of shell-model single-particle states, then the purpose of stripping experiments is only to determine "spectroscopic coefficients" in the linear combinations that make up the many-body nuclear wavefunctions (5.117). For definiteness let us consider the reaction $A(d, p)B$. The insertion of (5.117) in (5.109) shows that a spectroscopic coefficient is precisely the projection of ψ_B onto the vector-coupled product of ψ_A and a single-particle state:

$$\mathscr{S}^{1/2}(\alpha_A J_A, lj \mid \alpha_B J_B) = \sqrt{N_A + 1}\langle\{\psi_{nlj}(n_0), \psi_{\alpha_A J_A}\}_{J_n M_n} \mid \psi_{\alpha_B J_B M_B}\rangle. \quad (8.1)$$

Here N_A is the total number of neutrons in nucleus A. We recall [see (5.124)] that comparison of the DW theory with experiment allows $\mathscr{S}$ values to be measured. Let us now consider the $\mathscr{S}$ values that correspond to a number of simple nuclear wavefunctions. The standard treatment of this subject was given by Macfarlane and French [1] and by French [2, 3]. Helpful reviews of applications have been given by Cohen [4].

It is essential to keep in mind that $\psi_{\alpha_A J_A M_A}$ and $\psi_{\alpha_B J_B M_B}$ are normalized and antisymmetrized. Hence these wavefunctions are linear combinations of Slater determinants. There is a coefficient $\sqrt{N_A + 1}$ in (8.1) because (N_A) particles are antisymmetrized in nucleus A, and ($N_A + 1$) particles are antisymmetrized in nucleus B.

We now observe [1] that inert groups of nucleons, whose total angular momentum quantum numbers are zero, have no influence on $\mathscr{S}$ factors. "Inert groups" of nucleons are any common factors in all terms of both the target-nucleus and the product-nucleus wavefunctions. Inert groups that have $J = 0$ do not participate in the spectroscopy of either nucleus, and do not participate in the transfer process. Their only link with the "active" nucleons

* For consistency the particle-number formalism (*second quantization*) is not used in Sections 8.1–8.3, despite its suitability for this type of discussion. It is used in most of the references that are cited.

of the system is through the antisymmetrization requirement. The commonest inert groups are closed shells.

Suppose n neutrons of nucleus B are active and have the normalized, antisymmetrized wavefunction $\psi^a_{\alpha_B J_B M_B}(0, 1, \ldots, n-1)$. Suppose n_i neutrons are inert and have the normalized, antisymmetrized wavefunction $\psi^i(n, \ldots, N_A)$, which has $J = 0$. Here we have used $N_B = N_A + 1 = n + n_i$. The full wavefunction of the product nucleus is obtained by antisymmetrizing the product $\psi^a_{\alpha_B J_B M_B}\psi^i$, and is

$$\psi_{\alpha_B J_B M_B}(0, 1, \ldots, N_A) = \left[\frac{n!\, n_i!}{(N_A+1)!}\right]^{1/2} \sum_P (-)^P P \psi^a_{\alpha_B J_B M_B}\psi^i, \tag{8.2}$$

where the P are permutations that exchange neutrons between $\psi^a_{\alpha_B J_B M_B}$ and ψ^i, as in (3.65)–(3.67). Likewise, $(n-1)$ neutrons of nucleus A are active and have the normalized, antisymmetrized wavefunction $\psi^a_{\alpha_A J_A M_A}(1, \ldots, n-1)$, so that the full wavefunction of the target nucleus is

$$\psi_{\alpha_A J_A M_A}(1, \ldots, N_A) = \left[\frac{(n-1)!\, n_i!}{N_A!}\right]^{1/2} \sum_P (-)^P P \psi^a_{\alpha_A J_A M_A}\psi^i. \tag{8.3}$$

We calculate the $\mathscr{S}$ factor by inserting (8.2, 3) in (8.1). Because any one permutation in (8.3) has nonzero overlap with only one permutation in (8.2), the $\mathscr{S}$ factor is

$$\mathscr{S}^{1/2}(\alpha_A J_A, lj \mid \alpha_B J_B) = \sqrt{n}\,\langle\{\psi_{nlj}(0), \psi^a_{\alpha_A J_A}\}_{J_B M_B} \mid \psi^a_{\alpha_B J_B M_B}\rangle. \tag{8.4}$$

The inert nucleons are seen to be irrelevant. Apart from the change in the coefficient, (8.4) differs from (8.1) only in the superscripts a, which remind us that inert nucleons are to be omitted from the calculation of $\mathscr{S}$. Henceforth these superscripts are omitted.*

In the simplest stripping reaction nucleus A is entirely composed of inert closed shells. These shells are not excited in nucleus B. Hence $n = 1$, and we find $\mathscr{S} = 1$ in agreement with our preconceptions.

Simple results also follow if B is a closed-shell nucleus, and if A is formed from B by removing one nucleon ("adding one hole") without exciting any of the others. In this case the overlap integral in (8.4) has the value unity. Because $n = 2j + 1$, we find

$$\mathscr{S} = 2j + 1.$$

To understand this expression it is best to read (8.4) backwards: namely, the probability for removing one nucleon from a filled shell is proportional

* It should be apparent from the above analysis that the coefficient $\sqrt{n}$ reduces again when detailed calculations are performed. Under general conditions of configuration mixing the coefficient that survives is $\sqrt{n_{\text{equiv}}}$, where n_{equiv} is the number of active nucleons in ψ_B that have the quantum numbers nlj.

to the number of nucleons it contains. However, we also note that the angular momentum weight factors in the stripping cross section [see (5.124)] are related to the above large $\mathscr{S}$ factor. Because of this relationship the (d, p) cross section for inserting the last neutron in the j-shell is the same as the (p, d) cross section for removing the first. This is in agreement with our preconceptions.

Of course, the general expressions for $\mathscr{S}$ when a single j-shell is active are more complicated than the ones found above for the two endpoints of the shell. Let us consider target and product nucleus wavefunctions that have "maximum pairing," that is, those with minimum seniority [5]. The odd nucleus of the pair A, B then has seniority $v = 1$, and the even nucleus has seniority $v = 0$. For transitions between such states of minimum seniority we find

$$\begin{aligned} \mathscr{S} &= n, \qquad \text{for} \qquad n\text{-even}, \\ &= 1 - \frac{n-1}{2j+1}, \qquad \text{for} \qquad n\text{-odd}, \end{aligned} \tag{8.5}$$

where n, as before, is the number of equivalent neutrons in nucleus B. Equations 8.5 predict strong fluctuations of ground state $\mathscr{S}$ values as a given j-shell is filled; for example, for $j = \frac{7}{2}$ the sequence of $\mathscr{S}$ values is

$$\mathscr{S} = 1, 2, \tfrac{3}{4}, 4, \tfrac{1}{2}, 6, \tfrac{1}{4}, 8.$$

To derive (8.5) we construct wavefunctions of the required seniorities, and we insert these wavefunctions in (8.4). (A simple method for the construction of $v = 0, 1$ wavefunctions is given in Appendix 4 of the textbook by R. G. Sachs [6].) There is no need to give the detailed calculation here.

The above sequence of $\mathscr{S}$ values is deceptive in some ways, and it is helpful to consider the associated sequence of stripping cross sections (5.124). The cross sections depend on the product of $\mathscr{S}$ and the angular-momentum weight factor $(2J_B + 1)/(2J_A + 1)$. For a $v = 0$ nucleus $J = 0$, for a $v = 1$ nucleus $J = j$. Therefore, for $j = \frac{7}{2}$ the sequence of cross sections obeys

$$\left(\frac{2J_B + 1}{2J_A + 1}\right)\mathscr{S} = 8, \tfrac{1}{4}, 6, \tfrac{1}{2}, 4, \tfrac{3}{4}, 2, 1.$$

Here we see that small cross sections appear if the target nucleus is odd. This is appropriate, because there is a large variety of ways in which the added neutron can couple with the odd nucleus, and only one of these has $v = 0$.

We note a slight generalization of (8.5). It is clear that both stripping and pickup link an even nucleus that has pure seniority $v = 0$ only to $v = 1$ wavefunctions in the adjacent odd nuclei. Hence if the wavefunctions of the odd nuclei have mixed seniorities, the $\mathscr{S}$ factors are given by (8.5), multiplied by the probabilities of the $v = 1$ parts of these wavefunctions.

Another simple example concerns the addition of a nucleon in a state ψ_{nlj} that is *not equivalent* to any of the states in $\psi_{\alpha_A J_A M_A}$. In this case

$$\psi_{\alpha_B J_B M_B} = b_1 \mathscr{A}\{\psi_{nlj}(0), \psi_{\alpha_A J_A}\}_{J_B M_B} + b_2 \{\text{other configurations}\}, \quad (8.6)$$

where $\mathscr{A}$ antisymmetrizes the vector-coupled product. The second term in (8.6) does not contribute to $\mathscr{S}$; it only "dilutes" the wavefunction. In the first term we easily see that antisymmetrization of a product of inequivalent orbitals inserts a normalization constant $1/\sqrt{n}$, which cancels the coefficient in (8.4). We find

$$\mathscr{S} = (b_1)^2. \quad (8.7)$$

As a slight generalization of this example, we consider that $\psi_{\alpha_A J_A M_A}$ is a linear combination of two terms

$$\psi_{\alpha_A J_A M_A} = a_1 \psi^{(1)}_{J_A M_A} + a_2 \psi^{(2)}_{J_A M_A}, \quad (8.8)$$

and that interaction with the added nucleon alters the coefficients in this linear combination, to give

$$\psi_{\alpha_B J_B M_B} = b_1 \mathscr{A}\{\psi_{nlj}(0), \psi^{(1)}_{J_A}\}_{J_B M_B} + b_2 \mathscr{A}\{\psi_{nlj}(0), \psi^{(2)}_{J_A}\}_{J_B M_B} + b_3 \{\text{other configurations}\}. \quad (8.9)$$

The substitution of (8.8) and (8.9) in (8.4) yields

$$\mathscr{S} = (a_1 b_1 + a_2 b_2)^2. \quad (8.10)$$

The two terms in (8.8) would often be two independent shell-model configurations that are mixed by the nuclear interaction. It would be quite typical for this mixture to be altered (polarized) by the addition of another nucleon, so that $a_1/a_2 \neq b_1/b_2$.

Equations 8.8–8.10 illustrate the general problem of configuration mixing among the active nucleons. In the general calculation ψ_A is a linear combination of orthonormal basis states, and ψ_B is formed by vector coupling these basis states with the added nucleon. When we come to calculate (8.4), each term of ψ_B overlaps with its basis state in ψ_A, and the full quantity $\mathscr{S}^{1/2}$ is the sum of the individual overlaps. Complications arise only if the added nucleon is equivalent to one or more nucleons in a basis state. The contributions that arise from basis states of this type must be evaluated by means of (8.5), or by means of similar formulas derived for higher seniorities.

The significance of basis states equivalent to the added neutron is most readily appreciated through the idea of *occupation numbers* of single-particle states. For brevity, let us denote the ψ_{nlj} single-particle state by the single symbol (j). Let us define normalized, antisymmetrized states $(j)^\nu$ that have ν equivalent particles in the j-shell and minimum seniority. Hence, for these states $v = 0$ if ν is even, $v = 1$ if ν is odd. We now consider as an example a target nucleus wavefunction of the form

$$\psi_{\alpha_A J_A M_A} = \sum_{k=0} a_k \mathscr{A}\{(j)^{2k+\delta}, \psi_k\}_{J_A M_A}, \tag{8.11}$$

where $\delta = 0, 1$ according to whether the target nucleus is even or odd. The functions ψ_k are normalized and antisymmetrized and contain orbitals not equivalent to (j). It is further understood that the ψ_k contain enough orbitals for there to be exactly $n - 1$ active nucleons in each term of (8.11), with $n - 1 \geqslant 2j$. We consider a corresponding product nucleus wavefunction of the form

$$\psi_{\alpha_B J_B M_B} = \sum_{k=0} a_k \mathscr{A}\{(j)^{2k+\delta+1}, \psi_k\}_{J_B M_B}, \tag{8.12}$$

where the coefficients a_k of (8.11) are used again. [Wavefunctions having the structure of (8.11) and (8.12) are important in pairing theory.] We define the *fullness* of the state (j) in these two wavefunctions to be the weighted averages

$$V_j^2(A) = (2j+1)^{-1} \sum_{k=0} a_k^2 (2k+\delta), \tag{8.13a}$$

$$V_j^2(B) = (2j+1)^{-1} \sum_{k=0} a_k^2 (2k+\delta+1). \tag{8.13b}$$

Evidently the fullness obeys $0 \leqslant V_j^2 \leqslant 1$. Corresponding quantities

$$U_j^2(A) = 1 - V_j^2(A), \tag{8.14a}$$

$$U_j^2(B) = 1 - V_j^2(B), \tag{8.14b}$$

are called the *emptiness* of the state (j) in the two given wavefunctions. We note that in the particle-number representation of shell-model wavefunctions (*second quantization*) the fullness V_j^2 is just $(2j+1)^{-1}$ times the expectation of the number operator for the single-particle state (j) in the given nuclear state.

The spectroscopic factor for the transition $A \to B$ is obtained by inserting (8.11) and (8.12) in (8.4) and making use of (8.5). We find

$$\mathscr{S}_j^{1/2} = \sum_{k=0} a_k^2 [2k+2]^{1/2}, \quad \text{for an odd target}, \tag{8.15a}$$

$$\mathscr{S}_j^{1/2} = \sum_{k=0} a_k^2 \left[1 - \frac{2k}{2j+1}\right]^{1/2}, \quad \text{for an even target}. \tag{8.15b}$$

To simplify these equations and to obtain convenient expressions for $\mathscr{S}_j$ itself we introduce approximations. It is necessary to assume that the non-vanishing values of a_k^2 concentrate closely in the vicinity of some $q = k_{\text{average}}$. We then expand the square roots to first order in $(k - q)$ and regroup terms, using the normalization condition

$$\sum_k a_k^2 = 1.$$

To first order in $(k - q)$ it is found that

$$\mathscr{S}_j \approx (2j + 1)V_j^2(B), \qquad \text{for an odd target,} \qquad (8.16a)$$

$$\mathscr{S}_j \approx U_j^2(A), \qquad \text{for an even target.} \qquad (8.16b)$$

Within the accuracy discussed, we see that measurements of spectroscopic factors for the single-particle state (j) give the probabilities that this state is occupied. Such knowledge of occupation probabilities gives interesting insights about nuclei A and B no matter how complicated their wavefunctions. The interpretation of $\mathscr{S}$ factors in terms of occupation probabilities has received widespread application [4, 7, 8].

The above derivation of the connection between $\mathscr{S}$ factors and occupation probabilities can be criticized at several steps. (a) Of course, to go from (8.15*a*, *b*) to (8.16*a*, *b*) we make an explicit assumption that the occupation of state (j) is large. By implication we also assume the occupation of state (j) is a slowly-varying function of mass number. (b) The assumption that nuclear properties vary slowly with mass number also appears at an earlier step, when a single set of coefficients a_i is used in (8.11) and (8.12). (c) The wavefunctions $\psi_{\alpha_A J_A M_A}$ and $\psi_{\alpha_B J_B M_B}$ can contain terms of other types than the ones used in (8.11) and (8.12). For example, ν equivalent particles in the j-shell can be coupled to give higher seniorities than the $v = 0, 1$ of the states $(j)^\nu$. Also, the numbers of equivalent particles in the various terms of one nuclear wavefunction need not be all odd or all even. Although effects of type (c) are known to be weak, they should not be overlooked.

The above criticisms suggest that (8.16*a*, *b*) may be so rough that they should be used only for preliminary analyses of experiments, in advance of careful theoretical treatments of nuclei A and B. Before arriving at such a harsh conclusion, we note that if the structures of nuclei A and B are described by pairing theory, the relations between $\mathscr{S}$ factors and occupation probabilities [7] are exactly those of (8.16*a*, *b*), without need for any further approximations. The fact that these relations are exact consequences of pairing theory is consistent with the rough approximations we used. Pairing theory itself is rough, and is based on approximations similar to those used to derive (8.16*a*, *b*). In conclusion, although (8.16*a*, *b*) are approximate, the widespread interest in pairing theory lends them great value. We see in the next section that sum rules yield more precise determinations of occupation probabilities.

8.2 Sum Rules

Individual $\mathscr{S}$ factors link pairs of nuclear states, with the result that a given $\mathscr{S}$ factor depends on the detailed structural properties of two nuclei. Hence a table of observed $\mathscr{S}$ factors is a complicated body of information about pairs of nuclei. By use of the completeness relations of quantum mechanics we develop sum rules that arrange this information into more useable forms. Sum rule methods, as French emphasizes [3], allow us to get at, "· · · those things that are of real interest, such as the strengths, centroid energies, and widths of various excitations of nuclei." Furthermore, sum rules generate quantities that pertain to one nucleus at a time.

Sum rules come about because $\mathscr{S}$ factors are quadratic in the matrix elements that link nucleus A with nucleus B. Hence the eigenfunctions of either nucleus enter in an $\mathscr{S}$ sum as a complete set of *intermediate states* in a ground-state matrix element of the other nucleus. We note that sum rules for pickup and for stripping yield different information: By summing with respect to stripping final states we obtain properties of nucleus A. By summing with respect to pickup final states we obtain properties of nucleus B.

By suitable insertion of weight factors it is possible to develop $\mathscr{S}$ sums for the evaluation of ground-state expectations of a variety of interesting operators. The weight factors may depend on the excitation energies of the states over which we sum, or they may depend on the angular momenta of these states. Such weighted sums are discussed later. It is convenient to begin with unweighted sums over stripping final states $\psi_{\alpha_B J_B M_B}$.

One typical sum over the $\mathscr{S}$ factors of (8.4) is

$$n \sum_{\alpha_B J_B M_B} \langle \{\psi_{nlj}, \psi_{\alpha_A J_A}\}_{J_B M_B} \mid \psi_{\alpha_B J_B M_B} \rangle \langle \psi_{\alpha_B J_B M_B} \mid \{\psi_{nlj}, \psi_{\alpha_A J_A}\}_{J_B M_B} \rangle$$

$$= \sum_{\alpha_B J_B M_B} \mathscr{S}(\alpha_A J_A, lj \mid \alpha_B J_B)$$

$$= \sum_{\alpha_B J_B} (2J_B + 1) \mathscr{S}(\alpha_A J_A, lj \mid \alpha_B J_B), \quad (8.17)$$

where the last form follows because the $\mathscr{S}$ factors do not depend on M_B. Here we sum over all J_B reached by coupling j with J_A, and over all energy eigenstates of nucleus B (labelled by the index α_B) with angular momentum J_B. The sum over J_B has more than one term only if $J_A \neq 0$. However, it is typical for the sum over α_B to include many terms. This comes about because the target is polarizable, that is, because the product term $\{\psi_{nlj}, \psi_{\alpha_A J_A}\}_{J_B M_B}$ usually is only one of many parents that contribute to the eigenfunctions of nucleus B. Because the added nucleon interacts with the nucleons of the target nucleus, this particular parent can be distributed among many eigenfunctions $\psi_{\alpha_B J_B M_B}$, and this distribution allows many nonvanishing $\mathscr{S}$ factors for the ψ_{nlj} orbital. A trivial example of such splitting of the spectroscopic strength is given in (8.8)–(8.10) and the accompanying discussion. In that example the strength is split among all states ψ_B that contain the terms $\psi^{(1)}_{J_A M_A}$ and $\psi^{(2)}_{J_A M_A}$.

The sum over the states $\psi_{\alpha_B J_B M_B}$ in (8.17) cannot, as it stands, be replaced by a unit operator. The states $\psi_{\alpha_B J_B M_B}$ are fully antisymmetrized. Therefore they span a smaller space than the products $\psi_{nlj}\psi_{\alpha_A J_A}$. This question of antisymmetrization complicates the calculation of the unweighted sums; it is also the origin of much of their interest.

The sum over the $\psi_{\alpha_B J_B M_B}$ can be replaced by a unit operator provided we first insert a projection operator $\mathscr{P}_A$ to select the antisymmetric parts of the products $\psi_{nlj}\psi_{\alpha_A J_A}$. Because the unit operator may be expressed in any representation, we introduce the convenient uncoupled representation, and obtain

$$\sum_{\alpha_B J_B} (2J_B + 1)\mathscr{S}(\alpha_A J_A, lj \mid \alpha_B J_B)$$
$$= n \sum_{m, M_A} \langle \psi_{nljm}\psi_{\alpha_A J_A M_A} \mid \mathscr{P}_A \psi_{nljm}\psi_{\alpha_A J_A M_A}\rangle. \quad (8.18)$$

Equation 8.18 is especially simple if ψ_{nlj} is not equivalent to any of the orbitals in $\psi_{\alpha_A J_A}$. In this case the operator $\mathscr{P}_A$ generates a factor $(1/n)$ that cancels the coefficient in (8.18). However, the sums over m and M_A are unrestricted. Hence we obtain

$$\sum_{\alpha_B J_B} (2J_B + 1)\mathscr{S}(\alpha_A J_A, lj \mid \alpha_B J_B) = (2j + 1)(2J_A + 1). \quad (8.19)$$

It is noteworthy that the sum in (8.19) contains the same statistical factor that appears in the stripping cross section. Therefore the sum rule may be applied even if J_B is unknown. French stresses [1, 2, 3] that the statistical factors always possess this convenient property.

When ψ_{nlj} is equivalent to orbitals already occupied in $\psi_{\alpha_A J_A}$, the $\mathscr{P}_A$ operator introduces a relation between the m sum and the M_A sum. Let us suppose $\psi_{\alpha_A J_A}$ is a linear combination of terms that contain various numbers of particles in the j-shell, as in (8.11). It is clear from (8.18) that the orthogonality of these terms causes them to contribute independently to the $\mathscr{S}$ sum.* Therefore we are free to examine the contribution in (8.18) of one term in $\psi_{\alpha_A J_A}$ with a definite number of particles q in the j-shell. Whatever the angular momentum coupling in this term, it may be expressed as a linear combination of Slater determinants, with each Slater determinant containing a definite set of projection quantum numbers for the q orbitals $\psi_{nlj\mu}$. Hence the operator $\mathscr{P}_A$ excludes q values of m from the sum in (8.18). We obtain

$$\sum_{\alpha_B J_B} (2J_B + 1)\mathscr{S}(\alpha_A J_A, lj \mid \alpha_B J_B) = (2J_A + 1)\sum_{q=0} a_q^2[2j + 1 - q], \quad (8.20)$$

* Independent terms in ψ_A always [1] contribute independently to the $\mathscr{S}$ sum, i.e., cross terms always cancel.

where the sum extends over all (odd and even) values of q. Equation 8.20 is conveniently expressed in terms of the emptiness parameter U_j^2 [see (8.14) for a definition of U_j^2 in the special case q-even] to give the standard result

$$\sum_{\alpha_B J_B} \left(\frac{2J_B + 1}{2J_A + 1}\right) \mathscr{S}(\alpha_A J_A, lj \mid \alpha_B J_B) = (2j + 1)U_j^2(A). \tag{8.21}$$

Equation 8.21 is exact, and is not based on pairing theory or on approximations related to those used in pairing theory, in contrast to the discussion of individual $\mathscr{S}$ factors in Section 8.1. Thus, summation over *all states* reached by inserting a particle in the j-shell yields quite precise information about ψ_A. Equation 8.19 is of course a special case of (8.21).

Equation 8.21 is the general nonenergy-weighted (NEW) monopole sum rule for stripping transitions, provided only nucleons of one type (neutrons or protons) are active. It has been extensively applied for the spectroscopic analysis of stripping experiments (e.g., see [4, 8]). One rather obvious caution about (8.21) should be made: Because of close two-body correlations in nuclei, part of the "strength" for inserting a particle in the j-shell may be displaced to very high excitation energies ($\sim$100 MeV), where it cannot be identified by stripping experiments. Since this effect may be partly compensated by corresponding ground-state correlations, the net resulting modification of the $\mathscr{S}$ sums is not obvious.

Equation 8.21 bears a close relation to the $\mathscr{S}$ factor of (8.16b), for the ground-state transition based on a seniority zero, even target nucleus. For this case pairing theory predicts that the entire $\mathscr{S}$ sum is concentrated on a single final state. However, (8.21) is very different from the $\mathscr{S}$ factor of (8.16a), for the ground state transition based on a seniority one, odd target nucleus. In this case the single, seniority-zero ground state of the product nucleus is not representative of the states in the complete $\mathscr{S}$ sum.

The monopole sum rule for pickup transitions is obtained from the $\mathscr{S}$ factors of (8.4) by constructing the sum

$$\sum_{\alpha_A J_A} \mathscr{S}(\alpha_A J_A, lj \mid \alpha_B J_B)$$
$$= n \sum_{\alpha_A J_A} \langle \psi_{\alpha_B J_B M_B} \mid \{\psi_{nlj}, \psi_{\alpha_A J_A}\}_{J_B M_B}\rangle \langle \{\psi_{nlj}, \psi_{\alpha_A J_A}\}_{J_B M_B} \mid \psi_{\alpha_B J_B M_B}\rangle. \tag{8.22}$$

To sum over the states of nucleus A it is necessary to undo the angular momentum coupling of the product $\{\psi_{nlj}, \psi_{\alpha_A J_A}\}_{J_B M_B}$. This is accomplished by extending (8.22) to the form

$$\sum_{\alpha_A J_A} \mathscr{S}(\alpha_A J_A, lj \mid \alpha_B J_B)$$
$$= n \sum_{\alpha_A J_A J' M'} \langle \psi_{\alpha_B J_B M_B} \mid \{\psi_{nlj}, \psi_{\alpha_A J_A}\}_{J'M'}\rangle \langle \{\psi_{nlj}, \psi_{\alpha_A J_A}\}_{J'M'} \mid \psi_{\alpha_B J_B M_B}\rangle, \tag{8.23}$$

where the additional terms that have $J'M' \neq J_BM_B$ do not contribute to the matrix elements. A change of representation is now introduced in (8.23), to transform the sum over $J'M'$ to a sum over the projections of j and J_A, and thereby to allow the completeness of the states $\psi_{\alpha_A J_A M_A}$ to be exploited. This procedure gives

$$\sum_{\alpha_A J_A} \mathscr{S}(\alpha_A J_A, lj \mid \alpha_B J_B) = n \sum_m \langle \psi_{\alpha_B J_B M_B} \mid \psi_{nljm}(n_0) \rangle \langle \psi_{nljm}(n_0) \mid \psi_{\alpha_B J_B M_B} \rangle. \tag{8.24}$$

Here the remaining sum over single-particle intermediate states serves as a projection operator, to select from $\psi_{\alpha_B J_B M_B}$ the terms in which particle n_0 occupies the j-shell. The sum over m picks up as many nonvanishing contributions as there are particles in the j-shell, with the result that (8.24) reduces to

$$\sum_{\alpha_A J_A} \mathscr{S}(\alpha_A J_A, lj \mid \alpha_B J_B) = (2j + 1)\, V_j^2(B), \tag{8.25}$$

where V_j^2 is the j-shell fullness parameter. Equation 8.25 is the general NEW monopole sum rule for pickup transitions. Once again, this equation is exact, and is not in any sense obtained from pairing theory. Once again the sum rule may be used in conjunction with measured cross sections, even if not all values of J_A are known, because the same angular momentum weight factors (none at all!) appear in (8.25) and in the pickup cross section.

Cohen has emphasized [4] that use of (8.25) in conjunction with (8.21) allows accurate determinations of U_j^2/V_j^2, a parameter that tends to be sensitive to the value of j.

Macfarlane and French [1] give modified forms of (8.21) and (8.25) that are applicable for cases in which neutrons and protons with the same nlj quantum numbers are active simultaneously. These modified NEW monopole sums have interesting applications; for example, for nuclei with a neutron excess. In this case insertion of a proton or removal of a neutron generally excites two widely separated groups of states that have different i-spin. Individual sums over these two groups bear simple relations to the numbers of active neutrons and protons [9]. These sum rules show that particle transfer reactions preferentially excite the states of lower i-spin. The sum rules have been used to resolve uncertainties of DW calculations [10].

Energy-weighted sum rules are obtained by constructing sums in which the $\mathscr{S}$ factors are multiplied by powers of the difference of the total energies of nuclei A and B. Such sums reduce to ground-state matrix elements, inasmuch as powers of $E_B - E_A$ may be absorbed into the $\mathscr{S}$ factors as powers of $H_B - H_A$. Here H_B is the full many-body Hamiltonian for the N_B particles of nucleus B, and H_A is the many-body Hamiltonian for the $N_A = N_B - 1$ particles of nucleus A. Now as the power of $E_B - E_A$ is increased, the weighted sums rapidly become more sensitive to experimental and theoretical uncertainties. The latter sensitivity comes about because matrix elements of powers of H depend on correlations in the ground-state wavefunctions. For these reasons the only weight factors that have received practical applications have been $E_B - E_A$ and $(E_B - E_A)^2$.

The linear-energy-weighted (LEW) stripping sum rule is constructed by inserting the coefficient $E_B - E_A$ in (8.17) and absorbing this coefficient as an operator in the matrix elements. This procedure gives

$$\sum_{\alpha_B J_B} (2J_B + 1)(E_B - E_A)\mathscr{S}(\alpha_A J_A, lj \mid \alpha_B J_B) = n \sum_{\alpha_B J_B M_B} \langle \{\psi_{nlj}, \psi_{\alpha_A J_A}\}_{J_B M_B} (H_B - E_A) \mid \psi_{\alpha_B J_B M_B}\rangle \times \langle \psi_{\alpha_B J_B M_B} \mid \{\psi_{nlj}, \psi_{\alpha_A J_A}\}_{J_B M_B}\rangle. \tag{8.26}$$

We treat the states $\psi_{\alpha_B J_B M_B}$ as a complete set, as before, taking care to insert a projection operator $\mathscr{P}_A$ to isolate the antisymmetric parts of the product states. The sum reduces to

$$\sum_{\alpha_B J_B} (2J_B + 1)(E_B - E_A)\mathscr{S}(\alpha_A J_A, lj \mid \alpha_B J_B) = n \sum_{m, M_A} \langle \mathscr{P}_A \psi_{nljm} \psi_{\alpha_A J_A M_A} \mid (H_B - E_A) \mathscr{P}_A \psi_{nljm} \psi_{\alpha_A J_A M_A}\rangle. \tag{8.27}$$

A more explicit expression is obtained by omitting $\mathscr{P}_A$ on one side of the matrix element and replacing E_A by a Hamiltonian H_A that operates on the coordinates in $\psi_{\alpha_A J_A M_A}$:

$$\sum_{\alpha_B J_B} (2J_B + 1)(E_B - E_A)\mathscr{S}(\alpha_A J_A, lj \mid \alpha_B J_B) = n \sum_{m, M_A} \langle \mathscr{P}_A \psi_{nljm} \psi_{\alpha_A J_A M_A} \mid (H_B - H_A) \psi_{nljm} \psi_{\alpha_A J_A M_A}\rangle. \tag{8.28}$$

Equations 8.27 and 8.28 are equivalent. These expressions give the expectation of the energy of a particle in the nlj state that interacts with the target nucleus in its ground state. This expectation takes full account of the complexities of the individual interactions and of the particle distributions in the target nucleus, no matter how complicated these may be.

If $\psi_{\alpha_A J_A M_A}$ is a closed shell nucleus, the sum role expressions of (8.27) and (8.28) give the single-particle energy ε_{nlj} of the state ψ_{nlj}. When several particles are active, (8.27) and (8.28) contain additional contributions from the residual interactions among the active particles [3, 8, 11]. It is found from pairing theory that (8.27) and (8.28) yield in good approximation the energy of a one "quasi-particle excitation" based on the state ψ_{nlj} [4, 8].

The linear-energy-weighted pickup sum rule is constructed by inserting $E_B - E_A$ in (8.22) and by absorbing this coefficient as an operator in the matrix elements. Angular momentum is decoupled, as in (8.23) and (8.24), and the sum over $\alpha_A J_A$ is performed. This procedure gives

$$\sum_{\alpha_A J_A} (E_B - E_A)\mathscr{S}(\alpha_A J_A, lj \mid \alpha_B J_B)$$
$$= n \sum_m \langle \psi_{\alpha_B J_B M_B} \mid \psi_{nljm}(n_0)\rangle\langle \psi_{nljm}(n_0) \mid (H_B - H_A)\psi_{\alpha_B J_B M_B}\rangle, \quad (8.29)$$

where H_A operates on all coordinates other than n_0. Once again, the remaining sum over single-particle intermediate states serves as a projection operator, which selects the parts of $\psi_{\alpha_B J_B M_B}$ in which particle n_0 occupies the j-shell. Once again, interactions among the active nucleons of nucleus B keep (8.29) from being quite the single-particle energy of ψ_{nlj}. It is interesting [3, 8, 9] that several discrepancies between the sum rules and the single-particle energies cancel if (8.28) and (8.29) are added together.*

The sum rules discussed thus far are based on unweighted sums over intermediate-state angular momenta. As a result they reduce to ground-state expectations of monopole operators. In effect, these unweighted sum rules extract only monopole information from a table of experimental $\mathscr{S}$ factors. On the other hand, French has observed [13, 3, 11, 12] that information about higher multipoles can be extracted, provided $J_A \neq 0$ in stripping experiments, or provided $J_B \neq 0$ in pickup experiments. Under these circumstances $\mathscr{S}$ sums with J-dependent weight factors can be constructed, such that the weighted sums reduce to ground-state expectations of operators of higher multipolarity. For example, sums with dipole weight factors reduce to ground-state expectations of the angular momentum operator or of the magnetic moment operator [14]. Suitable sum rules allow the isolation of contributions of particular orbitals to these nuclear properties.

As with energy-weighted sum rules, the multipole-weighted sum rules rapidly lose accuracy as the complexity of the weight factor increases. The weight factors in the multipole sum rules have both positive and negative signs.

8.3 Collective Nuclei

We now consider transitions among bound states ψ_A and ψ_B that are derived from rotational or vibrational collective models. An inclusive treatment of $\mathscr{S}$ factors for such cases was given by Satchler ([15], see also [1]).

* Equations 8.27–8.29 diverge if H_B and H_A contain "realistic" two-body interactions, which have hard cores. The divergence of these sums is correct [12]; it comes about because short-range correlations induced by the hard core displace part of the stripping strength to energies $E_B - E_A$ that are infinite. On the other hand, stripping experiments detect only the part of the strength that lies at low excitation. Presumably sums over this part of the strength should be associated with "effective" two-body interactions, which do not contain hard cores.

The present discussion is based on his treatment. A useful discussion of applications of these $\mathscr{S}$ factors was given by Macfarlane [16].

Nuclear deformation affects both the bound wavefunctions and the continuum wavefunctions. Modifications of the continuum wavefunctions are considered in Section 6.5, where they are discussed in terms of generalized distorted waves. Such generalized distorted waves incorporate inelastic couplings among states of the target nucleus or among states of the product nucleus. It is seen that the indirect stripping transitions allowed by this procedure tend to be weak.

Modifications of the bound wavefunctions affect the $\mathscr{S}$ factors. These effects of deformation must be treated, whether or not we use generalized distorted waves. It is seen that deformation effects in the $\mathscr{S}$ factors tend to be characteristic of the collective models.

Rotational models are characterized by deformed intrinsic wavefunctions whose orientation rotates adiabatically, so that all the states of one rotational band share the same intrinsic wavefunction. Under such a model, when a nucleon is inserted in a nucleus it is inserted in the intrinsic wavefunction; this insertion generates a new intrinsic wavefunction that is then the basis of a new band system. To calculate $\mathscr{S}$ factors it evidently is necessary to proceed in two steps: first the overlap of the intrinsic wavefunctions is computed, second the angular momenta of the individual states are utilized to transform to the space-fixed coordinates in which the reaction analysis takes place. It is important that deformed intrinsic wavefunctions contain a mixture of angular momenta. The $\mathscr{S}$ factors for individual states of a rotational band are proportional to individual components of this mixture. Hence stripping experiments can give rather direct determinations of intrinsic wavefunctions.

To calculate $\mathscr{S}$ factors for rotational nuclei, wavefunctions of the usual adiabatic rotational model [17] are inserted in (8.4). For cylindrically-symmetric deformations these wavefunctions are of the form

$$\psi_{KJM} = \left[\frac{(2J+1)}{16\pi^2(1+\delta_{K,0})}\right]^{1/2} \psi_{vib} \times \{\mathscr{D}_{MK}{}^{J}(\theta_i)\chi_K + \text{reflection}\}. \quad (8.30)$$

Here χ_K is the deformed intrinsic wavefunction, described in body-fixed coordinates, the θ_i are the Euler angles of the body-fixed axes relative to space-fixed axes, and the $\mathscr{D}_{MK}{}^{J}$ are rotation matrices. The second term of (8.30) is inserted to ensure definite parity. It is generated by reflecting the first term in the equatorial plane of the deformed nucleus. The quantum number K is the projection of the angular momentum of the intrinsic wavefunction on the nuclear symmetry axis. Because the nucleus is cylindrically symmetric, this projection has a well-defined value, even though the total angular momentum of the deformed intrinsic wavefunction is not well defined. Finally, the wavefunction ϕ_{vib} describes vibrations of the shape of the deformed rotator.

The intrinsic functions of the target and residual nuclei are written in greater detail as

$$\chi_{A,K_A}(y'), \qquad \chi_{B,K_B}(y', n_0'),$$

where y' denotes the coordinates of the nucleons in the target nucleus, relative to body-fixed axes, and n_0' denotes the coordinates of the transferred nucleon, relative to the body-fixed axes. These intrinsic functions generally are regarded as single Slater determinants of deformed single-particle orbitals, the possibility of linear combinations being ignored. In this picture χ_A and χ_B differ in the orbital of only one nucleon. The overlaps between the two terms of ψ_A and the two terms of ψ_B select for the orbital of the transferred nucleon the two possibilities

$$\int \chi^*_{A,\pm K_A} \chi_{B,\pm K_B} \, dy' = \chi_\Omega(n_0'), \tag{8.31}$$

where $|\Omega| = |K_A \pm K_B|$. The upper sign in Ω occurs if the single-particle orbitals in χ_A and χ_B are not introduced with maximum pairing, hence it occurs only for odd-odd nuclei or for highly excited bands.

A deformed single-particle orbital $\chi_{\alpha\Omega}(n_0')$, labelled by α, may be expanded in a spherical basis,* in terms of functions that have definite j:

$$\chi_{\alpha\Omega}(n_0') = \sum_{nlj} c_{nlj}(\alpha, \Omega) \psi_{nlj\Omega}(n_0'). \tag{8.32}$$

We now see that a given stripping transition picks out one component from the expansion of (8.32).

Insertion of the above expressions in (8.4) yields the $\mathscr{S}$ factors for transitions between states of cylindrically-symmetric deformed rotators:

$$\mathscr{S}(K_A J_A, lj \mid K_B J_B) = g^2 \left(\frac{2J_A + 1}{2J_B + 1} \right) \langle J_B K_B \mid J_A j; \mp K_A, K_B \pm K_A \rangle^2$$
$$\times \, c^2_{nlj}(K_B \pm K_A) \langle \phi_A \mid \phi_B \rangle^2, \tag{8.33}$$

where

$$g^2 = 1 + \delta(K_A, 0) + \delta(K_B, 0). \tag{8.34}$$

The Clebsch-Gordan factor in (8.33) represents the probability of finding the rotating target and the incident nucleon in correct relative alignment for capture. The coefficient c^2_{nlj} represents the probability that the orbital of the

* In Nilsson's calculations [18] a representation different from (8.32) is used. His basis functions are labelled by $nl\Lambda$, where Λ is the projection of l, and his coefficients are related to those above by a Clebsch-Gordan transformation

$$c_{nlj}(\alpha, \Omega) = \sum_{\Lambda} a_{nl\Lambda} \langle l\tfrac{1}{2}; \Lambda, \Omega - \Lambda \mid j\Omega \rangle.$$

transferred nucleon appears in the intrinsic wavefunction. The vibrational-overlap factor is more difficult to treat with accuracy. It tends to be absorbed in an "intrinsic $\mathscr{S}$ factor" if more complete models of χ_A and χ_B are developed [16]. Short of such more complete discussions we merely note [15, 1] that for transitions between ground-state bands the vibrational factor tends to be of the order unity. This factor conveniently cancels when $\mathscr{S}$ factors for the various states of a rotational band are compared.

The orthonormality properties of the Clebsch-Gordan factor in (8.33) allow us to deduce a sum rule. For the final states of a *given rotational band*, summation over all states J_B reached by capture with given l and j gives

$$\sum_{J_B}(2J_B + 1)\mathscr{S}(K_AJ_A, lj \mid K_BJ_B) = g^2(2J_A + 1)c_{nlj}^2(K_B \pm K_A). \quad (8.34$$

If (8.34) is now further summed over the contributions from different lj in the same band of levels, the normalization condition $\sum_{lj} c_{nlj}^2 = 1$ of the inserted orbital gives

$$\sum_{J_Blj}(2J_B + 1)\mathscr{S}(K_AJ_A, lj \mid K_BJ_B) = g^2(2J_A + 1). \quad (8.35)$$

Equation 8.35 may be used, for example, to check the normalization of DW calculations. We also observe in (8.35) that for most levels of the product nucleus we expect $J_B \gg J_A$. Therefore, it is clear, especially if many different l and j contribute to the sum, that individual rotational $\mathscr{S}$ factors must be small. Sheline and Shelton [19, 20], for example, noted that particle transfer excites levels above the pairing gap much more strongly than it excites the rotational levels in the gap.

A summary of early (d, p) experiments with rotational nuclei was given by Macfarlane [16]. Among targets for which $\mathscr{S}$ factors were studied we may list: at 15 MeV, Er^{167}, W^{182}, W^{183} by Isoya [21]; at 12 MeV, Dy^{163} by Shelton and Sheline [20]; at 12 MeV, Yb^{176}, Hf^{178} by Vergnes and Sheline [22]; at 7 MeV, Ta^{181} by Erskine and Buechner [23]; at 12 MeV, U^{235}, U^{238} by Macefield and Middleton [24]; at 12 MeV, W^{182}, W^{184}, W^{186} by Erskine [25]. On the whole, for even-even targets the $\mathscr{S}$ factors for levels of the ground state band agree within factors of 2 or 3 with the predictions of (8.33), using coefficients c_{nlj}^2 obtained from Nilsson's single-particle wavefunctions. Some indication that the discrepancies indicate deficiencies of the Nilsson wavefunctions and not of the reaction theory is contributed by calculation [22] of the decoupling parameter a. A calculation that uses the measured values of c_{nlj}^2 gives a more satisfactory value of a than one that uses Nilsson's parameter values. It is also of interest that predictions of the $\mathscr{S}$ factors for W^{183} must use [25] wavefunctions that incorporate band mixing.

A similar attempt to apply band-mixing ideas to the reaction $Ta^{181}(d, p)Ta^{182}$ was less successful [23]. Ta^{182} is an odd-odd nucleus. It seems that the stripping experiment is a severe test of our understanding of the structure of this nucleus.

The stripping experiments are performed at sufficiently high energies to excite a large portion of the spectrum of the residual nucleus, both in and above the pairing gap. Bands built on vibrations seem to be excited with intensities comparable to that of the ground state band [21]. An adequate theory of the $\mathscr{S}$ factors for these bands requires a more complete treatment of the intrinsic wavefunction than is given in the derivation of (8.33); for example, Kern, Mikoshiba, Sheline, Udagawa, and Yoshida [26] derive the intrinsic wavefunctions of γ-vibrational bands by applying pairing theory to Nilsson's deformed single-particle orbitals.

Satchler also developed expressions [15, 1] for stripping to excited levels of *spherical vibrating nuclei*. However, it has since been realized, as remarked by Kumar and Baranger [17], that the simple vibrational model seldom is applicable. Therefore there is no need to review this part of Satchler's analysis. In any case the analysis is simple enough. The vibration and the inserted nucleon are considered to be independent degrees of freedom of the product nucleus. (The exclusion principle is ignored!) The angular momenta carried by the nucleon and the vibration are appropriately vector coupled. Then the crucial result of the calculation, as we anticipate, is that $\mathscr{S} \neq 0$ only if the target and product nuclei are at least to some extent in the same vibrational state. Hence $\mathscr{S}$ factors for excited states measure vibrational admixtures. It is not surprising that Kenefick and Sheline [27] found that $\mathscr{S}$ factors in the 12 MeV (d, p) reactions on $Sm^{147,149}$ are not in agreement with this simple theory. It seems more pertinent to employ microscopic analyses of the vibrational states [8].

8.4 Radial Form Factors

The concept of spectroscopic factors is meaningful only if single-particle wavefunctions are assumed to have the same radial shapes in all nuclear energy states. If this condition is fulfilled, nuclear structure theory is purely a matter of constructing and understanding linear combinations of small numbers of single-particle wavefunctions, and $\mathscr{S}$ factors are the meeting ground between reaction theory and nuclear structure theory.

Single-particle functions can have altered radial shapes only if several nucleons are active, and if the residual interactions among these nucleons mix basis functions that have different principal quantum numbers. Throughout most of the nuclear volume it seems correct to ignore this effect and to assume that single-particle radial wavefunctions are not altered by residual interactions. This is the normal assumption of nuclear structure calculations. However, this assumption obviously breaks down at the nuclear surface.

Different energies are required to remove a nucleon with quantum numbers lj from the different eigenstates of nucleus B. Because the separation energy varies from state to state, the lj radial wavefunction in the different eigenstates must have different shapes in the surface region, even if in all these states it has the same shape in the nuclear interior.

Unfortunately (Chapter 5), the nucleon-transfer process is sensitive to details of shape of radial wavefunctions in the nuclear surface region. Therefore, even though the structure calculations from which we extract $\mathscr{S}$ factors are a good guide to the wavefunction of the transferred nucleon in the nuclear interior, we see that to predict the magnitude of stripping cross sections it is necessary to develop further information about the radial shape of this function.

The lj radial wavefunction must be treated with particular care whenever this orbital is only a small admixture in some nuclear energy state, or whenever the energy of the nuclear state is far removed from the lj single-particle energy.

The wavefunction of the transferred nucleon is defined in general [see (5.109, 5.111)] as the overlap between the target nucleus eigenfunction ψ_A, and the product nucleus eigenfunction ψ_B. The radial part of this wavefunction is given by

$$\mathscr{R}_{lj}^{\alpha_A J_A}(r_0) = \langle \{\mathscr{Y}_{l\frac{1}{2}j}(\hat{n}_0), \psi_{\alpha_A J_A}\}_{J_B M_B}, \psi_{\alpha_B J_B M_B}\rangle, \tag{8.36}$$

where the comma in the scalar product denotes integration only over the internal coordinates of $\psi_{\alpha_A J_A}$ and over the coordinates in the spin-angle function $\mathscr{Y}_{l\frac{1}{2}j}(\hat{n}_0)$. The variable r_0 is the radial displacement of n_0 from the center of mass of A.

For large enough r_0, nucleon n_0 is not in interaction with A, and the shape of $\mathscr{R}_{lj}^{\alpha_A J_A}$ therefore is controlled by the separation energy $(E_B - E_A)$. In this region of r_0 the shape of $\mathscr{R}_{lj}^{\alpha_A J_A}$ is that of a Hankel function (for neutrons) or that of a Coulomb-modified Hankel function (for protons). To construct the radial wavefunction for all values of r_0, one popular technique computes $\mathscr{R}_{lj}^{\alpha_A J_A}$ as a single-particle eigenfunction in a Woods-Saxon potential well, with the depth of the potential adjusted to give the radial function the correct shape at large r_0. This technique is known as the "well-depth (WD) method." For a single nucleon outside a closed-shell core the transfer process excites only one state, and the WD method may be regarded as exact. It is subject to criticism only insofar as the shape of a Woods-Saxon potential may not be an adequate approximation to the shape of the Hartree-Fock potential.* On the other hand, if many nucleons are active the WD method loses

* Sharpey-Schafer [27] discusses an improved technique for choosing the parameters in the Woods-Saxon potential.

its validity. In particular, because the WD method ignores the many-body aspects of the transfer process, there is no reason to believe that it generates radial functions whose shapes are correct at small r_0. Because the WD radial wavefunctions may be wrong at small r_0, the process of normalizing these wavefunctions may lead to wrong amplitudes at large r_0, hence to wrong stripping cross sections.

Many authors have discussed improved methods for the calculation of the radial wavefunction of the transferred nucleon [28–36]. From a qualitative point of view, what is required of any improved calculation of $\mathscr{R}_{lj}^{\alpha_A J_A}$ is that it generate a function whose shape at small r_0 is very nearly that of a shell-model basis function, and whose shape at large r_0 is controlled by the separation energy. It is interesting that this qualitative point of view received direct application in early work by Tobocman [37]; however his method was in other respects too crude, and it was replaced by the WD method. Later, Sherr and collaborators [10] encountered apparent inconsistencies among $\mathscr{S}$ factors extracted from experiment with WD wavefunctions, and they were led to suggest the extreme view that radial wavefunctions should be calculated without any reference to individual separation energies. However, their approach was also too crude.

Pinkston and Satchler [29] proposed the most complete method for the calculation of $\mathscr{R}_{lj}^{\alpha_A J_A}$. In their approach the eigenfunction $\psi_{\alpha_B J_B M_B}$ is expanded in a complete set of antisymmetric energy eigenstates $\psi_{\alpha_P J_P M_P}$ of the nucleus with one less nucleon. The coefficients in this expansion are the relative wavefunctions*

$$\mathscr{R}_{l'j'}^{\alpha_P J_P}(n_0) = \langle \{\mathscr{Y}_{l' \frac{1}{2} j'}(\hat{n}_0), \psi_{\alpha_P J_P}\}_{J_B M_B}, \psi_{\alpha_B J_B M_B} \rangle. \tag{8.37}$$

The Schrödinger equation for $\psi_{\alpha_B J_B M_B}$ then takes the form of an infinite set of coupled differential equations for these relative wavefunctions. The one desired relative wavefunction $\mathscr{R}_{lj}^{\alpha_A J_A}$ is obtained by finding some adequate approximation in which to solve the coupled equations for the entire set of functions $\mathscr{R}_{l'j'}^{\alpha_P J_P}$.

The Pinkston-Satchler equations can be solved by iteration. To zero order in the coupling, the $\mathscr{R}_{l'j'}^{\alpha_P J_P}$ can be taken as multiples of shell-model basis functions, with undetermined coefficients. In this approximation the coupled differential equations reduce to a set of linear algebraic equations for the undetermined coefficients. These algebraic equations are precisely the equations of usual shell-model calculations. Hence it is possible to insert shell-model wavefunctions into the coupling terms of the Pinkston-Satchler

* It should be realized that the $\mathscr{R}_{l'j'}^{\alpha_P J_P}$ are defined in terms of eigenstates of two different nuclei. Therefore their quantum-mechanical properties are not simple. In particular, these functions are not orthogonal and cannot be approximated as Hartree-Fock eigenfunctions [30, 32].

equations, and to use these equations to generate first-order improvements of the relative wavefunctions. This method was applied by Prakash and Austern [32].

A more compact derivation of the first-order iterative solution was given by Kawai and Yazaki [34], using a method that is not based on the Pinkston-Satchler equations.* In the Kawai-Yazaki derivation we take the Schrödinger equation for $\psi_{\alpha_B J_B M_B}$ in the form

$$0 = \left[H_A - \frac{\hbar^2}{2M}\nabla_0^{\,2} + U(r_0) + V_{n_0 A} - E_B\right]\psi_{\alpha_B J_B M_B}, \qquad (8.38)$$

where $U(r_0)$ is a single-particle potential for particle n_0 and V_{n_0A} is the sum of two-body residual interactions between n_0 and the nucleons in A. Equation 8.38 then is left multiplied by

$$\{\mathscr{Y}_{l\frac{1}{2}j}(\hat{n}_0), \psi_{\alpha_A J_A}\}_{J_B M_B},$$

and is integrated over $\hat{n}_0$ and the coordinates of A, to give

$$0 = \left\langle \{\mathscr{Y}_{l\frac{1}{2}j}(\hat{n}_0), \psi_{\alpha_A J_A}\}_{J_B M_B}, \left[-\frac{\hbar^2}{2M}\frac{1}{r_0}\frac{\partial^2}{\partial r_0^{\,2}}r_0 + \frac{\hbar^2}{2M}\frac{l(l+1)}{r_0^{\,2}} + U(r_0) + E_A - E_B + V_{n_0 A}\right]\psi_{\alpha_B J_B M_B}\right\rangle. \qquad (8.39)$$

Most of the operators in (8.39) depend only on r_0, and may be removed from the matrix element, to give

$$\left[-\frac{\hbar^2}{2M}\frac{1}{r_0}\frac{d^2}{dr_0^{\,2}}r_0 + \frac{\hbar^2}{2M}\frac{l(l+1)}{r_0^{\,2}} + U(r_0) + E_A - E_B\right]\mathscr{R}_{lj}^{\alpha_A J_A}(r_0) = -\langle\{\mathscr{Y}_{l\frac{1}{2}j}(\hat{n}_0), \psi_{\alpha_A J_A}\}_{J_B M_B}, V_{n_0 A}\psi_{\alpha_B J_B M_B}\rangle, \qquad (8.40)$$

where the definition of the relative radial wavefunction (8.36) has been inserted. Equation 8.40 allows us to compute the required radial wavefunction as a solution of a linear ordinary differential equation if the inhomogeneous term on the right hand side (RHS) of the equation is known. We note that *the exact separation energy $E_B - E_A$ appears in* (8.40). Therefore solutions of this equation necessarily have correct shapes at large r_0.

The Pinkston-Satchler equations can be recovered from (8.40) by expanding $\psi_{\alpha_B J_B M_B}$ in a complete set, as in (8.37). However, it is more convenient and flexible to perform analyses of $\psi_{\alpha_B J_B M_B}$ in connection with particular applications of (8.40).

* The method of Kawai and Yazaki can be adapted rather easily to the analysis of form factors for two-nucleon transfer (M. Kawai, private communication).

In practical applications of (8.40) we substitute the shell-model wavefunction $\psi^{SM}_{\alpha_B J_B M_B}$ for $\psi_{\alpha_B J_B M_B}$. Although the shell-model wavefunction is defective in the region of the nuclear surface, it yields useful results in (8.40) because it is multiplied by the short-range operator $V_{n_0 A}$. This iterative method was applied by Prakash and Austern [32] in a study of the (p, d) and (d, t) reactions on Ni^{58}. In the (p, d) calculations the relative radial wavefunctions they obtained improved the fit to the angular distribution for the $f_{5/2}$ pickup leading to the 0.78 MeV $(5/2^-)$ state of Ni^{57}. In the (d, t) calculations the altered normalization of the wavefunctions at large r_0 led to as much as fifty percent alterations of cross sections.

A comprehensive study of form factors, largely on the basis of iterative solutions of (8.40), has been presented by Philpott, Pinkston, and Satchler [39].

We note at this point that (8.40) intermingles the calculation of the shape and of the normalization of $\mathscr{R}^{\alpha_A J_A}_{lj}$. No $\mathscr{S}$ factor is displayed. Although some separation of shape and normalization can be obtained by isolating diagonal matrix elements on the right hand side of the equation, it is clear that our improved calculation of radial wavefunctions to some extent spoils the convenience of stripping theory. The calculation of radial wavefunctions adequate for a careful reduction of stripping data may require considerable interplay between structure analyses and reaction analyses. As one step in this interplay, experimental values of configuration-mixing coefficients might be introduced in the shell-model wavefunction in (8.40).

We also note that the magnitude of $\mathscr{R}^{\alpha_A J_A}_{lj}$ in (8.40) is proportional to $V_{n_0 A}$. This observation emphasizes that correct normalizations can be obtained only if the residual interactions used in (8.40) are identical with the ones used to compute the shell-model wavefunction $\psi^{SM}_{\alpha_B J_B M_B}$.

Often the coupling term on the RHS of (8.40) is localized in the nuclear-surface region, near some radius R. When this occurs, a simple estimate of R immediately fixes the shape of $\mathscr{R}^{\alpha_A J_A}_{lj}$ in terms of the potential and the separation energy (this is measured) on the LHS of the equation [32]. Indeed, the shape of $\mathscr{R}^{\alpha_A J_A}_{lj}$ is identical with the shape of the Green's function for the LHS. This surface-delta-function approximation may yield estimates of form factors with as much ease as the WD method and with far greater accuracy. We note that the surface-delta approximation resembles a method [39], in which the *radius* of the single-particle potential is adjusted to give $\mathscr{R}^{\alpha_A J_A}_{lj}$ the correct separation-energy shape at large r_0.

Sometimes a perturbative solution of (8.40), by iteration of the shell-model solution, is not of sufficient accuracy. Greater accuracy is obtained if the more strongly coupled orbitals in $\psi_{\alpha_B J_B M_B}$ are treated as having undetermined radial shapes, and if coupled equations for these radial wavefunctions are solved. This procedure may be regarded as a partial return to the equations of Pinkston and Satchler, with due attention now being given to the truncation of the equations.

Huby and Hutton [31] used a coupled-equations approach in a study of form factors for the reaction $Ni^{58}(p, d)$. They regarded Ni^{58} as composed of an inert Ni^{56} closed-shell core, plus two active neutrons in the wavefunction

$$\psi(Ni^{58}) \approx u_a(r_1)u_a(r_2)\,|(2p_{3/2})_0{}^2\rangle + u_b(r_1)u_b(r_2)\,|(1f_{5/2})_0{}^2\rangle, \tag{8.41}$$

where the kets represent normalized spin-angle wavefunctions and the second term of (8.41) is much weaker than the first. Diagonalization of H with the above wavefunction yields coupled equations for u_a and u_b. The coupling of u_a and u_b reflects the long tail of the p function into the calculation of the f function, with the result that the main peak of the (d, p) angular distribution for the $f_{5/2}$ orbital is narrowed. A similar result was obtained in the perturbative calculation by Prakash and Austern [32]. Satisfactory agreement with experiment is obtained if the effect caused by improved form factors is combined with the effect caused by inclusion of the deuteron D-state [38].

Coupled equations can be of significance even if form factors for nucleon transfer are computed without explicit introduction of residual interactions with other active nucleons. Often we are interested in eigenfunctions of single-particle Schrödinger equations in which the potential is deformed, or in which the potential includes some dependence on i-spin. A single-particle orbital that is governed by a deformed potential is expanded in a spherical basis

$$\chi(\mathbf{r}) = r^{-1} \sum_{lj\Omega} u_{lj\Omega}(r)\,|lj\Omega\rangle, \tag{8.42}$$

and the radial functions for the different spherical components are found to obey coupled differential equations. Rost shows [33] that in the region of the nuclear surface exact solutions of these coupled equations have amplitudes as much as fifteen percent different from amplitudes obtained by zero-order degenerate perturbation theory.

More interesting results arise in the analysis of form factors for (p, d) reactions that lead to isobaric analogue states [10, 33, 36]. In this case an i-spin dependent term is included in the single-particle potential, and the form factors are computed by solving the Lane equations. We express the wavefunction of the target nucleus as a linear combination of two terms

$$\psi_B = \psi_C\psi_p + \psi_A\psi_n. \tag{8.43}$$

Here ψ_C is a low-lying state of excitation of the nucleus formed by removing a proton from B, and ψ_p is the wavefunction for the C, p relative motion. The state ψ_A is the isobaric analogue of ψ_C. Because ψ_A has a higher excitation energy than ψ_C, the relative function ψ_n must be more strongly damped asymptotically than ψ_p. The WD approximation for ψ_n and ψ_p would place great importance on this difference between the asymptotic energies. On the other hand, the entire interest of the second term of (8.43) hinges upon its coupling with the first term. This coupling causes ψ_p and ψ_n to have similar shapes in the region of the nuclear surface where the coupling is strong. As a result, although ψ_p and ψ_n have different shapes far outside the nucleus, the magnitude of the tail of ψ_n is much greater than predicted by the WD approximation. Coupling to the tail of the proton wavefunction helps lead the neutron out of the nucleus, and enhances the cross section for neutron pickup by as much as a factor of two [36].

References

[1] M. H. Macfarlane and J. B. French, *loc. cit.*, Chap. 5, Ref. 105.
[2] J. B. French, *loc. cit.*, Chap. 5, Ref. 104.
[3] J. B. French, in *Proceedings of the International School of Physics "Enrico Fermi"* Course XXXVI, C. Bloch, Ed., Academic, New York, 1966.
[4] B. L. Cohen, in *Proceedings of the Third Nordic-Dutch Accelerator Symposium*, University of Helsinki, 1964 (unpublished); also in *Nuclear Spin-Parity Assignments*, N. B. Gove and R. L. Robinson, Eds., Academic, New York, 1966.
[5] A. de-Shalit and I. Talmi, *loc. cit.*, Chap. 5, Ref. 168.
[6] R. G. Sachs, *loc. cit.*, Chap. 2, Ref. 30.
[7] S. Yoshida, *Phys. Rev.* **123,** 2122 (1961).
[8] S. Yoshida, *Nucl. Phys.* **38,** 380 (1962).
[9] J. B. French and M. H. Macfarlane, *Nucl. Phys.* **26,** 168 (1961).
[10] R. Sherr, B. F. Bayman, E. Rost, M. E. Rickey, and C. G. Hoot, *Phys. Rev.* **139,** B1272 (1965).
[11] M. A. Moinester, *Nucl. Phys.* **A94,** 81 (1967).
[12] J. B. French, in *Proceedings of the Summer Study Group*, Brookhaven Laboratory Report BNL 948, 165 (unpublished).
[13] J. B. French, *Phys. Letters* **13,** 249 (1964).
[14] M. E. DeLópez, M. Mazari, T. A. Belote, W. E. Dorenbusch, and O. Hansen, *Nucl. Phys.* **A94,** 673 (1967).
[15] G. R. Satchler, *Ann. Phys.* **3,** 275 (1958).
[16] M. H. Macfarlane, in *Nuclear Spectroscopy with Direct Reactions*, edited by F. E. Throw, Argonne National Laboratory report ANL-6878, 1964, (unpublished).
[17] A. Bohr and B. Mottelson, *loc. cit.*, Chap. 5, Ref. 49. See also, A. Kerman, in *Nuclear Reactions*, P. M. Endt and M. Demeur, Eds., North-Holland, Amsterdam, 1959; J. P. Elliott, lectures at University of Rochester, NYO-2271, 1958, (unpublished); K. Kumar and M. Baranger, *Nucl. Phys.* **A92,** 608 (1967).

[18] S. G. Nilsson, *Kgl. Danske Videnskab. Selskab, Mat.-fys. Medd.* **29,** No. 16 (1955).
[19] R. K. Sheline, W. N. Shelton, and R. A. Kenefick, *Phys. Letters* **5,** 129 (1963).
[20] W. N. Shelton and R. K. Sheline, *Phys. Rev.* **133,** B624 (1964).
[21] A. Isoya, *Phys. Rev.* **130,** 234 (1963).
[22] M. N. Vergnes and R. K. Sheline, *Phys. Rev.* **132,** 1736 (1963).
[23] J. R. Erskine and W. W. Buechner, *Phys. Rev.* **133,** B370 (1964).
[24] B. E. F. Macefield and R. Middleton, *Nucl. Phys.* **59,** 561 (1964).
[25] J. R. Erskine, *Phys. Rev.* **138,** B66 (1965).
[26] J. Kern, O. Mikoshiba, R. K. Sheline, T. Udagawa, and S. Yoshida, *Nucl. Phys.* **A104,** 642 (1967).
[27] J. F. Sharpey-Schafer, *Phys. Letters* **26B,** 652 (1968).
[28] N. Austern, *Phys. Rev.* **136,** B1743 (1964).
[29] W. T. Pinkston and G. R. Satchler, *Nucl. Phys.* **72,** 641 (1965).
[30] T. Berggren, *Nucl. Phys.* **72,** 337 (1965).
[31] R. Huby and J. L. Hutton, *Phys. Letters* **19,** 660 (1966).
[32] A Prakash, Ph.D. thesis, University of Pittsburgh 1966, (unpublished), *Phys. Rev. Letters* **20,** 864 (1968); A. Prakash and N. Austern, *Ann. Phys.* **51,** 418 (1969).
[33] E. Rost, *Phys. Rev.* **154,** 994 (1967).
[34] M. Kawai and K. Yazaki, *Progr. Theoret. Phys.* **37,** 638 (1967); and subsequent publications.
[35] K. Sugawara, *Nucl. Phys.* **A110,** 305 (1968).
[36] R. Stock and T. Tamura, *Phys. Letters* **22,** 304 (1966).
[37] W. Tobocman, *Phys. Rev.* **115,** 98 (1959).
[38] R. C. Johnson and F. D. Santos, *loc. cit.*, Chap. 5, Ref. 135.
[39] R. J. Philpott, W. T. Pinkston, and G. R. Satchler, *Nucl. Phys.* **A119,** 241 (1968), *ibid,* **A125,** 176 (1969).

CHAPTER 9

Polarization and Angular Correlations

Elsewhere in this book only the differential cross sections of nuclear reactions are discussed. However, differential cross sections entail averages over all initial and final spin projections. These averages are seen, for example, in the basic formulas of the DW method (4.59, 4.61). A complete labelling of the amplitude $T_{\beta\alpha}{}^{\mathrm{DW}}$ in these equations is

$$T^{\mathrm{DW}}(\mathbf{k}_\beta, m_b, M_B; \mathbf{k}_\alpha, m_a, M_A), \tag{9.1}$$

where the initial and final spin projections and relative momenta are now indicated.* Then the differential cross section is proportional to $|T_{\beta\alpha}{}^{\mathrm{DW}}|^2$, averaged over all the spin projections m_a, M_A, m_b, M_B. Because of this averaging the differential cross section is only a function of the bombarding energy and of the angle between $\mathbf{k}_\alpha$ and $\mathbf{k}_\beta$.

More complete information about reactions can be obtained by measuring cross sections that are not so severely averaged with respect to the spin projections. Polarization experiments achieve this by directly measuring the distribution of the cross section among the possible spin projections of one of the projectiles. Angular correlation experiments concern reactions in which one of the product nuclei is formed in an excited state. By measuring the angular distribution of secondary radiation from this excited state, it is possible to deduce the relative populations of the spin projections produced in the primary reaction. Such studies of secondary radiation may be regarded as indirect polarization experiments.

Thus the term "polarization" is used in general to denote any ordering of the nuclear spins. Devons and Goldfarb [1] surveyed the theory of polarization in nuclear reactions. Satchler [2] gave an excellent summary of applications of this theory in DW analyses of direct reactions; much of the discussion in this chapter parallels his article. A similar summary of DW applications was given by Goldfarb and Johnson [3].

* Throughout Chapter 9 the order of the indices on the transition amplitude is opposite to that used elsewhere in this book. The present ordering, which is widely used in the literature, facilitates the construction of matrix equations.

In principle it is possible to perform enough different polarization experiments to obtain an almost complete description of the reaction amplitude. In the fundamental studies of nucleon-nucleon scattering considerable efforts are made to achieve just such a level of completeness [4, 5], and these efforts have been well rewarded. However, it is not clear to what extent comparable efforts may be relevant in studies of direct reactions. Accurate data that bear on this question are only beginning to be available, largely since the development of polarized ion sources.

We must not forget that DI theories concern only parts of the full physical wavefunction for a nuclear reaction. (See Chapters 2, 4, 11). Very sensitive tests of the reaction, as in polarization experiments, therefore encounter DI-CN interference effects that may cause fundamental limitations of interpretation. Indeed, polarization cross sections for light nuclei are known to fluctuate with respect to bombarding energy.

The fact that polarization cross sections are sensitive to CN contributions led to suggestions that such cross sections be used to distinguish DI and CN reaction mechanisms. These suggestions were based on the observation that DI reaction mechanisms possess special simplicities that limit the complexity of polarization cross sections, (e.g., see [6]). These DI simplicities could be used to derive certain parameter-independent *equations of redundancy*. By checking whether these equations are fulfilled one could check the purity of the DI reaction mechanism. Unfortunately, the derivations of the equations of redundancy tended to be based on oversimplified DI theories. Theories that carry all normal complications, such as spin-orbit terms of optical potentials, do not limit the polarization cross sections to a sufficient extent to give equations of redundancy that are of much experimental interest [3].

A more specific way to distinguish CN and DI reaction mechanisms is found in angular correlation experiments in which the primary radiation is emitted at 0° or 180°. For such angles of emission the intermediate excited nucleus can be formed in only a limited number of magnetic substates, and the populations of these substates can be deduced with great certainty by measuring the angular distribution of the secondary radiation [7]. One can then determine whether these populations obey the further limitations imposed by DI theories [8, 9]. It has been remarked that CN admixtures in the amplitude should cause especially strong fluctuations in the excitation functions of the individual substates selected in this fashion, and these fluctuations might be studied [9].

However, the value of polarization experiments is far more likely to lie in the quantitative improvement of well-established DI theories than in speculative tests of reaction mechanisms. Much of the work to be done by this means has to do with nucleon spins. DI differential cross sections are not sensitive to spin-dependent interactions or to the way spins are coupled up in wavefunctions; for example, deuteron stripping experiments have difficulty in determining the total angular momentum transfer. Polarization experiments allow great progress with such questions. They are required for the determination of spin-dependent terms in optical potentials, they are sensitive to spin-flip terms in the nuclear form factors (hence to the coupling scheme of bound wavefunctions), they give promise of allowing explicit measurements of the total j transfer in stripping. Systematic study of all these applications of polarization experiments is just beginning.

In the present chapter, Section 9.1 discusses the density and efficiency matrices used to describe polarizations and polarization measurements. Section 9.2 discusses the relation between the initial and final density matrices in a collision experiment, and derives formulas that express cross sections in terms of scattered amplitudes. Section 9.3 discusses polarizations derived from the DW theory, and gives several comparisons with experiment. Section 9.4 discusses angular correlations.

9.1 Description of a Polarized System

To discuss the theory of polarization it is necessary to develop quantum mechanical techniques for the description of systems that are partially polarized. We recognize that any single wavefunction describes a pure state, which is likely to be completely polarized along some suitably chosen axis of projection. Partial polarization is, in general, a property of statistically uncorrelated ensembles of particles, like those produced in beams from accelerators, and it must be described by sets of wavefunctions with random relative phases.

One way to describe the polarization of an ensemble is to introduce a system of coordinate axes and to discuss the number of systems in each magnetic substate defined with respect to these axes. All experiments utilize this description. However, this simple approach places too much emphasis on the choice of the axes of projection. As a result it is difficult to relate measurements along different axes, it is difficult to relate different kinds of experiments, and it is difficult to exploit modern methods of angular momentum summation.

The density matrix [1, 4, 10–13] is the preferred starting point for analysis. It provides a compact and flexible description of all aspects of polarization, and it allows the easy calculation of explicit theoretical results for special cases. To develop the theory of the density matrix, let us consider an ensemble whose systems are distributed over the set of incoherent, normalized states $\lambda^{(K)}$. These $\lambda^{(K)}$ are the various independent states in which the systems are prepared, and are not necessarily orthogonal. Let $w(K)$ be the number of systems in state $\lambda^{(K)}$. Then the average value of some operator $\mathcal{O}$ is obtained by computing its expectations in the individual states $\lambda^{(K)}$, and averaging these expectations over the ensemble:

$$\bar{\mathcal{O}} = \frac{\sum_K w(K)(\lambda^{(K)}, \mathcal{O}\lambda^{(K)})}{\sum_K w(K)(\lambda^{(K)}, \lambda^{(K)})}. \tag{9.2}$$

To perform the calculations in (9.2), it is generally necessary to introduce a set of orthonormal basis functions ϕ_m. In terms of these functions

$$\lambda^{(K)} = \sum_m a_m^{(K)} \phi_m, \tag{9.3}$$

and (9.2) becomes

$$\bar{\mathcal{O}} = \frac{\sum_{m,n} \sum_K w(K) a_m^{(K)*} \mathcal{O}_{mn} a_n^{(K)}}{\sum_m \sum_K w(K) a_m^{(K)*} a_m^{(K)}}. \tag{9.4}$$

If we now define the density matrix ρ to be the operator whose matrix elements in the basis ϕ_m are

$$\rho_{nm} \equiv \sum_K w(K) a_n^{(K)} a_m^{(K)*}, \tag{9.5}$$

then (9.4) is seen to be of the form

$$\bar{\mathcal{O}} = \frac{\text{trace}\,\{\rho\mathcal{O}\}}{\text{trace}\,\{\rho\}}. \tag{9.6}$$

All information about the polarization of the ensemble is contained in the matrix ρ. In a rough sort of way the matrix elements ρ_{mn} are the probabilities for states ϕ_m and ϕ_n to occur coherently [14] in the polarized ensemble. The benefits of the density-matrix formalism may be noted:

1. The trace of any operator is independent of its representation. Therefore we are free to formulate the matrix ρ in any representation in which the probabilities of independent states are known, and we are free to transform to any other representation that may be convenient for calculation or for comparison with experiment. The choice of a quantization axis for angular momentum might be one such change of representation.

2. In polarization discussions the matrix ρ is constructed in a product basis, in which ϕ_m and ϕ_n have definite projections of angular momentum. However, by suitable change of representation it is possible to express ρ as a linear combination of "statistical tensors" that are irreducible under rotations and are convenient for calculations. Because ρ is a scalar operator (it describes a physical state), the coefficients of the statistical tensors are tensor parameters that have immediate experimental interpretations.

3. In polarization discussions ρ is of finite rank, say N. Furthermore, ρ is Hermitian by construction. Therefore the maximum number of independent

parameters in ρ is N^2, and this sets an immediate upper limit to the complexity of polarization phenomena. A complete measurement of the state of polarization need only measure the N^2 parameters of the density matrix.

As an example, let us consider an ensemble of systems that have angular momentum j, and that all have projection $\tilde{m}$ along some given axis. In a jm representation with respect to this axis, the matrix elements of ρ are

$$\rho_{mm'} = \delta_{m\tilde{m}}\,\delta_{m'\tilde{m}}. \tag{9.7}$$

This is the density matrix for a completely polarized, pure state.

As another example, let us consider the density matrix for an ensemble with angular momentum j and zero polarization. Zero polarization implies that all members of a set of basis states that span the spin space have equal probabilities. With the use of a jm representation in (9.5), the matrix elements of the normalized density matrix are found to be

$$\rho_{mm'} = N^{-1}\,\delta_{mm'}, \tag{9.8}$$

where $N = 2j + 1$ is the rank of the matrix. To verify that (9.8) describes zero polarization we substitute it in (9.6). It is found that the average expectations of the components of $\mathbf{J}$ all vanish, as required, and that the average expectations of all traceless tensors formed from these components (see below) also vanish. We see that (9.8) is invariant under change of representation, and therefore an unpolarized ensemble is described equally well as a random assemblage of plane-polarized systems, or circularly-polarized systems, etc.

Collision experiments require descriptions of the state of polarization of the entire set of noninteracting collision partners. It is clear that the overall initial density matrix is the direct product of the density matrices for the individual particles.

As another interesting and useful example, let us derive the complete density matrix for spin-$\frac{1}{2}$ particles. In this case the matrix is a Hermitian operator in the 2×2 spin space, and it contains four real parameters. These four parameters may be introduced as the coefficients needed to express ρ as a linear combination of the four independent Hermitian operators 1, σ_x, σ_y, σ_z, where 1 is the unit matrix. Because the spin matrices transform under rotations as the components of a pseudovector, and because ρ is a scalar, it is clear that the coefficients of the three spin matrices are proportional to the components of a pseudovector observable. This observable must be the polarization. Because the spin matrices are traceless it is clear that the coefficient of the unit matrix must be the total intensity of the beam. A complete expression for the density matrix then is

$$\rho = \tfrac{1}{2}I[1 + (\mathbf{P}\cdot\boldsymbol{\sigma})], \tag{9.9}$$

where I is the total beam intensity and the coefficient of $\boldsymbol{\sigma}$ has been normalized so that $\mathbf{P}$ is the polarization vector. The four parameters in the above expression for ρ are I and the three components of $\mathbf{P}$. To verify that $\mathbf{P}$ in (9.9) is indeed the polarization vector we recognize that a measurement of the projection of the polarization along some axis $\hat{n}$ is equivalent to the measurement of the average expectation of $(\boldsymbol{\sigma} \cdot \hat{n})$. From (9.6) this expectation is found to be

$$\overline{(\boldsymbol{\sigma} \cdot \hat{n})} = \frac{\operatorname{tr}\{\frac{1}{2}I[1 + (\mathbf{P} \cdot \boldsymbol{\sigma})](\boldsymbol{\sigma} \cdot \hat{n})\}}{\operatorname{tr}\{\frac{1}{2}I[1 + (\mathbf{P} \cdot \boldsymbol{\sigma})]\}},$$
$$= (\mathbf{P} \cdot \hat{n}).$$

Evidently $\overline{(\boldsymbol{\sigma} \cdot \hat{n})}$ is maximized if $\hat{n}$ lies along $\mathbf{P}$, and the maximum value is P as required.

For particles of spin greater than $\frac{1}{2}$ a full parameterization of the density matrix requires information that goes beyond the mere specification of the polarization vector [1, 12–18]. Let us take as an example spin-1 particles, such as deuterons. In this case ρ is a Hermitian operator in a 3×3 spin space, and it contains nine real parameters. Four of these parameters are the coefficients of the Hermitian matrices, 1, S_x, S_y, S_z, where $\hbar\mathbf{S}$ is the angular momentum operator. As with spin-$\frac{1}{2}$ these four parameters are the total intensity and the three components of the polarization vector. Five additional parameters enter as coefficients of the five linearly-independent symmetric products of pairs of the operators S_x, S_y, S_z. These five parameters are the components of a traceless symmetric tensor, that describes "tensor polarization" of a deuteron beam. The complete density matrix for spin-1 then is*

$$\rho = \tfrac{1}{3}I[1 + \tfrac{3}{2}(\mathbf{P} \cdot \mathbf{S}) + (\mathbf{Q} : \mathbf{O})], \tag{9.10}$$

where $\mathbf{Q}$ is the five-component, tensor polarization and $\mathbf{O}$ is the tensor operator whose Cartesian components are

$$O_{ij} = \tfrac{1}{2}(S_i S_j + S_j S_i) - \tfrac{1}{3}(S)^2 \delta_{ij}. \tag{9.11}$$

We note that no tensor operator of the form of (9.11) exists for spin-$\frac{1}{2}$, because products of the Pauli spin matrices can be linearized. Likewise, for spin-1 it is not possible to form tensors of higher than second rank. We also note that in practical calculations $\mathbf{Q}$ and $\mathbf{O}$ usually are handled in spherical components.

For a simple illustration of the content of (9.10), we consider a deuteron beam prepared in definite states of projection $m = 1, 0, -1$ with respect to some given axis. Let the probabilities of these three states be $w(+)$, $w(0)$, and $w(-)$, where $w(+) + w(0) + w(-) = 1$. Then, from (9.5) the density

* The normalization of the vector term in (9.10) is discussed in connection with (9.29).

matrix in *m*-representation is

$$\rho = \begin{pmatrix} w(+) & 0 & 0 \\ 0 & w(0) & 0 \\ 0 & 0 & w(-) \end{pmatrix}.$$

We put this matrix in the form of (9.10) by expressing it as the linear combination

$$\rho = \tfrac{1}{3}\begin{pmatrix} 1 & 0 & 0 \\ 0 & 1 & 0 \\ 0 & 0 & 1 \end{pmatrix}$$

$$+ \tfrac{1}{2}[w(+) - w(-)]\begin{pmatrix} 1 & 0 & 0 \\ 0 & 0 & 0 \\ 0 & 0 & -1 \end{pmatrix}$$

$$+ \tfrac{1}{2}[w(0) - \tfrac{1}{3}]\begin{pmatrix} -1 & 0 & 0 \\ 0 & 2 & 0 \\ 0 & 0 & -1 \end{pmatrix}.$$

Thus for a beam polarized with respect to a single axis the vector polarization is determined by $[w(+) - w(-)]$, and the tensor polarization is determined by $[w(0) - (\frac{1}{3})]$. Pure vector polarization requires $w(0) = \frac{1}{3}$.

The above discussion shows that the density matrix is a mathematical device that provides a complete and compact description of all aspects of the polarization of a physical system. For a collision experiment the initial density matrix describes the polarizations with which the collision partners are prepared. The final density matrix describes the polarizations that emerge from the collision, as determined by the properties of the scattering amplitudes (Section 9.2).

Coester and Jauch [19] suggested the introduction of another mathematical device, the "efficiency matrix," to provide a corresponding complete and compact description of all aspects of the measurements in a given experiment. Measurement implies finding the probabilities for projecting the system onto a set of independent (incoherent) states $\Pi^{(Q)}$. From the original definition of the density matrix in terms of the probabilities of incoherent states $\lambda^{(K)}$, we recognize that the probability that the system is found in a single state $\Pi^{(Q)}$ is $\langle Q| \rho |Q\rangle$. The probability that an experiment detects a given ensemble of the states $\Pi^{(Q)}$ then is

$$W = \sum_Q \varepsilon_Q \langle Q| \rho |Q\rangle, \tag{9.12}$$

where ε_Q is the *efficiency* with which given states in the ensemble are detected. (The usual values for ε_Q are $\varepsilon_Q = 0, 1$.) If we express the states $\Pi^{(Q)}$ in an arbitrary representation, as in the earlier discussion, then

$$\Pi^{(Q)} = \sum_m b_m^{(Q)} \phi_m, \tag{9.13}$$

and

$$W = \text{tr}\,\{\varepsilon\rho\}, \tag{9.14}$$

where the general form of the efficiency matrix ε is

$$\varepsilon_{mn} = \sum_Q \varepsilon_Q b_m^{(Q)} b_n^{(Q)*}. \tag{9.15}$$

Calculations based on (9.14) are independent of representation, and enable one to derive any required correlation in a systematic fashion, following a few straightforward rules [14]. The efficiency matrix can be expanded in tensors [1, 13, 14, 19] to allow an orderly comparison with the density matrix.

The efficiency matrix for a reaction that produces several particles or degrees of freedom is the direct product of the efficiency matrices for detecting each individual particle or degree of freedom. In this direct product the efficiency matrix for any unobserved spin or angular variable is a delta function in the eigenvalues of that variable. This form of the efficiency matrix for an unobserved variable stands in close analogy with (9.8), which gives the density matrix for an unpolarized variable.

9.2 Polarization in Collision Experiments

In a collision experiment the initial density matrix ρ^i describes an incident beam prepared in a set of incoherent spin states $\lambda^{(K)}$ with probabilities $w(K)$. These incoherent states are expanded in terms of channel wavefunctions ψ_α [see (1.1)], with the result that the matrix elements of ρ^i are

$$\rho^i_{\alpha\alpha'} = \sum_K \langle\alpha \mid \lambda^{(K)}\rangle w(K) \langle\lambda^{(K)} \mid \alpha'\rangle. \tag{9.16}$$

We now require the final density matrix ρ^f that describes the properties of particles emitted from the collision. As with ρ^i, the matrix ρ^f is used to compute expectation values of given operators $\mathcal{O}$, averaged over the incoherent states $\lambda^{(K)}$. However, the expectations computed with ρ^f concern particles that have been emitted from the collision, and have attained asymptotically large displacements from each other. Expectations computed for these particles involve overlaps of the scattered waves generated from the incoherent states. These scattered waves are expanded in terms of the channel wavefunctions ψ_β for the exit channels, and have magnitudes given by the scattered amplitudes $T_{\beta\alpha}$. Then the average expectation of an operator $\mathcal{O}$, computed in the final states derived from the $\lambda^{(K)}$, is found to be

$$\bar{\mathcal{O}} = \sum_K w(K) \sum_{\alpha\beta\alpha'\beta'} \langle \lambda^{(K)} | \alpha \rangle T^*_{\alpha\beta} \langle \beta | \mathcal{O} | \beta' \rangle T_{\beta'\alpha'} \langle \alpha' | \lambda^{(K)} \rangle. \tag{9.17}$$

The insertion of (9.16) gives a reduced form of (9.17):

$$\bar{\mathcal{O}} = \sum_{\alpha\beta\alpha'\beta'} \langle \beta | \mathcal{O} | \beta' \rangle T_{\beta'\alpha'} \rho^i_{\alpha'\alpha} T^*_{\alpha\beta}, \tag{9.18}$$

which may in turn be written

$$\bar{\mathcal{O}} = \text{trace}\,\{\mathcal{O}\rho^f\} = \text{trace}\,\{\rho^f\mathcal{O}\}, \tag{9.19}$$

where

$$\rho^f = T\rho^i T^\dagger. \tag{9.20}$$

We see that the T-matrix transforms ρ^i into ρ^f.

A general expression for the experimental correlation function, suitable as a framework for the analysis of all possible polarization experiments, is obtained by combining (9.14) and (9.20), giving

$$W_{fi} = \text{tr}\,\{\varepsilon^f T\rho^i T^\dagger\}. \tag{9.21}$$

Here ε^f is the efficiency matrix that describes the conditions of detection, and ρ^i is the density matrix that describes the conditions of beam and target preparation. Then the differential cross section for a two-body reaction $a + A \rightarrow b + B$ may be written as a simple multiple of W_{fi},

$$\frac{d\sigma_{fi}}{d\Omega_f} = \frac{\mu_\alpha\mu_\beta}{(2\pi\hbar^2)^2}\left(\frac{k_\beta}{k_\alpha}\right) W_{fi}, \tag{9.22}$$

where our usual normalization of the amplitudes $T_{\beta\alpha}$ [see (1.32)] is employed. The matrices ρ^i and ε^f give the weights with which various spin states affect the cross section.

If all final spin states are detected with equal probability, the efficiency matrix ε^f has the matrix elements

$$\varepsilon^f_{m_b M_B, m_b' M_B'} = \delta_{m_b m_b'}\, \delta_{M_B M_B'}. \tag{9.23}$$

If all incident spin states are prepared with equal probability, the normalized density matrix ρ^i has the matrix elements

$$\rho^i_{m_a M_A, m_a' M_A'} = (2s_a + 1)^{-1}(2J_A + 1)^{-1}\, \delta_{m_a m_a'}\, \delta_{M_A M_A'}. \tag{9.24}$$

The usual unpolarized differential cross section

$$\left(\frac{d\sigma_{fi}}{d\Omega_f}\right)_0 = \frac{\mu_\alpha\mu_\beta}{(2\pi\hbar^2)^2}\left(\frac{k_\beta}{k_\alpha}\right)\frac{1}{(2s_a + 1)(2J_A + 1)} \sum_{\substack{m_a M_A \\ m_b M_B}} |T_{\beta\alpha}|^2, \tag{9.25}$$

is obtained if both (9.23) and (9.24) are substituted in (9.21) and (9.22).

Detection of the vector polarization of product particle b implies measurement of the average projection of $\mathbf{s}_b$ along some axis $\hat{n}$. The efficiency matrix that describes this measurement is

$$\varepsilon_f = s_b^{-1}(\mathbf{s}_b \cdot \hat{n}), \tag{9.26}$$

in which a unit matrix in the spin space of particle B is an implied factor. (For particles that have spin $s_b = \frac{1}{2}$, for example, ε^f has the expectations $+1, -1$ for spins that are parallel, anti-parallel to $\hat{n}$). Then the cross section for the production of polarized particles, with an unpolarized beam incident on an unpolarized target, is derived by substituting (9.24) and (9.26) in (9.21) and (9.22):

$$\left(\frac{d\sigma_{fi}}{d\Omega_f}\right)_{\text{pol}} = \frac{\mu_\alpha \mu_\beta}{(2\pi\hbar^2)^2}\left(\frac{k_\beta}{k_\alpha}\right)\frac{1}{s_b(2s_a+1)(2J_A+1)} \operatorname{tr}\{(\mathbf{s}_b \cdot \hat{n})TT^\dagger\}. \tag{9.27}$$

The usual "polarization" of the product particles is the ratio of (9.25) and (9.27),

$$\begin{aligned}(\mathbf{p}_b \cdot \hat{n}) &= \frac{(d\sigma_{fi}/d\Omega_f)_{\text{pol}}}{(d\sigma_{fi}/d\Omega_f)_0}, \\ &= \frac{s_b^{-1}\operatorname{tr}\{(\mathbf{s}_b \cdot \hat{n})TT^\dagger\}}{\operatorname{tr}\{TT^\dagger\}}.\end{aligned} \tag{9.28}$$

It is clear that the polarization vector $\mathbf{p}_b$ is perpendicular to the scattering plane. To see this we recognize that $\mathbf{s}_a$, $\mathbf{J}_A$, $\mathbf{J}_B$ are unpolarized and do not define directions in space. Therefore the direction of $\mathbf{p}_b$ must be determined by the directions of the momenta $\mathbf{k}_\alpha$, $\mathbf{k}_\beta$. In particular, $\mathbf{p}_b$ must be parallel to the one pseudovector $[\mathbf{k}_\alpha \times \mathbf{k}_\beta]$ that can be formed from $\mathbf{k}_\alpha$ and $\mathbf{k}_\beta$. In more general discussions this result is derived as a special case of theorems concerning the expectations of statistical tensors in particular coordinate systems. It is standard (*Basel convention*) to define the magnitude of $\mathbf{p}_b$ to be positive if $\mathbf{p}_b$ has the same sense as $[\mathbf{k}_\alpha \times \mathbf{k}_\beta]$.

If the incident particle a has vector polarization $\mathbf{P}_a$, the normalized incident density matrix is

$$\rho^i = \frac{1}{(2s_a+1)(2J_A+1)}\left[1 + \left(\frac{3}{s_a+1}\right)(\mathbf{P}_a \cdot \mathbf{s}_a)\right], \tag{9.29}$$

in which a unit matrix in the spin space of particle A is an implied factor. The normalization factor $3/(s_a + 1)$ is inserted in order that the product of ρ^i by the corresponding efficiency matrix for the detection of $\mathbf{p}_a$ have trace unity. Then the cross section measured by spin-independent detectors, if a polarized beam impinges on an unpolarized target, is derived by substituting (9.23) and (9.29) in (9.21) and (9.22):

$$\left(\frac{d\sigma_{fi}}{d\Omega_f}\right) = \frac{\mu_\alpha \mu_\beta}{(2\pi\hbar^2)^2}\left(\frac{k_\beta}{k_\alpha}\right)\frac{1}{(2s_a+1)(2J_A+1)} \times \operatorname{tr}\left\{T\left[1+\left(\frac{3}{s_a+1}\right)(\mathbf{P}_a\cdot\mathbf{s}_a)\right]T^\dagger\right\}. \tag{9.30}$$

The first term of (9.30) is the usual unpolarized cross section (9.25). The second term of (9.30) is affected by the incident polarization; this term has an asymmetry $\mathbf{A}_b$ that we define quantitatively in terms of the ratio

$$(\mathbf{P}_a\cdot\mathbf{A}_b) \equiv \left(\frac{3}{s_a+1}\right)\frac{\operatorname{tr}\{T(\mathbf{P}_a\cdot\mathbf{s}_a)T^\dagger\}}{\operatorname{tr}\{TT^\dagger\}}. \tag{9.31}$$

This asymmetry* is a pseudovector, which lies in a direction determined by the average expectation $\operatorname{tr}\{T\mathbf{s}_a T^\dagger\}$. As with the polarization, the direction of the asymmetry must be parallel to $[\mathbf{k}_\alpha \times \mathbf{k}_\beta]$, the normal to the scattering plane. For the present discussion it is important that the direction of this normal reverses, by definition, according to whether $\mathbf{k}_\beta$ lies to the left or right of the incident beam. Hence, for a given value of the scattering angle, the sign of the asymmetric term in the cross section of (9.30) reverses, according to whether we detect left or right scattering. The magnitude of A_b is measured by the left-right ratio of cross sections

$$P_a A_b = \frac{(d\sigma_{fi}/d\Omega_f)_L - (d\sigma_{fi}/d\Omega_f)_R}{2(d\sigma_{fi}/d\Omega_f)_0}, \tag{9.32}$$

where the direction of the incident polarization $\mathbf{P}_a$ is chosen perpendicular to the scattering plane. Measurement of the left-right cross section ratio enables A_b to be determined if P_a is known, or enables P_a to be determined if A_b is known.

There is a close resemblance between the polarization in (9.28) and the asymmetry in (9.31). Vector expressions for these two quantities are

$$\mathbf{p}_b = \frac{s_b^{-1}\operatorname{tr}\{\mathbf{s}_b TT^\dagger\}}{\operatorname{tr}\{TT^\dagger\}}, \tag{9.28$'$}$$

$$\mathbf{A}_b = \left(\frac{3}{s_a+1}\right)\frac{\operatorname{tr}\{T\mathbf{s}_a T^\dagger\}}{\operatorname{tr}\{TT^\dagger\}}. \tag{9.31$'$}$$

* The asymmetry $\mathbf{A}_b$ is also sometimes called the "analyzing power" of the reaction. With polarized ion sources, it is fairly easy to measure the asymmetry.

Both these quantities concern measurements on the reaction $a + A \rightarrow b + B$. Even closer resemblances are found if these two quantities are compared with the corresponding quantities in the inverse reaction $b + B \rightarrow a + A$. For example, the polarization produced in the inverse reaction is

$$\mathbf{p}_a = \frac{s_a^{-1}\,\mathrm{tr}\,\{\mathbf{s}_a TT^\dagger\}}{\mathrm{tr}\,\{TT^\dagger\}}, \tag{9.33}$$

where the unitarity of the S-matrix guarantees the presence of the same T-matrix as in (9.28′) and (9.31′). We see that (9.31′) and (9.33) differ primarily by the order of the operators in the numerators. However, it was discovered very early [11, 12, 20] that the time-reversal invariance of the nuclear Hamiltonian causes the S-matrix to be symmetric, and therefore causes the orders of the operators in (9.31′) and (9.33) to be equivalent. This leads to the relation

$$\mathbf{A}_b = \left(\frac{3s_a}{s_a + 1}\right)\mathbf{p}_a. \tag{9.34}$$

Equation 9.34 is known as the "polarization-asymmetry theorem." Corresponding relations have been derived for statistical tensors of higher rank [21].

Biedenharn contributed the important remark [22] that the polarization-asymmetry theorem remains valid even for DI calculations that contain complex optical potentials. Such complex potentials violate time-reversal invariance. However, this time-reversal violation only represents the transfer of flux into channels omitted from the DI calculations, and does not affect the symmetry of the S-matrix for the retained channels. Satchler generalized Biedenharn's observation by adding the further remark [18] that for spin-dependent optical potentials we must take care that any terms odd under time reversal should also be anti-Hermitian. This requirement limits the types of tensor terms that can appear in the optical potential for deuterons.

With the aid of the polarization-asymmetry theorem, (9.32) for the left-right cross section ratio assumes the simple form

$$P_a p_a = \left(\frac{3s_a}{s_a + 1}\right)\frac{[(d\sigma_{fi}/d\Omega_f)_L \quad (d\sigma_{fi}/d\Omega_f)_R]}{2(d\sigma_{fi}/d\Omega_f)_0}. \tag{9.35}$$

Thus the cross section ratio is given by the product of the incident polarization in the reaction $a + A \rightarrow b + B$, multiplied by the polarization that would be produced in the inverse reaction.

When the reaction $a + A \rightarrow b + B$ produces nucleus B in an excited state, an experimental investigation of the inverse reaction would require use of a target that contains nuclei in this excited state. Since such targets are unlikely, authors occasionally disregard this requirement, and compare a reaction that produces an excited state with an "inverse" reaction based on the corresponding ground state. Even under these conditions, the relations expressed by the polarization-asymmetry theorem often are fulfilled fairly well [3, 23, 24].

The polarizations of the target and residual nuclei satisfy relations that resemble the polarization-asymmetry theorem for the light projectiles. The implications of these relations are discussed by Goldfarb and Bromley [25].

We note above that the pseudovector $[\mathbf{k}_\alpha \times \mathbf{k}_\beta]$, the normal to the scattering plane, plays a special role in polarization discussions. Many general expressions are simplified if one of the coordinate axes is taken to lie along this normal. Thorough discussions of the symmetries found by use of such special coordinate systems are given in [2, 3]. These discussions are not presented in this book. For inelastic scattering with small energy loss another interesting special coordinate system is indicated by "adiabatic symmetry." This symmetry is discussed in Section 9.4.

9.3 Polarization in the DW Theory

To compute polarization cross sections for some particular reaction it is necessary to substitute the amplitude for that reaction into the appropriate formulas of Section 9.2. We now consider the polarization effects associated with amplitudes derived from the important DW theory. Amplitudes derived from other theories may be treated in a similar fashion.

Because the DW theory requires information about spin-dependent terms in optical potentials, we begin with a discussion of *elastic scattering*: Polarization measurements for nucleons have long been performed using methods of double scattering; these measurements have determined the strength of the spin-orbit term in the optical potential and have shown that its range is smaller than that of the central term (see Section 5.2). Corresponding studies of the deuteron potential are complicated by the possible presence of tensor terms [18]. However, with the development of polarized ion sources it has been shown [26, 27] that the tensor terms are in fact much weaker than the $\mathbf{L} \cdot \mathbf{S}$ vector term. The strength and shape of the vector term have been measured, and its radius and diffuseness have been shown to be smaller [27] than previously suspected [28]. It is interesting that DW calculations that use the improved deuteron potentials derived from these polarization studies seem to fit [29] the back-angle j-dependence in the differential cross sections of (d, p) stripping reactions (see Section 5.8).

Many authors have noted that the spin-dependent terms of optical potentials are weak enough so that perturbation theory may be used to compute the polarization in elastic scattering experiments. This is not to say that small polarizations need be associated with these weak potentials. At deep minima in differential cross sections even small displacements between the amplitudes for spin-up and spin-down scattering give rise to large polarizations, which show rapid changes of sign. Nevertheless, first-order calculations of the amplitudes give adequate descriptions of these fluctuating polarizations. A rough first-order theory by Rodberg [30] developed this idea, and suggested that the angular distribution of the polarization should be proportional to the derivative of the differential cross section with respect to scattering angle. A more accurate first-order theory was developed subsequently by Goldfarb, Greenlees, and Hooper [31]. A semiclassical first-order theory was discussed by de-Shalit and Hufner, and by Gubkin [32].

Let us now substitute the DW amplitude into the formulas of Section 9.2, to obtain explicit expressions for the polarization and asymmetry in reaction experiments. Although spin-orbit terms of optical potentials cause important modifications, they are omitted from the analysis, and the results obtained using spin-orbit interactions are only quoted when required. Hence it is the non-spin-orbit DW amplitude of (4.59) that is used in the analysis. We follow Satchler [2] and combine the spectroscopic coefficient A and reduced amplitude β into a single symbol B

$$B_{sj}^{\ lm} \equiv A_{lsj}\beta_{sj}^{\ lm}. \tag{9.36}$$

The polarization of particle b is found from (9.28). If we choose a coordinate system that has its z-axis along $\mathbf{k}_\alpha$ and its y-axis along $[\mathbf{k}_\alpha \times \mathbf{k}_\beta]$, then the polarization $\mathbf{p}_b$ lies along the y-axis and has the magnitude

$$\begin{aligned} p_b(\Theta) = &\left\{\sum_{lsjm} |B_{sj}^{\ lm}|^2\right\}^{-1} \sum_{ll'ss'j} \hat{s}\hat{s}'\hat{l}'\hat{s}_b\left[\frac{2(s_b+1)}{s_b}\right]^{1/2} \\ &\times (-)^{j-l-s_a+s_b} W(ss'll';1j)W(s_bs_bss';1s_a) \\ &\times \sum_m \langle l'1; m+1, -1 \mid lm\rangle \operatorname{Im}[B_{sj}^{\ lm} B_{s'j}^{l'm+1*}], \end{aligned} \tag{9.37}$$

where $\hat{x} = (2x+1)^{1/2}$. The asymmetry caused by a vector-polarized incident beam is found from (9.31). It may also be computed from (9.37) by use of the polarization-asymmetry theorem. The explicit expression is obtained by interchanging s_a with s_b and multiplying by $(-)^{s-s'}$. Expressions for the asymmetries caused by incident tensor polarizations of higher rank may be found in the literature [2, 3].

It is of interest that the sum on j in (9.37) is incoherent, just as in the differential cross section of (4.62). This j is the total angular momentum transferred by the residual interaction $V_\beta - U_\beta$; it is the difference between the angular momenta of nuclei A and B. Because the polarization of particle b is averaged over the spin orientations of nuclei A and B, the incoherence on j might seem rather obvious. On the other hand, if the spins of these two nuclei are to be decoupled from the rest of the calculation it is essential that the DW matrix element contain only a single state of each nucleus. Hence, the incoherence on j requires the factored form of the DW wavefunctions; it is a consequence of the basic structure of the DW theory [2]. It is clear from this discussion that the incoherence would not be destroyed by the use of spin-dependent optical potentials, provided that these potentials only depend on the spins of the light projectiles, a and b (as is usual, see Section 5.2).

Our omission of spin-dependent distortion does limit the interference among different values of the orbital angular momentum transfer l, l'. The Clebsch-Gordan coefficient in the last line of (9.37) indicates, that under this condition, l and l' are directly linked by the polarization, a spherical tensor of rank one. This gives $l' = l, l \pm 1$. This restriction is of particular importance for *normal-parity* transitions, because for these it causes the double sum on l, l' to reduce to an incoherent single sum. Transitions of this type (see Chapter 5) include nearly all reactions initiated by projectiles of low mass, such as inelastic scattering, stripping from S states, or any other transitions in which the zero-range approximation is valid. The normal parity gives $(l + l')$ even; as a result $l' = l$, uniquely, and (9.37) becomes

$$p_b(\Theta) = \left\{\sum_{lsjm} |B_{sj}{}^{lm}|^2\right\}^{-1} \sum_{lss'j} \hat{s}\hat{s}'\hat{l}\hat{s}_b \left[\frac{(s_b + 1)}{l(l + 1)s_b}\right]^{1/2}$$
$$\times (-)^{j-l-s_a+s_b} W(ss'll; 1j) W(s_b s_b s s'; 1 s_a)$$
$$\times \sum_m [(l - m)(l + m + 1)]^{1/2} \operatorname{Im} [B_{sj}{}^{lm} B_{s'j}^{lm+1*}]. \qquad (9.38)$$

The disappearance of l, l' interference greatly reduces the consequences of any l-mixing. However, (9.38) still contains interference between different spin-transfers, s.

Further discussion of polarization requires consideration of individual reactions, and requires use of the explicit form factors derived in Chapter 5. It is clear that calculations of polarizations are natural byproducts of all usual DW calculations of differential cross sections.

For nucleon inelastic scattering, the projectile spins have the values $s_a = s_b = \frac{1}{2}$, and the possible spin-transfers are $s = 0$ or $s = 1$, corresponding to "non-spin-flip" and "spin-flip" transitions, respectively. Although the $s = 1$ transitions are believed to be fairly weak (Section 5.5) they are important for the polarization. To see this, we note that terms in (9.38) with $s = s' = 0$ vanish. Hence nonvanishing polarization requires either spin-dependent distortions or spin-flip nuclear transitions. Much of the investigation of polarization in (p, p') reactions concerns the determination of the relative importance of these two effects.

Accurate measurements of polarization angular distributions in (p, p') reactions have become more numerous since the development of polarized ion sources. In many cases, DW calculations of the (p, p') polarizations seem to be as satisfactory as for the differential cross sections [33–37]. Since the experiments generally treat excitations that show strong collective enhancement, it is appropriate that most of the DW calculations are based on the macroscopic collective model (see Chapter 5) of the excitation process. In this model the projectile is pictured as interacting with a deformed optical potential whose deformation depends on the internal coordinates of the target nucleus. Occasional calculations that employ microscopic models of the excitation process are less successful [36, 37], apparently for the reasons discussed in Chapter 5. Figures 9.1 and 9.2 give an example of the medium-energy (p, p') polarization data now available, and show how DW calculations of the differential cross section and polarization compare with experiment. To obtain theoretical results of the quality in Figs. 9.1 and 9.2 it is necessary to adjust the optical potential to fit the elastic differential cross section and polarization to the largest possible scattering angles. Of course, the (p, p') polarization is more sensitive to small parameter variations than the differential cross section.

The successful DW description of the (p, p') polarization is obtained as a natural outcome of the collective model of inelastic scattering, which thus is seen to fit a remarkably wide range of phenomena. It is even more impressive to realize that the polarization calculations require consistent, careful application of the collective model. This is emphasized by the treatment of the spin-orbit term of the optical potential. The polarization results are not satisfactory unless a suitable deformation of the spin-orbit term is included in the residual interaction that excites the nucleus.

Because the deformed spin-orbit potential does not depend on the spin of the target nucleus, or on the coordinates of individual nucleons in the target nucleus, the residual interaction given by this potential can only have spin-transfer $s = 0$, with $j = l$, in the sense of the standard multipole expansion of the DW theory. On the other hand, although there is no spin-flip interaction between the projectile and the target nucleus, there is a triple spin-flip interaction that involves the spin and orbital angular momentum of the projectile in the presence of the target nucleus. Standard DW codes must be altered somewhat to accommodate this interaction.

Small effects can cause large modifications of the polarization. For this reason it is of interest that Fricke, Gross and Zucker [35] show that coupled-channels calculations give much the same (p, p') polarizations as the DW calculations.

In (p, p') polarization studies for bombarding energies above 100 MeV it is usually possible to treat the spin-orbit distorting potential in perturbation approximation, either in a full DW calculation or in WKB approximation [24, 38–41]. Then the spin-orbit distortion and the spin-flip term of the nuclear form factor make additive contributions to the net polarization. Both

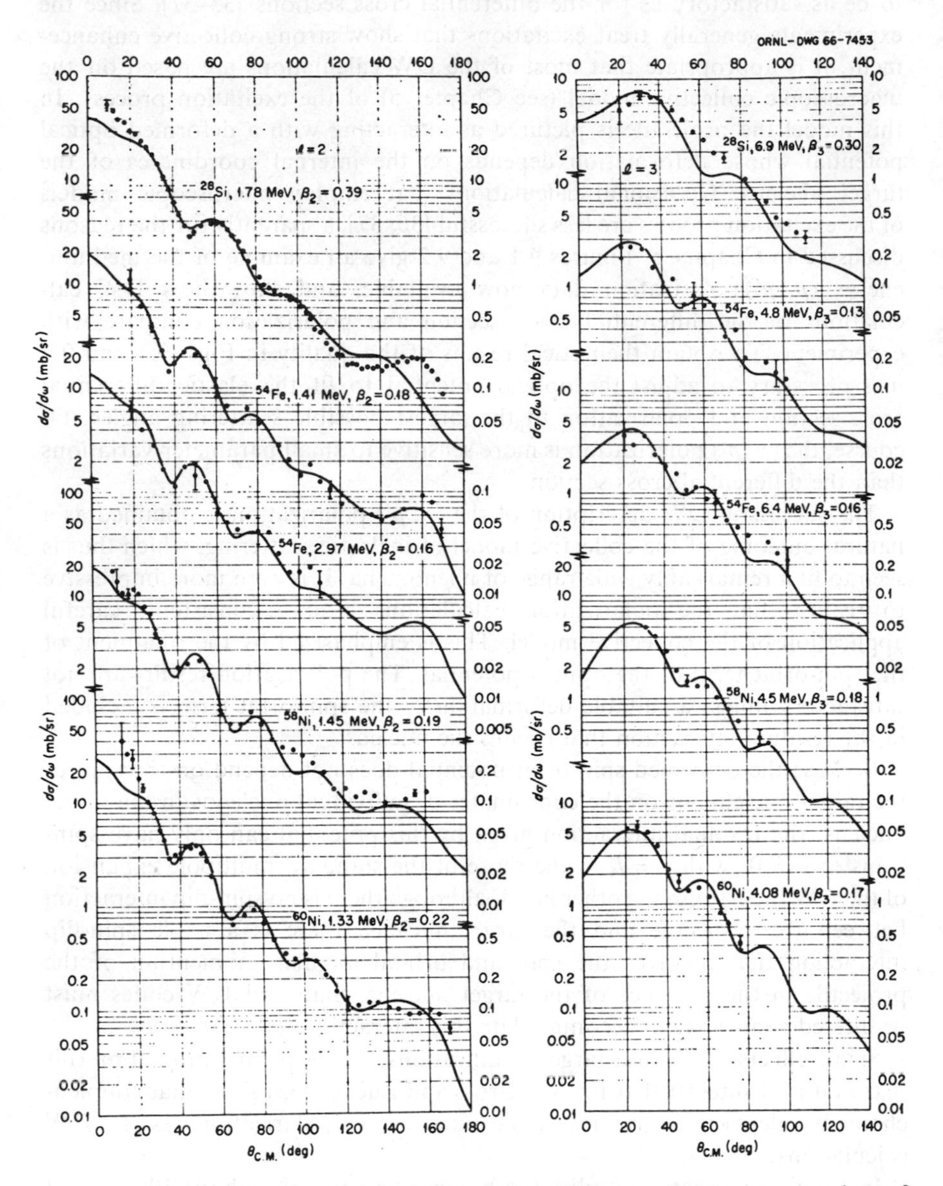

Fig. 9.1 Cross-section data and DW predictions, for the inelastic scattering of 40 MeV protons by the indicated nuclei. (From [35], Fig. 10.)

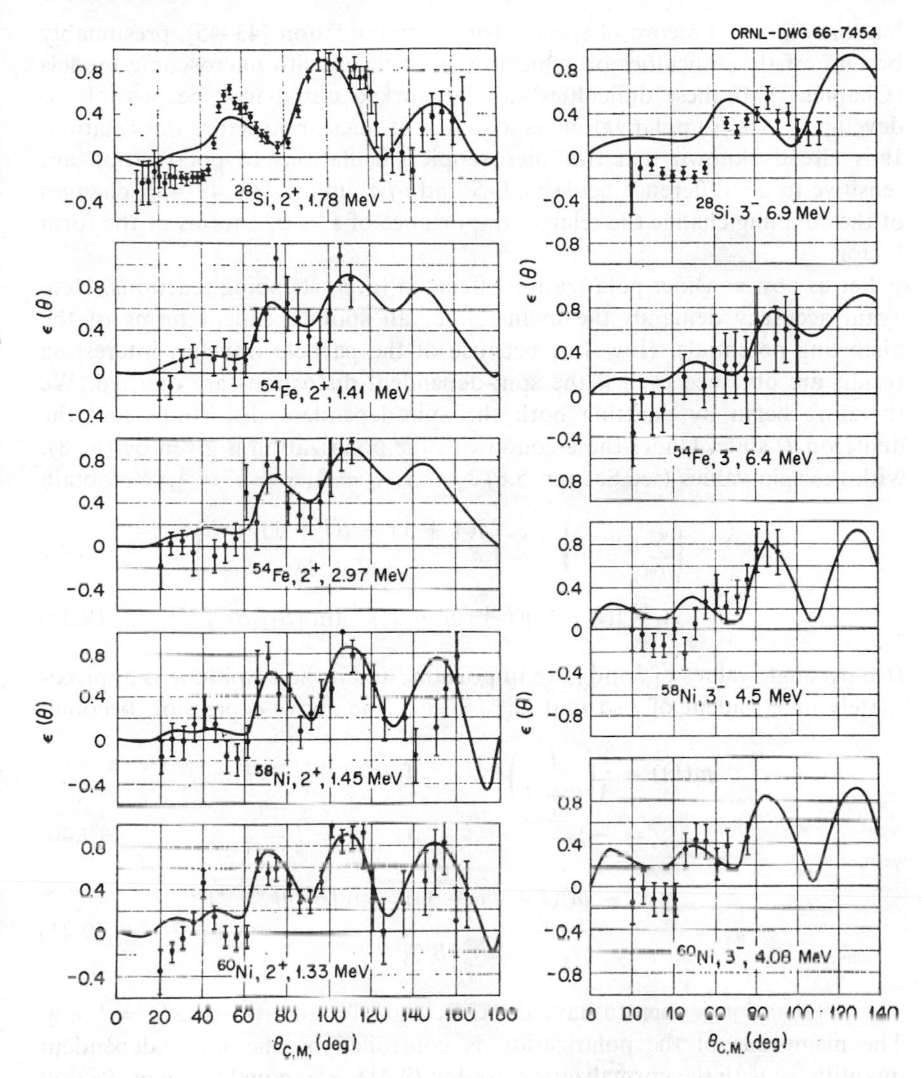

Fig. 9.2 Inelastic asymmetry data and DW predictions associated with the differential cross sections of Fig. 9.1. (From [35], Fig. 11.)

contributions are important, contrary to earlier suggestions by Köhler [42]. The first-order contribution due to the form factor can be computed from (9.38), and it possesses all the simple properties of that expression.

At high energy it would seem possible to treat the (p, p') residual interaction by distorted-waves impulse approximation (Section 5.6) in conjunction with microscopic models of the nuclear excitation. This important approach has had only a mixed record of success for the polarization [43–45], presumably because of the difficulties of achieving consistency with microscopic models (Chapter 5). If these difficulties can be worked out it may be possible to develop the (p, p') polarization as a source of nuclear structure information. It is already known that the microscopic calculations of polarization are sensitive to the difference between L-S and j-j coupling [38, 43–45]; changes of the coupling change the relative importance of $s = 0, 1$ terms of the form factor.

Let us now consider polarization effects in (d, p) stripping reactions. Here again accuracy demands the inclusion of all spin-dependent terms of the distorting potentials. However, because of the particle transfer, interesting results are obtained even if the spin-dependent distortions are omitted. We therefore begin by omitting both the spin-dependent distortions and the deuteron D-state. Under these conditions the polarization is given by (9.38), with the spin values (see Section 5.8) $s_a = 1$, $s_b = \frac{1}{2}$, $s = s' = \frac{1}{2}$. We obtain

$$p_b(\Theta) = \left\{\sum_{jlm} |B^{lm}_{\frac{1}{2}j}|^2\right\}^{-1} \sum_{lj} \left[\frac{j(j+1) - l(l+1) - \frac{3}{4}}{3l(l+1)}\right] \times \sum_m [(l-m)(l+m+1)]^{\frac{1}{2}} \operatorname{Im}[B^{lm}_{\frac{1}{2}j} B^{lm+1*}_{\frac{1}{2}j}]. \tag{9.39}$$

If only single values of l and j are important, and if the amplitude is approximately independent of j so that $B^{lm}_{\frac{1}{2}j} \approx B^{lm}$, the above expression becomes

$$\begin{aligned} p_b(\Theta) &= \frac{1}{3}\left(\frac{l}{l+1}\right)\pi_l, \quad &\text{if} \quad j = l + \tfrac{1}{2}, \\ &= -\tfrac{1}{3}\pi_l, \quad &\text{if} \quad j = l - \tfrac{1}{2}, \end{aligned} \tag{9.40}$$

where

$$\pi_l = \frac{\sum_m [(l-m)(l+m+1)]^{\frac{1}{2}} \operatorname{Im}[B^{lm}B^{lm+1*}]}{l \sum_m |B^{lm}|^2}. \tag{9.41}$$

The polarization is seen to have opposite signs for the two cases $j = l \pm \frac{1}{2}$. The magnitude of the polarization is controlled by the spin-independent quantity π_l; with the normalization used in (9.41), π_l is equal to the projection of the vector $\mathbf{l}$ on the direction $[\mathbf{k}_\alpha \times \mathbf{k}_\beta]$,

$$\pi_l = \frac{\langle l_y \rangle}{l}, \tag{9.42}$$

and thus cannot exceed unity Therefore, the magnitude of the polarization has the upper limit $|p_b| \leqslant \frac{1}{3}$. These results were first given by Newns [46].

Semiclassical interpretations of the above results were given by Newns [46] and by Biedenharn and Satchler [47]. The semiclassical analysis begins with the observation that π_l would vanish in the absence of distortion (e.g., in the Butler theory), for in this case the probability that the neutron is captured does not depend on its location in the target nucleus; as a result the average impact parameter of the captured neutron has the value zero. Distortions (even if spin-independent) then must tend to upset the uniform distribution of impact parameters and cause the average $\langle \mathbf{l} \rangle$ to be nonvanishing. With $\langle \mathbf{l} \rangle$ nonvanishing the spins of neutrons captured with $j = l + \frac{1}{2}$ tend to lie parallel to the preferred direction of $\langle \mathbf{l} \rangle$, while the spins of neutrons captured with $j = l - \frac{1}{2}$ tend to lie opposite to $\langle \mathbf{l} \rangle$. Polarization of the outgoing proton comes about because the nucleons in the incident deuteron tend to have parallel spins. The limit $|p_b| \leqslant \frac{1}{3}$ is a measure of the probability that these nucleons do in fact have parallel spins [47].

The above results for the proton polarization also apply for the asymmetry in (d, p) reactions initiated by vector-polarized deuterons, provided that spin-dependent distortions may be ignored. The asymmetry is computed from (9.38) by interchanging the proton and deuteron spins. This step only affects a numerical multiplier, with the result that the proton asymmetry produced by polarized deuterons is exactly three times the proton polarization produced by unpolarized deuterons [48]. The factor 3 implies that polarized deuteron beams can cause asymmetries of up to one hundred percent, presumably because the nucleons in such deuteron beams can have parallel spins.

Equation 9.40 has interesting implications for nuclear spectroscopy because it indicates that measurement of the sign of the polarization should suffice to distinguish transitions having $j = l \pm \frac{1}{2}$. Early investigations of (d, p) polarization did not substantiate this rule [49], and it was presumed that the omission of spin-dependent distortions rendered (9.40) invalid. However, these early experiments treated light nuclei, for which CN complications are likely, and they could not compare $j = l \pm \frac{1}{2}$ measurements for the same nucleus. More recent experimental work has emphasized medium-mass nuclei, and has treated cases for which pairs of known $j = l \pm \frac{1}{2}$ transitions could be compared [29, 50]. This work suggests that under some conditions (9.40) may be quantitatively reliable in the region of the main stripping peak. Yule and Haeberli [29] were able to study asymmetries for the following cases: eight $l = 1$ transitions with $j = \frac{1}{2}, \frac{3}{2}$ in $Mg^{24}(d, p)$, $Ca^{40}(d, p)$, and $Cr^{52}(d, p)$; two $l = 2$ transitions with $j = \frac{3}{2}, \frac{5}{2}$ in $Mg^{24}(d, p)$; four $l = 3$ transitions with $j = \frac{5}{2}, \frac{7}{2}$ in $Ca^{40}(d, p)$ and $Cr^{52}(d, p)$. The experiments show that the asymmetries for the two j values associated with a given l are reliably of opposite sign. Yule and Haeberli [29] showed further that DW calculations agreed with these experimental results and were not overly sensitive to small parameter variations in the vicinity of physically-reasonable values. Near the main stripping peak the gross features of the polarization also were not greatly influenced by the introduction of spin-dependent distortions. Yule and Haeberli attribute the stability of their DW calculations to the L-space localization discussed by Hooper [51].

Spin-dependent distortions seem to be more important at other energies and in other mass regions than those treated by Yule and Haeberli. Not only do DW calculations of the polarization frequently show great sensitivity to spin-dependent distortions, but measured polarizations often disagree markedly with the predictions of (9.40). Thus, transitions with $l = 0$ are observed to have nonzero polarizations [49] whereas (9.40) predicts $p_b = 0$. Furthermore transitions with $l \neq 0$ sometimes have polarizations that exceed the limit $|p_b| \leqslant \frac{1}{3}$. It appears to be necessary to build up a body of experience with spin-dependent distortions before their full significance can be judged. Calculations that treat these interactions in first order [52, 53] may be helpful for the assessment of trends; for example it is known that to first order, tensor terms in the deuteron optical potential do not affect the vector polarization in the (p, d) reaction [48, 52]. In any case, the basic step in the study of the spin-dependent interactions must be the measurement of polarizations in elastic scattering.

The deuteron D-state causes further complications of the polarization in (d, p) reactions. Particular difficulties arise because D-state effects are enhanced by interference with the effects of spin-dependent distortions [54, 55]. However, we recall that even in the absence of spin-dependent distortions (Section 5.8) the D-state introduces transitions that have spin-transfer $s = \frac{3}{2}$, and it causes l-mixing. These effects already cause a certain amount of tensor polarization in (p, d) reactions [54] of the order found experimentally [56]. It is of interest that D-state effects increase with l.

9.4 Angular Correlations

Angular correlation theory concerns the relation between the angles of emission of two, successive *dynamically-independent* radiations from a nuclear system. By assumption, the second such radiation has such a small influence on the conditions of emission of the first, that the intermediate nucleus may be considered to be in a sharp state. For nuclear reactions followed by gamma ray emission, such as we consider in the present discussion, the assumption of dynamical independence tends to be fulfilled very accurately. It is obvious that dynamically-independent radiations are correlated only because the first radiation may leave the intermediate nucleus partially polarized, so that the second radiation is emitted from a polarized source.

There is a vast literature on angular correlations (e.g., see [1, 7, 13, 14, 57–59]), most of which does not concern nuclear reactions and is ignored.

Here we only attempt a brief introduction to the use of gamma ray correlations to understand direct reactions. This aspect of correlations has thus far received fairly limited application, largely because of the difficulty of coincidence experiments with medium-energy cyclotrons. However, new instrumentation gives promise of rapid improvements.

Because we assume dynamical independence, the overall amplitude for a reaction followed by gamma ray emission has the product form

$$\mathscr{T}_{\gamma\alpha} = \sum_{\beta} \Gamma_{\gamma\beta} T_{\beta\alpha}, \tag{9.43}$$

where $\Gamma_{\gamma\beta}$ is the gamma ray amplitude. From (9.20), the overall final density matrix is

$$\rho^f = \Gamma(T\rho^i T^\dagger)\Gamma^\dagger. \tag{9.44}$$

From (9.21), the overall experimental correlation then has the factored form

$$W = \mathrm{tr}\,\{\varepsilon^f \Gamma(T\rho^i T^\dagger)\Gamma^\dagger\} = \mathrm{tr}\,\{\rho\Lambda\}, \tag{9.45}$$

where

$$\rho = T\rho^i T^\dagger \tag{9.46}$$

is the density matrix for the particle transition, and

$$\Lambda = \Gamma^\dagger \varepsilon^f \Gamma \tag{9.47}$$

is the "statistical matrix" for the subsequent gamma ray transition. The factored form of W allows the two transitions to be studied separately, and it permits the use of matrices ρ and Λ that are taken over from studies of other reactions.

In correlation experiments, polarizations generally are neither prepared nor detected. As a consequence, ρ^i and ε^f in (9.46) and (9.47) are delta functions in the spin projections of the incident and emergent radiations, and the matrices ρ and Λ reduce to

$$\rho(M_B, M'_B) = \sum_{m_a M_A m_b} T(\mathbf{k}_\beta, m_b, M_B; \mathbf{k}_\alpha, m_a, M_A) \times T^*(\mathbf{k}_\beta, m_b, M'_B; \mathbf{k}_\alpha, m_a, M_A), \tag{9.48}$$

and

$$\Lambda(M'_B, M_B) = \sum_{M_C} \Gamma^*(\mathbf{k}_\gamma, M_C; M'_B)\Gamma(\mathbf{k}_\gamma, M_C; M_B), \tag{9.49}$$

where $\mathbf{k}_\gamma$ is the wave vector of the gamma ray, and where the nuclear state produced by gamma ray emission has angular momentum J_C, M_C.

The angular distribution of the coincidence gamma radiation is determined by the $\mathbf{k}_\gamma$-dependence of Λ, and by the matrix elements of the ρ-matrix. The latter matrix elements enter as parameters that determine the relative weights of the various M_B-states from which the gamma radiation emanates. We note that under some experimental arrangements [7] the values of these parameters are largely independent of the mechanism of the reaction that produces the ρ-matrix. Such experiments emphasize the role of the angular momenta in the Λ-matrix. Our present interests concern both these angular momenta and also the use of the Λ-matrix to test mechanism-dependent properties of the ρ-matrix.

The density matrix $\rho(M_B, M_B')$ is defined in the spin space of nucleus B, hence in general it contains $(2J_B + 1)^2$ real parameters. Interesting simplifications of this density matrix appear in the DW theory [2, 6, 8], because in this theory the distribution among the projections of J_B is produced simply by vector coupling the target spin J_A with the distribution among the projections of the transfer angular momenta lsj. Special selection rules on the latter distribution then lead to identities on the $\rho(M_B, M_B')$ distribution. One interesting example of these identities is Goldfarb's demonstration [8] that in (d, p) reactions the ρ-matrix for protons emitted at 0° or 180° depends on only *one parameter*. This parameter may be used to measure the relative importance of CN and DI reaction mechanisms.

We also note that DW calculations of the ρ-matrix are not expected to show much sensitivity to spin-dependent distorting potentials. Such potentials directly involve the spins of the light projectiles, a and b, and therefore affect the polarizations of these particles. However, the angular momenta of nuclei A and B do not enter DW calculations through the distorting potentials, but only through the spin-flip terms of the residual interaction $(V_\beta - U_\beta)$.

To put the Λ-matrix and the correlation function into standard form, we expand the ρ-matrix in spherical tensors:

$$\rho(M_B, M_B') = \sum_{KQ} \rho_{KQ}(B)(-)^{K-J_B-M_B}\langle J_B J_B; -M_B M_B' \mid KQ\rangle. \quad (9.50)$$

The insertion of (9.50) in (9.45) then gives the correlation function

$$W = \sum_{KQ} \rho_{KQ}(B)\left[\frac{4\pi}{2K+1}\right]^{1/2} Y_K{}^Q(\hat{k}_\gamma)F_K, \quad (9.51)$$

in terms of the extensively-tabulated [57, 59] correlation coefficients

$$F_K = \sum_{LL'} C_L C_{L'} F_K(LL'J_C J_B). \quad (9.52)$$

Here C_L, $C_{L'}$ are the probability amplitudes for emission of gamma rays of multipole orders L, L'. The rapid convergence of the multipole series for nuclear gamma radiation causes only the lowest nonvanishing terms in (9.52) to be of any importance. At worst it may be necessary to carry interference between electric and magnetic multipoles of adjacent orders, however, even this complication is sometimes suppressed by selection rules, with the result that the sum in (9.52) reduces to a single term. This simplification limits the complication of the correlation function, to the extent that the sum on K in (9.51) sometimes terminates well before the maximum $K \leqslant 2J_B$ allowed by (9.50).

It is also important that the F_K in (9.52) vanish if K is odd. This property expresses the expected symmetry of the gamma ray angular distribution with respect to reflection in the $\mathbf{k}_\alpha$, $\mathbf{k}_\beta$ plane, the reaction plane. Because K must be even, reactions with $J_B = \frac{1}{2}$ have $K = 0$ and give isotropic radiation patterns. It is interesting that Gemmell et al. [60] could use this simple fact to distinguish $j = \frac{1}{2}, \frac{3}{2}$ transitions in $Fe^{54}(d, p)$ reactions with $l = 1$. Transitions that gave non-isotropic decay radiation could be identified as $j = \frac{3}{2}$.

Another interesting symmetry is found if the distorted waves $\chi_\alpha^{(+)}$ and $\chi_\beta^{(-)}$ are replaced by the corresponding plane waves. This step introduces a close relation between the linear momentum transfer $\mathbf{q} = \mathbf{k}_\alpha - \mathbf{k}_\beta$ and the orbital angular momentum transfer $\mathbf{l}$, with the result that $\mathbf{q}$ becomes an axis of symmetry of the decay radiation. The introduction of distorted waves sometimes leaves vestiges of this symmetry.

Gamma-ray correlations in inelastic scattering were studied in the pioneer DW calculations by Levinson and Banerjee [61]. Their successful description of $C^{12}(p, p'\gamma)$ data at 16 and 19 MeV gave strong support to the direct reaction interpretation of the experiments. Investigations of $(p, p'\gamma)$ reactions at higher energies seem to show that the gamma-ray correlation is a sensitive indicator of spin-flip terms in the nuclear interaction [40, 62]. Distortions seem to affect only the overall magnitude of the cross section, and not the gamma-ray correlation pattern.

The gamma radiation that follows inelastic scattering possesses a special symmetry if the nuclear reaction can be treated by the *adiabatic approximation* (see Section 6.3). In this approximation the internal variables of the target nucleus change so slowly that they are effectively constant during the period of interaction with the projectile. Inelastic scattering then may be treated as a special case of elastic scattering, the nuclear excitations being derived from the impulse the scattered projectile delivers to the target nucleus. In this adiabatic approximation the reaction is symmetric with respect to interchange of the entrance and exit channels [63–65], and the impulse delivered to the nucleus is constrained to lie along the adiabatic recoil direction $\mathbf{q} = \mathbf{k}_\alpha - \mathbf{k}_\beta$ where $|\mathbf{k}_\alpha| = |\mathbf{k}_\beta|$. Therefore $\mathbf{q}$ is an axis of symmetry, both for the inelastic amplitude and for the ρ-matrix that parameterizes the gamma-ray correlation. If $\mathbf{J}_B$ is quantized along this adiabatic symmetry axis, then the matrix elements $\rho(M_B, M_B')$ are nonvanishing only if M_B, M_B' are even.* Radiation

* Of course, the condition $|\mathbf{k}_\alpha| = |\mathbf{k}_\beta|$ distinguishes the adiabatic symmetry axis from the plane-wave symmetry axis mentioned earlier. The difference is especially marked at small scattering angles.

patterns observed in $(p, p'\gamma)$ and $(\alpha, \alpha'\gamma)$ experiments often show such symmetry about the adiabatic recoil direction, with the axis of symmetry shifting correctly as the correlated proton or alpha particle is detected at different scattering angles. The DW calculations of Levinson and Banerjee also tend to show adiabatic symmetry [64].

In some experiments the gamma-ray symmetry axis shows striking departures from the adiabatic prediction, even in cases of (α, α') excitation of low-lying collective levels, which should conform very accurately to the adiabatic approximation [66]. Near minima of the alpha-particle angular distribution the gamma-ray symmetry angle frequently is a rapidly varying function of the alpha-particle scattering angle, and it exhibits a spectacular tangent-like behavior. Inglis has given extensive semi-classical interpretations of this phenomenon [67]. Straightforward DW calculations of the inelastic scattering tend to agree with the observed oscillations of the gamma-ray symmetry angle; however, the calculated oscillations are found to be very sensitive to details of distorting potentials [66]. Unfortunately, this sensitivity suggests that the oscillations are a special phenomenon, which is not likely to yield much information about other aspects of the reaction. Similar results were obtained in abbreviated DW calculations [68], of the type mentioned in Chapter 7.

References

[1] S. Devons and L. J. B. Goldfarb, *Handbuch der Physik*, Vol. 42, S. Flügge, Ed., Springer Verlag, Berlin, 1957.
[2] G. R. Satchler, *loc. cit.*, Chap. 4, Ref. 16.
[3] L. J. B. Goldfarb and R. C. Johnson, *Nucl. Phys.* **18,** 353 (1960).
[4] L. Wolfenstein, in *Annual Review of Nuclear Science*, Vol. 6, J. G. Beckerley, M. D. Kamen, and L. I. Schiff, Eds., Annual Reviews, Palo Alto, 1956.
[5] R. Wilson, *The Nucleon-Nucleon Interaction*, Interscience, New York, 1963; P. Signell, in *Advances in Nuclear Physics*, Vol. 2, M. Baranger and E. Vogt, Eds., Plenum, New York, 1969.
[6] R. Huby, M. Y. Refai, and G. R. Satchler, *Nucl. Phys.* **9,** 94 (1958); see also J. P. Martin, K. S. Quisenberry, and C. A. Low, Jr., *Phys. Rev.* **120,** 492 (1960).
[7] A. E. Litherland and A. J. Ferguson, *loc. cit.*, Chap. 2, Ref. 31.
[8] L. J. B. Goldfarb, *loc. cit.*, Chap. 2, Ref. 32.
[9] J. M. O'Dell et al., *loc. cit.*, Chap. 2, Ref. 33.
[10] R. Tolman, *Principles of Statistical Mechanics*, Chap. 9, Oxford University Press, Oxford, 1938.
[11] L. Wolfenstein and J. Ashkin, *Phys. Rev.* **85,** 947 (1952).
[12] R. H. Dalitz, *Proc. Phys. Soc.* (London) **A65,** 175 (1952).

[13] D. M. Brink and G. R. Satchler, *Angular Momentum*, pp. 107, et seq., Oxford University Press, Oxford, 1962.
[14] A. M. Lane, lectures at the Harwell Laboratory, May, 1961, (unpublished).
[15] W. Lakin, *Phys. Rev.* **98,** 139 (1955).
[16] L. C. Biedenharn, *Ann. Phys.* **4,** 104 (1958).
[17] L. J. B. Goldfarb, *Nucl. Phys.* **7,** 622 (1958).
[18] G. R. Satchler, *Nucl. Phys.* **21,** 116 (1960).
[19] F. Coester and J. M. Jauch, *Helv. Phys. Acta* **26,** 3 (1953).
[20] R. J. Blin-Stoyle, *Proc. Phys. Soc.* **A65,** 452 (1952).
[21] G. R. Satchler, *Nucl. Phys.* **8,** 65 (1958).
[22] L. C. Biedenharn, *Nucl. Phys.* **10,** 620 (1959).
[23] L. J. B. Goldfarb and R. C. Johnson, *Nucl. Phys.* **21,** 462 (1960).
[24] C. T. Tindle, *Nucl. Phys.* **A110,** 193 (1968).
[25] L. J. B. Goldfarb and D. A. Bromley, *Nucl. Phys.* **39,** 408 (1962).
[26] J. Raynal, *Phys. Letters* **3,** 331 (1963); *ibid,* **7,** 281 (1964).
[27] P. Schwandt and W. Haeberli, *Nucl. Phys.* **A110,** 585 (1968).
[28] R. H. Bassel, R. M. Drisko, G. R. Satchler, L. L. Lee, Jr., J. P. Schiffer, and B. Zeidman, *Phys. Rev.* **136B,** 960 (1964).
[29] T. J. Yule and W. Haeberli, *Nucl. Phys.* **A117,** 1 (1968).
[30] L. S. Rodberg, *Nucl. Phys.* **15,** 72 (1960).
[31] L. J. B. Goldfarb, G. W. Greenlees, and M. B. Hooper, *Phys. Rev.* **144,** 829 (1966).
[32] A. de-Shalit and I. Hufner, *Phys. Letters* **15,** 52 (1965); I. A. Gubkin, *Nucl. Phys.* **A111,** 605 (1968).
[33] R. M. Craig, J. C. Dore, G. W. Greenlees, J. Lowe, and D. L. Watson, *Nucl. Phys.* **83,** 493 (1966).
[34] D. J. Baugh, M. J. Kenny, J. Lowe, D. L. Watson, and H. Wojciechowski, *Nucl. Phys.* **A99,** 203 (1967).
[35] M. P. Fricke, E. E. Gross, and A. Zucker, *Phys. Rev.* **163,** 1153 (1967).
[36] C. Glashausser, R. de Swiniarski, J. Thirion, and A. D. Hill, *Phys. Rev.* **164,** 1437 (1967).
[37] S. A. Fulling and G. R. Satchler, *Nucl. Phys.* **A111,** 81 (1968).
[38] E. J. Squires, *Nucl. Phys.* **6,** 504 (1958).
[39] D. J. Hooton and N. Ashcroft, *Proc. Phys. Soc.* (London) **81,** 193 (1963).
[40] D. J. Rowe, G. L. Salmon, A. B. Clegg, and D. Newton, *Nucl. Phys.* **54,** 193 (1964).
[41] R. M. Haybron, H. McManus, A. Werner, R. M. Drisko, and G. R. Satchler, *Phys. Rev. Letters* **12,** 249 (1964).
[42] H. S. Köhler, *Nucl. Phys.* **9,** 49 (1958).
[43] R. M. Haybron and H. McManus, *loc. cit.*, Chap. 5, Ref. 76.
[44] R. M. Haybron and H. McManus, *loc. cit.*, Chap. 5, Ref. 77.
[45] H. K. Lee and H. McManus, *loc. cit.*, Chap. 7, Ref. 40.
[46] H. C. Newns, *Proc. Phys. Soc.* (London) **B66,** 477 (1953).

[47] L. C. Biedenharn and G. R. Satchler, *Helv. Phys. Acta*, Supplement VI, 372 (1960).

[48] G. R. Satchler, *Nucl. Phys.* **6,** 543 (1958); **13,** 697 (1959).

[49] L. H. Reber and J. X. Saladin, *Phys. Rev.* **133,** B1155 (1964); S. A. Hjorth, J. X. Saladin, and G. R. Satchler, *Phys. Rev.* **138,** B1425 (1965).

[50] A. A. Rollefson, P. F. Brown, J. A. Burke, P. A. Crowley, and J. X. Saladin, *Phys. Rev.* **154,** 1088 (1967).

[51] M. B. Hooper, *loc. cit.*, Chap. 5, Ref. 122.

[52] R. C. Johnson, *Nucl. Phys.* **35,** 654 (1962).

[53] R. Cirelli and P. Gulmanelli, *Il Nuovo Cimento* **26,** 194 (1962); R. Cirelli, A. Marini, and P. Gulmanelli, *Il Nuovo Cimento* **47B,** 39 (1967).

[54] R. C. Johnson, *loc. cit.*, Chap. 5, Ref. 133.

[55] R. C. Johnson and F. Santos, *loc. cit.*, Chap. 5, Ref. 135.

[56] S. E. Darden and A. J. Froelich, *Phys. Rev.* **140,** B69 (1965).

[57] L. C. Biedenharn and M. E. Rose, *Rev. Mod. Phys.* **25,** 729 (1953); K. Alder, B. Stech, and A. Winther, *Phys. Rev.* **112,** 2029 (1958).

[58] M. E. Rose, *Elementary Theory of Angular Momentum*, Wiley, New York, 1957.

[59] H. Frauenfelder and R. M. Steffen, in *Alpha-, Beta-, and Gamma-Ray Spectroscopy*, K. Siegbahn, Ed., North-Holland, Amsterdam, 1965.

[60] D. S. Gemmell, L. L. Lee, Jr., A. Marinov, and J. P. Schiffer, *Phys. Rev.* **144,** 923 (1966).

[61] C. A. Levinson and M. K. Banerjee, *loc. cit.*, Chap. 5, Ref. 4.

[62] A. B. Clegg and G. R. Satchler, *Nucl. Phys.* **27,** 431 (1961).

[63] G. R. Satchler, *Nucl. Phys.* **18,** 110 (1960).

[64] J. S. Blair and L. Wilets, *Phys. Rev.* **121,** 1493 (1961).

[65] J. S. Blair, *loc. cit.*, Chap. 7, Ref. 23.

[66] D. K. McDaniels, D. L. Hendrie, R. H. Bassel, and G. R. Satchler, *Phys. Letters* **1,** 295 (1962); W. W. Eidson, J. G. Cramer, Jr., D. E. Blatchely, and R. D. Bent, *Nucl. Phys.* **55,** 613 (1964); D. E. Blatchely and R. D. Bent, *Nucl. Phys.* **61,** 641 (1965); W. W. Eidson, J. G. Cramer, Jr., and G. P. Eckley, *Phys. Letters* **18,** 34 (1965).

[67] D. R. Inglis, *Phys. Letters* **10,** 336 (1964); *Preludes in Theoretical Physics*, A. de-Shalit, H. Feshbach, and L. Van Hove, Eds., North-Holland, Amsterdam, 1966; *Phys. Rev.* **142,** 591 (1966); *Phys. Rev.* **157,** 873 (1967).

[68] E. V. Inopin and S. Shehata, *Nucl. Phys.* **50,** 317 (1964); J. G. Wills and J. G. Cramer, Jr., *Proc. Conf. on Nuclear Spectroscopy with Direct Reactions* (Chicago), Argonne National Lab. report ANL-6848 (1964) p. 147.

CHAPTER 10

DW Exchange and Recoil Effects

Banerjee and Pal [1] remarked in 1965 that the theory of exchange effects in direct reaction processes was in a "state of some confusion." Little has happened since to lessen this confusion.

In part the confusion is of a purely technical origin. It is seen in Section 5.12 that the exchange terms in the DW theory contain form factors of extremely long range. As a result the integrations in these terms are subject to considerable cancellations, and are sensitive to details of wavefunctions. Furthermore, the difficult six-dimensional numerical integration procedure of Section 5.13 must be used. It is unfortunate that most calculations of the exchange terms have been careless about these technical questions, often to the extent of approximating the distorted waves by plane waves! Thereby, misleading results have been obtained. One example of such technical carelessness is found in calculations that associate exchange terms with back-angle peaks in angular distributions, but that overlook the back-angle peaks (see Chapter 5) that frequently appear in accurate evaluations of the DW direct term.

Additional confusion is inherent in the initial formulation of the DW theory, and may be seen already in the classification of DW exchange terms in (4.91–4.93) and (4.97–4.99). Two terms in this classification, the *heavy-particle stripping term* and the *distorting-potential exchange term*, have especially been subjects of controversy [2]. In these "core-exchange terms" a valence particle interacts with the core and is ejected, while at the same time the incident particle changes its state from free to bound without experiencing any interaction. One could well suspect the significance of these terms, on the grounds that the interaction between a valence particle and the core should primarily serve as a distorting potential and should not cause the valence particle to change its state. This suspicion seems to be borne out by the mathematical structure of the core-exchange terms. We observe that the bound and free wavefunctions for the incident particle tend to be orthogonal, and therefore the HPS term and the distorting-potential term tend to vanish. Furthermore, even if these two terms were not small individually they would tend to cancel each other, because the final-state distorting potential is nearly the same as the interaction of the emerging particle with the core. The orthogonality effect and the cancellation of the core interaction by the distorting potential were discovered in the period 1960–62 by Bassel and Gerjuoy, by Day, Rodberg, Snow and Sucher, and by Levin [3]. These effects had been overlooked in earlier work because earlier authors had used plane waves for the free particles.*

It is somewhat strange that heavy-particle stripping terms should seem of such little importance, in view of the qualitative analogy with the direct stripping process that originally motivated Madansky and Owen [2] to consider the HPS process. To see this analogy we note that an HPS contribution to a reaction $A(a, b)B$ has by definition the property that the (light) particle a interacts with the (heavy) core of A and thereby shakes off the (light) particle b. In effect, particle a strips the core from nucleus A. This process resembles the reaction $A(d, p)B$ in which nucleus A strips the neutron from the incident deuteron, and thereby liberates the proton. However, although these two processes are subject to similar qualitative descriptions they are quantitatively very dissimilar. Corresponding steps of the two processes involve very different mass ratios and ratios of interaction strengths. The direct stripping process uses as a matter of course wavefunctions that take into account exactly the strong interaction of both light particles with the core; only the weak interaction between the light particles is treated as a perturbation, and it induces the light particles to change their state of motion in the field of the core. In HPS, on the other hand, the particle we picture as being transferred (the core) is the one that dominates the wavefunctions of the system. Both light particles move in the field of this "transferred particle" and they faithfully follow its motion, because it moves slowly and is a seat of strong interactions. Under these conditions, it is clear that HPS must be weak. (Edwards [2] visualized an HPS process $A(a, b)B$ in which projectile a is presumably more massive than the core of A, and is the seat of the distorting potentials. His process does not conform with usual qualitative ideas about HPS.)

However, it is not clear that secondary effects left after the introduction of accurate distorted waves are small enough to allow the complete neglect of the core-exchange terms. The orthogonality of bound and free wavefunctions is broken both by core recoil and by the virtual excitation of internal coordinates of the core. The latter effect, especially, allows the single-particle Hamiltonians that govern the bound and free wavefunctions to be slightly

* For clarity, we note here that the core matrix element that Day et al. demonstrate to vanish is not present in the classification in Chapter 4. Their matrix element is the asymptotic amplitude of the *homogeneous solution* of the distorted-waves differential equation, and it is omitted in the preliminary discussions in Chapter 3 because there are no incident waves in rearrangement channels. Day et al. discuss this point in their footnote 6.

different. This effect is related to the partial cancellation of the core interaction by the final-state distorting potential, and is also related to the well-known energy dependence of the optical potential. We are thus led to realize that a proper analysis of the core-exchange terms requires careful analysis of the derivation of the optical potential for antisymmetrized wavefunctions. Particular care is required, lest approximations tolerable in the direct term introduce errors comparable in magnitude to the entire core-exchange effect.

Antisymmetrization of the optical model has a long history (for reviews see [4, 5]), much of which is not relevant to the present discussion. In some applications [4] the optical potential is entirely derived by calculation, as an intermediate step toward the calculation of cross sections. In such applications antisymmetrization is of great importance because it controls the intermediate states used in the derivation. However, the optical potentials in nuclear physics are only in part derived. In part they also are fitted to experiment and they thereby automatically incorporate many effects of antisymmetrization. These questions are discussed in Section 10.1. The analysis is combined with a discussion of the overcompleteness of the DI model for antisymmetrized wavefunctions.

The core-exchange terms are also discussed in Section 10.1, under the assumption of no recoil. Recoil is introduced in Section 10.2.

A *knockon exchange term* also appears in the classification in Chapter 4, (4.91, 4.97). Because this term contains an interaction between the incident and ejected particles it is not affected by the near orthogonality of bound and free single-particle wavefunctions. Hence, while this term is subject to the technical difficulties mentioned earlier, its evaluation can at least proceed from a well-defined starting point. It is discussed in Section 10.3.

Some exchange terms that are important in collisions of heavy ions are described in Section 10.4

10.1 Superfluous Wavefunctions, Core-exchange

Exact wavefunctions can be antisymmetrized either before or after solving the Schrödinger equation, as noted in Chapter 4. However, DI calculations are performed with severely truncated model wavefunctions, and we cannot be sure such truncated wavefunctions are associated with symmetrical model Hamiltonians unless antisymmetrization is inserted at the first stage of calculation. We therefore treat the antisymmetrized model wavefunction

$$\Psi_{\text{model}} = \mathscr{A} \sum_{\gamma} \xi_{\gamma}(0)\psi_{\gamma}(1, \ldots, n). \tag{10.1}$$

For definiteness and simplicity it will be considered that (10.1) describes inelastic scattering of spinless nucleons. Hence ψ_{γ} are bound antisymmetric energy eigenstates of the target nucleus, and ξ_{γ} are the associated functions that describe the motion of a projectile relative to this nucleus. We now treat the influence of antisymmetrization on the calculation of the functions ξ_γ. In this discussion we assume $n \gg 1$, so that effects of recoil may be put aside. (These effects will be considered in the following section.) We also ignore the possibility (see Section 4.9) that the exchange terms in (10.1) may be less important than other, omitted rearrangement terms of similar mathematical structure.

To calculate the functions ξ_γ we follow the procedure of Section 4.2, and require Ψ_{model} to be a solution of the model Schrodinger equation

$$(E - H_{PP} - \bar{\mathcal{O}})\Psi_{\text{model}} = 0. \tag{10.2}$$

Here H_{PP} and $\bar{\mathcal{O}}$ are symmetrical operators in the truncated antisymmetric part of Hilbert space. The operator $\bar{\mathcal{O}}$, as usual, corrects for the coupling to parts of Hilbert space that are not contained in the DI model.

Equation 10.2 is made explicit, as in Section 4.9, by left multiplication by each of the internal functions $\psi_\gamma^*(1, \ldots, n)$, followed by integration over the coordinates $1, \ldots, n$. In this fashion (10.2) is reduced to a set of coupled equations for the relative wavefunctions $\xi_\gamma(0)$. Because projection on the limited set of internal functions ψ_γ is equivalent to specification of the DI model, the matrix elements of H_{PP} in the coupled equations simply are matrix elements of the full H, and the coupled equations may be written

$$\sum_\gamma \langle \psi_\beta(1, \ldots, n)| \, E - H - \bar{\mathcal{O}} \, |\mathscr{A}\psi_\gamma(1, \ldots, n)\rangle \xi_\gamma(0) = 0, \tag{10.3}$$

where the bra-ket notation indicates integration over the coordinates $1, \ldots, n$. Clearly, each bracket symbol in (10.3) represents an operator that acts on the coordinate 0 in $\xi_\gamma(0)$:—Two aspects of the above set of coupled equations still remain obscure. One is that the exchange terms in (10.3) generate highly nonlocal operations on the functions $\xi_\gamma(0)$, and the significance of these operations must be understood before practical calculations based on (10.3) can be developed. The other obscure point is that no symmetric optical operators of the kind required for $\bar{\mathcal{O}}$ have ever been developed. Therefore, some method for relating $\bar{\mathcal{O}}$ to usual optical model ideas must be found.

The complications of the exchange terms in (10.3) are tied in with the nonuniqueness that is well known [5, 6] for the coefficients ξ_γ in antisymmetrized expressions of the form of (10.1). As an example of this nonuniqueness we note that if one of the ψ_γ should be a determinant of orthogonal single-particle orbitals, then any linear combination of these orbitals could be added to the associated ξ_γ without affecting the expression $\mathscr{A}\xi_\gamma\psi_\gamma$. In general the ψ_γ are not so simple as in this example, and it is more difficult to describe the superfluous parts of the ξ_γ. Indeed, in general the antisymmetrized terms in (10.1) are not even orthogonal, despite the orthogonality of the parent states ψ_γ, and a description of the superfluous parts requires simultaneous analysis of all the terms in the wavefunction. A full analysis of the superfluity problem is obtained [5, 6] in terms of the density matrix $\mathscr{K}$, whose matrix elements are

$$K_{\gamma\gamma'}(\mathbf{r}, \mathbf{r}') \equiv n\langle\psi_\gamma(\mathbf{r}, 2, 3, \ldots, n), \psi_{\gamma'}(\mathbf{r}', 2, 3, \ldots, n)\rangle. \tag{10.4}$$

This analysis shows that eigenfunctions of $\mathscr{K}$ that have eigenvalue unity may be added to the ξ_γ without affecting (10.1). Let us denote the ith $\mathscr{K}$-matrix eigenfunction by the symbol $\boldsymbol{\sigma}_i(0)$. Then for each i in this (orthonormal) set of superfluous functions the component of $\boldsymbol{\sigma}_i(0)$ in the relative wavefunction $\xi_\gamma(0)$ is $\sigma_{i\gamma}(0)$, and the entire set of components satisfies

$$\mathscr{A} \sum_\gamma \sigma_{i\gamma}(0)\psi_\gamma(1, \ldots, n) = 0. \tag{10.5}$$

Although the superfluous eigenfunctions only are defined above in terms of the antisymmetrization operations in (10.1), the $\sigma_{i\gamma}(0)$ so defined nevertheless are solutions of the coupled set of (10.3). This comes about [7], because the matrix elements in these equations are antisymmetrized with respect to the same set of bound states ψ_γ that determine the superfluous functions. In effect, the combination of the direct and exchange terms in the coupled equations contains a projection operator that annihilates any superfluous parts in the $\xi_\gamma(0)$. Hence the superfluous and nonsuperfluous parts of the $\xi_\gamma(0)$ are separately solutions of the coupled equations; the coupled equations do not link these two parts of the wavefunction. Again, a simple example of this situation is found if the ψ_γ all are single determinants, composed of the single-particle orbitals governed by a given Hartree-Fock potential. In this case the superfluous parts of the ξ_γ are precisely the single-particle orbitals occupied in the bound states ψ_γ, and the above discussion merely states that HF orbitals can be made orthogonal. *Similar results follow for collisions between composite projectiles* [7]

When constructing nuclear reaction theories, it clearly is advantageous to exclude the superfluous parts of the $\xi_\gamma(0)$ at an early stage of calculation. Use of only nonsuperfluous parts of the $\xi_\gamma(0)$ certainly minimizes cancellation between direct and exchange terms when Ψ_{model} is applied for the calculation of the transition amplitude. Hence a calculation based on the nonsuperfluous wavefunctions should have minimum sensitivity to approximations. Moreover, as Saito has emphasized [7], the relative wavefunctions $\xi_\gamma(0)$ are easier to interpret if the superfluous parts are first removed. This is especially helpful in studies of the scattering of composite projectiles.

However, it is in the development of approximate coupled equations to replace (10.3) that the systematic exclusion of superfluous parts is most advantageous. For simplicity, let us specialize to the discussion of elastic scattering, so that the superfluous functions $\sigma_{i\gamma}(0)$ may be treated as single-index quantities $\sigma_i(0)$. The antisymmetrized wavefunction for elastic scattering is

$$\Psi = \mathscr{A}\xi(0)\psi(1, \ldots, n), \tag{10.6}$$

where ψ is the wavefunction of the target-nucleus ground state. The superfluous functions $\sigma_i(0)$ are derived from $\psi(1, \ldots, n)$ according to the discussion following (10.4). We now use the (orthonormal) set of functions $\sigma_i(0)$ to construct an idempotent projection operator, $P^2 = P$, of the form

$$P = 1 - \sum_i |\sigma_i(0)\rangle\langle\sigma_i(0)|, \tag{10.7}$$

This operator [5–7] excludes superfluous parts from $\xi(0)$. To apply the operator P we note that for elastic scattering the coupled set of (10.3) reduces to the single integrodifferential equation

$$\langle\psi(1, \ldots, n)|\, E - H - \overline{\mathcal{O}}\, |\mathscr{A}\psi(1, \ldots, n)\rangle\xi(0) = 0. \tag{10.8}$$

In view of the discussion given earlier, the nonsuperfluous part of $\xi(0)$ is in itself a solution of (10.8), and therefore insertion of the projection operator gives an equation

$$P\langle\psi(1, \ldots, n)|\, E - H - \overline{\mathcal{O}}\, |\mathscr{A}\psi(1, \ldots, n)\rangle P\xi(0) = 0, \tag{10.9}$$

whose nonsuperfluous solutions are identical with those of (10.8). It is easier to introduce approximations in (10.9) than in (10.8).

The projection operator in (10.9) allows as its first consequence a modified treatment of the core-exchange terms. To see this, we write out (10.9) more explicitly, as in Chapter 4, to obtain

$$\begin{aligned} 0 = {} & P\langle\psi(1, \ldots, n)|\, E - \varepsilon - T_0 - V_{0A} - \overline{\mathcal{O}}\, |\psi(1, \ldots, n)\rangle P\xi(0) \\ & - nP\langle\psi(1, \ldots, n)|\, V(10)\, |\psi(0, 2, \ldots, n) P\xi(1)\rangle \\ & - nP\langle\psi(1, \ldots, n)|\, E - \varepsilon - T_0 - \sum_{i=2}^{n} V(i0) - \overline{\mathcal{O}}\, |\psi(0, 2, \ldots, n) P\xi(1)\rangle, \end{aligned} \tag{10.10}$$

where ε is the binding energy of the target nucleus ground state, and V_{0A} is the net interaction between nucleon 0 and the (labelled) nucleons of the target nucleus. Here the knockon exchange term has been isolated in the second line and the core-exchange terms have been isolated in the third line. It is clear that if ψ were a single determinant, all the orbitals in this determinant would be orthogonal to $P\xi$, and the third line of (10.10) would vanish identically, as in the Hartree-Fock theory. This result is much simpler than would be found in (10.8), for example, where the core-exchange terms play the important role of compelling the superfluous functions $\sigma_i(0)$ to be solutions. Hence the difference between (10.8) and (10.9, 10.10) is precisely that in the two latter equations the superfluous functions are dealt with by the projection operator, and therefore the core-exchange terms are irrelevant. Even if ψ is not a single determinant, it is an excellent approximation [7] to omit the core-exchange terms from (10.10). Introduction of the projection operator greatly reduces the sensitivity to these complicated terms.

Another consequence of the form of (10.10) is that the role of the optical operator $\bar{\mathcal{O}}$ can now be clarified. Because the projection operator excludes coupling to superfluous parts of Hilbert space, it is not essential in (10.10) to treat $\bar{\mathcal{O}}$ as a symmetric operator. In fact, except for the core-exchange terms, $\bar{\mathcal{O}}$ appears only in the rather simple direct matrix element

$$\langle\psi(1, \ldots, n)|\ \bar{\mathcal{O}}\ |\psi(1, \ldots, n)\rangle$$

and it is straightforward to treat this matrix element as a phenomenological operator that acts on the coordinate 0, in line with usual optical model procedures.

In summary, a practical calculation of the relative wavefunction $P\xi$, for an antisymmetric treatment of eleastic scattering, may be based on (10.10) in the following fashion: (a) The last line of (10.10) is omitted entirely. (b) The interactions in the direct term of (10.10) are replaced by a complex phenomenological potential

$$U(0) \equiv \langle\psi(1, \ldots, n)|\ V_{0A} + \bar{\mathcal{O}}\ |\psi(1, \ldots, n)\rangle. \tag{10.11}$$

(c) Depending on the properties of the wavefunction ψ, the knockon exchange term in (10.10) either is carried explicitly, as would be necessary if ψ contains very loosely-bound orbitals, or it is absorbed implicitly as a part of the phenomenological potential $U(0)$. Hence the proper treatment of this knockon term may tend to be somewhat case dependent.

According to the above discussion, (10.10) is replaced by the approximate equivalent

$$0 \approx P[E - \varepsilon - T_0 - U(0)]P\xi(0)$$
$$-nP\langle\psi(1, \ldots, n)|\ V(01)\ |\psi(0, 2, \ldots, n)P\xi(1)\rangle, \tag{10.12}$$

where the knockon term has been left explicit. To treat the projection operator in (10.12) we now take advantage of the separable form of the operator $(1 - P)$ and recast (10.12) as

$$[E - \varepsilon - T_0 - U(0)]P\xi(0) - n\langle\psi(1, \ldots, n)|\ V(01)\ |\psi(0, 2, \ldots, n)P\xi(1)\rangle$$
$$= \sum_i \lambda_i \sigma_i(0). \tag{10.13}$$

Here the parameters λ_i represent definite integrals, derived by left multiplying the expression $[E - \varepsilon - T_0 - U(0)]P\xi(0)$ by the operator $(1 - P)$.

Equation 10.12 is most easily solved by treating it as an inhomogeneous differential equation that is integrated numerically for arbitrary sets of values of the parameters λ_i. From the set of solutions thus obtained, the one that is orthogonal to all the superfluous functions $\sigma_i(0)$ is selected. Thus the parameters λ_i are handled like Lagrange multipliers.

The orthogonality terms on the RHS of (10.13) eliminate the superfluous parts of $P\xi(0)$, and ensure that spurious intermediate states do not enter in the calculation of this wavefunction. Many authors have advocated the introduction of such orthogonality terms when solving the Schrodinger equation for antisymmetrized wavefunctions [4, 7, 8]. At low bombarding energy the orthogonality terms cause considerable scattering, irrespective of the interactions in (10.13).

In practical calculations with (10.13), the difficulty in computing accurate $\mathscr{K}$-matrix eigenfunctions must undoubtedly compel the use of greatly simplified versions of the functions $\sigma_i(0)$, at the cost of some superfluity in the relative wavefunction $P\xi(0)$. A practical approximate set of superfluous functions might ordinarily be just the principal bound single-particle states occupied in the wavefunction ψ.

From the standpoint of nuclear reaction theory, the above discussion of elastic scattering has the interest that it provides an antisymmetric procedure for the calculation of $\chi_\alpha^{(+)}$ and $\chi_\beta^{(-)}$, the distorted waves in DW calculations. Presumably all antisymmetrized DW calculations should use distorted waves calculated by this procedure. This has never yet been done.

Let us now return to the discussion of core-exchange contributions to nuclear reactions that was initiated in the Introduction to this chapter. A typical core-exchange term in the DW amplitude for nucleon inelastic scattering is the HPS term of (4.92):

$$\langle \psi_\beta(1, \ldots, n)\chi_\beta^{(-)}(0)| \sum_{i=2}^{n} V(i0)\, |\psi_\alpha(0, 2, \ldots, n)\chi_\alpha^{(+)}(1).\rangle \tag{10.14}$$

Unfortunately, it would appear that our analysis of antisymmetrized elastic scattering yields no more understanding of (10.14) than we already had in the Introduction. Insofar as ψ_α and ψ_β are made up of single-particle orbitals governed by a single, given Hartree-Fock potential, the distorted wave $\chi_\alpha^{(+)}$ will be orthogonal to all the single-particle orbitals in ψ_β, and the HPS amplitude will vanish. To obtain a nonvanishing HPS amplitude it would appear necessary to carry the core-exchange effects to higher order, and to include them in the determination of the relative wavefunctions. It probably also is necessary to go to the full superfluity discussion, based on (10.4), and to determine and use superfluous functions $\sigma_{i\gamma}(0)$ that take account of channels α, β in a self-consistent fashion. Evidently a thorough analysis of the core-exchange terms requires the study of explicit models, and remains a problem for the future.

10.2 Recoil Effects

In an attempt to find a physical basis for the core-exchange process (heavy-particle stripping), the discussion in Section 10.1 considered several "polarizibility effects," that might break the orthogonality of bound and continuum single-particle wavefunctions. However, within the constraint of antisymmetrization, it was difficult to treat these effects well enough to evaluate their contribution to the HPS process.

Within the distorted waves context, recoil is treated merely by introducing correct, nonredundant relative coordinates in the equations of Chapter 4. Let us consider as an example recoil effects in the antisymmetrized DW amplitude for nucleon inelastic scattering (4.89). We denote nucleons 2, . . . , n by the inclusive symbol C. Then suitable relative coordinates are $\mathbf{r}_{1C}$, the displacement of 1 from the center of mass of C, and $\mathbf{r}_{0A}$, the displacement of 0 from the center of mass of C and 1. (The kinetic-energy operator separates in these coordinates.) We find

$$\mathbf{r}_{0A} = \mathbf{r}_{0C} - n^{-1}\mathbf{r}_{1C}, \tag{10.16}$$

where $\mathbf{r}_{1C}$ is the displacement of 1 from the center of mass of C. The coordinates $\mathbf{r}_{1C}$ and $\mathbf{r}_{0A}$ are natural variables in the exit channel and in the direct term in the entrance channel. Upon introducing the above explicit coordinates in (4.89), the amplitude for inelastic scattering becomes

$$T_{\alpha\beta}{}^{\mathrm{DW}} = \big(\chi_\beta^{(-)}(\mathbf{r}_{0A})\psi_B(\mathbf{r}_{1C}, C), [V(10) + V_{C0} - U_\beta(r_{0A})] \times [1 - nP_{01}]\chi_\alpha^{(+)}(\mathbf{r}_{0A})\psi_A(\mathbf{r}_{1C}, C)\big), \tag{10.17}$$

where $V_{C0} = \sum_{i=2}^{n} V(i0)$ is the sum of all interactions of particle 0 with the core.

The introduction of explicit relative coordinates evidently does not imply any alteration of the direct term in (10.17).

The exchange term in (10.17) is

$$T_{\alpha\beta}{}^{\mathrm{DW}}(EX) = -n\big(\chi_\beta^{(-)}(\mathbf{r}_{0A})\psi_B(\mathbf{r}_{1C}, C), [V(10) + V_{C0} - U_\beta(r_{0A})] \times \chi_\alpha^{(+)}(\mathbf{r}_{1C} - n^{-1}\mathbf{r}_{0C})\psi_A(\mathbf{r}_{0C}, C)\big). \tag{10.18}$$

For simplicity, let us assume ψ_B and ψ_A to be single determinants that differ by the excitation of one orbital. Then only one term of each determinant contributes to nonvanishing core overlap, and (10.18) reduces to

$$T_{\alpha\beta}{}^{\mathrm{DW}}(EX) = -\big(\chi_\beta^{(-)}(\mathbf{r}_{0A})\phi_B(\mathbf{r}_{1C}), [V(10) + \bar{V}(r_{0C}) - U_\beta(r_{0A})] \times \chi_\alpha^{(+)}(\mathbf{r}_{1C} - n^{-1}\mathbf{r}_{0C})\phi_A(\mathbf{r}_{0C})\big), \tag{10.19}$$

where the coefficient n in (10.18) has been cancelled by the normalization coefficients of ψ_A and ψ_B. The interaction $\bar{V}(r_{0C})$ is the expectation of V_{0C} with respect to the internal wavefunction of the core, hence it is a function of the displacement of 0 from the center of mass of the core. The functions

ϕ_B and ϕ_A are the single-nucleon orbitals by which ψ_B and ψ_A differ. Upon introduction of the standard variables $\mathbf{r}_{0A}$ and $\mathbf{r}_{1C}$, (10.19) becomes

$$T_{\alpha\beta}{}^{\mathrm{DW}}(EX) = -(\chi_\beta^{(-)}(\mathbf{r}_{0A})\phi_B(\mathbf{r}_{1C}), [V(10) + \bar{V}(|\mathbf{r}_{0A} + n^{-1}\mathbf{r}_{1C}|) - U_\beta(r_{0A})] \times \chi_\alpha^{(+)}([1 - n^{-2}]\mathbf{r}_{1C} - n^{-1}\mathbf{r}_{0A})\, \phi_A(\mathbf{r}_{0A} + n^{-1}\mathbf{r}_{1C})). \quad (10.20)$$

Recoil corrections in (10.20) appear in the arguments of $\bar{V}$, $\chi_\alpha^{(+)}$, and $\chi_\beta^{(-)}$. These corrections cause major modifications of the core-exchange terms, but are of only secondary importance for the knockon term, based on $V(10)$. Hence recoil corrections to the knockon term are not discussed further.

Let us now assume the core-exchange terms in (10.20) would vanish in the absence of recoil. In other words, let us assume the bound and continuum wavefunctions are governed by the same single-particle Hamiltonian, and would be orthogonal if they had the same argument. Then, to first order in n^{-1}, there is a nonvanishing recoil contribution because $\bar{V}$ and U_β do not have the same argument. This contribution is

$$T_{\alpha\beta}{}^{\mathrm{DW}}(\mathrm{RECOIL}) \approx -n^{-1}(\chi_\beta^{(-)}(\mathbf{r}_{0A})\, \phi_B(\mathbf{r}_{1C}), (\mathbf{r}_{1C} \cdot \nabla\bar{V}(r_{0A})) \times \chi_\alpha^{(+)}(\mathbf{r}_{1C})\, \phi_A(\mathbf{r}_{0A})). \quad (10.21)$$

Additional nonvanishing contributions would arise from the recoil corrections in $\chi_\alpha^{(+)}$ and ϕ_A if $\bar{V}(r_{0A}) \neq U_\beta(r_{0A})$. While normal DW considerations allow $\bar{V} \neq U_\beta$ (see Section 4.5), our present assumption that bound and continuum wavefunctions are governed by the same Hamiltonian removes this freedom. Hence (10.21) probably is the principal recoil contribution to heavy-particle stripping. In any case, the structure of (10.21) typifies the other nonvanishing terms of (10.20).

It is very reasonable that the interaction of a projectile with a core of finite mass cannot be cancelled exactly by the optical interaction with the entire nucleus, inasmuch as the two interactions involve displacements from different centers. This simple effect is the origin of (10.21).

The cross section given by (10.21) was studied by R. C. Johnson and N. Austern (unpublished) and by L. S. Rodberg [9]. We now use the methods of these authors to develop qualitative understanding of the recoil amplitude. We first introduce a double commutator of $\mathbf{r}_{1C}$ with the single-particle Hamiltonian, to replace the long-range operator $\mathbf{r}_{1C}$ by the localized operator $\nabla\bar{V}$. Here $\bar{V}$ is the average interaction of particle 1 with the core; it is the same as the $\bar{V}$ introduced previously, because particles 1 and 0 in the present example are identical. Equation 10.21 then assumes the more symmetrical form

$$T_{\alpha\beta}{}^{\mathrm{DW}}(\mathrm{RECOIL}) \approx -n^{-1}\left(\frac{\hbar^2}{M}\right)(E_\alpha + \varepsilon_B)^{-2} \times (\chi_\beta^{(-)}, \nabla\bar{V}\phi_A) \cdot (\phi_B, \nabla\bar{V}\chi_\alpha^{(+)}), \quad (10.22)$$

where M is the nucleon mass, E_α is the (positive) kinetic energy in channel α, and ε_B is the (positive) binding energy of ϕ_B. Equation 10.22 decreases with E_α in the fashion appropriate for a dipole matrix element.

The interactions $\nabla \bar{V}$ in (10.22) are typical of the interactions in the direct DW amplitude for inelastic scattering (see Chapter 5). Hence the n^{-1} coefficient immediately implies that the recoil amplitude is much smaller than the direct amplitude. Some further reduction of the magnitude of (10.22) comes about because the $\nabla \bar{V}$ operators link the distorted waves to bound states rather than to each other. Hence only one or two partial waves of $\chi_\alpha^{(+)}$ and $\chi_\beta^{(-)}$ participate in (10.22), and these partial waves probably have low angular momentum and are strongly attenuated by absorption. Thus we can conclude that the recoil amplitude is much smaller than the direct amplitude, and is negligible except for light nuclei and under unusual circumstances. It is likely that recoil makes a smaller contribution to the core-exchange process than the somewhat vaguer effects treated in Section 10.1.

Not only is the recoil amplitude small, but we see in (10.22) that this amplitude does not give narrow back-angle peaks in angular distributions. The correlation between the directions of the asymptotic momenta of $\chi_\alpha^{(+)}$ and $\chi_\beta^{(-)}$ comes about only through the vector operator $\nabla \bar{V}$, which links these distorted waves with the bound states ϕ_B and ϕ_A. As a result, the recoil amplitude can only contain zero order and second order Legendre functions, and it must be fore-aft symmetric and vary slowly with scattering angle, irrespective of the details of (10.22). Under these conditions the recoil amplitude has none of the properties associated with analyses of experiment [2] in terms of heavy-particle stripping.

10.3 Knockon

We see in Sections 10.1, 2 that the core-exchange terms associated with heavy-particle stripping are very small unless the core is strongly polarizable. However, no certain general estimate of the polarizibility effect is available, short of the introduction of explicit models for individual cases.

The knockon exchange term of (4.91) and (4.97) and (10.17–10.20) has more definite properties and it is frequently important. Because its contributions in inelastic scattering are much larger than in rearrangement collisions, as noted in Chapter 4, we treat these two types of reactions separately. In both cases careful attention to details of wavefunctions and to finite-range effects is required, as noted in the Introduction to Chapter 10. In addition, both the direct and knockon terms in these two cases should presumably be computed with optical wavefunctions from which spurious parts are eliminated, as described in Section 10.1.

The full DW amplitude for inelastic scattering, aside from core-exchange effects, is obtained by combining the direct and knockon terms. We take advantage of the identity of nucleons to express both terms in the same form, and we thereby obtain for the full DW amplitude

$$T_{\alpha\beta}{}^{\mathrm{DW}} = \left(\chi_\beta^{(-)}(0)\psi_B(1, \ldots, n), \sum_{i=1}^{n} V(i0)[1 - P_{i0}]\, \chi_\alpha^{(+)}(0)\psi_A(1, \ldots, n)\right). \tag{10.23}$$

Here the distorting potential is dropped from the direct term, as in Chapter 5, because the bound wavefunctions ψ_A and ψ_B are orthogonal.

We see in (10.23) that introduction of the knockon term antisymmetrizes the motion of the projectile relative to each valence nucleon with which it engages in a two-body interaction. It is of immediate interest that this amount of antisymmetrization appears automatically in calculations that apply the distorted waves impulse approximation (Section 5.6) to the study of inelastic scattering at high energy. This comes about because DWIA is based on the experimentally-observed two-body scattering amplitude, and this amplitude refers to antisymmetrized pairs of nucleons. For bombarding energies much below 100 MeV we know the DWIA fails because the intermediate states accessible to nucleons inside a nucleus are not the same as in free space. Probably the failure of DWIA appears first in the knockon term because this term is sensitive to details of wavefunctions.

A number of DW calculations at low and medium bombarding energies have been based on the straightforward evaluation of (10.23), using effective two-nucleon interactions of the sort given in (5.75); for example, the original study of $C^{12}(p, p')$ by Banerjee and Levinson [10] was of this type. Some more recent calculations [11–13] indicate that exchange effects are appreciable, while others [14] indicate they are negligible. Unfortunately, the significance of all these ostensibly antisymmetrized calculations still is uncertain, and they probably should be regarded as only preliminary attempts to gauge the importance of the knockon effect. In none of these calculations were spurious parts removed from the distorted waves, and in no case were the distorted waves fitted to elastic scattering data in an antisymmetrized fashion, as described in Section 10.1. While the effects described there tend to be negligible except at very low energies, it is difficult to draw final conclusions in advance of numerical investigations.

The direct and exchange amplitudes in (10.23) have easily distinguishable roles only if we are careful to express the interaction $V(i0)$ entirely in terms of spin-exchange and charge-exchange operators, as explained in Section 5.5. Under these conditions the knockon exchange term contains all the space-exchange effects in the calculation, and our development of the properties of

(10.23) becomes applicable both for inelastic scattering and for charge-exchange reactions. In both cases complicated spatial overlaps are isolated in the knockon term. Thus far only occasional authors have utilized this orderly classification of exchange effects.

Under the above organization of exchange effects the knockon term interchanges the spatial coordinates of interacting bound and free nucleons. Hence it is appreciable only if $V(i0)$ has appreciable matrix elements for large momentum transfers. The importance of the knockon term evidently depends on the range of $V(i0)$. If $V(i0)$ is of zero range the direct and knockon terms in (10.23) are identical up to a multiplicative factor* of order unity, and exchange effects must be very large. However, as $V(i0)$ becomes of longer range it is less able to mediate large momentum transfers, and the emerging particle is less able to receive the full momentum of the incident particle. For the glancing collisions that then occur, as Rodberg emphasized [9], the knockon amplitude is strongly reduced and it loses any tendency for forward peaking (contrary to common belief). However, practical values for the range of $V(i0)$ probably lie closer to the short-range extreme [13].

Additional properties of the knockon term may be seen by carrying the formal development of (10.23) a little further. Because all terms in the sum over i give identical contributions, we take $i = 1$, and perform a parentage expansion of ψ_A and ψ_B to isolate the dependence on particle 1. A typical term from this expansion gives (the notation is simplified and we ignore spins)

$$T_{\alpha\beta}(\text{KNOCKON}) = -(\chi_\beta^{(-)}(\mathbf{r}_0)\,\phi_{\lambda\mu}(\mathbf{r}_1),\, V(10)\chi_\alpha^{(+)}(\mathbf{r}_1)\,\phi'_{\lambda'\mu'}(\mathbf{r}_0)). \quad (10.24)$$

The interaction $V(10)$ is now expanded in multipoles, as in (5.76) and (5.78). Omitting the coefficient of the scalar function $g(r_{01})$, the multipole expansion yields

$$T_{\alpha\beta}(\text{KNOCKON}) = -4\pi \sum_{lm} (\chi_\beta^{(-)}(\mathbf{r}_0)\,\phi_{\lambda\mu}(\mathbf{r}_1),\, g_l(r_0, r_1) \times Y_l^m(\hat{r}_1) Y_l^{m*}(\hat{r}_0)\chi_\alpha^{(+)}(\mathbf{r}_1)\,\phi'_{\lambda'\mu'}(\mathbf{r}_0)). \quad (10.25)$$

Because the distorted waves contain all values of angular momenta, (10.25) does not select only one or two multipoles of the interaction, in the fashion characteristic of the direct DW term. Instead, we must rely on the convergence of the multipole series itself to limit the angular momenta in (10.25). For an interaction of not too short range this series converges well, and the g_l are large only for small l; therefore, the important angular momenta in $\chi_\beta^{(-)}$ lie near λ' and the important angular momenta in $\chi_\alpha^{(+)}$ lie near λ. We now see another effect that tends to reduce the knockon term: If λ and λ' are small (as is often true of nuclei) then interactions of low multipolarity only couple

* This factor is associated with charge-exchange and spin-exchange parts of $V(i0)$.

the bound states to the strongly-absorbed low partial waves in $\chi_\alpha^{(+)}$ and $\chi_\beta^{(-)}$.

An effect of particular interest appears if the angular momentum transfer in the transition $\psi_A \rightarrow \psi_B$ is large. In this case the direct amplitude in (10.23) is small because it uniquely selects from the multipole series a term that has high multipolarity, that therefore has small magnitude. By contrast, because all multipoles participate in the exchange amplitude, irrespective of angular momentum transfer, it can be large. Atkinson and Madsen [13] stress that for transitions of high angular momentum and for $V(i0)$ of not too short range the exchange amplitude can easily be much larger than the direct amplitude. This effect immediately vitiates attempts to fit $V(i0)$ to inelastic scattering data, multipole by multipole, in the simple fashion pursued by Satchler and collaborators (see Section 5.5).*

Evidently the importance of the knockon effect is very dependent on circumstances. We may also consider its role in collective excitations. In these cases the inelastic transition $\psi_A \rightarrow \psi_B$ is strong, because many single-particle excitations combine constructively. Many terms from the parentage expansions of ψ_A and ψ_B must be taken. Because of the complications of the knockon term one might suppose [15] it would not have the same interference structure as the direct term, and one might suppose its participation in collective excitations would be inhibited. However, Atkinson and Madsen [13] give evidence that this idea is false, and that the direct and exchange terms contain similar constructive interference.

Finally, we note the obvious point that the knockon effect may be enhanced for states ϕ_λ and $\phi'_{\lambda'}$ that are weakly bound. In this case the tails of the bound wavefunctions extend to large radii and can overlap with continuum partial waves that are not strongly absorbed. In addition, the multipole expansion of the interaction converges slowly at large radii, and more multipoles can participate in the reaction.

To compare knockon effects in rearrangement collisions with those in inelastic scattering, let us take the (d, p) reaction as an example for discussion. The DW expression for the (d, p) knockon amplitude was previously given in (4.97), (5.153) and (5.155). We see that (d, p) knockon entails capture of a deuteron and ejection of a proton, so that three nucleons change their state of motion. However, (p, p') knockon entails capture of one proton and ejection of another, so that only two nucleons change their state of motion. Clearly the knockon term in inelastic scattering is inherently less complicated, and for this reason alone it may be expected to be larger. The (d, p) knockon term must in addition be much smaller than the (d, p) direct term because it requires both nucleons of the incident deuteron to have close overlap with

* Excellent additional studies of this effect are presented by W. G. Love et al., *Phys. Letters* **29B**, 478 (1969).

the target nucleon, whereas the direct term allows one nucleon to pass by at large radius as a spectator.

Another difference between inelastic scattering and rearrangement is seen by recognizing that, unlike the situation in inelastic scattering, the introduction of zero-range two-nucleon interactions does not reduce the (d, p) knockon term to the same form as the direct term. The nuclear interactions are with individual nucleons in the deuteron, therefore to achieve close resemblance between the (d, p) direct and knockon terms the deuteron wavefunction would have to have zero range. However, the deuteron wavefunction is large and loose. Therefore the capture of the deuteron can only be mediated by those matrix elements of the individual interactions that supply small momentum to the relative motion of the two nucleons of the deuteron. It is as if the individual two-nucleon interactions were averaged over the deuteron wavefunction to give an "effective interaction" that then mediates capture of the deuteron. In view of the long range of this effective interaction, the knockon amplitude must tend to be weak and isotropic.

Some enhancement of the (d, p) knockon term may be possible if there is strong two-nucleon clustering in the surface region of the residual nucleus. Although this effect would provide a large "reduced width" for deuteron capture (good overlap), it would not, of course, relax the limitations of momentum matching mentioned in the preceding paragraph. It is amusing that the structure of the matrix element for (d, p) knockon resembles that for the (He^3, p) stripping reaction. In each case a deuteron is captured and a proton changes its state of motion. This resemblance may be exploitable, to provide a measure of the knockon term.

Despite the above arguments that the (d, p) knockon term is small, it has special interest because it obeys selection rules different from the (d, p) direct term. This comes about because the direct term selectively inserts into the nucleus a nucleon of definite lj. This property imposes a configuration selection rule that is generally more restrictive than the mere conservation of overall angular momentum and parity to which the knockon term is subject. An interesting example is the much discussed reaction $B^{10}(d, p_1)B^{11}$, from the 3^+ ground state of B^{10} to the $(\frac{1}{2})^-$ first-excited state of B^{11}. This reaction cannot proceed by capture of a p-shell neutron unless there is simultaneous strong spin flip of both the incident deuteron and the outgoing proton. However it is allowed for knockon. Barz [16] estimates that knockon can well be responsible for the $\sim$0.1 mb/steradian cross section that is observed.*

* A review of earlier analyses of the $B^{10}(d, p_1)$ reaction may be found in the author's article in *Fast Neutron Physics*, Vol. 2. This reaction had excited attention because early measurements gave an angular distribution similar to that for ordinary $l = 1$ stripping. However, this was fortuitous.

Other cases have been studied, in which normal stripping is forbidden but knockon is allowed. In none of these cases are large cross sections seen.

On the whole, the knockon contribution to stripping seems to be comparable in magnitude with that of core excitation, discussed in Section 6.5. As with that process, interference between direct and knockon terms may sometimes be important [16].

10.4 Heavy Ions

In reactions initiated by the collision of two heavy ions the leading exchange term often is comparable in importance with the direct term and is conveniently reinterpreted as the amplitude for some other, simple transfer process. This comes about because the masses of the colliding nuclei are nearly equal, hence permutations that exchange the cores of these nuclei are equivalent to the transfers of small clusters of valence nucleons. To see this let us treat as an example a reaction $A + B \rightarrow C + D$, in which A, B, and D, respectively, contain one, two, three more nucleons than C. Let us regard C as the "core" of A, B, and D. Let us also use C in an inclusive sense to denote the set of labelled nucleons in nucleus C, and let us use C' to denote the distinct set of labelled nucleons in the core of D. Then a partially antisymmetrized expression for the DW transition amplitude, following Sections 3.5 and 4.8, is

$$T_{\alpha\beta}{}^{\mathrm{DW}} = (\psi_C(C)\psi_D(C', 1, 2, 3)\, \chi_\beta^{(-)}(\mathbf{r}_\beta), [V_\beta - U_\beta]\{\psi_A(C, 1)\psi_B(C', 2, 3)\chi_\alpha^{(+)}(\mathbf{r}_\alpha) \pm \psi_A(C', 1)\psi_B(C, 2, 3)\, \chi_\alpha^{(+)}(\mathbf{r}_\alpha')\}), \quad (10.26)$$

where 1, 2, and 3 denote the coordinates of the three valence nucleons. The channel coordinate $\mathbf{r}_\beta$ is the displacement of the center of mass of C from the center of mass of $\{C', 1, 2, 3\}$; the coordinate $\mathbf{r}_\alpha$ is the displacement of the center of mass of $\{C, 1\}$ from $\{C', 2, 3\}$; the coordinate $\mathbf{r}_\alpha'$ is the displacement of the center of mass of $\{C', 1\}$ from $\{C, 2, 3\}$. We note that (10.26) is only partially antisymmetrized, because permutations that exchange only parts of the cores of A and B are omitted. In comparing the direct and exchange terms in (10.26), it is now evident that the direct term describes the transfer of nucleon 1 from A to B, whereas the exchange term describes the transfer of nucleons 2 and 3 from B to A. While the relative importance of these two processes depends on details of parentage coefficients, they should surely often be of comparable importance.

The direct and exchange terms in (10.26) tend to peak strongly at forward and backward scattering angles, respectively, hence there is reasonable hope for independent experimental investigations of these terms. The peaking is strong because the momenta are large and the projectiles are strongly absorbing (see Chapters 5, 7). Under these conditions the direct term is large when the momenta associated with $\mathbf{r}_\beta$ and $\mathbf{r}_\alpha$ are parallel, that is, when nucleus C emerges in the incident direction of A. The exchange term is large when the momenta associated with $\mathbf{r}_\beta$ and $\mathbf{r}'_\alpha$ are parallel, that is, when nucleus C emerges in the incident direction of B.

Quantitative understanding of (10.26) is assisted by the fact that the distorted waves are dominated by the interaction of the cores, C and C'. Hence it is correct to use the same distorted waves in both the direct and exchange terms.

Particularly clear examples of exchange in heavy-ion collisions are obtained in elastic scattering when nuclei C, D are identical with nuclei A, B, as shown by von Oertzen, Gutbrod, Müller, Voos, and Bock [17]. Let A be less massive than B, so that B consists of $A + a$, where a denotes a nucleon or a small cluster of nucleons. Then the DW amplitude for A, B elastic scattering is

$$\begin{aligned} T_{el}{}^{\mathrm{DW}} &= (\chi_\beta^{(-)}(\mathbf{r}_\beta), U_\beta e^{i(\mathbf{k}_\alpha\cdot\mathbf{r}_\beta)}) \\ &\quad + (\psi_A(A)\psi_B(A', a)\,\chi_\beta^{(-)}(\mathbf{r}_\beta), [V_\beta - U_\beta]\{\psi_A(A)\psi_B(A', a)\,\chi_\alpha^{(+)}(\mathbf{r}_\alpha) \\ &\qquad \pm \psi_A(A')\psi_B(A, a)\,\chi_\alpha^{(+)}(\mathbf{r}'_\alpha)\}). \qquad (10.27) \end{aligned}$$

We now choose the distorting potential so that the direct DW matrix element of $[V_\beta - U_\beta]$ is zero, so that the first term of (10.27) generates the entire small-angle elastic scattering. Then (10.27) simplifies to

$$\begin{aligned} T_{el}{}^{\mathrm{DW}} &= (\chi_\beta^{(-)}(\mathbf{r}_\beta), U_\beta e^{i(\mathbf{k}_\alpha\cdot\mathbf{r}_\beta)}) \\ &\quad \pm (\psi_A(A)\psi_B(A', a)\,\chi_\beta^{(-)}(\mathbf{r}_\beta), [V_\beta - U_\beta]\psi_A(A')\psi_B(A, a)\,\chi_\alpha^{(+)}(\mathbf{r}'_\alpha)). \qquad (10.28) \end{aligned}$$

Here the exchange term has the form of the amplitude for transfer of a from B to A. The distorted waves in this exchange term are identical with the elastic-scattering wavefunction in the forward-angle direct term. Diffraction approximations can be used for this wavefunction [17].

Figure 10.1 shows the differential cross sections for the three elastic scattering processes $C^{12} + O^{16}$ at 35 MeV, $C^{12} + F^{19}$ at 40 MeV, $C^{12} + B^{10}$ at 18 MeV. Striking back-angle structure appears for $C^{12} + O^{16}$ and for $C^{12} + B^{10}$ because the exchange terms in these cases correspond to the very easy transfers of an alpha particle and of a deuteron, respectively. The angular distribution of this structure can be associated quantitatively with the forward-angle elastic scattering, to the extent that the forward-angle and backward-angle oscillatory structures both even have the same dependence on bombarding energy. The $C^{12} + F^{19}$ collision gives no significant back-angle structure, presumably because in this case the exchange term requires transfer of a Li^7 cluster.

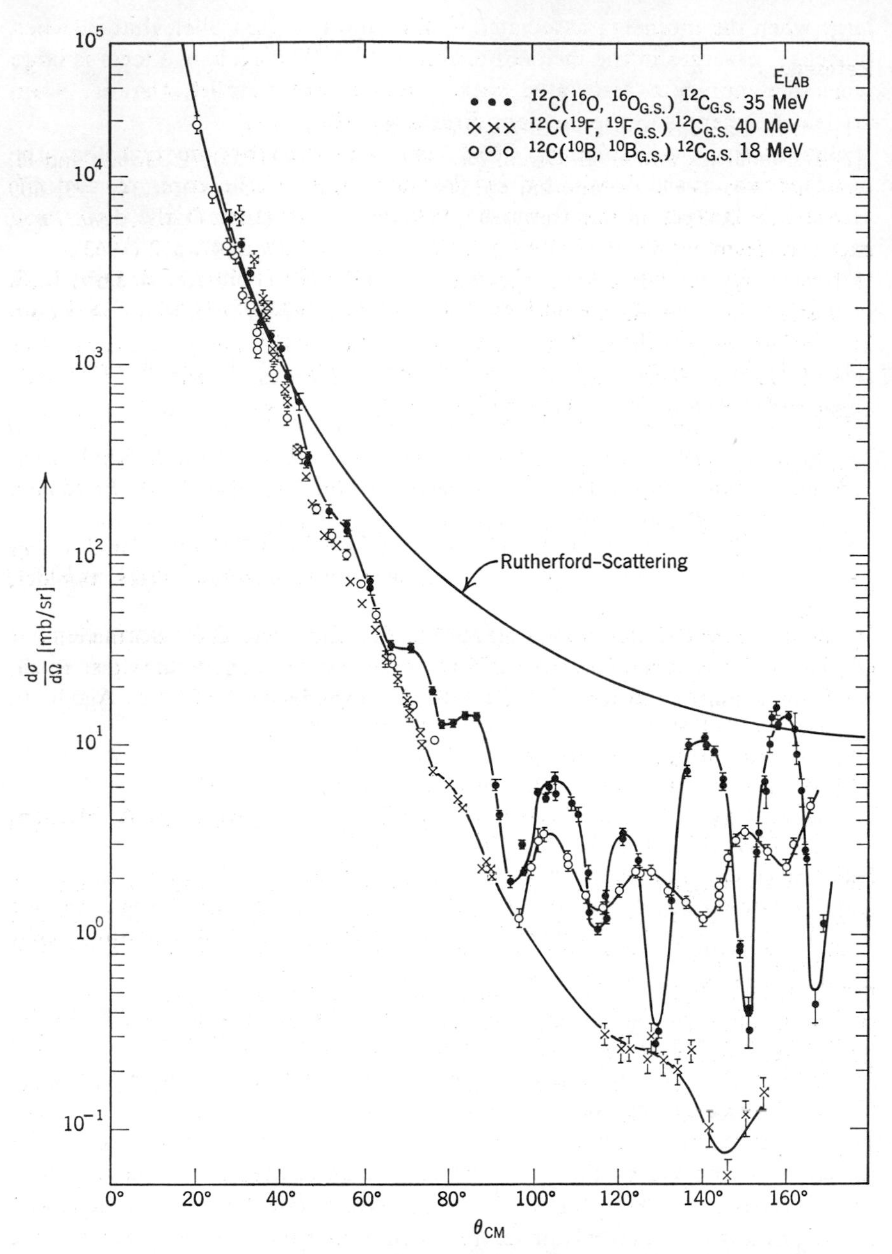

Fig. 10.1 Differential cross sections for heavy-ion elastic scattering as indicated on the graph. (From [17], Fig. 1.)

References

[1] M. K. Banerjee and D. Pal, *Nucl. Phys.* **83,** 575 (1966).
[2] L. Madansky and G. E. Owen, *Phys. Rev.* **99,** 1608 (1955); T. Honda and H. Ui, *Prog. Theoret. Phys.* **25,** 613, 635 (1961), and *Nucl. Phys.* **34,** 593, 609 (1962); D. Robson, *Nucl. Phys.* **33,** 594 (1962), *ibid*, **42** 592 (1963), *Proc. Phys. Soc.* (London) **80,** 1067 (1962); S. Edwards, *Nucl. Phys.* **47,** 652 (1963).
[3] R. H. Bassel and E. Gerjuoy, *Phys. Rev.* **117,** 749 (1960); T. B. Day, L. S. Rodberg, G. A. Snow, and J. Sucher, *Phys. Rev.* **123,** 1051 (1961); F. S. Levin, *Nucl. Phys.* **36,** 119 (1962).
[4] M. H. Mittleman, in *Advances in Theoretical Physics*, Vol. 1, K. A. Brueckner, Ed., Academic, New York, 1965.
[5] W. A. Friedman and H. Feshbach, in *Spectroscopic and Group Theoretical Methods in Physics*, Eds. F. Bloch, S. G. Cohen, A. de Shalit, S. Sambrusky, and I. Talmi, North-Holland, Amsterdam 1968; see also W. A. Friedman, *Ann. Phys.* **45,** 265 (1967).
[6] H. Feshbach, *Ann. Phys.* **23,** 47 (1963), see also A. K. Kerman, in *Lectures in Theoretical Physics*, Vol. VIII C, University of Colorado Press, Boulder, 1966.
[7] S. Saito, *Prog. Theoret. Phys.* **41,** 705 (1969), and references given therein.
[8] R. G. Sachs, *Phys. Rev.* **95,** 1065 (1954); L. M. Frantz, R. L. Mills, R. G. Newton, and A. M. Sessler, *Phys. Rev. Letters* **1,** 340 (1958); A. Agodi, F. Catara, and M. Di Toro, *Ann. Phys.* **49,** 445 (1968).
[9] L. S. Rodberg, *Nucl. Phys.* **47,** 1 (1963).
[10] M. K. Banerjee and C. A. Levinson, *loc. cit.*, Chap. 5, Ref. 4.
[11] K. A. Amos, *Nucl. Phys.* **A103,** 657 (1967); also K. A. Amos, V. A. Madsen, and I. E. McCarthy, *loc. cit.*, Chap. 5, Ref. 68.
[12] A. Agodi and G. Schiffrer, *loc. cit.*, Chap. 5, Ref. 67.
[13] J. Atkinson and V. A. Madsen, *Phys. Rev. Letters* **21,** 295 (1968), see also D. Agassi and R. Schaeffer, *Phys. Letters* **26B,** 703 (1968) and subsequent publications.
[14] T. Une, S. Yamaji, and H. Yoshida, *loc. cit.*, Chap. 5, Ref. 92.
[15] N. K. Glendenning and M. Veneroni, *loc. cit.*, Chap. 5, Ref. 46.
[16] H. W. Barz, *Nucl. Phys.* **A91,** 262 (1967).
[17] W. von Oertzen, H. H. Gutbrod, M. Müller, U. Voos and R. Bock, *Phys. Letters* **26B,** 291 (1968).

CHAPTER 11

Unified Theories, Dispersion Theories

Chapters 1–10 deal with the standard theoretical approaches on which nearly all practical direct reaction calculations are based. Two topics of a less standard nature are now presented. The topic of "unified theories" resumes the discussion of the relation between CN and DI reaction mechanisms that is initiated in Chapters 2 and 4. The topic of "pology" treats reactions by means of an alternative formulation of quantum mechanics that emphasizes analytic properties of the scattering matrix, rather than wavefunctions.

11.1 Unified Theories

A detailed formulation of direct reaction theories is given in Chapter 4. These theories are in some sense supposed to concern simple aspects of the nuclear dynamics, aspects in which only a few degrees of freedom are excited. This qualitative picture leads us to construct DI theories by dividing the overall wavefunction,

$$\Psi = \Psi_P + \Psi_Q, \tag{11.1}$$

into a part Ψ_P in which complicated nuclear excitations are suppressed, and a part Ψ_Q in which these excitations are retained. Because any reaction theory must at least carry enough degrees of freedom to describe the open channels, we are led to identify Ψ_P with Ψ_{open} of Chapter 2, and to express it as the sum of products

$$\Psi_P = \sum_{\alpha} \psi_{1\alpha}\psi_{2\alpha}\xi_\alpha(\mathbf{r}_{1\alpha} - \mathbf{r}_{2\alpha}), \tag{11.2}$$

where $\psi_{1\alpha}$ and $\psi_{2\alpha}$ are the internal wavefunctions of the colliding nuclei in channel α, and where ξ_α is the relative wavefunction in channel α. Our "simple" direct reaction theories then concern the coupling among the various terms of (11.2).

The function Ψ_Q, except for minor qualifications, may be identified with Ψ_{closed} of Chapter 2. Thus Ψ_Q is considered to contain such complicated and extensive internal excitations that no one degree of freedom has sufficient energy for breakup.

The nuclear interaction couples Ψ_P with Ψ_Q, and thereby it allows the complications of Ψ_Q to affect Ψ_P. One effect of these complications then is to cause the term that couples Ψ_Q with Ψ_P to fluctuate with energy; this in turn causes the properties of Ψ_P to fluctuate with energy. Unfortunately, these fluctuations prevent theories of the full open-channel function Ψ_P from having the simplicity we require of DI theories, even though Ψ_P is itself constructed to be simple. To overcome this problem the discussion in Chapter 4 goes on to define the DI part of Ψ as the part of Ψ_P that is generated by carrying only the energy-averaged coupling to Ψ_Q. We thereby arrive at the wavefunction Ψ_{model} of (4.34) and (4.35). All the discussion in Chapters 5–10 then concerns Ψ_{model}. Because Ψ_{model} is only a fragment of the overall wavefunction, it is governed by non-Hermitian interactions.

Because we assumed Ψ_Q is entirely closed (it exhibits no flux asymptotically), this function only influences scattering amplitudes through its coupling to Ψ_P. Our analysis distinguishes two kinds of influence exercised by Ψ_Q: One influence is through the energy-averaged coupling between Ψ_Q and Ψ_P. This energy-averaged coupling affects Ψ_{model}, and through Ψ_{model} it affects parts of the scattering amplitude that have slow energy variations. The other influence of Ψ_Q is through the fluctuating part of the coupling to Ψ_P, the part that remains after the energy-averaged coupling has been removed. This fluctuating coupling has a zero mean. It is the source of a fluctuating part of Ψ_P,

$$\Psi_{\text{fluct}} = \Psi_P - \Psi_{\text{model}}, \tag{11.3}$$

which is, in turn, the source of parts of the scattering amplitude that fluctuate with zero mean.* Hence, by distinguishing between the average and fluctuating parts of the coupling we have precisely distinguished between the average and fluctuating parts of the scattering amplitudes, as discussed in Section 2.1. Hence the wish to construct DI theories that are simple at *all stages* has compelled us to identify these theories with the energy-averaged parts of the scattering amplitudes. We recall (2.3) that the average and fluctuating parts of the amplitudes make noninterfering contributions to energy-averaged cross sections.

Because the analysis summarized above is primarily intended to introduce our use of simple DI theories, it avoids several interesting questions about the relation between DI and CN. Further discussion of these questions is now given. We see how the choice of explicit basis functions affects the relation between DI and CN, and we see how DI and CN interfere when cross sections are not averaged over broad intervals of energy. To some extent this discussion reopens the question of how best to define DI.

* Because Ψ_{fluct} only possesses outgoing waves, its contribution to the scattering amplitudes adds linearly to the contribution from Ψ_{model}.

Most detailed formal developments of unified theories employ a shell-model basis [1–6], and treat elastic and inelastic scattering of nucleons. Only for this case do the formal theories seem to give results that go beyond the insights of our general qualitative discussion. We now develop an analysis of this case. This analysis makes reference to the study of direct reactions by Hüfner, Mahaux, and Weidenmüller [1], who explored a shell-model theory of nucleon scattering. Several references to the work of HMW are given in Chapter 2.

The shell-model analysis uses single-nucleon orbitals ϕ_i that are governed by a Hermitian shell-model potential $U(\mathbf{r})$. This potential possesses a "realistic" shape, of the sort discussed in earlier chapters. In particular, $U(\mathbf{r})$ has finite depth, therefore the spectrum of the orbitals ϕ_i includes both a set of discrete states and a single-nucleon continuum. We build basis states for analyses of the A-nucleon system by constructing antisymmetrized products of the ϕ_i. The states constructed in this way form a complete set, and are eigenstates of the shell-model Hamiltonian

$$H_0 = \sum_{j=1}^{A} [T(j) + U(\mathbf{r}_j)], \tag{11.4}$$

where $T(j)$ is the kinetic energy operator for nucleon j.

The shell-model analysis is given definiteness by truncating the above basis so that the retained basis functions consist of only those products in which at most one ϕ_i is chosen from the single-nucleon continuum. Under this truncation the retained basis functions are of two types. One type is the finite set of functions Φ_μ that are entirely products of bound orbitals ϕ_i. Because the number of nucleons is large, functions of the type Φ_μ are abundant up to very high excitation energies, the total energy merely being distributed among the single-nucleon orbitals so that no one orbital is in the single-nucleon continuum. The functions Φ_μ possess eigenenergies ε_μ, so that

$$H_0\Phi_\mu = \varepsilon_\mu \Phi_\mu. \tag{11.5}$$

Basis functions of the second type are the $\chi_E{}^\lambda$. These are antisymmetrized products formed from one continuum orbital that has energy $E - \varepsilon_\lambda \geqslant 0$ multiplied by* $A - 1$ other orbitals that are bound and that have a total energy ε_λ. Thus

$$H_0\chi_E{}^\lambda = E\chi_E{}^\lambda. \tag{11.6}$$

* HMW use functions $\chi_E{}^\lambda$ that have a closer correspondence with elastic scattering. Their $\chi_E{}^\lambda$ are formed by multiplying a continuum orbital by actual bound eigenfunctions of the target nucleus. The transformation to this basis [1, 4, 7, 8] complicates the analysis, somewhat along the lines of the discussion in Section 10.1. Fortunately our present interests do not require this much care. At each stage we regard the $\chi_E{}^\lambda$ as being built either from single determinants or from target-nucleus eigenstates, whichever allows the simpler procedure.

The functions described above are orthonormal, such that

$$\langle \Phi_\mu \mid \Phi_\nu \rangle = \delta_{\mu\nu}, \qquad \langle \Phi_\mu \mid \chi_E{}^\lambda \rangle = 0,$$

and

$$\langle \chi_E{}^\mu \mid \chi_{E'}{}^\nu \rangle = \delta_{\mu\nu}\, \delta(E - E').$$

Clearly the spectrum of eigenenergies of the bound functions Φ_μ overlaps the continuous spectra of the functions $\chi_E{}^\lambda$. Hence we are led to speak of "bound states in the continuum."

With the above truncated basis, not more than one nucleon at a time is allowed to have positive energy, and therefore we are limited to the study of nucleon scattering reactions. In fact, this limitation is required for mathematical reasons, for it is awkward to form the wavefunctions of composite projectiles, such as deuterons, by building linear combinations of products of positive-energy shell-model orbitals. Within the above basis the first stage of our DI $-$ CN division is now no longer flexible. The functions $\chi_E{}^\lambda$ correspond asymptotically to the open channels for nucleon scattering, and must be taken as the basis functions of the P-space. Therefore Ψ_P is a linear combination of the $\chi_E{}^\lambda$. Likewise the basis functions Φ_μ are closed and must be taken to span the Q-space.*

In principle there is still some remaining flexibility in the DI $-$ CN division, because no precise specification of the shell-model potential $U(\mathbf{r})$ has yet been given. However, we know that by suitable choice of $U(\mathbf{r})$ it is possible to get shell-model wavefunctions that give good zero-order descriptions both of nuclear bound states and of nucleon elastic scattering. With $U(\mathbf{r})$ thus chosen the full Hamiltonian of the system is then formed by adding to H_0 a *residual interaction* V,

$$H = H_0 + V, \tag{11.7}$$

that can, in some sense, be treated perturbatively. Hence, by accepting the $U(\mathbf{r})$ that gives good zero-order wavefunctions we have some hope of using the shell model to develop at least semiquantitative theories of nucleon scattering, starting from known nuclear interactions.

The benefits derived from use of an explicit shell-model basis are now clear. First, there is a very simple separation of P-space from Q-space, because Φ_μ and $\chi_E{}^\lambda$ are orthogonal. Second, because of the close association of the $\chi_E{}^\lambda$ with elastic scattering, there is the easy plausibility of using these functions as a basis for DI inelastic scattering analyses. Third, because so much of the nuclear interaction is built into the basis functions, there is the possibility that $(H - H_0)$ can be treated perturbatively. Hence it may be possible to derive the effective interactions of the reduced DI theory, and

* Later we consider including a limited number of the Φ_μ in the P-space. These are called "doorway states."

thereby to go beyond mere phenomenological assumptions about those interactions. It may also be possible to derive the statistical properties of the accompanying CN theory. Finally, there is the advantage that the Pauli principle is incorporated at the very first step, because antisymmetrized basis functions are used.

The shell-model theory of scattering may be developed in detail as a simplified special case of the projection operator analysis in Section 4.2. In this development, because H_0 is diagonal in the set of basis states that define P-space and Q-space, it is only necessary to use projection operators to classify the matrix elements of the residual interaction, V. We obtain

$$\begin{aligned} H &= H_0 + V, \\ &= H_0 + V_{PP} + V_{PQ} + V_{QP} + V_{QQ}, \end{aligned}$$

as in Section 4.2, where V_{PP} is the part of V that links states $\chi_E{}^\lambda$ with each other; V_{PQ} and V_{QP} are the parts of V that link $\chi_E{}^\lambda$ with Φ_μ; and V_{QQ} links the Φ_μ with each other. HMW [1] collect these parts of V into "compound" and "direct" interactions, respectively,

$$V_1 = V_{PQ} + V_{QP} + V_{QQ}, \tag{11.8}$$

$$V_2 = V_{PP}. \tag{11.9}$$

Here V_1 not only governs the eigenstates of Q-space but also links these eigenstates to P-space. However, V_2 only links the various $\chi_E{}^\lambda$ of P-space with each other, without coupling any of them to the closed states Φ_μ that make up Q-space. If V_2 were absent the open channels would only be linked through the Φ_μ, and we would have a pure CN theory. Therefore V_2 causes the DI aspects of the shell-model theory.

The entire subsequent analysis now concerns the extent to which V_1 and V_2 can or cannot be placed in one-to-one correspondence with CN and DI parts of the scattering amplitudes. Because the relations between interactions and amplitudes are not linear, only approximate correspondences can exist. Therefore the explicit structure of the shell-model theory is once again helpful, because it allows these correspondences to be studied numerically [1]. Perturbative methods may be used.

Although the shell-model theory regards both V_1 and V_2 as small corrections to H_0, it is neither necessary nor desirable to treat V_1 as a perturbation. The interaction V_1 causes resonances. We therefore seek to treat V_1 exactly and to treat only V_2 as a perturbation. It may be shown [1] that under most conditions a theory constructed in this fashion converges quite well, provided averages over intervals of a few hundred keV are used.

The technique by which V_1 is treated exactly may be seen by omitting V_2 and considering how to diagonalize the Hamiltonian

$$H_1 = H_0 + V_1. \tag{11.10}$$

In the restricted set of basis functions $\chi_E{}^\lambda$, Φ_μ we recognize that H_0 is already diagonal and that V_1 only couples the $\chi_E{}^\lambda$ to the *finite set of intermediate states* Φ_μ. Under these conditions V_1 reduces to a separable interaction. As in usual shell-model work, the diagonalization then requires inversion of a finite matrix whose rank equals the total number of bound functions Φ_μ. Naturally, this matrix can be very large. However, it is susceptible of being discussed by normal methods of nuclear structure analysis.

One formal method by which to carry through the matrix diagonalization of H_1, described above, is to return to the coupled equations with which the projection operator formalism begins, (4.10, 4.11), and to eliminate Ψ_Q between the two equations. We find

$$(E - H_{PP})\Psi_P = V_{PQ}(E - H_{QQ})^{-1}V_{QP}\Psi_P, \tag{11.11}$$

where, of course, in the present discussion $V_{PP} = V_2 = 0$, and therefore $H_{PP} = H_0$. The complicated effective interaction on the RHS of (11.11) requires inversion of only a finite matrix. Because this interaction is separable, the condition $H_{PP} = H_0$ allows (11.11) to be solved by algebraic methods.

Before proceeding with the solution of (11.11) it is best to rework the effective interaction into a more convenient form. The form in (11.11) displays singularities that apparently lie on the real E axis. However, these singularities are cancelled by corresponding zeroes in Ψ_P, and the singularities that in fact do affect Ψ_P are displaced below the real axis. The actual singularities are joint properties of H_{QQ} and H_{PP}; they are also (through H_{PP}) affected by the asymptotic boundary conditions in P-space. Under these circumstances it is best to rearrange the effective interaction to show more clearly the influence of H_{PP} on the singularities. The structure of the singularities in (11.11) is seen in (11.17–19). To obtain these rearranged equations we modify the method by which Ψ_Q is eliminated from the coupled equations. We return again to the method of Section 4.2; we begin [see (4.13)] by solving the first of the coupled equations to obtain Ψ_P in terms of Ψ_Q:

$$\Psi_P = \mathring{\Psi}_P + (E^+ - H_{PP})^{-1}V_{PQ}\Psi_Q, \tag{11.12}$$

where $\mathring{\Psi}_P$ is the homogeneous solution determined by H_{PP} in the absence of coupling. This solution is then substituted in the second equation to obtain an uncoupled inhomogeneous equation for Ψ_Q in terms of $\mathring{\Psi}_P$ [see (4.15)]

$$[E - H_{QQ} - V_{QP}(E^+ - H_{PP})^{-1}V_{PQ}]\Psi_Q = V_{QP}\mathring{\Psi}_P. \tag{11.13}$$

Equation 11.13 is then used to eliminate Ψ_Q from the first of the original coupled equations, and also to eliminate Ψ_Q from (11.12). These equations become

$$(E - H_{PP})\Psi_P = V_{PQ}\,\Lambda\, V_{QP}\mathring{\Psi}_P, \tag{11.14}$$

$$\Psi_P = \{1 + (E^+ - H_{PP})^{-1}V_{PQ}\,\Lambda\, V_{QP}\}\mathring{\Psi}_P, \tag{11.15}$$

where

$$\Lambda \equiv [E - H_{QQ} - V_{QP}(E^+ - H_{PP})^{-1}V_{QP}]^{-1}. \tag{11.16}$$

As a final step, $\mathring{\Psi}_P$ is then eliminated between (11.14) and (11.15), to give a homogeneous reduced Schrödinger equation for Ψ_P:

$$(E - H_{PP})\Psi_P = M\Psi_P, \tag{11.17}$$

$$M \equiv V_{PQ}\,\Lambda\, V_{QP}\{1 + (E^+ - H_{PP})^{-1}V_{PQ}\,\Lambda\, V_{QP}\}^{-1}. \tag{11.18}$$

(An alternative way to transform from (11.11) to (11.17) is given in Appendix 2 of an article by Toledo Piza and Kerman [9].) We once again see in (11.18) that the effective interaction can be handled by algebraic methods, provided that Q-space is finite and $H_{PP} = H_0$.

Equation 11.18 shows the nature of an approximation used in Chapter 4 to go from the inhomogeneous (4.34) to the homogeneous (4.36). That approximation implies that the energy average of the last factor in (11.18) may be replaced by the value unity. Although the shell-model context of the present discussion allows explicit calculation of the last factor, our qualitative understanding of (11.17) and (11.18) is assisted if this factor is once again omitted. Hence we approximate

$$M \approx V_{PQ}[E - H_{QQ} - V_{QP}(E^+ - H_{PP})^{-1}V_{QP}]^{-1}V_{QP}. \tag{11.19}$$

The effective interaction M now only differs from the interaction in (11.11) because the denominator in (11.19) contains a decay width operator

$$V_{QP}(E^+ - H_{PP})^{-1}V_{PQ}. \tag{11.20}$$

This operator is previously discussed in Section 4.2. The interaction M still has poles in the complex E-plane, which lie near the discrete eigenvalues of H_{QQ}. However the decay width shifts and broadens these poles. Each pole in M causes a resonance in the scattering amplitudes derived from Ψ_P.

We illustrate the structure of (11.19) by taking Q-space to consist of only one bound state Φ, whose energy is given by $H_{QQ}\Phi = \varepsilon\Phi$: In this case, returning to the notation of HMW [1], (11.19) gives

$$M = V_1\,|\Phi\rangle[E - \varepsilon - \langle\Phi|\, V_1(E^+ - H_0)^{-1}V_1\,|\Phi\rangle]^{-1}\langle\Phi|\, V_1. \tag{11.21}$$

The factor in brackets in (11.21) is the usual Breit-Wigner denominator; it is a c-number. Upon introducing (11.21) into (11.17) or (11.14), the contribution that M causes in the scattering amplitudes is found to be

$$\frac{\langle \mathring{\Psi}_{P'}^{(-)}| V_1 |\Phi\rangle\langle\Phi| V_1 |\mathring{\Psi}_P^{(+)}\rangle}{E - \varepsilon - \langle\Phi| V_1(E^+ - H_0)^{-1}V_1 |\Phi\rangle}. \tag{11.22}$$

Equation 11.22 is the Breit-Wigner amplitude. The matrix elements in the numerator of (11.22) are the resonance widths. Usual penetration factors enter through the channel wavefunctions $\mathring{\Psi}_{P'}^{(-)}$ and $\mathring{\Psi}_P^{(+)}$.

The above formalism shows how to handle the CN interactions V_1. It also shows how these interactions cause resonances by coupling the open channels to the bound states Φ_μ. (Of course, near thresholds there also may be *single-particle resonances* that originate in the open channels themselves.) We may now return to the principal subject of our analysis and see how the DI interaction V_2 modifies the formalism.

The interaction V_2 is included at all stages in the equations already derived. It is only noted from time to time in discussions of these equations how algebraic solutions could be obtained if V_2 were omitted. Hence to exhibit the role of V_2 we merely recopy the reduced Schrödinger equation of (11.17), and insert the full P-space Hamiltonian, $H_{PP} = H_0 + V_2$. For the effective interaction we use the approximate form given in (11.19), and obtain

$$(E - H_0 - V_2)\Psi_P \approx V_{PQ}[E - H_{QQ} - V_{QP}(E^+ - H_0 - V_2)^{-1}V_{PQ}]^{-1}V_{QP}\Psi_P. \tag{11.23}$$

Thus V_2 appears in the reduced Schrödinger equation in two places. It is an additional interaction that directly affects Ψ_P, and it modifies the widths and shifts of the resonances caused by coupling to Ψ_Q.

It is understandable that straightforward attempts to obtain solutions of (11.23) by treating V_2 as a perturbation are likely to show poor convergence [7]. The presence of V_2 produces minor shifts of the resonances caused by V_1. With sharp resonances, very small shifts of this nature can easily cause major local modifications of scattering amplitudes.

It is also understandable that energy averaging creates the conditions under which perturbative treatments of V_2 become possible [1]. Small displacements of resonances are irrelevant for energy-averaged quantities. We derive the energy-averaged equivalent of (11.23) by averaging the effective interaction M, to obtain

$$\bar{M} = V_{PQ}[E + iI - H_{QQ} - V_{QP}(E + iI - H_0 - V_2)^{-1}V_{PQ}]^{-1}V_{QP}, \tag{11.24}$$

as in Chapter 4. This averaged interaction then is inserted in (11.23), to give the reduced Schrödinger equation

$$(E - H_0 - V_2 - \bar{M})\Psi_{\text{model}} = 0. \tag{11.25}$$

This equation is identical with (4.36), on which all the DI analysis of the present book is based. For strong interactions V_2, (11.25) must be solved by the method of coupled channels; for fairly weak interactions V_2, (11.25) may be solved by the DW method. Both these methods are predicated on an initial step of energy averaging.

At this stage the shell-model analysis has returned to the earlier view that direct reactions should be identified with energy-averaged reduced dynamics, and therefore they should concern energy-averaged amplitudes. Moreover, an additional reason for energy averaging has appeared; it is now seen as a condition for convergence of DI calculational methods.

We are now ready to discuss the CN processes that accompany the DI aspects of reactions. This discussion includes some consideration of DI-CN "interference". HMW [1] identify three different circumstances for which unified discussions of DI and CN processes may be of interest. These are: (a) It is satisfactory to define DI in an energy-averaged sense, as above, but we wish to analyze the *law of distribution* of the fluctuations caused by the CN couplings. (b) Details of individual, isolated resonances are of interest, so that energy averaging must be avoided altogether. (c) A fine resonance structure that we wish to suppress by energy averaging is superimposed on a coarse resonance structure that we wish to study in detail. Clearly, cases (a) and (b) require very different methods of analysis, while case (c) requires a combination of the methods used in cases (a) and (b).

Case (b) is primarily of interest for reactions in which low-energy projectiles are scattered by light nuclei. In this case the energies of the closed eigenstates of H_{QQ} are well separated and easily resolved experimentally, and the number of these states is so small that it is practical to carry them on an equal footing with the wavefunctions of the open channels. In the context of the present formalism we might proceed by calculating solutions of (11.23) as it stands. Because such a procedure combines DI and CN processes in one inclusive Schrödinger equation, it necessarily preserves unitarity of the S-matrix; this contrasts (Chapter 2) with procedures that merely construct linear combinations of DI and CN amplitudes. However, we note that under the conditions of case (b) unitarity can also be preserved by more phenomenological procedures; for example, because there is no energy averaging, the Schrödinger equation for this case does not contain complex potentials; therefore we can treat DI and CN as independent Hermitian contributions to the K-matrix [6, 10].

In most reaction experiments the bombarding energies and masses are so large that a detailed study of the fine structure in excitation functions is neither practical nor interesting. Under these conditions case (a) is relevant. We therefore identify DI processes with energy-averaged scattering amplitudes, in the sense of (11.25), and proceed to investigate the properties of the fluctuating amplitudes,

$$f_{\text{CN}} \equiv f - f_{\text{DI}}, \tag{11.26}$$

left over after the energy-averaged parts have been removed. In Chapter 2 we see that CN and DI amplitudes defined in this fashion contribute non-interfering parts to energy-averaged cross sections, because the fluctuations of f_{CN} have a zero mean. The next level of investigation, as Ericson stressed [11], concerns how the fluctuations of f_{CN} are distributed about this mean, and how they affect cross sections. In investigations at this level it is usually assumed [11] that f_{CN} can be treated as a linear combination of pole terms (simple resonances) that have statistically uncorrelated energies and widths. On the other hand, in the shell-model theory the fluctuations of f_{CN} are caused by the fluctuations of the effective interaction M that stands in (11.23). Hence in this theory we should properly derive the distribution of the fluctuations of f_{CN} from our knowledge of the fluctuations of M. There have been very few attempts to accomplish such derivations [1, 12].

To derive the distribution of fluctuations of f_{CN} it is necessary to face the fact that nuclei are not statistical systems, and that at all stages their dynamical properties are well-defined and determinate. Statistical theories are relevant only because at some level of accuracy we choose to become vague about details of the dynamics. At this level we replace dynamical models by assumed distribution laws, and from these assumed distributions we derive the distributions of other quantities of physical interest. Hence the central problem of any statistical theory is to find at what stage in the investigation of the dynamics it is most plausible to insert the initial statistical assumption.

HMW [1] contend that in the shell-model theory the quantities that should most properly be treated as statistically uncorrelated are the matrix elements that couple the $\chi_E{}^\lambda$ with the eigenstates of H_{QQ}. It is argued in Chapter 2 that the complications of the CN eigenfunctions are so great that it is plausible to treat these matrix elements of V_{PQ} as random variables. Moldauer [12] introduces his basic statistical assumption at a similar stage, but in the context of a boundary-matching theory.

Provided $V_2 = 0$, the HMW statistical assumption implies that the amplitudes calculated from (11.23) can be expanded in pole terms with statistically-uncorrelated energies and widths. Much the same set of poles probably is obtained if $V_2 \neq 0$. However, how does the introduction of $V_2 \neq 0$ affect the distributions of widths and energies of these poles? Unfortunately, no conclusive answer to this question has been obtained. HMW feel that if an amplitude is expressed as the sum of a background term plus pole terms, then with $V_2 \neq 0$ the parameters of the pole terms must be correlated. Such correlations would interfere seriously with attempts to separate the DI and CN amplitudes experimentally by studies of statistical fluctuations. Further discussion of this question is beyond the scope of this book.

Finally we come to case (c), in which fine resonance structure and coarse resonance structure are superimposed in the same reaction. In reactions that display such "intermediate structure" it is interesting to consider averaging the effective interaction M over an energy interval that is large enough to suppress the fine structure but small enough to give good resolution of the coarse structure. Under such intermediate averaging (11.23) goes over to a reduced Schrödinger equation that contains both complex potentials and explicit resonance terms.

The required transformation of (11.23) is obtained by recognizing [13] that coarse energy-dependent structure appears if the Q-space for the reaction under consideration contains a limited number of states (either a subset of the Φ_μ or a subset of collective linear combinations of the Φ_μ) that are especially strongly coupled to the P-space. Such strongly-coupled states are called "doorway states," to suggest that the most favored way to couple P-space with the remainder of Q-space may be through these doorways. In the shell-model theory, doorway states can be formed by inserting single particle-hole excitations in the $\chi_E{}^\lambda$. Clearly, although the doorways are strongly favored CN *excitations*, they are not necessarily eigenstates of H_{QQ}, therefore they do not have an immediate one-to-one correspondence with observed resonances. Intermediate averaging establishes such a correspondence.

To remove the doorway excitations from the energy-averaging process, we alter the division between P-space and Q-space. This may be done either by transferring the doorway states from Q-space to P-space, or by replacing (11.1) by a three-term separation of the wavefunction.* Under the latter procedure

$$\Psi = \Psi_P + \Psi_d + \Psi_q, \tag{11.27}$$

where d labels the doorway part of Q-space and q labels the remainder of Q-space. Because P-space, d-space, and q-space are mutually orthogonal, the

* The flexibility with which we undertake these modifications should make it quite clear that, in the end, a DI model of a reaction is whatever part of the dynamics we choose to treat by explicitly solving a Schrödinger equation.

Schrödinger equation may be expressed as a set of three coupled equations

$$(E - H_{PP})\Psi_P = H_{Pd}\Psi_d + H_{Pq}\Psi_q, \tag{11.28}$$

$$(E - H_{dd})\Psi_d = H_{dP}\Psi_P + H_{dq}\Psi_q, \tag{11.29}$$

$$(E - H_{qq})\Psi_q = H_{qP}\Psi_P + H_{qd}\Psi_d. \tag{11.30}$$

We reduce these three equations by eliminating q-space, in the same fashion in which (4.10) and (4.11) are reduced [see (11.11–11.23)] by eliminating Q-space. The algebra is very much simplified by introduction of the "pure doorway state" assumption, that is, by assuming $H_{Pq} = 0$ so that P-space is coupled to q-space only through intermediate doorway excitations. Under this assumption the coupled equations are

$$(E - H_{PP})\Psi_P = H_{Pd}\Psi_d, \tag{11.28'}$$

$$(E - H_{dd})\Psi_d = H_{dP}\Psi_P + H_{dq}\Psi_q, \tag{11.29'}$$

$$(E - H_{qq})\Psi_q = H_{qd}\Psi_d. \tag{11.30'}$$

It is sufficient to solve (11.30′) for Ψ_q and to substitute this solution in (11.29′). Then the coupled system becomes

$$(E - H_{PP})\Psi_P = H_{Pd}\Psi_d, \tag{11.28'}$$

$$(E - H_{dd} - M_{dd})\Psi_d = H_{dP}\Psi_P, \tag{11.31}$$

with

$$M_{dd} \equiv H_{dq}(E - H_{qq})^{-1}H_{qd}. \tag{11.32}$$

The nature of the singularities in the effective interaction may be clarified by a further transformation of M_{dd}, in the same fashion in which (11.11) is transformed into (11.18). In this case the techniques of [9] would be especially convenient. However, the results of this transformation are not needed here, and it is omitted.

The hypothesis of intermediate structure suggests that the density of states in q-space is much greater than that in d-space. Therefore the fluctuations of M_{dd} are much more closely spaced than the eigenvalues of H_{dd}, therefore intermediate averaging primarily affects M_{dd}. Averaging converts M_{dd} to

$$\bar{M}_{dd} = H_{dq}(E + iI - H_{qq})^{-1}H_{qd}, \tag{11.33}$$

a non-Hermitian effective interaction that contains a negative imaginary part. The imaginary part of $\bar{M}_{dd}$ expresses damping of the doorway excitations by excitation of states in q-space. This is called *downward damping* [13] because it carries us deeper into the compound nucleus.

Thus, intermediate averaging converts (11.28′) and (11.31) into a coupled pair of equations for $\bar{\Psi}_P$, $\bar{\Psi}_d$ that are very much like the coupled pair for

Ψ_P, Ψ_Q from which the more general discussion started. The doorway states in these coupled equations cause resonances. Because the doorway states are strongly coupled to $\bar{\Psi}_P$, the doorway resonances are broad and strong. Only minor modifications are caused by the non-Hermitian interaction $\bar{M}_{dd}$ that results from the averaging. It introduces an extra width in the doorway resonances, which adds to the escape width caused by coupling to $\bar{\Psi}_P$.

A single reduced Schrödinger equation for the open-channel wavefunction of the averaged problem may be obtained by substituting $\bar{M}_{dd}$ in (11.31) and then substituting the solution of (11.31) in (11.28′). This procedure gives

$$[E - H_{PP} - H_{Pd}(E - H_{dd} - \bar{M}_{dd})^{-1}H_{dP}]\bar{\Psi}_P = 0. \qquad (11.34)$$

Equation 11.34 is identical with (2.88) of [13]. It is instructive to compare (11.34) with (11.25), the basic equation of energy-averaged DI theories.

Isobar analogue resonances provide interesting experimental examples of intermediate structure. Such resonances are narrow and well resolved from each other. However their breadths are much greater than the spacing of the fluctuations of the underlying fine structure. Applications of ideas of intermediate structure for this case are abundant in the literature.

Other experimental studies of DI-CN competition and interference are less clearcut [14]. Attempts have been made to gather data in the vicinity of isolated resonances in stripping experiments or in elastic scattering experiments. The theories used with these experiments are based on simple addition of DI and CN amplitudes, an approach that may not even preserve unitarity, and that is criticized in a number of places in this book. Despite these criticisms, the simplicity of this approach suggests that it be given further attention, due care being taken about conserved geometrical quantum numbers. To do much better it would be necessary to go over to difficult nonperturbative solutions of (11.23) or (11.34). (However in some cases it may be possible to go a little beyond the linear theory by combining DI and CN in the K-matrix.)

11.2 Dispersion Theories

Theories of elementary particle processes tend to avoid the use of wavefunctions and matrix elements; they instead work directly with amplitudes for the processes. These theories attempt to establish relations that allow unknown amplitudes to be expressed as functions of known amplitudes. This technique was developed originally as a means to avoid calculating with strong, poorly known interactions. Amplitudes possess the convenience that they generally are finite. In addition, theories that work only with amplitudes are independent of the introduction of speculative intermediate quantities. Indeed, enthusiasts of this method have often claimed that older methods of quantum mechanics could be replaced entirely by a body of suitable relations among amplitudes. Now, it is noteworthy that a central feature of such theories consists of analyses of the amplitudes as analytic functions of a variety of complex variables. The theories then tend to devolve into studies of the singularities of these functions.

Similar approaches to nuclear reaction theory have been attempted by a number of authors [15–21]. Once again the wish to avoid perturbative calculations with the strong nuclear force has served as partial motivation for the introduction of dispersion theories. Another motivation has been the hope that individual direct reaction processes could be placed in close correspondence with individual singularities of the amplitudes.

However, dispersion methods have not been as successful in nuclear reaction theory as in elementary particle theory. This is, in part, because in nuclear reaction theory the difficulties caused by the strong nuclear force can be dealt with to a sufficient extent by the introduction of "effective forces." Frequently these are derived by reaction matrix methods. Another reason for the failure of dispersion methods is that in nuclear reactions the various singularities of the amplitudes are not as well separated as in elementary particle processes. Therefore no one singularity is uniquely close to the physical region of the complex variables, to the extent that it can dominate the amplitude. Finally, the most important reason why dispersion methods have not replaced calculations with wavefunctions probably lies in the fact that a great deal is known about nuclei. There is a large body of well-established experience with theories of nuclear structure. Nearly all this experience is formulated in terms of configuration space models and wavefunctions, and is poorly adapted for use in dispersion theories. Under these conditions, while dispersion methods are excellent for the investigation of processes among the poorly understood elementary particles, they seem to be of marginal interest for working out the consequences of DI theories of nuclear reactions. Still, it is possible that dispersion methods will become of greater interest as reaction experiments are extended to higher energies [21].

Interest in the application of dispersion methods to nuclear reactions was initiated by Amado's observation [15] that the amplitude for a stripping reaction possesses a pole in the complex momentum-transfer plane, and by his further observation that at this pole the amplitude is independent of all distortion effects. The existence of this pole would seem to characterize the "mechanism" of a stripping reaction. Under suitable experimental conditions the stripping pole lies very near the physical region of the momentum transfer variable, and it has a good chance to dominate the amplitude. Hence it may be possible to analyze stripping experiments in terms of this pole, and it may be possible to extrapolate stripping data to the pole to obtain accurate $\mathscr{S}$ factors. There has been considerable investigation of these ideas [20, 22–24].

Both the origin of the stripping pole and the reason why it in fact seldom dominates the amplitude can be understood by application of standard wavefunction methods. Let us go back to (3.20), in which the exact transition amplitude is expressed as an overlap of wavefunctions. For the reaction $A + a \rightarrow B + b$ this overlap has the form

$$T_{\alpha\beta} = \int e^{-i(\mathbf{k}_\beta \cdot \mathbf{r}_\beta)} \langle \psi_B \psi_b | V_{bB} | \Psi^{(+)} \rangle \, d^3 r_\beta, \tag{11.35}$$

where V_{Bb} is the sum of all interactions between the particles of B and the particles of b. For definiteness let us suppose $a = b + x$, so that the reaction proceeds by the transfer of x from a to A to form B. Let us then consider the exact $\Psi^{(+)}$ to consist of an incident plane wave plus a remainder term:

$$\Psi^{(+)} = \psi_A \psi_a e^{i(\mathbf{k}_\alpha \cdot \mathbf{r}_\alpha)} + \text{remainder}; \tag{11.36}$$

and let us examine the part of $T_{\alpha\beta}$ that is obtained from only the plane wave part of $\Psi^{(+)}$,

$$T_{\alpha\beta}{}^0 = \int e^{-i(\mathbf{k}_\beta \cdot \mathbf{r}_\beta)} \langle \psi_B \psi_b | V_{bB} + V_{bx} | \psi_A \psi_a \rangle e^{i(\mathbf{k}_\alpha \cdot \mathbf{r}_\alpha)} d^3 r_\alpha \, d^3 r_\beta. \tag{11.37}$$

At this stage the bra-ket notation indicates integrations over coordinates other than $\mathbf{r}_\alpha$ and $\mathbf{r}_\beta$. Suitable transformation of coordinates in (11.37) then gives

$$T_{\alpha\beta}{}^0 = \int e^{i(\mathbf{q} \cdot \mathbf{r}_x)} \langle \psi_B \psi_b | V_{bB} + V_{bx} | \psi_A \psi_a \rangle e^{i(\mathbf{K} \cdot \mathbf{r})} \, d^3 r_x \, d^3 r, \tag{11.38}$$

where

$$\mathbf{q} \equiv \mathbf{k}_\alpha - \frac{m_A}{m_B} \mathbf{k}_\beta, \tag{11.39}$$

$$\mathbf{K} \equiv \frac{m_b}{m_a} \mathbf{k}_\alpha - \mathbf{k}_\beta, \tag{11.40}$$

and where $\mathbf{r}_x$ is the displacement of x from A and $\mathbf{r}$ is the displacement of b from x.

Equation 11.38 is not symmetrical in the two variables $\mathbf{r}$ and $\mathbf{r}_x$. As $r \rightarrow \infty$ the integral converges because the interactions V_{bB} and V_{bx} go very rapidly to zero. However, as $r_x \rightarrow \infty$ it is only the V_{bB} term in (11.38) whose convergence is controlled by a short-ranged interaction; the convergence of the V_{bx} term is controlled by the bound-state wavefunction ψ_B. Hence the amplitude can be very large if ψ_B is weakly bound and $q \approx 0$. This possibility is noted in Section 5.9, in the discussion of Coulomb stripping. What is new in the present discussion is the observation that for a suitable *imaginary* value of q the integration with respect to r_x always diverges, no matter how strongly

ψ_B is bound. Hence the Born amplitude $\overset{\circ}{T}_{\alpha\beta}$ possesses a singularity at this (nonphysical) imaginary value of the momentum transfer.

The nature and location of the singularity in the V_{bx} term of (11.38) may be seen by carrying through the integrations. The interaction V_{bx} may be eliminated by use of the Schrödinger equation for ψ_a, to give

$$\overset{\circ}{T}_{\alpha\beta}\,(\text{sing}) = \left(-B_a - \frac{\hbar^2 K^2}{2\mu_r}\right)\int e^{i(\mathbf{q}\cdot\mathbf{r}_x)}\langle\psi_B \mid \psi_A\rangle\, d^3r_x \times \int e^{i(\mathbf{K}\cdot\mathbf{r})}\langle\psi_b \mid \psi_a\rangle\, d^3r, \quad (11.41)$$

where μ_r is the reduced mass in the variable r. This expression has a characteristic product form [21]; it is composed of a propagator, multiplied by vertex functions that are amplitudes for the two elementary processes $B \to A + x$, $a \to b + x$. The entire q-dependence of (11.41) is contained in the amplitude for the process $B \to A + x$,

$$\int e^{i(\mathbf{q}\cdot\mathbf{r}_x)}\langle\psi_B \mid \psi_A\rangle\, d^3r_x = i^l(2l+1)\int_0^\infty r_x^2\, dr_x j_l(qr_x)\int d\Omega_x P_l(\hat{q}\cdot\hat{r}_x)\langle\psi_B \mid \psi_A\rangle, \quad (11.42)$$

where l is the angular momentum of the transferred particle in the overlap $\langle\psi_B \mid \psi_A\rangle$. Now the singularity of (11.42) is contributed by asymptotically large values of r_x, and for such radii the overlap reduces to a spherical Hankel function

$$i^l(2l+1)\int P_l(\hat{q}\cdot\hat{r}_x)\langle\psi_B \mid \psi_A\rangle\, d\Omega_x \to \kappa N_l h_l^{(1)}(i\kappa r_x), \quad (11.43)$$

where N_l is the asymptotic normalization factor and $\hbar^2\kappa^2/2\mu_x$ is the energy required to separate x from B. The insertion of (11.43) in (11.42) at last shows the singularity to have the form

$$\int e^{i(\mathbf{q}\cdot\mathbf{r}_x)}\langle\psi_B \mid \psi_A\rangle\, d^3r_x = \frac{qN_l}{q^2+\kappa^2} + \text{nonsingular part}. \quad (11.44)$$

The singularity is seen to be a simple pole at the location $q = i\kappa$. This is the stripping pole. At this pole the integrand in the r_x-integration in (11.42) reduces asymptotically to a constant, and it is for this reason that the integral diverges. The residue at the stripping pole measures N_l, the asymptotic normalization of the wavefunction of the transferred particle.

It is clear from the above discussion that (11.44) gives the entire singular part of the Born amplitude, $\overset{\circ}{T}_{\alpha\beta}$. The approximations used in our evaluation of this amplitude make only finite contributions and do not affect the residue.

We may now return to the non-Born part of the amplitude, and ask what contributions arise from the remainder term in (11.36). Asymptotically this remainder term is composed only of outgoing waves. These outgoing waves represent scattering in channels α, β, and rearrangement in channels $\gamma \neq \alpha$, β. The rearrangement terms are finite and do not affect the stripping pole. The scattered waves are of greater importance; however, they decrease asymptotically as r_α^{-1} and r_β^{-1}, respectively, therefore their contributions to the exact stripping amplitude also remain finite as $q \to i\kappa$. It is for this reason that distortion effects do not affect the residue at the stripping pole.

However, we recall that for small r_α, r_β the scattered waves in channels α, β are as important as the plane waves. It is only because the Born amplitude receives divergent contributions from very large values of r_x that the plane waves have enhanced importance at the stripping pole. Because the scattered waves are themselves rather slowly decreasing functions of r_x, the enhanced role of the plane waves reduces very rapidly if q departs from the pole. For $q = i\kappa + \varepsilon$ the radial integral in the Born amplitude cuts off for $r_x \approx |\varepsilon|^{-1}$; therefore, for $|\varepsilon| \gtrsim 0.02\ \text{fm}^{-1}$, say, it is no longer likely that the Born term dominates the contributions from the scattered waves. This discussion shows that the stripping pole normally only dominates the amplitude within a region of momentum that is small compared with the displacement of the pole from momenta of experimental concern. Under these conditions it is extremely unlikely that experimentally determined cross sections can ever be extrapolated to the stripping pole. Idealized "computer experiments" have repeatedly verified this analysis [22–24].

More searching analyses of the singularities of the stripping amplitude show that the scattered waves in channels α, β are to be associated with branch points that lie near the stripping pole. The branch point associated with Coulomb scattering actually coincides with the pole. (With regard to this latter singularity, it is pertinent to remark that the use of Born amplitudes that incorporate exact treatments of Coulomb scattering [20, 23, 24] yields no significant improvement of the extrapolations.)

We recall that the stripping pole corresponds to a Born amplitude that is a product of amplitudes for elementary processes (11.41). Hence the pole is associated with a Feynman diagram that contains two vertices, at which *independent processes* take place. This observation suggests the possibility that the assumption of pole dominance may be tested experimentally by checking whether the kinematic variables in a reaction are distributed statistically in the fashion allowed by two successive, independent processes. A test of this type was developed by Treiman and Yang [17, 20, 21]. It is applicable for reactions that yield three or more product particles.

More sophisticated dispersion analyses that go beyond the mere location of singularities in the transition amplitude have been developed. In these analyses distortion effects are introduced by the derivation of integral equations in which the branch cuts that arise from elastic scattering are treated on equal terms with the Born singularity. A major starting point for derivations of these equations is the unitarity condition [16–21], a quadratic matrix equation that expresses necessary relations among amplitudes. Of course, only a few elementary singularities can be incorporated in the integral equations, therefore these equations must be regarded as models of the reaction process, somewhat in the same spirit as the DW method. It is not clear whether the integral equations express models that go beyond the standard DW method. In any case, it has been difficult to develop useful results from these equations.

References

[1] J. Hüfner, C. Mahaux, and H. A. Weidenmüller, *loc. cit.*, Chap. 2, Ref. 3.
[2] R. H. Lemmer and C. M. Shakin, *loc. cit.*, Chap. 6, Ref. 20.
[3] B. Buck and A. D. Hill, *loc. cit.*, Chap. 6, Ref. 22.
[4] C. Bloch and collaborators, *loc. cit.*, Chap. 6, Ref. 23.
[5] J. Raynal, M. A. Melkanoff, and T. Sawada, *loc. cit.*, Chap. 6, Ref. 24.
[6] W. MacDonald and A. Mekjian, *Phys. Rev.* **160,** 730 (1967).
[7] W. Glöckle, J. Hüfner, and H. A. Weidenmüller, *Nucl. Phys.* **A90,** 481 (1967).
[8] H. A. Weidenmüller and K. Dietrich, *Nucl. Phys.* **83,** 332 (1966), Section 4.
[9] A. F. R. de Toledo Piza and A. K. Kerman, *Ann. Phys.* **43,** 363 (1967).
[10] L. Garside and W. Tobocman, *Phys. Rev.* **173,** 1047 (1968).
[11] T. Ericson et al., *loc. cit.*, Chap. 2, Ref. 7.
[12] P. A. Moldauer, *loc. cit.*, Chapter 2, Ref. 6, N. Ullah, *Phys. Letters* **28B,** 240 (1968).
[13] H. Feshbach, A. K. Kerman, and R. H. Lemmer, *Ann. Phys.* **41,** 230 (1967), and other references cited there.
[14] J. B. Marion and G. Weber, *Phys. Rev.* **102,** 1355 (1956); J. A. Evans, T. A Kuehner, and E. Almqvist, *Phys. Rev.* **131,** 1632, 1642 (1963); W. W. Daehnick, *Phys. Rev.* **135,** B1168 (1964); W. P. Alford, L. M. Blau, and D. Cline, *Nucl. Phys.* **61,** 368 (1965); E. F. Pessoa, R. L. Dangle, N. Veta, and O. Sala, *Nucl. Phys.* **68,** 337 (1965); J. P. Schapira, J. O. Newton, R. S. Blake, and D. J. Jacobs, *Nucl. Phys.* **80,** 565 (1966); T. Gudenhus, M. Cosack, R. Felst, and H. Wahl, *Nucl. Phys.* **80,** 577 (1966); P. E. Hodgson and D. Wilmore, *Proc. Phys. Soc.* **90,** 361 (1967); A. Budzanowski, K. Grotowski, L. Jarczyk, H. Niewodniczanski and A. Strzatkowski (1966) to be published; H. W. Fulbright, J. A. Robbins, R. West, D. P. Saylor, J. W. Verba, *Nucl. Phys.* **A94,** 214 (1967); M. Fazio, P. Guazzoni, S. Micheletti, M. Pignanelli, and L. Zetta, *Nucl. Phys.* **A111,** 255 (1968).
[15] R. D. Amado, *Phys. Rev. Letters* **2,** 399 (1959).

[16] I. S. Shapiro, *J. Exptl. Theoret. Phys.* (USSR) **41,** 1616 (1961); [Eng. translation, *Sov. Phys. JETP* **14,** 1148 (1962)]; *Nucl. Phys.* **28,** 244 (1961).
[17] I. S. Shapiro, in *Selected Topics in Nuclear Theory*, F. Janouch, Ed., IAEA (Vienna, 1962).
[18] I. S. Shapiro, in *Proceedings of Int. Conf. on Nuclear Physics*, Vol. 1, Centre National de la Recherche Scientifique, Paris, 1964.
[19] A. Dar and W. Tobocman, *Phys. Rev. Letters* **12,** 511 (1964).
[20] H. J. Schnitzer, *Rev. Mod. Phys.* **37,** 666 (1964).
[21] I. S. Shapiro, *Proceedings of the International School of Physics*, "*Enrico Fermi*," Course XXXVIII, T. E. O. Ericson, Ed., Academic, New York, 1967.
[22] N. R. Gibbs and W. Tobocman, *Phys. Rev.* **124,** 1496 (1961).
[23] C. Dullemond and H. J. Schnitzer, *Phys. Rev.* **129,** 821 (1963).
[24] W. K. Bertram and L. J. Tassie, *Phys. Rev.* **166,** 1029 (1968).

General References

The following references may be consulted as alternatives to portions of this book. The material from several of these references is used freely, often without explicit citation.

W. Tobocman, *Theory of Direct Nuclear Reactions*, Oxford University Press, London, 1961.

N. Austern, in *Fast Neutron Physics*, Vol. II, J. B. Marion and J. L. Fowler, Eds., Interscience, New York, 1963.

N. Austern, in *Selected Topics in Nuclear Theory*, F. Janouch, Ed., IAEA, Vienna, 1963.

N. K. Glendenning, in *Annual Reviews of Nuclear Science*, Vol. 13, Annual Reviews, Palo Alto, 1963.

Direct Interactions and Nuclear Reaction Mechanisms, E. Clementel and C. Villi, Eds., Gordon and Breach, New York, 1963.

G. R. Satchler, in *Lectures in Theoretical Physics*, Vol. VIII C, P. D. Kunz, D. A. Lind, W. E. Brittin, Eds., University of Colorado Press, Boulder, 1966.

C. Bloch, in *International School of Physics*, "*Enrico Fermi*," Course XXXVI, C. Bloch, Ed., Academic, New York, 1966.

N. K. Glendenning, in *International School of Physics* "*Enrico Fermi*," Course XL, M. Jean and A. Ricci, Eds., Academic, New York, 1969

Index*

*Subtopic entries are in some cases listed in the order of importance.